Design Computing and Cognition '04

# Design Computing and Cognition '04

*Edited by*

John S. Gero

*Key Centre of Design Computing and Cognition,*
*University of Sydney,*
*Sydney, New South Wales, Australia*

KLUWER ACADEMIC PUBLISHERS
DORDRECHT / BOSTON / LONDON

A C.I.P. Catalogue record for this book is available from the Library of Congress.

ISBN 978-90-481-6650-3 (PB)
ISBN 978-1-4020-2393-4 (e-book) ISBN 1-4020-2393-6 (e-book)

---

Published by Kluwer Academic Publishers,
P.O. Box 17, 3300 AA Dordrecht, The Netherlands.

Sold and distributed in North, Central and South America
by Kluwer Academic Publishers,
101 Philip Drive, Norwell, MA 02061, U.S.A.

In all other countries, sold and distributed
by Kluwer Academic Publishers,
P.O. Box 322, 3300 AH Dordrecht, The Netherlands.

*Printed on acid-free paper*

## TABLE OF CONTENTS

## List of Referees

Henri Achten, Eindhoven University of Technology, Netherlands
Tom Addis, University of Portsmouth, UK
Omer Akin, Carnegie Mellon University, USA
Tom Arciszewski, George Mason University, USA
Jean-Paul Barthes, Université de Compiègne, France
Kirsty Beilharz, University of Sydney, Australia
Peter Bentley, University College London, UK
Bo-Christer Bjork, Royal Institute of Technology, Sweden
Lucienne Blessing, Technical University Berlin, Germany
Frances Brazier, Vrije Universiteit Amsterdam, Netherlands
Alan Bridges, University of Strathclyde, UK
Dave Brown, Worcester Polytechnic Institite, USA
Jon Cagan, Carnegie Mellon University, USA
Luisa Caldas, Instituto Superior Technico, Portugal
Scott Chase, University of Strathclyde, UK
Nancy Cheng, University of Oregon, USA
Maolin Chiu, National Cheng-Kung University, Taiwan
John Clarkson, Cambridge University, UK
Nigel Cross, The Open University, UK
Bharat Dave, University of Melbourne, Australia
Bauke de Vries, Eindhoven University of Technology, Netherlands
Ellen Yi-Luen Do, University of Washington, USA
Wolfgang Dokonal, University of Technology Graz, Austria
Dirk Donath, Bauhaus-Universitaet Weimar, Germany
Andy Dong, University of Sydney, Australia
Jose Duarte, Instituto Superior Technico, Portugal
Alex Duffy, University of Strathclyde, UK
Chris Earl, The Open University, UK
Chuck Eastman, Georgia Institute of Technology, USA
Martin Fischer, Stanford University, USA
Renate Fruchter, Stanford University, USA
Haruyuki Fujii, Tokyo Institute of Technology, Japan
Issam Fujita, Iwate Prefectural University, Japan
Esther Gelle, Swiss Federal Institute of Technology, Switzerland
John Gero, University of Sydney, Australia
Alberto Giretti, Universita di Ancona, Italy
Gabi Goldschmidt, Technion - Israel Institute of Technology, Israel
Mark Gross, University of Washington, USA
David Gunaratnam, University of Sydney, Australia
John Haymaker, Stanford University, USA
Tony Holden, University of Cambridge, UK
Koichi Hori, University of Tokyo, Japan
Ludger Hovestadt, Federal Institute of Technology, Zurich, Switzerland
TaySheng Jeng, National Cheng Kung University, Taiwan
Richard Junge, Technical University Munich, Germany
Julie Jupp, University of Sydney, Australia
Udo Kannengiesser, University of Sydney, Australia
Vladimir Kazakov, University of Sydney, Australia
Ruediger Klein, DaimlerChrysler, Germany
Terry Knight, Massachusetts Institute of Technology, USA
Branko Kolarevic, University of Pennsylvania, USA
Alex Koutamanis, Delft University of Technology, Netherlands
Ramesh Krishnamurti, Carnegie Mellon University, USA
Sourav Kundu, Knowledge Solutions Group, Japan
Tom Kvan, University of Hong Kong, Hong Kong
Bryan Lawson, Sheffield University, UK
Pierre Leclerq, University of Liège, Belgium
Andrew Li, Chinese University of Hong Kong, Hong Kong
Udo Lindemann, Technical University of Munich, Germany
Hod Lipson, Cornell University, USA
Ardeshir Mahdavi, Vienna University of Technology, Austria
Mary Lou, Maher, University of Sydney, Australia
Dorian Marjanovic, University of Zagreb, Croatia
Bob Martens, Vienna University of Technology, Austria
Hari Narayanan, Auburn University, USA
Barry O'Sullivan, University College Cork, Ireland
Rivka Oxman, Technion Israel Institute of Technology, Israel
Jens Pohl, California Polytechnic State University, USA
Sattiraju Prabhakar, Information Science Institute, USA
Rabee Reffat, University of Sydney, Australia
Yoram Reich, Tel Aviv University, Israel
Michael Rosenman, University of Sydney, Australia
Stephan Rudolf, University of Stuttgart, Germany
Rob Saunders, City University London, UK
Linda Schmidt, University of Maryland, USA
Gerhard Schmitt, Federal Institute of Technology, Switzerland
Thorsten Schnier, University of Birmingham, UK
Kristi Shea, Cambridge University, UK
Greg Smith, University of Sydney, Australia
Ian Smith, Federal Institute of Technology, Lausanne, Switzerland
Tim Smithers, VICOMTech, Spain
Ricardo Sosa, University of Sydney, Australia
Ram Sriram, NIST, USA
Louis Steinberg, Rutgers University, USA
George Stiny, Massachusetts Institute of Technology, USA
Rudi Stouffs, University of Technology Delft, Netherlands
Carole Strohecker, Media Lab Europe, Ireland
Masaki Suwa, Chukyo University, Japan
Tapio Takala, Helsinki University of Technology, Finland
Hideaki Takeda, Nara Institute of Science and Technology, Japan
Hsien Hui Tang, Chang Gung University, Taiwan
Wade Troxell, Colorado State University, USA
Jin-Yeu Tsou, Chinese University of Hong Kong, Hong Kong
Bige Tuncer, Delft University of Technology, Netherlands
Ziga Turk, University of Ljubljana, Slovenia
Barbara Tversky, Stanford University, USA
Jerzy Wojtowicz, University of British Columbia, Canada
Rob Woodbury, Simon Fraser University, Canada
Stefan Wrona, Warsaw University of Technology, Poland

# Preface

Modern design research has a history that commenced around forty-five years ago. Early design research made use of concepts associated with formal methods and was driven by the potential of the then novel digital computer. Whilst much has happened during the intervening years including developments in design simulation, design optimization, design representation, design databases, documentation of designs, 2D and then later 3D modeling, and design generation, it took the introduction of artificial intelligence into design research a little more than twenty years ago to provide renewed vigour to the field. What artificial intelligence provides includes new ways of representing designs, new ways of simulating designs and new ways of generating designs all based on symbolic computation.

Artificial intelligence provides an environmentally rich paradigm within which design research based on computational constructions can be carried out. This has been one of the foundations for the developing field called *design computing*.

Recently, there has been a growing interest in what designers do when they design and what they do when they design using computational tools. Much of this interest is driven by the need to have a better understanding of human designers in order to create better tools for them. Part of the reason, however, is the need to better understand designers as part of the development of design science. This studying of designers is the basis of a newly emergent field called *design cognition* that draws part of its source from cognitive science.

This new conference series aims at providing a bridge between the two fields of design computing and design cognition. The confluence of these two fields is likely to provide the foundation for further advances in each of them.

The papers in this volume are from the *First International Conference on Design Computing and Cognition (DCC'04)* held at the Massachusetts Institute of Technology, USA. They represent the state-of-the-art of research and development in design computing and cognition. They are of particular interest to researchers, developers and users of advanced computation in design and those who need to gain a better understanding of designing.

In these proceedings the papers are grouped under the following nine headings, describing both advances in theory and application and demonstrating the depth and breadth of design computing and the still burgeoning field of design cognition:

Conceptual Design
Design Cognition
Design Generation
Representation in Design
Patterns in Design
Designing with Shapes and Features
Agents in Design
Words and Rational in Design
Interaction: Sentience and Sketching

Over 140 papers were submitted to the conference. Each paper was extensively reviewed by three referees drawn from the international panel of ninety-nine active referees listed. Thanks go to them, for the quality of these papers depends on their efforts. The reviewers' recommendations were then assessed before the final recommendation was made.

Lai Chui Looi deserves particular thanks for it was she who took the authors' final submissions and turned them into a uniform whole. The final quality of the manuscript bears her mark.

**John S Gero**
University of Sydney
April 2004

## CONCEPTUAL DESIGN

JS Gero (ed), *Design Computing and Cognition'04*, 3-22


# AN INTELLIGENT ASSISTANT FOR CONCEPTUAL DESIGN

## *Informed Search Using a Mapping of Abstract Qualities to Physical Form*

KIMBERLE KOILE
*Massachusetts Institute of Technology, USA*

**Abstract.** In early stages of design, the language used is often very abstract. In architectural design, for example, architects and their clients use experiential terms such as "private" or "open" to describe spaces. The Architect's Collaborator (TAC) is a prototype design assistant that supports iterative design refinement using abstract, experiential terms. TAC explores the space of possible designs in search of solutions satisfying specified abstract goals by employing a strategy we call dependency-directed redesign: It evaluates a design with respect to a set of goals, uses an explanation of the evaluation to guide proposal and refinement of design repair suggestions, then carries out the repair suggestions to create new designs.

## 1. Introduction

In early stages of design, the language used is often very abstract. Engineers might talk about designing a piece of equipment that is "easy to maintain". Clothing designers talk of "baggy" clothing. Architects and their clients use experiential terms such as "private" and "open". Throughout the design process these abstract terms are operationalized and translated into physical characteristics of the artifact being designed. The design process can be viewed as one of exploration, trying to turn goals, often articulated only in very abstract terms at the beginning of the process, into an artifact that realizes those goals.

If we are to build programs that help designers during the early stages of design, often termed conceptual design, we must give those programs rich vocabularies and the capability to represent and reason with abstract concepts. The hypothesis put forth in this paper is the following: Computational tools can support conceptual design by providing a mapping of abstract terms to measurable design features and by using that mapping in an informed, exploratory search of a design space. The Architect's

Collaborator (TAC) is a prototype design support system that illustrates these ideas in the domain of architecture. TAC employs techniques from artificial intelligence to explore a space of designs using a technique we call *dependency-directed redesign.* TAC is an intelligent design assistant that focuses on design refinement using abstract terms, leaving to the designer the tasks of providing a starting design, specifying and respecifying goals, and ranking designs.

This paper begins by describing a design problem that TAC solved, then discusses how TAC works and gives results from an experiment with a Frank Lloyd Wright Prairie house and from a real-world design example. The paper then discusses related work, future work, and contributions.

## 2. An Architectural Example

Architectural design is well-suited to research on conceptual design for several reasons. Most design problems exhibit the difficulties mentioned earlier: They are exploratory in nature and involve the use of terms representing abstract, experiential qualities. Such experiential qualities—e.g., openness, spaciousness, privacy—are not easily articulated or formalized. Yet they are an essential part of the architectural design process: Architects and their clients often describe desired spaces in terms of these qualities; architects use their knowledge from past experiences, from environment behavior research, and from their own theories to create physical form that manifests such qualities. This knowledge can be articulated and structured as general design knowledge (Wright 1954; Alexander et al. 1977; Zeisel and Welch 1981; Hertzberger 1993). As illustrated in this paper, this design knowledge can be operationalized and used as the basis for a conceptual design support system that reasons about abstract qualities and physical form.

To illustrate the above idea, TAC was given the design of an existing house that the owners and their architects were redesigning. Several problems with the house were identified, one of which is illustrated in Figure 1, the living and dining rooms felt small and isolated from each other.

One way to solve the size problem is to make the rooms larger. Another way is to make the rooms feel larger by creating views to neighboring spaces. Creating views also helps with the feeling of isolation. Given a goal of having the dining room not feel small and isolated, TAC used its knowledge base of architectural concepts to translate this goal into having the dining room visually open from the living room. It calculated a visual openness value, Figure 2, and determined that the value was insufficient.

TAC proposed making the rooms feel larger and less isolated by increasing the visual openness of the dining room. It suggested design

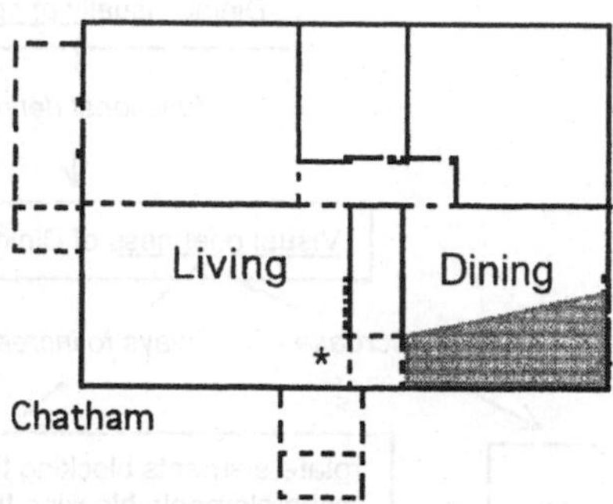


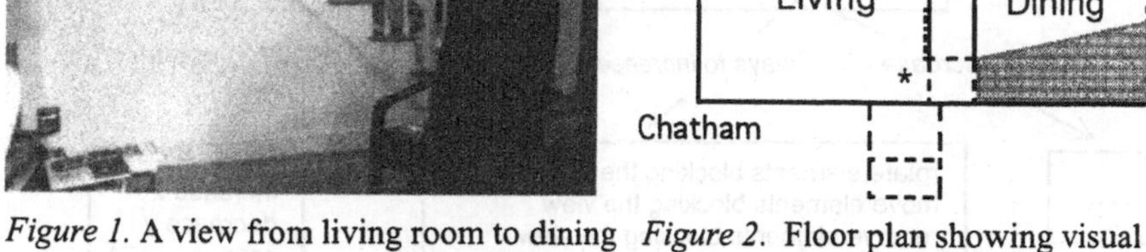

*Figure 1.* A view from living room to dining

*Figure 2.* Floor plan showing visual openness of dining from viewpoint *; value is 0.42. Shaded region is visible. Dotted lines are open edges.

modifications to achieve this increase, and created seven new designs by: rotating the stair 90 degrees, rotating the stair 270 degrees, moving the stair to three different exterior edges, removing the stair, replacing the stair wall with a screen. The last solution, with the stair wall "screenified", was implemented by the owners, Figures 3 and 4.

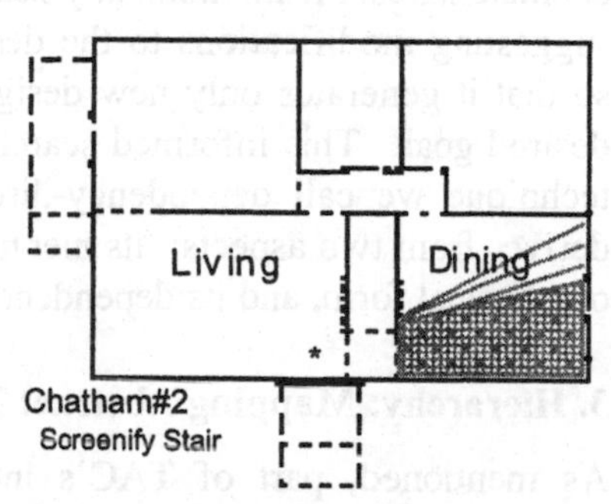


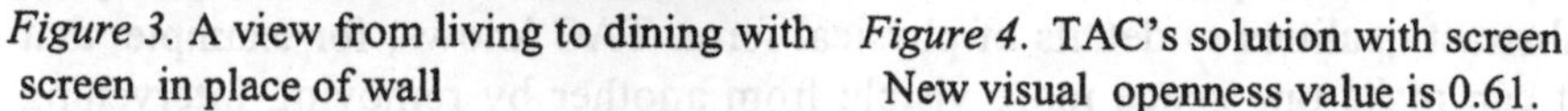

*Figure 3.* A view from living to dining with screen in place of wall

*Figure 4.* TAC's solution with screen New visual openness value is 0.61.

TAC creates new designs with a visually-open dining room by using a mapping between abstract qualities and operators on physical form, Figure 5. It locates the function that relates visually-open to visual openness, and finds that one territory is visually open to another if at least .6 of its area is visible. Finding this not true, TAC uses a general rule about making an expression of the form "x greater than y" true by increasing y, and proposes increasing visual openness (the value of y in this case). It then finds in its knowledge base techniques for increasing visual openness by modifying the things blocking the view. It determines that the stair blocks the view, then applies each of the techniques to the original design, producing the new designs.

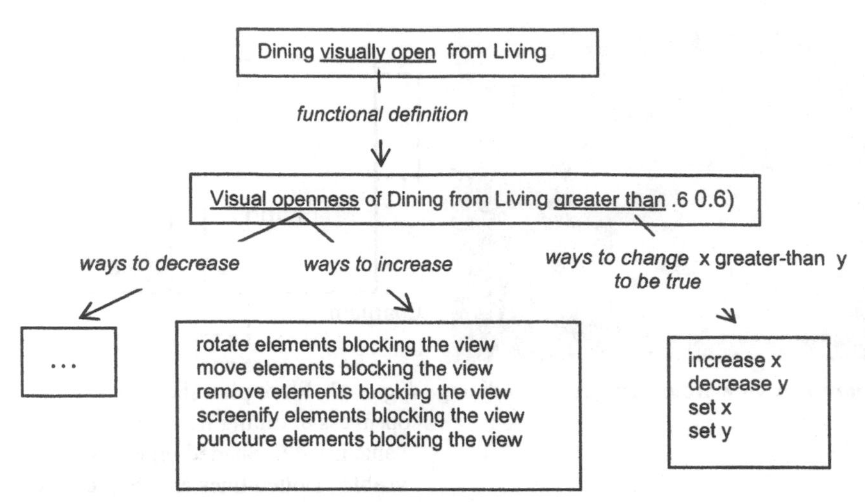


*Figure 5.* Portion of TAC's mapping of abstract terms to operators on physical form

This example illustrates TAC's behavior: It translates a goal stated in terms of abstract qualities into operators on physical form. It performs this translation using a hierarchy that maps abstract terms to physically measurable design characteristics and design operators for achieving those characteristics. It methodically searches the space of possible solutions by suggesting modifications to the design, pruning suggestions when possible so that it generates only new designs with a good chance of satisfying the desired goal. This informed search of a design space is performed using a technique we call dependency-directed redesign. TAC's intelligence thus derives from two aspects: its hierarchy that maps abstract terms to operators on physical form, and its dependency-directed redesign strategy.

## 3. Hierarchy: Mapping Abstract Terms to Physical Terms

As mentioned, part of TAC's intelligence derives from its mapping of abstract qualities to details of physical form. TAC knows, for example, that it can make one space more visible from another by removing intervening walls. It knows that it can make a space feel more private by making less of it visible or by making the path to it from a front door less direct. TAC also knows about characteristics of Frank Lloyd Wright's Prairie houses.

TAC represents architectural knowledge—general knowledge as well as a designer's or client's particular preferences—using constructs called *design characteristics*. TAC also contains domain-independent knowledge from geometry, arithmetic, logic, and computation, which it represents using what we call *TAC-functions*. The following sections describe TAC's representation for designs, design characteristics, and TAC-functions, and illustrate how these constructs are used to map abstract qualities to details of physical form.

### 3.1. REPRESENTING DESIGNS

TAC represents a design as a set of five models, each capturing a different aspect of a design. The *design element model* contains size and location information for walls, windows, etc.; it can be thought of as a primitive computer-aided design (CAD) model. The *edge model* is a two-dimensional geometric abstraction of the design element model, containing points and non-overlapping edges. Edges are either one-dimensional abstractions of design elements (e.g., walls), or one-dimensional projections of design elements. Projections, also called projected edges, are "invisible" edges that extend in a parallel or perpendicular direction from design element edges and help bound two-dimensional regions we call territories (Kincaid 1997). Territories are grouped into a *territory model*, another geometric abstraction of a design element model. See Figures 2 and 4 for examples. A *use space model* pairs territories with uses specified by the designer. Finally, a *circulation model* is a graph representing paths between doorways.[1]

TAC's representation for a design differs from most other representations of architectural designs in three significant ways. First, the fundamental vocabulary is that of design elements—walls, windows, etc. Most other knowledge-based architectural design systems that generate new designs represent only spaces, and thus cannot reason about physical form. Second, territories, often called spaces in other systems, are derived from the design elements, not specified independently. Finally, most other systems do not have separate representations for territories and use; a notable exception is Simoff and Maher (1998). Representing use separately from territories enables TAC to reason about physical form independently of intended use.

### 3.2. DEFINING DESIGN CHARACTERISTICS

As mentioned above, design characteristics represent architectural properties of a design, including such concepts as visual openness, physical accessibility, and floor plan area. Some design characteristics can be computed directly from design elements, while others are derived from computed design characteristics and are related to physical form via those characteristics. Design characteristics form a decomposition hierarchy, with characteristics computed from physical form at the bottom and those derived from them higher up. In this way experiential qualities are mapped into details of physical form. Four design characteristics, which appear in examples throughout this paper, illustrate this mapping. The decomposition hierarchy for these characteristics is shown at the end of this section.

[1] Design elements and their edges are entered by hand using a 2D design editor; projected edges, territories, and circulation paths are computed automatically. Visualization capabilities more sophisticated than 2D floor plans are possible, but are outside the scope of this research, which is focused on intelligent exploration of design space.

Example 1: Visual-openness is quantitative and measures the portion of a territory visible from another territory. Visual-openness is an example of a design characteristic that is computed from physical form elements; its evaluation function is a "black box" computational geometry routine. Figures 2 and 4 show the results of visual-openness calculations.

Example 2: Visually-open is boolean-valued and defined in terms of visual-openness by using a threshold: A territory is considered visually-open from another territory if at least 0.6 of its area is visible from the other territory. Visually-open is an example of a derived design characteristic. Its *evaluation function* is defined in terms of visual-openness using a Lisp-like expression: (gt (visual-openness x from y) 0.6).

Example 3: Perceived-main-entryness is vector-valued and gives a measure of the perception of an exterior door as a main entry. Characteristics that influence a visitor's choice of door when approaching a house are *components* of perceived-main-entryness. These include distance between door and street, straightness of path between door and street, and formality of door. Perceived-main-entryness is a derived characteristic; its evaluation function collects all components into a vector. Perceived-main-entryness also has *necessary conditions*: In order to have a perceived-main-entryness value, for example, an exterior door must be visible from the street.

Example 4: The design characteristic perceived-main-entry is defined in terms of perceived-main-entryness. Its value is the exterior door most likely to be perceived as the main entry. TAC constructs a partial order that ranks exterior doors by their perceived-main-entryness values, and returns the top of the partial order as the value of perceived-main-entry. Alternatively, the evaluation function for perceived-main-entry could combine the components of the perceived-main-entryness vector into a single value and choose the door with the highest value. Notice, however, that the components are incommensurate, and it is not necessarily meaningful, nor obvious how to combine them into a single value.

These examples of design characteristics illustrate TAC's decomposition hierarchy of characteristics. The means by which a design characteristic's evaluation function is defined determines the characteristic's place in the hierarchy. A characteristic that is considered to be directly related to physical form is at the bottom of the hierarchy and has an evaluation function that is a predefined "black box" that operates on one or more models representing a design. A derived design characteristic is higher up in the hierarchy and has an evaluation function constructed using one of three different methods: *evaluation function body* (e.g., as visually-open), *components*, or *components and necessary conditions* (e.g., as perceived-main-entryness). These methods provide the means for constructing the hierarchy, as shown in Figure 6.

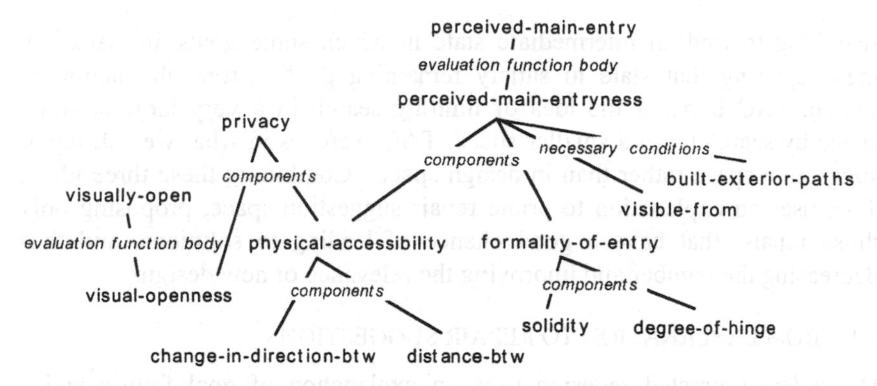


*Figure 6.* Dependency links for some of TAC's design characteristics

### 3.3. COMPLETENESS

TAC's knowledge base is complete enough to solve interesting, simple two-dimensional redesign problems, as with the Chatham house design problems described in sections 2 and 5. It contains 30 TAC-functions which represent arithmetic relations, logical relations, computational constructs, and set concepts. These TAC-functions form a basic set of domain independent functions out of which new design characteristics can be built for other architectural design problems. The remaining ten TAC-functions represent geometric concepts, e.g., distance between two things. More geometric concepts could be added, e.g., alignment, overlap (Cui and Randell 1992).

TAC contains 62 design characteristics, which represent architectural concepts such as privacy, visual openness, paths between two design elements. Forty of these proved sufficient for the Chatham design problems. The remaining 22 characteristics were added for a Frank Lloyd Wright Prairie house experiment and included Wright-specific characteristics such as circuitous path, place of prospect, and place of refuge (Hildebrand 1991). More design characteristics could be added easily for design problems involving other architectural types or other architects.

## 4. Dependency-Directed Redesign

As mentioned earlier, TAC's intelligence derives from its design characteristic hierarchy, which maps abstract concepts to details of physical form, and from its informed search using that hierarchy. Its informed search employs a technique we call *dependency-directed redesign*, which is inspired by artificial intelligence work on dependency-directed backtracking (Stallman and Sussman 1977), plan repair (Sussman 1975; Simmons 1992), and abstraction in search (Sacerdoti 1974). From dependency-directed backtracking, TAC borrows the idea of using an explanation of goal failure to guide search for a solution. From plan repair, TAC borrows the idea of searching to find an intermediate state in which some goals are satisfied, then repairing that state to satisfy remaining goals. From abstraction in search, TAC borrows the idea of limiting search in a very large solution space by searching in a smaller space: TAC searches in what we call repair suggestion space rather than in design space. Combining these three ideas, TAS uses an explanation to prune repair suggestion space, proposing only those repairs that have a good chance of leading to solutions, and thus decreasing the number and improving the relevance of new designs.

## 4.1. FROM EXPLANATION TO REPAIR SUGGESTIONS

Dependency-directed redesign uses an explanation of goal failure and a knowledge base of repair strategies to propose suggestions for modifying a design: Given an initial design and a set of goals, TAC evaluates a design with respect to the goals and uses the resulting explanation to propose repair suggestions for any goals not satisfied. It then prunes and refines suggestions, and creates new designs for the remaining suggestions.

Returning to the Chatham house prior to remodeling, Figures 1 and 2, consider the goal of having the dining room visually open from the living room. The goal is represented by the expression (visually-open Dining from Living). TAC evaluates this goal, determines that it is not satisfied, and produces an explanation of the failure in the form of a tree that represents a trace of the goal expression's evaluation, Figure 7. By walking down the tree, TAC can determine why a goal was not satisfied and then use that information to propose suggestions for design repair.

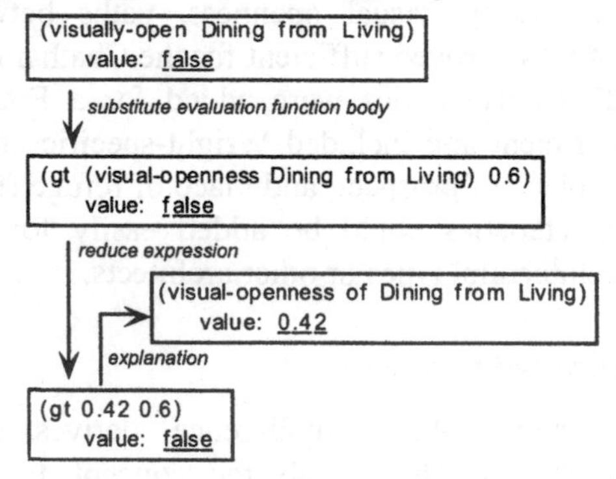


*Figure 7.* Explanation for (visually-open Dining from Living) for Chatham example

In particular, TAC identifies opportunities to repair the cause of failure by looking for expressions whose value it knows how to change via domain independent routines called *fixers*. Fixers reason about how to get from a current value to a desired value; they propose increasing, decreasing, or setting values. They rely on a characteristic's *increasers*, *decreasers*, and *setters*—expressions that when evaluated modify a design, thereby changing the value of the characteristic. (Examples of fixers and increasers were shown in Figure 5, labeled as "ways to change ..." and "ways to increase", respectively.) Whether a design characteristic or TAC-function has a fixer, increasers, decreasers, or setters depends on the nature of the characteristic or function. Some characteristics, such as the color of a design element, are directly settable and have setters that change a value. Others, such as visual-openness, are not directly settable, and hence do not have setters; instead their values are changed by modifying the design. Thus, instead of setters, the characteristic visual-openness has increasers, since certain modification operators, e.g., removing an intervening design element, have a good chance of increasing its value.

To repair the Chatham design so that the dining room is visually open from the living room, TAC traverses the explanation shown in Figure 7 until it finds methods for "fixing" a node's expression. When it gets to the gt node, it finds that it knows how to fix a (gt x y) expression: it can set y to be greater than x, decrease y to be less than x, set x to be greater than y, or increase x to be greater than y. In the current expression, y is a constant, 0.6., and cannot be decreased or set; x is the visual-openness expression, and visual-openness cannot be directly set. One option remains: increasing x. TAC checks its knowledge base and finds that it knows how to increase the value of visual-openness by means of increasers associated with that characteristic. So it proposes increasing the value of visual-openness:

```
(increase-value of (visual-openness Dining from Living)
                   until visual-openness greater than 0.6)
```

TAC then retrieves increasers (see Figure 5), which are written in terms of operators on design elements that block the view between things, e.g.,

```
(remove blocking-elts-btw x y)        (screenify blocking-elts-btw x y)
```

Substituting arguments of Dining and Living from the original goal expression, TAC then proposes specific repair suggestions, e.g.,

```
(remove blocking-elts-btw Dining Living)   (screenify blocking-elts-btw Dining Living)
```

TAC now checks the design to identify design elements that block the view, finds the stair, substitutes it into the repair expressions, and proposes, e.g.,

```
(remove Stair)                        (screenify Stair)
```

For each suggestion, TAC then creates new designs, one of which was shown in Figure 4.

## 4.2. FROM REPAIR SUGGESTIONS FOR GOALS TO NEW DESIGNS

The Chatham house example illustrates how TAC works with a single goal, translating a goal expression into operators on physical form, carrying out those operators to create new designs. More realistic design problems have multiple, often conflicting goals. TAC deals with this situation by using a generate-and-test control structure, generating intermediate designs that satisfy a subset of the goals, then iteratively repairing those designs to satisfy remaining goals. An enabling assumption for this approach is that some goals will be independent, so that working on one goal does not always undo a previously satisfied goal (Sussman 1975). For the goals that do interact, some amount of work to reevaluate and resatisfy goals is necessary. TAC limits the amount of work in two ways. First, as previously described, it separates the proposing of repairs from the performing of repairs, thereby enabling it to avoid designs that it knows will not satisfy goals. Second, its generate-and-test control structure includes a lookahead step: When proposing repair suggestions for a particular goal, it "looks ahead" for potential goal interactions. It looks both for conflict, i.e., when satisfying a goal will undo an already satisfied goal, and synergy, i.e., when a modification will satisfy more than one goal. Three kinds of conflict and synergy were identified: obvious, predictable, unpredictable. TAC handles the first two of these: Obvious interactions are detected by comparing goals, predictable interactions are detected by comparing repair suggestions for goals. Being able to reason about obvious and predictable interactions enables pruning of repair suggestions before creating designs, which helps control search and increases the chances that intermediate designs are closer to solutions. Unpredictable interactions, by definition, cannot be detected ahead of time, and lead to the need for generate-and-test.

An example of TAC's reasoning with multiple goals is shown for one of Frank Lloyd Wright's Prairie houses, the Horner house, Figure 8.

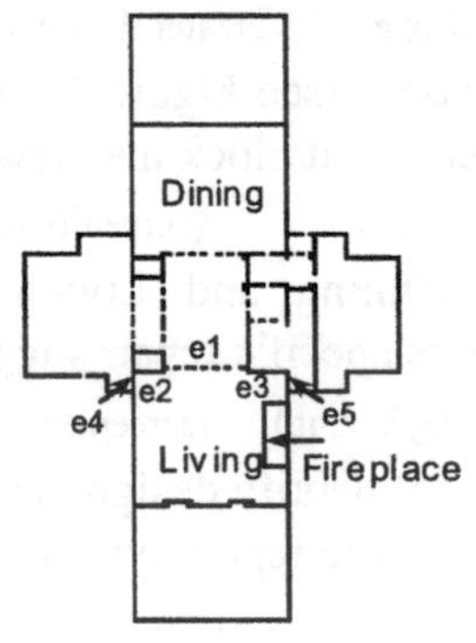


*Figure 8.* Edge model for Horner house; e1 to e5 are edges

TAC was asked to evaluate the house with respect to five goals usually satisfied in Prairie houses: the center of the Living room visible from the Dining room, the Living room visually open from the Dining room, one fireplace in the Living room and one in the entire design, and the fireplace on an interior edge. The first four goals are satisfied already, but the last goal is not: the fireplace is not on an interior edge. Attempting to satisfy this goal illustrates some interesting goal interactions.

Figure 9 illustrates TAC's behavior given the five goals. Starting with the unsatisfied goal, TAC proposes six suggestions (s1 to s6): move the fireplace to any of five interior edges (e1 to e5) or add a new fireplace on an interior edge. It notices that adding a fireplace conflicts with the goal of having one fireplace, so it prunes that suggestion. It then creates five new designs, each with a fireplace on one of the specified edges. It checks these designs and finds that in D4 and D5 the fireplace is not entirely on an interior edge, so it discards these two designs. It rechecks the other four goals for the remaining designs, finding that D2 and D3 are solutions. It determines that moving the fireplace to e1 (in design D1) has caused the visibility goals to become unsatisfied: the fireplace has blocked the view between living and dining territories. So TAC proposes removing or puncturing[2] the fireplace. It notices that removing the fireplace will conflict with keeping the number of fireplaces at one, so it prunes that suggestion. It carries out the fireplace puncture operation and creates design D6. It checks the visually-open goal and finds it now satisfied, so D6 is a solution. It has no more designs to check, so it stops, returning solutions D2, D3, and D6: Horner#2, Horner#3, and Horner#1#1 in Figure 10.

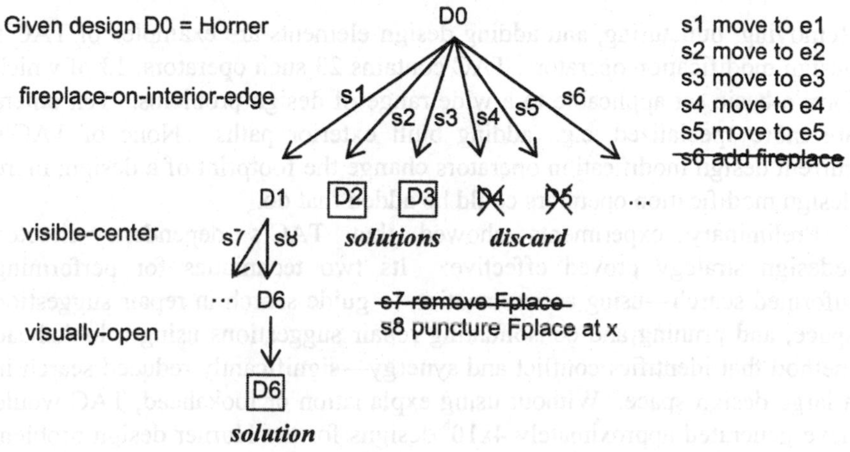


*Figure 9.* Control structure for Horner design example

The goals in this example exhibited several different kinds of interaction. Obvious synergy was exhibited by the two fireplace count goals: having one fireplace in the living territory also satisfied having one fireplace in the entire design. Predictable conflict occurred between the goal of having one fireplace and a suggestion to remove the fireplace. Note that the goals themselves in this case were not in conflict, but rather one goal was in

[2] "Puncturing" a fireplace is a technique Wright used in the Robie house.

conflict with a particular repair suggestion proposed for another goal. Unpredictable synergy occurred between the visible-center goal and the visually-open goal: puncturing the fireplace to make the living territory center visible also caused the living territory to be visually open.

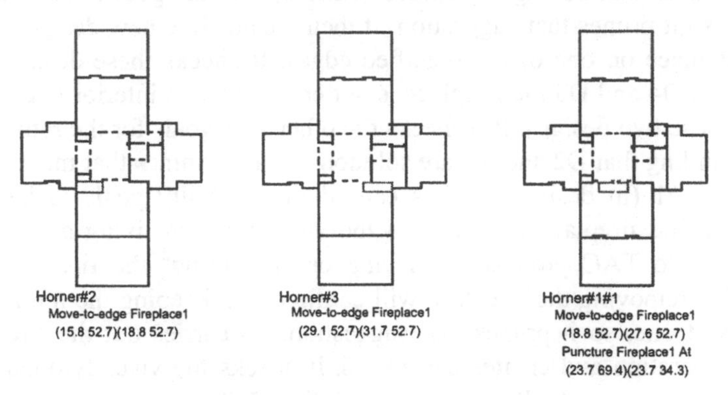
Horner#2
Move-to-edge Fireplace1
(15.8 52.7)(18.8 52.7)
Horner#3
Move-to-edge Fireplace1
(29.1 52.7)(31.7 52.7)
Horner#1#1
Move-to-edge Fireplace1
(18.8 52.7)(27.6 52.7)
Puncture Fireplace1 At
(23.7 69.4)(23.7 34.3)

*Figure 10.* Solutions for Horner design problem

### 4.3. EFFECTIVENESS

Removing, puncturing, and adding design elements are examples of TAC's design modification operators. TAC contains 23 such operators, 13 of which form a basic set applicable to a wide range of design problems. Ten others are more specialized, e.g., adding built exterior paths. None of TAC's current design modification operators change the footprint of a design; more design modification operators could be added that do.

Preliminary experiments showed that TAC's dependency-directed redesign strategy proved effective: Its two techniques for performing informed search—using an explanation to guide search in repair suggestion space, and pruning and consolidating repair suggestions using a lookahead method that identifies conflict and synergy—significantly reduced search in a large design space. Without using explanation or lookahead, TAC would have generated approximately $4x10^8$ designs for the Horner design problem described in this section: 23 operators, 10 producing at least 4 new designs each, yields 53 new designs for each of 5 goals, or $53^5$, approximately $4x10^8$.

The tables below summarize control structure experiments for the Horner design problem. Five goals were specified in each of two orders, optimal and nonoptimal.[3] Table 1 gives results using explanation to guide search; Table 2 gives results using both explanation and lookahead.

[3] An optimal goal order is one in which goals with synergistic operators, i.e., that will satisfy more than one goal, precede goals with which they interact. See Koile (2001) for details.

TABLE 1. Five goals, Horner design problem, explanation used, no lookahead

| Goal Order | # solutions | # designs | # repair cycles |
|---|---|---|---|
| optimal | 4 | 16 | 5 |
| nonoptimal | 37 | 339 | 48 |

TABLE 2. Five goals, Horner design problem, explanation and lookahead used

| Goal Order | # solutions | # designs | # repair cycles |
|---|---|---|---|
| optimal | 4 | 8 | 3 |
| nonoptimal | 11 | 47 | 5 |

Using an explanation to guide search reduced the number of designs generated to 339 for a nonoptimal goal order, and to 16 for an optimal goal order, which is considerably better than $4x10^8$. Adding the lookahead mechanism further reduced the number of designs generated for nonoptimal goal order to 47, and for optimal goal order to 8.

The optimal goal order, both with and without lookahead, resulted in the same four solutions. The nonoptimal goal order, however, resulted in additional solutions. Most of these solutions were very similar to those found with optimal order. They might have punctured the fireplace in a slightly different location, for example. Several of the solutions found without lookahead, however, were significantly different, because designs were created that violated goals that were not the current focus—a situation not uncommon in search problems. TAC then repaired those designs, creating additional solutions. For this reason, the best control structure for generating solutions when goals interact would include an option for relaxing lookahead when desired.

## 5. A Real-World Design Problem

A system such as TAC can be used by architects as both a design tool and an analysis tool. This section illustrates TAC's utility as a design tool in an experiment using the Chatham house discussed in the opening example.[4]

The Chatham house was being redesigned at the same time that TAC was under development. The architects and TAC thus were able to work in tandem on the same design problems. TAC was given a model of the house and a set of design goals defined by the owners and their architects, and in response proposed new designs. Several of the designs are presented here to show that TAC finds plausible solutions to a real architectural design problem, and that is does so with breadth and generality.

---

[4] See Koile (2001) for discussion of TAC's utility in analysing designs and definitions of architectural type.

The Chatham house, floor plan, and approach paths are shown in Figure 11 and 12.

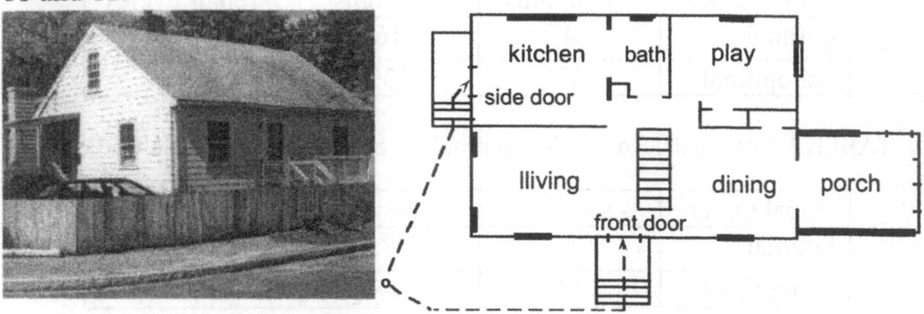


*Figure 11.* The Chatham house

*Figure 12.* Chatham house first floor and approach paths to exterior doors
The usual approach point is marked by o.

Four problems with the house were identified:

- site: visitors approaching the house are not sure which door to use
- entry: living room is not private with respect to the front door
- territories: main living spaces feel isolated from one another
- use: kitchen activity is too far from the dining activity

We phrased goals for TAC in terms of physical access and visual openness:

- site: one perceived main entry
- entry: living room visually semi-open and physically semi-accessible (i.e., reached via somewhat crooked path) from the perceived main entry
- territories: main living spaces visually open from one another
- use: kitchen activity next to the dining activity

Figure 13 shows one of the designs produced by the architects, along with a similar design proposed by TAC. In both designs, TAC and the architects solved the problem of having more than one perceived main entry by removing the front door and making the side door the new front door. Moving the front door also increased the change in direction, and thus the crookedness of the path, between the entrance and the living territory, and decreased the visibility of the living territory from the entrance. The visibility was decreased too much, so both TAC and the architects removed a section of wall between the front door and the living territory, a modification that also makes the living territory more easily accessible from the entrance. TAC and the architects turned the stair to increase visual openness between the dining and living territories. They exchanged the playroom and kitchen activities so that the kitchen activity would be adjacent to the dining activity.

The designs also show differences, some of which result from the architects' working with a larger goal set than TAC. Some of these goals were not given to TAC because they would not have illustrated new TAC behavior, e.g., making the kitchen territory more visually open from the dining territory. Other goals were outside the scope of TAC's current operators, which do not change a design's footprint, e.g., enlarging the entry porch. Other differences between TAC's designs and the architects' are due to both unspecified goals and lack of information in TAC's knowledge base. When the dining territory became smaller as a result of turning the stair, for example, TAC did not enlarge the territory at the expense of the porch, as the architects did: TAC did not know of an implicit assumption that the dining territory would not be smaller, nor that a territory can be enlarged by borrowing area from a neighboring territory.

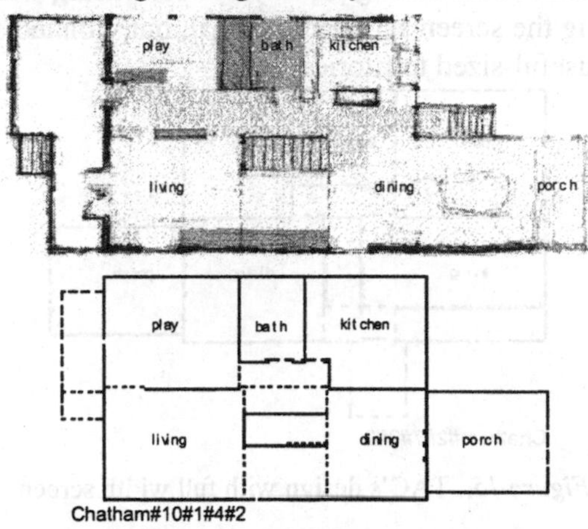


*Figure 13.* Architects' design (top) and TAC's design; labels are activities

An alternate design produced by the architects and a similar design proposed by TAC are shown in Figure 14.

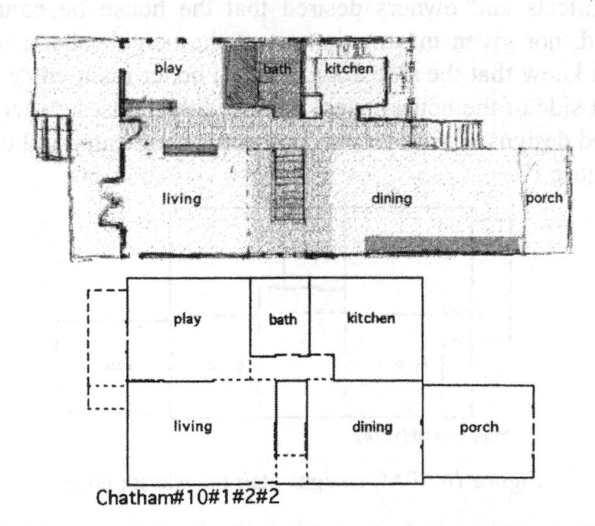


*Figure 14.* Alternate design by the architects (top) and TAC's similar design

In the designs shown in Figure 14, TAC and the architects again have made the side door the new front door, removed a section of wall between the entrance and the living territory, and exchanged playroom and kitchen activities. Instead of turning the stair, however, they have replaced the solid wall of the stair with a screen, e.g., as shown in Figure 3.

TAC came up with designs that differed significantly from the architects' designs. TAC's design in Figure 15 uses a screen the full width of the living territory as a means of increasing privacy by decreasing visibility from the front door. Adding the screen satisfied the goal, but violated an implicit goal of creating only useful-sized territories.

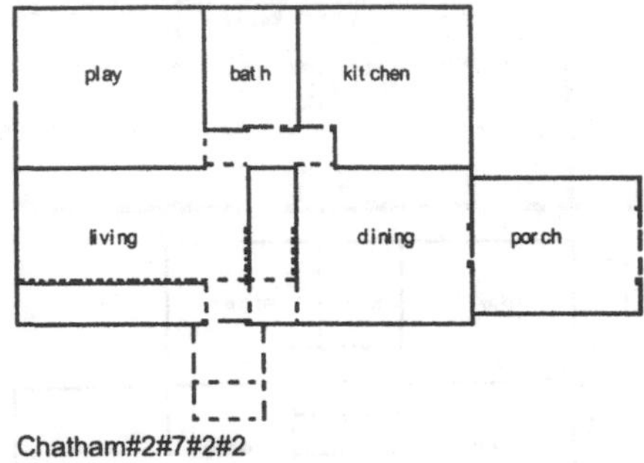


*Figure 15.* TAC's design with full width screen

TAC also was not told that the owners preferred the stair in a central location, so it suggested moving the stair to exterior walls, creating plausible designs but not what the owners had in mind, Figure 16. TAC was not told that the architects and owners desired that the house be connected to a neighborhood, nor given information about the neighborhood. As a result, TAC did not know that the side door makes a better main entry because the street on that side of the house is less busy and the houses closer together. It thus produced designs with the front door as the main entry and the side door removed, Figure 17.

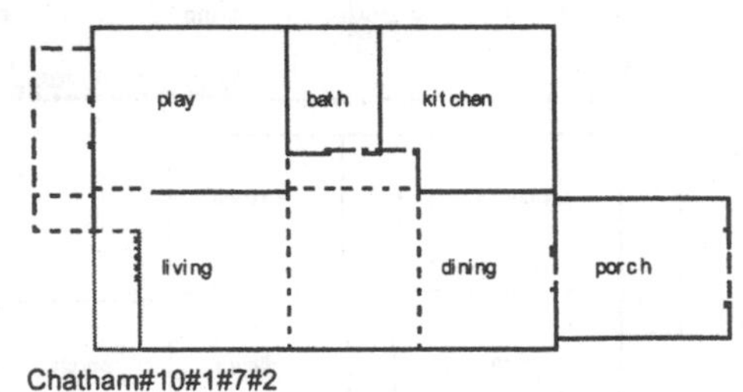


*Figure 16.* TAC design: stair on exterior edge

Some of TAC's novel designs are quite plausible and result from its ability as a computational tool to carry out transformations easily and quickly: It produces many variations on a theme, a task an architect would find very tedious. In some cases, TAC's designs may be redundant or even bad, as in Figure 15, but they can be easily set aside by the designer as she focuses on the designs that meet specified and unspecified criteria. TAC has utility as a brainstorming tool and can help a designer and client elucidate goals by calling attention to desired or undesired features.

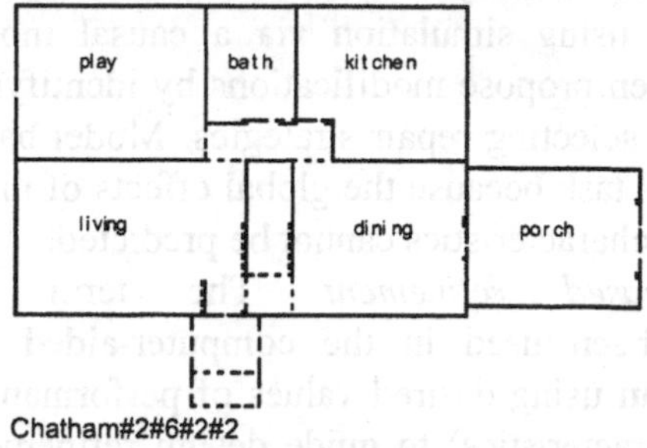


*Figure 17.* TAC design: screen at front door

## 6. Related Work

There is a vast literature on computational tools for conceptual design. Tools that are most similar to the work reported in this paper either reason with similar experiential knowledge or employ similar reasoning techniques. An earlier paper, (Koile 1997), surveyed systems that evaluated designs with respect to experiential qualities. The discussion here is confined to work that shares features with TAC's dependency-directed redesign strategy, especially in the field of architecture.

Two methodologies that share features with TAC's redesign strategy are case adaptation in case-based reasoning, and performance-based refinement.

*Case-base reasoning*: Case adaptation methods employed in case-based reasoning systems are similar in spirit to TAC's repair mechanism. Given a design case, they modify it to meet specified design goals. Indeed, TAC's modification operators can be thought of as a "taxonomy for design adaptation" (Oxman 1996). See Voss and Oxman (1996) for a survey.

Constraint satisfaction techniques have been used to adapt architectural design cases. Some systems first adapt a case's topology using graph algorithms, then adapt geometry using constraint satisfaction techniques (Smith et. al. 1996; Hua et. al. 1996). Design knowledge may be represented implicitly in the systems' parameters and constraints (Smith et. al. 1996), or explicitly using techniques such as hierarchies of object types (Giretti and Spalazzi 1997). Constraint satisfaction techniques are not appropriate for TAC's repair problem: Since particular design element arrangements for realizing abstract design characteristics are not known *a priori*, specific constraints between design elements cannot be specified.

Model-based reasoning techniques have been used to adapt cases, though typically for engineering fields in which qualitative models of device behavior can be built. Even though not in the domain of architecture, the systems described in Goel (1991) and Prabhakar and Goel (1998) are worthy of mention as examples of using explanation of failure (case mismatch) to guide iterative repair. The systems retrieve a mechanical design case, and evaluate the case using simulation via a causal model of the device's behavior. They then propose modifications by identifying the source of the device failure and selecting repair strategies. Model-based reasoning is not possible for TAC's task because the global effects of modification operators on abstract design characteristics cannot be predicted.

*Performance-based refinement.* The term "performance-based refinement" has been used in the computer-aided architectural design community to mean using desired values of performance variables (akin to TAC's design characteristics) to guide design refinement—just what TAC does. As Flemming and Mahdavi (1993) suggest, most performance-based refinement tools only evaluate performance variables; the designer must "guess" at likely design modifications for affecting desired values. The work of Mahdavi (1997; 1998) is an exception. GESTALT, described in Mahdavi (1997), employs an "intelligent" generate-and-test method to iteratively modify a design using knowledge of functional relationships between physical form and performance variables. It maps experiential qualities, such as light quality, to methods for changing them, as TAC does. Such qualities in GESTALT are quantitative (e.g., a five point scale of light quality) and can be mathematically modeled or formalized through regression analysis. Hence, optimization techniques can be used to select particular values for desired characteristics. TAC's power would be enhanced by employing this technique when possible, rather than always assuming monotonic relationships. Many design characteristics in architectural design are not quantitative, however, so TAC's qualitative reasoning cannot be replaced completely with optimization methods.

## 7. Future Work and Contributions

TAC's representations form a good foundation for the development of rich knowledge bases of architectural design knowledge. As with all systems that rely on knowledge bases, however, acquiring the knowledge is nontrivial. If a designer can assemble a set of designs that exhibit a particular characteristic, machine learning techniques may able to help with the knowledge acquisition task. In addition, explanation-based learning techniques may be useful in adding knowledge of discovered synergies and conflicts to the knowledge base.

TAC's knowledge base could be extended to include knowledge of materials and light, both of which affect the experiential qualities of a space. The knowledge base also could be extended to include sociological influences on physical form (Wright 1954; Hillier and Hanson 1984). Changing attitudes about domestic life, for example, transformed the front and back parlors of Victorian times into the modern-day living room.

Focus to date has been on TAC's representation and reasoning capabilities, with little time spent addressing user interface issues. Thinking about what constitutes an appropriate interface for designers opens up a number of intriguing possibilities. TAC would benefit, for example, from integration with a sketching tool, e.g., (Gross 1996), so that a designer could move between sketching and TAC's evaluation and repair steps. TAC might also benefit from an interface that allowed a user to increase or decrease values of design characteristics and observe the resulting changes in physical form. A similar idea is proposed in (Flemming and Mahdavi 1993).

TAC's control structure could be extended to support goal specification and refinement. As design goals evolve along with a design solution, a designer might want to interrupt one of TAC's evaluation and repair cycles, redefine goals, then have TAC continue. This extension would be straightforward. TAC also could be extended to assist a designer in specifying goals by suggesting some goals automatically. If a design has a second floor, for example, TAC could suggest that the design needs a stair. The issue of how complete the goal set needs to be and whether goals could be inferred are open research questions.

TAC demonstrates that it is possible to construct a prototype intelligent assistant that supports conceptual design via iterative design refinement, representing and reasoning about how experiential qualities are manifested in physical form. Its hierarchy of design characteristics provides a means for operationalizing abstract qualities. Its dependency-directed redesign mechanism provides a means for exploring a design space using abstract qualities. Its use in finding plausible solutions to a real architectural design problem demonstrates the real-world potential of these ideas.

## Acknowledgments

This research was funded by a National Science Foundation Graduate Fellowship. The author thanks Randall Davis, Howard Shrobe, Patrick Winston, Tomás Lozano-Peréz, and John Aspinall for assistance in AI; and Aaron Fleisher, Richard Krauss, Duncan Kincaid, and Mark Gross for assistance in architecture. The drawings in Figures 13 and 14 are courtesy of Duncan Kincaid and Daniel Gorini.

## References

Alexander, C, Ishikawa, S, Silverstein, M, Jacobsen, M, Fiksdahl-King, I and Angel, S: 1977, *A Pattern Language*, Oxford University Press, New York.

Cui, Z and Randell, D: 1992, Qualitative simulation based on a logical formalism of space and time, *AAAI '92*, pp. 679-684.

Flemming, U and Mahdavi, A: 1993, Simultaneous form generation and performance evaluation: A `two-way' inference approach, *in* U Flemming and S Van Wyk (eds), *CAAD Futures '93*, North-Holland, 161-174.

Giretti, A and Spalazzi, L: 1997, ASA: A conceptual design-support system, *Engineering Applications of Artificial Intelligence* **10**(1): 99-111.

Goel, AK: 1991, A model-based approach to case adaptation, *Proceedings of the Thirteenth Annual Conference of the Cognitive Science Society*, Lawrence Erlbaum, pp. 143-148.

Gross, MD: 1996, The electronic cocktail napkin--a computational environment for working with design diagrams, *Design Studies* **17**: 53-69.

Hertzberger, H: 1993, *Lessons for Students in Architecture*, Uitgeverij Publishers, Rotterdam.

Hillier, B and Hanson, J: 1984, *The Social Logic of Space*, Cambridge University Press.

Hildebrand, G: 1991, *The Wright Space: Pattern and Meaning in Frank Lloyd Wright's Houses*, University of Washington Press, Seattle.

Hua, K, Faltings, B and Smith, I: 1996, CADRE: Case-based geometric design, *Artificial Intelligence in Engineering* **10**: 171-183.

Kincaid, DS: 1997, *An Arithmetical Model of Spatial Definition*, Master of Architecture Thesis, Dept. of Department of Architecture, Massachusetts Institute of Technology.

Koile, K: 1997, Design conversations with your computer: evaluating experiential qualities of physical form, *in* R Junge (ed), *CAAD Futures '97*, Kluwer, pp. 203-218.

Koile, K: 2001, *The Architect's Collaborator: Toward Intelligent Tools for Conceptual Design*, PhD Thesis, Dept. of EECS, MIT.

Mahdavi, A and Suter, G: 1997, On implementing a computational facade design support tool, *Environment and Planning B* **24**: 493-508.

Mahdavi, A and Suter, G: 1998, On the implications of design process views for the development of computational design support tools, *Automation and Construction* **7**(2-3): 189-204.

Oxman, R: 1996, Design by re-representation: A model of visual reasoning in design, *Design Studies* **18**(4): 329-347.

Prabhakar, S and Goel, AK: 1998, Functional modeling for enabling adaptive design of devices for new environments, *Artificial Intelligence in Engineering* **12**: 417-444.

Saderdoti, E: 1974, Planning in a hierarchy of abstraction spaces, *Artificial Intelligence* **5**(2): 115-135.

Simmons, RG: 1992, The roles of associational and causal reasoning in problem solving, *Artificial Intelligence* **53**(2-3): 159-208.

Simoff, SJ and Maher, ML: 1998, Designing with the activity/space ontology, *in* JS Gero and F Sudweeks (eds), *Artificial Intelligence in Design '98*, Kluwer, 23-43.

Smith, I, Stalker, R and Lottaz, C: 1996, Creating design objects from cases for interactive spatial composition, *in* JS Gero (ed), *Artificial Intelligence in Design '96*, Kluwer, 97-116.

Stallman, R and Sussman, G: 1977, Forward reasoning and dependency-directed backtracking in a system for computer-aided circuit analysis, *Artificial Intelligence* **9**: 135-196.

Sussman, GJ: 1975, *A Computer Model of Skill Acquisition*, American Elsevier, New York.

Voss, A and Oxman, R: 1996, A study of case adaptation systems, *in* JS Gero (ed), *Artificial Intelligence in Design '96*, Kluwer, 173-189.

Wright, FL: 1954, *The Natural House*, Horizon Press, New York.

Zeisel, J and Welch, P: 1981, *Housing Designed for Families: A Summary of Research,* Joint Center for Urban Studies of MIT and Harvard University, Cambridge, MA.

JS Gero (ed), *Design Computing and Cognition'04*, 23-36


# FEATURE NODES: AN INTERACTION CONSTRUCT FOR SHARING INITIATIVE IN DESIGN EXPLORATION

SAMBIT DATTA
*Deakin University, Australia*

and

ROBERT F WOODBURY
*Simon Fraser University, Canada*

**Abstract.** Exploration with formal design systems comprises an iterative process of specifying problems, finding plausible and alternative solutions, judging the validity of solutions relative to problems and reformulating problems and solutions. Recent advances in formal generative design have developed the mathematics and algorithms to describe and perform conceptual design tasks. However, design remains a human enterprise: formalisms are part of a larger equation comprising human computer interaction. To support the user in designing with formal systems, shared representations that interleave initiative of the designer and the design formalism are necessary. The problem of devising representational structures in which initiative is sometimes taken by the designer and sometimes by a computer in working on a shared design task is reported in this paper. To address this problem, the requirements, representation and implementation of a shared interaction construct, *the feature node*, is described. The feature node facilitates the sharing of initiative in formulating and reformulating problems, generating solutions, making choices and navigating the history of exploration.

## 1. Introduction

Supporting computational design exploration requires flexible and extensible representational structures for supporting the iterative process of specifying problems, finding plausible and alternative solutions, judging the validity of solutions relative to problems and reformulating problems and solutions. Recent studies report representational frameworks that allow for flexible feature-based modelling (Leeuwen and Wagter 1998), sorts (Stouffs and Krishnamurti 2002) and design exemplars (Summers et al. 2002). However, design remains a human enterprise: to be scalable, representational formalisms need to be embedded within a broader framework of human computer interaction. To support the user in designing with formal systems, shared representations that integrate the role of the designer and generative formalisms are necessary.

Shared representational structures enable the designer to formulate problems, generate solutions, make choices and visually browse the history of exploration (alternatives, revisions) recorded in design space. Interaction over shared representations enables both user and formalism to exercise joint responsibility over domain goals. For example, Rich and Sidner (1998) report a domain level representation supporting collaboration between an agent and the user. The shared representation of Veloso et al. (1997) allows both automated and human planners are able to interact and construct plans jointly.

An exposition of the designer's actions in design space is captured in the metaphor of navigation (Chien and Flemming 1996). Navigation encapsulates the user's action in design space through the traversal of paths and landmarks defined over the navigation structure. A graphical notation supporting mixed-initiative human-computer dialogue is reported by Datta and Woodbury (2002). Flexible representational structures (Stouffs and Cumming 2003) and the visualization and interaction with such mixed representational data constructs allow users novel mechanisms for interaction.

We argue that, for sharing initiative between human and machine in design exploration, it is necessary to build intermediary representational structures shared between the designer's view of exploration and the symbol structures that represent them. We base our interaction construct on the symbol substrate for design space exploration reported in Woodbury et al. (1999). This formalism implements a formal mechanism for computing exploration in terms of types, features, descriptions and resolution algorithms (Burrow and Woodbury 1999; Woodbury et al. 2000). In particular it addresses the representational properties of information ordering, partiality, intensionality, structure sharing and satisfiability. The formalism (Woodbury et al. 1999) provides an incremental modelling approach based on data constructs supporting partiality and intentionality within the representational structures.

The view on representational flexibility and extensibility in design is addressed by the property that all data constructs are partial and intentional. In this paper, we report an interaction construct, the feature node, over this formalism. The interaction construct extends this view to provide a representation that is shared between a designer and the formalism. The requirements for such an interaction construct, its representation in terms of the symbol substrate and its implementation are described in the paper.

## 2. Requirements

Designing is a reflective conversation that involves the recursive processes of *seeing, moving* and *seeing* (Schön and Wiggins 1992). Choices, alternatives and versions emerge out of the interaction between designing (acting) and discovering (reflecting). Exploration encompasses the formulation of requirements and the generation of solutions based on these requirements. Problems and solutions co-evolve (Hybs and Gero 1992) and reformulation is an integral part of the problem definition (Smithers 2000). Problems and solutions in design are inherently partial (Woodbury et al. 1999). Generated solutions provide a large space of alternatives (Woodbury and Chang 1995) and representational models can reduce cognitive overload and facilitate choice making (Chien and Flemming 1997). Choices, their connections and the developing history of explicitly discovered design alternatives must be accessible to the designer through interaction with the *structure* of exploration. Exploration rationale (Smithers 2002) and design history (Burrow and Woodbury 2001) are significant tools for supporting exploration. The designer must be able to exploit this history through navigation and recombination of the paths of exploration.

The shared representation must provide a unified model for representing the set of problems, sub problems, problem revisions and associated reformulations that a designer actually considers. A designer should be able to make choices of alternative problems and solutions. Further, the representation must capture the history of problems, solutions and choices made by the designer during exploration. These requirements are addressed in our representation as follows:

- **Problems and Solutions** - A problem state corresponds to the designer's view of problem formulation. A solution state corresponds to the initial, intermediate and final designs satisfying a problem. The feature node must capture the connections between a problem and its possible solutions (partial or complete) uncovered in design space.
- **Choices** - Problems and requirements have multiple solutions. In design, a problem formulation may have no solutions, a finite number of solutions or an arbitrarily large collection of solutions. The intentional choices made by the designer in the reformulation of problems and commitment to generated solutions during exploration must be recorded in the feature node representation.
- **Exploration history** - Exploration history captures the rationale of exploration as a record of designer actions and formal moves. This history must be captured in a shared representation as a collection of ancestor and progeny feature nodes.

The requirements capture the generics of a shared interaction without describing the algorithmic and symbol level implications of the formal substrate. The mappings between the concepts and their relationships are modelled in using the unified modelling language, UML, notation (Jacobson et al. 1998).

The designer's view of exploration, shown in Figure 1, comprises problems, solutions, choices and history (their connections and the resulting explicit design space). The problem formulation and reformulation cycle, the solution generation and reuse cycle, the intentional choices made by the designer and the rationale of exploration history needs to be captured by the representation.

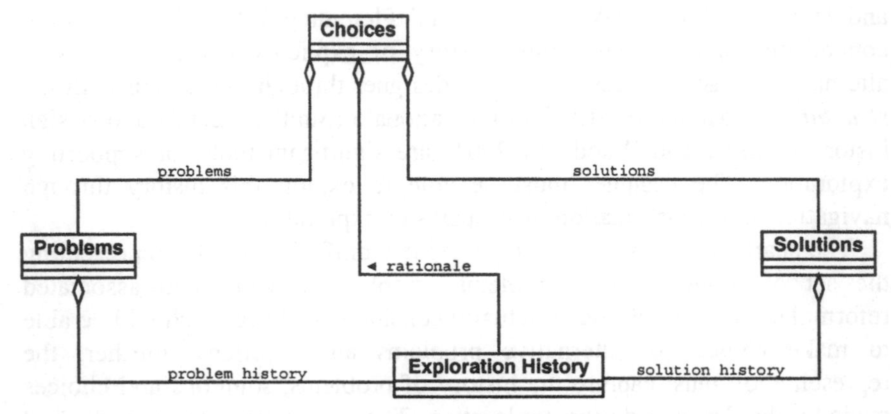


*Figure 1.* The designer's view of exploration can be captured through a representation of the following entities, problems, solutions, choices and history. The problem formulation and reformulation cycle, the solution generation and reuse cycle, the intentional choices of the designer and the rationale of exploration in the form of a history are captured in this view.

The representation must provide support for the designer in problem formulation, solution generation, choices of alternatives and interaction with exploration history. Problems need not be fixed. Designs can be partial or complete with respect to the initial problem formulation. The shared representation In the next section, we show how the requirements identified above are addressed through the *feature node*, a representational construct over the symbol substrate of design space exploration.

## 3. Feature Node Representation

The feature node, FNode, is an interface construct for composing the designer's view of exploration and the formal substrate into a shared representation. A feature node, encapsulates the designer's interaction with the formalism by coupling user actions with the elements of the underlying symbol level. The feature node connects to the formal substrate and supports problem formulation and reformulation, solution generation, choice-making over problems and solution alternatives and navigation of exploration history. The next sections describe each of the domain requirements and their mapping to feature nodes.

### 3.1. PROBLEMS, SOLUTIONS AND CHOICE

In our representation, a problem is represented as a PState. Problem formulations are defined through statements, Desc given in the query language of the description formalism (Woodbury et al. 1999). The problem fomulation process includes type specifications, description authoring and reformulation. The connection between a and a can be stated as follows:

$$\text{FNode} \equiv \bullet\ \text{FNode} \bullet \text{PState} : \text{Desc} \quad (1)$$

Equation 1 captures the feature node representation of the problem formulation and reformulation process. Problem specifications are specified through interaction with a FNode. A FNode, in turn composes a PState. These in turn compose disjunctive and non-disjunctive statements in the description language of typed feature feature structures, Desc. A partial or complete solution to a problem is represented as a solution state, SState. The solution generation process includes computing satisfaction and resolution of problem statements. The symbol level concept that implements the elements of a sequence of solution states is the PartialSatisfier. Through the FNode, the user and the formalism participate in a process of incremental generation of partial solutions of a problem statement.

User choices and actions on partial solutions are defined through a FNode. User guidance in the generative process is supported in the selection of a FNode from the collection of possible solutions and in the specification of the next step of resolution. The connection between a FNode and a SState is given in the path form as follows:

$$\text{FNode} \equiv \bullet\text{FNode} \bullet \text{SState} : \text{PartialSatisfier} \quad (2)$$

Equation 2 captures the partial satisfier that represents a partial solution, (the SState) as well as the type constraints that remain to be resolved. The composition of the relationship between a problem state and its partial satisfiers is shown in Figure 2.

The feature node, Fnode captures the relationship between a problem state, PState and an alternative design that is a partial solution to the problem, SState. In a Fnode, designs and problems, can be elaborated dynamically. This formulation supports the user in the problem formulation and solution generation process within the same interface construct.

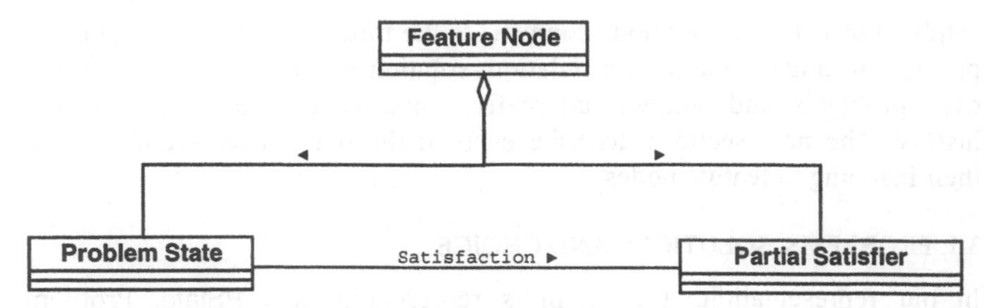


*Figure 2.* The feature node composes the relationship between the problem state and the partial satisfier

## 3.2. EXPLORATION HISTORY

Exploration history as uncovered by the designer's actions is recorded as the ordered collection of ancestor and progeny nodes, {FNode}. The exploration history is a tree where each branch is a path to a Fnode representing, either a problem, a solution or a choice point. The tree of feature nodes is the *satisfier space*, a record of all problem, solution and choice decisions made during exploration. The relationship between a satisfier space and its constituent feature nodes is shown in Figure 3.

## 3.3. OPERATIONS

The formal symbol level operators for supporting movement in design space are $\pi$-resolution, indexing and reuse, hysterical undo, design unification and design anti-unification (Burrow and Woodbury1999). At the interaction level, user-driven operations support the construction, navigation and synchronisation moves (Datta et al 2002) during exploration.

The feature node enables the designer to access both these types of operations as the Operations of exploration. Operations are further specialised into Intrinsics and Extrinsics. These are represented as attributes of a FNode. Given a feature node, Intrinsics allow the designer to access the formal (intrinsic) operators in the formal substrate. Extrinsics permit the manipulation of feature nodes with no analogue in the formal substrate.

These are cast as behvioural (extrinsic) attributes of a FNode. FNode•Intrinsics enable the user to access formal moves available in the symbol substrate. FNode•Extrinsics provide a hook to account for the contingent aspects of user interaction. The designer can access a partial satisfier and apply an operator from to extend the partial satisfier, PartialSatisfier.

### *3.3.1. Intrinsics*

In the FNode, formal moves are cast as *intrinsic attributes* of a feature node. They are intrinsic because they mirror the moves described in (Woodbury et al. 1999; Burrow and Woodbury 1999; Woodbury et al. 2000). Therefore, Intrinsics are Operations providing direct access to the underlying the arguments and operators of the design exploration machinery. With reference to a FNode and the formal operators in the substrate, Operations can be written as equivalent to the following path,

$$\text{Operations} \equiv \bullet\text{FNode} \bullet \text{Intrinsics} : \text{PartialSatisfier} \quad (3)$$

Through this formulation, FNode•Intrinsics capture design moves that mirror the operators, arguments and states already available in the formal substrate. The composition of Intrinsics as the attributes of a FNode is shown in Figure 4.

This connection to the underlying exploration machinery makes no claims about supporting the contingent intentionality of designer actions. To be truly mixed-initiative, feature nodes need to support operations that enable the designer to manipulate the underlying entities of a feature node.

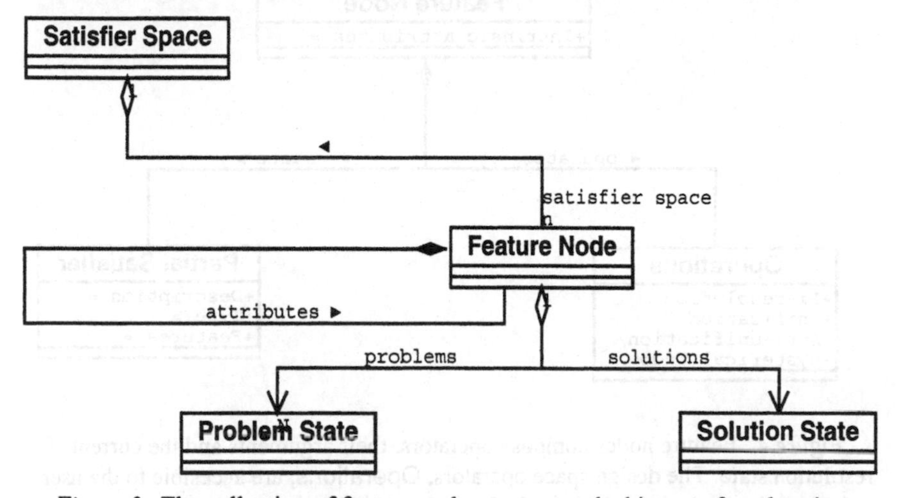


*Figure 3.* The collection of feature nodes, represent the history of exploration

### *3.3.2. Extrinsics*

The interactive manipulation of a feature node by the designer requires an additional set of operations. Extrinsics support contingent user interaction with no analogue in the formal substrate of the design space exploration. These operations are cast as *extrinsic* atttributes of a FNode.

Extrinsics capture the class of Operations that permit the user flexible and extensible interaction with the elements of a FNode. An example of an extrinsic operation is the direct manipulation of a feature node. Details of a graphical notation for incorporating extrinsics is reported in Datta et al (2002).

With reference to a FNode and the interaction operators external to the formal substrate, Extrinsics can be written as equivalent to the following path,

$$\text{Operations} \equiv \bullet\text{FNode} \bullet \text{Extrinsics} : \text{FNode} \tag{4}$$

In this manner, FNode•Extrinsics are defined recursively over feature nodes. The mapping of extrinsic attributes of a feature node is explained in Figure 5.

Summarising, the representation of a Fnode described above makes it possible to unify problem formulation, PState, intermediate and final solution generation, PState and the operators of exploration, Operations under a shared representation. An example implementation of the above representation in the framework of design space exploration is described in the next section.

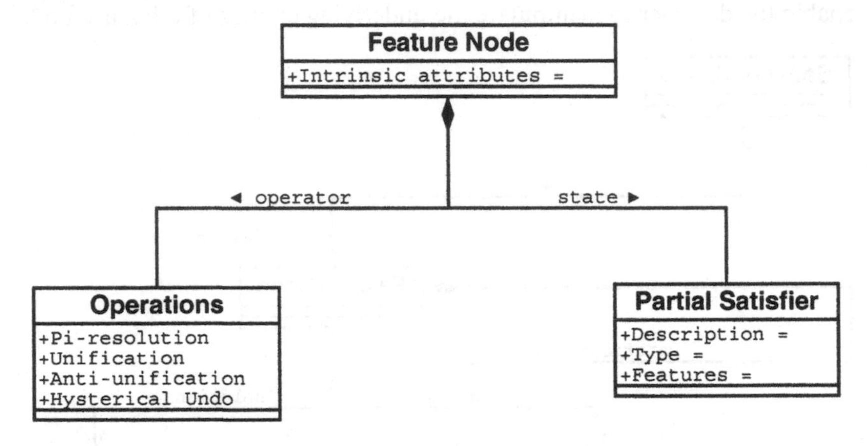


*Figure 4*. Feature nodes compose operators, their arguments and the current resolution state. The design space operators, Operations, are accesible to the user through the intrinsic attributes of a feature node. The current resolution state is represented by the partial satisfier.

## 4. Feature Node Implementation

The symbol substrate of design space exploration comprises an inheritance hierarchy of type definitions, *types*, a set of features, FEATS, constraints over types and features, *cons* and statements in a description language *Desc*. These components, are referred to as elements of the TypeSystem (Woodbury et al. 1999).

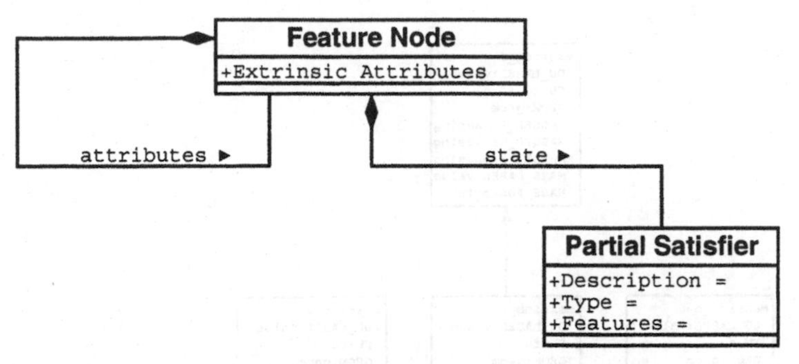


*Figure 5.* The extrinsic attributes of the feature node, FNode, represent the behavioural aspects of the feature node contingent upon user interaction but with no analogue in the formal substrate

In the TypeSystem, problems are specified through a type hierarchy and a description. The specification of a problem amounts to the construction of an inheritance hierarchy of types, a collection of feature declarations appropriate for those types and constraints on types.

In our interface construct, a PState composes the problem formulation and reformulation process. For example, the type *massing*, corresponding to a massing element, is a refinement of the type *configuration,* which is a subtype of type *du* or design_unit.

As shown in Figure 6, the type *massing*, inherits two features from *du*, DU_LABEL, GEOM. It introduces three types of features, namely MASS_LABEL, MASS_POS, FU. The feature MASS_LABEL serves to identify the geometric design_unit with a name. The feature MASS_POS provides the massing with a positional co-ordinate of type *point*. The feature FU is a hook to the functional roles that the massing element may play during exploration.

Massing configurations can recursively contain other massings. The type *massing* has a sequence of sub-types *massing_a, massing_b*... in which *massing_a* $\subseteq$ *massing_b*. Each massing of type *massing_n* introduces a feature MASSEL_N which denotes a sub-massing of *massing*. Thus a feature structure representing a massing of type *massing_f* is a design configuration of $a+b+c+d+e$ sub-massings.

Collection of path equations in the description language, appear in Table 1. These statements are analogous to the query, *What are the possible massing configurations in design space that satisfy the constraints on their features ?*

The PState is a compiled representation corresponding to the above description and is displayed in the interface through a feature node with the label, *Satspace Element* as shown in Figure 7. The generation process presents the user with all possible alternative combinations of satisfiers arising out of disjunctive dessriptions.

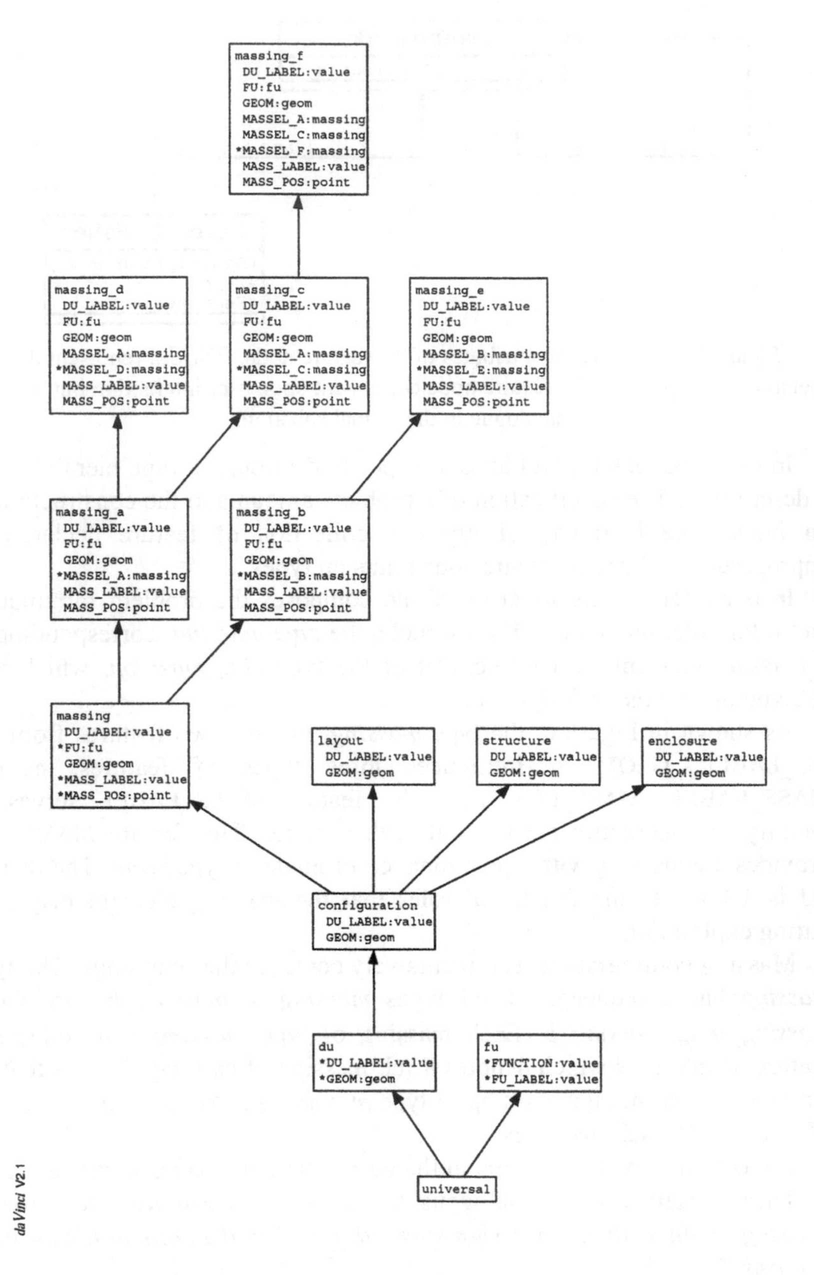


*Figure 6.* Representation of massing elements. A type hierarchy fragment showing their inheritance from the type *du* and the introduced features. The asterisk (*) symbol marks features that are introduced on that type.

TABLE 1. A collection of path equations in the form of a conjunctive description

```
(SFC_HOUSE GEOM POS : point &
SFC_HOUSE GEOM COMMAND : command) &
( FU : fu
& DU : du
& DU DU_LABEL == FU FU_LABEL
& DU GEOM COMMAND == SFC_HOUSE GEOM COMMAND
& PORCH SFC_PORCH : du
& ROOMROW ENTITY_A DU GEOM == SFC_HOUSE GEOM
& ROOMROW ENTITY_B DU GEOM == DU GEOM)
```

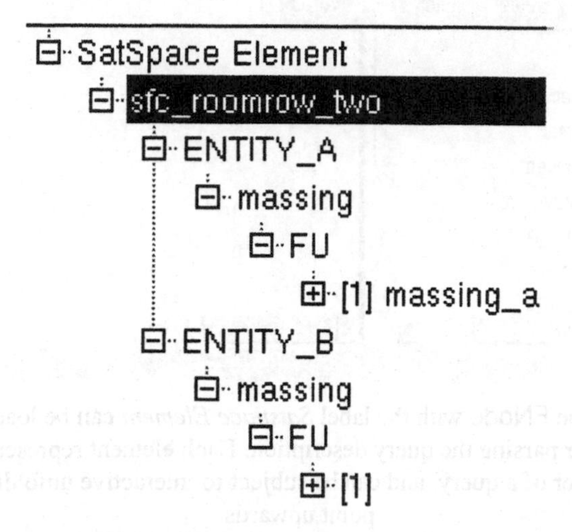


*Figure 7.* The label SatSpaceElement represents the feature node corresponding to the PState in Satisfier space. The figure shows a SState labelled by the type *sfc_roomrow_two* uncovered in the exploration of the problem.

Thus, a problem state, PState is the initial representation of a problem. It is on this structure that the process of modification and reformulation of problems occurs.

A *Satspace Element* labels a single exploration state, the FNode. An composes a and . The first element of a FNode is a PState as shown in Figure 8. By interaction with this element, the designer can either modify (reformulate) the PState or generate a new problem state. The rest of the elements of a FNode are entities representing SState nodes. These are the partial satisfiers (solution states) of the FNode.

By interaction with these elements, the designer can unfold the possible solution states of the problem. The label SatSpaceEl can be used interchangeably to represent feature nodes in satisfier space (see Section 3.2). Each SatSpaceEl captures the mapping between problem state, PState, their solutions as SState objects and the record of their connection to the underlying design space as uncovered by user choice. The substructures corresponding to the results of problem formulation and solution generation are returned by the formalism as feature nodes.

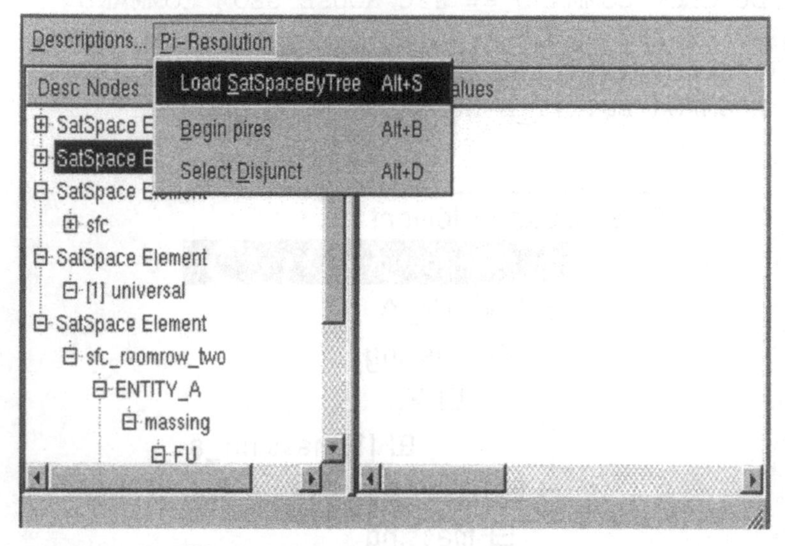


*Figure 8.* The FNode with the label *Satspace Element* can be loaded into the interface after parsing the query description. Each element represents the most general satisfier of a query, and can be subject to interactive unfolding from this point upwards.

The resolution procedure $\pi$-resolution (Burrow and Woodbury, 1999) acts incrementally by generating partially resolved structures. The process of generating partial satisfiers is expressed as a sequence of resolution steps, where each step records the resolution of a type constraint explicitly.

## 5. Discussion

The paper reports a shared representation for mixed-initiative design exploration. The feature node, FNode, is a representational construct integrating a designer's view of exploration comprising problems, solutions, choices and history over the symbol level representation of design space exploration. Through the FNode, *Problem state*, *Solution state*, *Choice* and *Operations* are explained.

First, the construct, FNode captures the dynamic and changing relationships between a problem state, PState and alternative designs (partial solutions) to the problem, SState. By interaction with this element, the designer can either modify (reformulate) the PState or generate a new problem state.

Second, the FNode records the intentional choices made by the designer. The elements of a FNode are entities representing nodes generated by the formalism. These are the partial satisfiers (solution states) of the FNode. By interaction with these elements, the designer can unfold the possible solution states of the current problem.

Third, the collection of ancestor and progeny feature nodes, { FNode} form a satisfier space. The satisfier space records problems, solutions and designer choices and therefore encapsulates the rationale of exploration.

## Acknowledgements

The authors gratefully acknowledge the contributions of Andrew Burrow, RMIT who made available his implementation of the representation framework of typed feature structures on which the massing example is based.

## References

Burrow, AL and Woodbury R: 2001, Design spaces–The forgotten artefact, *in* M Burry, T Dawson, J Rollo, and S Datta (eds), *Hand, Eye, Mind, Digital*, Deakin University,Geelong, pp. 56–62.

Burrow, AL and Woodbury R: 1999, π-resolution and design space exploration, *in* G Augenbroe and C Eastman (eds), *Computers in Building: Proceedings of the CAADF'99 Conference,* Kluwer , pp. 291–308.

Chien, S and Flemming U: 1996, *Design Space Navigation: An Annotated Bibliography*, Technical Report No EDRC 48-37-96, Engineering Design Research Center, Carnegie-Mellon University, Pittsburgh, PA, USA.

Chien, S and Flemming U: 1997, Information navigation in generative design systems, *in* Y Liu, J Tsou, and J Hou (eds), *CAADRIA 97, Vol. 2*, Hsinchu, Taiwan, National Chia Tung University, pp. 355–366.

Datta, S and Woodbury R: 2002, A graphical notation for mixed-initiative dialogue with generative design systems *in* JS Gero (ed), *Artificial Intelligence in Design'02*, Kluwer Academic Publishers, pp. 25–40.

Hybs, I and Gero JS: 1992, An evolutionary process model of design, *Design Studies* **13**(3): 273–290.

Jacobson, I, Booch G, and Rumbaugh J: 1998, *The Unified Modeling Language User Guide*, Reading, Addison Wesley, MA.

Leeuwen, JV and Wagter H: 1998, A features framework for architectural information: Dynamic models of design *in* JS Gero and F Sudweeks (eds), *Artificial Intelligence in Design '98*, Kluwer Academic Publishers, pp. 461–480

Rich, C and Sidner CL: 1998, COLLAGEN: A collaboration manager for software interface agents, *User Modeling and User-Adapted Interaction* **8**(3-4): 315–350.

Schon, DA and Wiggins G: 1992, Kinds of seeing and their functions in designing, *Design Studies* **13**(2): 135–156.

Smithers, T: 2000, Designing a font to test a theory, *in* JS Gero (ed), *Artificial Intelligence in Design '00*, Kluwer Academic Publishers, pp. 3–22.

Smithers, T: 2002, Synthesis in designing, *in* JS Gero (ed), *Artificial Intelligence in Design '02*, Kluwer Academic Publishers, pp. 3–24.

Stouffs, R and Cumming M: 2003, Querying design information through the visual manipulation of representational structures, *in* ML Chiu, JY. Tsou, T Kvan, M Morozumi, and TS Jeng (eds), *Digital Design: Research and Practice*, Kluwer Academic Publishers, pp. 41–53.

Stouffs, R and Krishnamurti R: 2002, Representational flexibility for design, *in* JS Gero (ed), *Artificial Intelligence in Design '02*, Kluwer Academic Publishers, pp. 105–128.

Summers, J, Lacroix Z, and J Shah: 2002, Case-based design facilitated by the design exemplar, *in* JS Gero (ed), *Artificial Intelligence in Design '02*, Kluwer Academic Publishers, pp. 453–476.

Veloso, MM, Mulvehill AM, and Cox MT: 1997, Rationale-supported mixed-initiative case-based planning, *Innovative Applications of Artificial Intelligence*, Providence, RI, pp. 1072–1077.

Woodbury, R, Burrow A, Datta S, and Chang T: 1999, Typed feature structures and design space exploration, special issue on generative design systems, *Artificial Intelligence in Design, Manufacturing and Engineering* **13**(4): 287–302.

Woodbury, R, Datta S, and Burrow A: 2000, Erasure in design space exploration, *in* JS Gero (ed) *Artificial Intelligence in Design'00*, Kluwer Academic Publishers pp. 521–544.

Woodbury, RF and Chang TW: 1995, Massing and enclosure design with SEED-Config., *ASCE Journal of Architectural Engineering* **1**(4): 170–178.

JS Gero (ed), *Design Computing and Cognition'04*, 37-55


# COMPUTER-AIDED CONCEPTUAL DESIGN OF BUILDING STRUCTURES

## *Geometric Modeling for the Synthesis Process*

RODRIGO MORA
*Concordia University, Canada*

and

HUGUES RIVARD, CLAUDE BÉDARD
*University of Quebec, Canada*

**Abstract.** This paper proposes a methodology for allowing engineers to synthesize structural systems more efficiently within a building architectural context. The methodology follows a top-down approach where the engineer focuses first on the overall structural implications of the building architecture, and lets the computer take care of more specific and time consuming tasks. The methodology is being developed in a three-phase project. The first phase is described in this paper. At the core of the methodology lies an integrated representation that is used by the engineer, with the assistance of synthesis algorithms, to create the design model. The representation describes structural and architectural concepts that are relevant during conceptual structural design and relies on geometric modeling techniques for allowing the engineer to reason about the topology and geometry of the model being created. A software prototype has been implemented and a test case demonstrates its capabilities in supporting design synthesis.

## 1. Introduction

There is a consensus among building design practitioners that better engineering feedback to the architect is required early on during the design process. Computer programs for performing structural analysis and detail design calculations have been in the market and used by practitioners for several years. However, there is still a scarcity of computer programs available to assist properly engineers in the conceptual design of building structures. The explanation for this lies in the fact that during conceptual design the engineer's work flow is highly dependent on the amount, quality and type of information that is exchanged with the architect. This in turn affects the quality and effectiveness of the structural engineering feedback provided to the architect. During conceptual design the engineer synthesizes alternative structural layouts while considering multiple conflicting building design criteria coming from the different participants involved in the building design process. This research project emphasizes only on the structural and architectural aspects.

Taking advantage of the recent advances in geometric modeling and parametric techniques, structural engineering packages now simplify the generation, visualization and manipulation of three-dimensional structural models. More recently, a group of applications has emerged that allows the configuration of the structure directly from the building architecture. The main drawbacks of these applications are the following: (1) they promote a constructive generation of the model of the structural system, element by element, in a bottom-up fashion. The constructive approach for conceptual structural design works well for the design of small buildings. However, this approach becomes tedious and error-prone for complex and larger buildings; (2) the integration with the building architecture is limited to selecting physical architectural entities and providing them with structural functionality. Other important factors are not considered, for example, the interdependency between architectural functional spaces and structural support configurations; (3) they lack geometrical reasoning capabilities and structural synthesis knowledge that allow the engineer to reason about the geometry and topology of the model being created while inspecting the building architecture and synthesizing structural solutions.

Structural systems in buildings are hierarchically organized into structural volumes, subsystems, assemblies and elements joined through connections. The main premise of this research project is that, in order to properly support conceptual design of building structures and enable timely engineering feedback to the architect, computers must allow engineers to synthesize the structural hierarchy within a building architectural context. A design methodology is presented to support this premise. The methodology follows a top-down refinement approach. The approach relies on a representation integrating structural and architectural conceptual design entities, and on synthesis algorithms that assist the engineer in synthesizing structural solutions.

The intend of this research project is to provide support for structural synthesis of most typical buildings made out of steel and/or concrete, such as office, apartment and educational buildings. This paper describes results from the first phase of a three-phase research project that aims at assisting engineers during conceptual design of building structures. Each phase builds

on the results from the preceding phases. The goals of these three phases are the following:
*Phase I: Synthesis based on geometric modeling and reasoning* - Provide assistance to the engineer in the inspection of the building architecture, the exploration of structural alternatives using his/her own knowledge, the integration of the structural system to the building architecture, and the verification of structural solutions. At this phase, the assistance is mostly based on geometric modeling and reasoning techniques.
*Phase II: Knowledge based synthesis* - Provide support for the same activities as described in the preceding phase. However, in addition to geometric modeling and reasoning, in this phase support is based on structural engineering knowledge regarding assembly types, loads, materials and element cross-sections, as well as constructability and cost concerns.
*Phase III: Changes and generation of alternatives* - Enable changes to the building architecture and the structural system as a result of two-way feedback between the architect and the engineer, as well as computer generation of feasible structural alternatives.

## 2. Literature Review

From a computer perspective, the problem of conceptual design of building structures can be treated as two closely coupled sub-problems: (1) develop a computer representation that properly describes the physical system being designed, (2) develop computational problem-solving methods that find satisfactory solutions to the design problem.
(1) Representation for the physical system - Several research and standardization efforts have been carried out over the last decade to develop standard representations for the structural system and the building architecture, such as CIMsteel standard and the Industry Foundation Classes. However, with few exceptions (Khemlani et al. 1997; Rivard and Fenves 2000a), the proposed representations are not geared towards supporting conceptual design.
(2) Problem solving methods - Over the last two decades researchers have focused on the second sub-problem by relying mostly on artificial intelligence (AI) techniques to explore conceptual design alternatives; relevant examples are: (1) expert systems (Maher 1985; Ravi and Bédard 1993) (2) formal logic and engineering first principles (Jain et al. 1991; Fuyama et. al 1997), (3) grammars (Meyer 1995), (4) case-based reasoning (CBR) systems (Maher and Zhang 1993; Maher and Balachandran 1994; Bailey and Smith 1994; Kumar and Raphael 1997; Rivard and Fenves 2000b), (5) fuzzy logic (Shen et al. 2001), (6) evolutionary algorithms (Grierson and Khajehpour 2002; Sisk et al. 2003; Rafiq et al. 2003) and (7)

hybrid systems such as a CBR system combined with a genetic algorithm for case adaptation (Soibelman and Peña-Mora 2000).

The main drawback of most AI-based research projects is that they tend to minimize the impact of the architectural design on the structural synthesis process. In addition, some of them simplify the representation of the physical system in order to adjust to the requirements of the selected problem solving technique(s). This research project aims at overcoming these limitations. In its first phase, it focuses on the representational aspects of the structural system and the building architecture, at the conceptual stage, while considering basic computational problem-solving aspects through synthesis algorithms. It is expected that more advanced computational techniques will benefit from the representation being proposed in this research project.

## 3. Design Methodology

The hierarchical organization of the structural system naturally influences the way engineers think while synthesizing structural configurations. Following this hierarchical organization, Lin and Stotesbury (1988) developed a, so called, total-system approach, which is essentially a top-down refinement approach. The total-system approach allows overall structural concepts to become contexts for thinking about local issues of detail component interactions and ensures compatibility between overall concepts and their constituent components. It also allows relating structural concepts at different levels of the structural hierarchy to architectural schemes, which translates in engineering feedback to the architect at each hierarchical level.

Rivard and Fenves (2000a) proposed a design model for the conceptual design of building structures, which is inspired by the total-system approach. In this model, the structural engineer is initially concerned with establishing three-dimensional structural schemes that respond to architectural space-form schemes (Lin and Stotesbury 1988). The hierarchical representation first breaks down the structural system into independent structural volumes that are assumed to behave as structural wholes. Independent Structural Volumes are in turn subdivided into smaller sub-volumes called structural zones. Structural zones are introduced in order to allow the definition of spatial structural requirements and loadings that correspond to architectural functions. Independent Structural Volumes are also decomposed into three structural subsystems, namely the foundation, the gravity, and the lateral subsystems. Each of these subsystems is further refined into structural assemblies. Finally, structural assemblies are decomposed into structural elements and their connections. The methodology proposed in this research project follows a total-system approach and uses the aforementioned top-down design model.

As illustrated in Figure 1, the engineer performs the synthesis process by following a sequence of steps numbered 1 to 5. In the diagram, thick vertical arrows oriented downwards indicate the top-down sequence of tasks performed by the engineer, while thin upwards arrows indicate backtracking between tasks. Activities numbered 2 and 4 have two sub-activities.

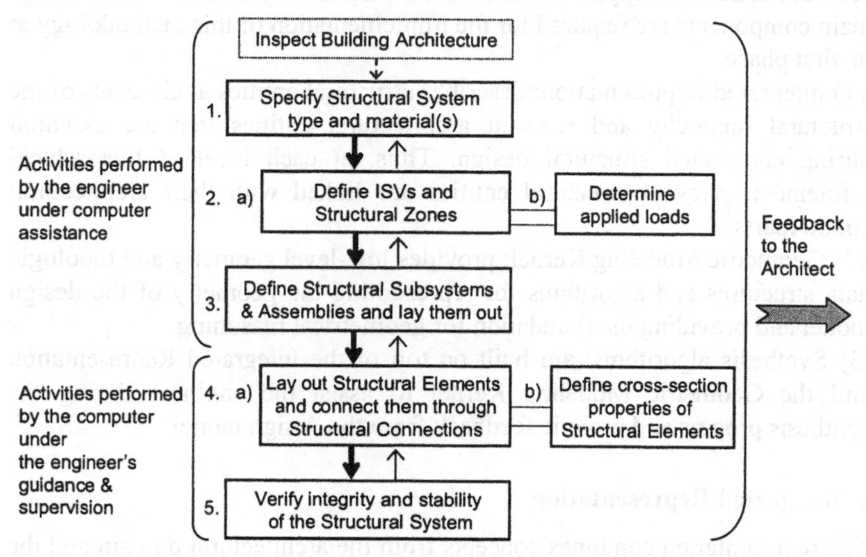


*Figure 1.* Design methodology for the synthesis stage in 5 steps

As shown in Figure 1, before beginning the actual synthesis process the engineer inspects the building architecture and possibly suggests global and/or local changes to the architecture (Meyer 1995). Although this activity actually takes place throughout the entire process, it is important however to make explicit this initial engineering feedback to the architect. In activity number 1 the engineer specifies the type of structural system based on the material, for example steel, concrete or composite structural system. Another important activity also takes place at the end of the process, which is the verification of the integrity and stability of structural systems being configured.

As indicated in the diagram, activities numbered 1, 2 and 3 are performed by the engineer with computer assistance whereas activities numbered 4 and 5 are performed by the computer under the guidance and supervision of the engineer. Therefore, the engineer is in charge of making strategic design decisions, which involve selecting structural system type and material(s), as well as subsystems and assemblies, and laying them out. While the computer takes care of more specific and time consuming tasks, such as arranging and connecting elements into assemblies, under the engineer's guidance and supervision. The big horizontal arrow pointing to the right indicates that feedback to the architect may be provided at any step during the process.

Phase I of this research project tackles activities 1 through 5, except for activities 2.b) and 4.b) that will be tackled in Phase II. While Phase III provides enhanced support for each aforementioned synthesis activity. Three main components are required for the implementation of this methodology at its first phase:

(1) Integrated Representation: describes structural entities at all levels of the structural hierarchy and relevant architectural entities that are essential during conceptual structural design. Thus, at each level of hierarchical refinement, relevant structural entities are linked with their architectural counterparts.

(2) Geometric Modeling Kernel: provides low-level geometry and topologic data structures and algorithms for representing the geometry of the design model and providing the foundation for geometrical reasoning.

(3) Synthesis algorithms: are built on top of the Integrated Representation and the Geometric Modeling Kernel to assist the engineer during the synthesis process and provide feedback from the design model.

## 4. Integrated Representation

The representation combines concepts from the architectural domain and the structural domain in a single representation scheme, Figure 2. The outcome of the conceptual design process is therefore a single building model. UML (Unified Modeling Language) notation is used in Figure 2 (Booch et al. 1997) to describe the entities of the representation along with their interactions. In the body of the text **bold** typeface is used to describe such entities. The more abstract entities in the representation, which are defined by the architect, are the **Site** and the **Building**. The **Site** may contain several **Buildings**. The **Building** includes a reference **Grid**. As shown in Figure 2, the **Building** has two more sub-entities: **Architectural_Model** and **Structural_System** that are in turn decomposed into their constituent entities. The central shaded portion of the figure shows key entities and relationships that act as links between the two disciplines.

*Architectural Domain Representation:*

The **Storey** is the most central concept in the **Architectural_Model**; all other entities in the **Architectural_Model** refer to the **Storey**. **Space** is another key concept; the **Architectural_Model** includes two types of **Space** entities: **Primary_Space** and **Secondary_Space**. A **Primary_Space** is the smallest space that is most explicitly established by physical objects such as surfaces, screens, etc. **Secondary Spaces** are groupings of **Primary Spaces,** for example, an entire floor could be defined as a **Secondary Space**. This space decomposition allows the later association of structural entities and function to architectural spaces. Two **ASlab** entities, i.e. the architectural view of a Slab, cover a **Storey**, one at the top (ceiling) and another at the bottom (floor). **Storeys** also have **AWall** and **AColumn** entities, which correspond to architectural views of **Walls and Columns**.

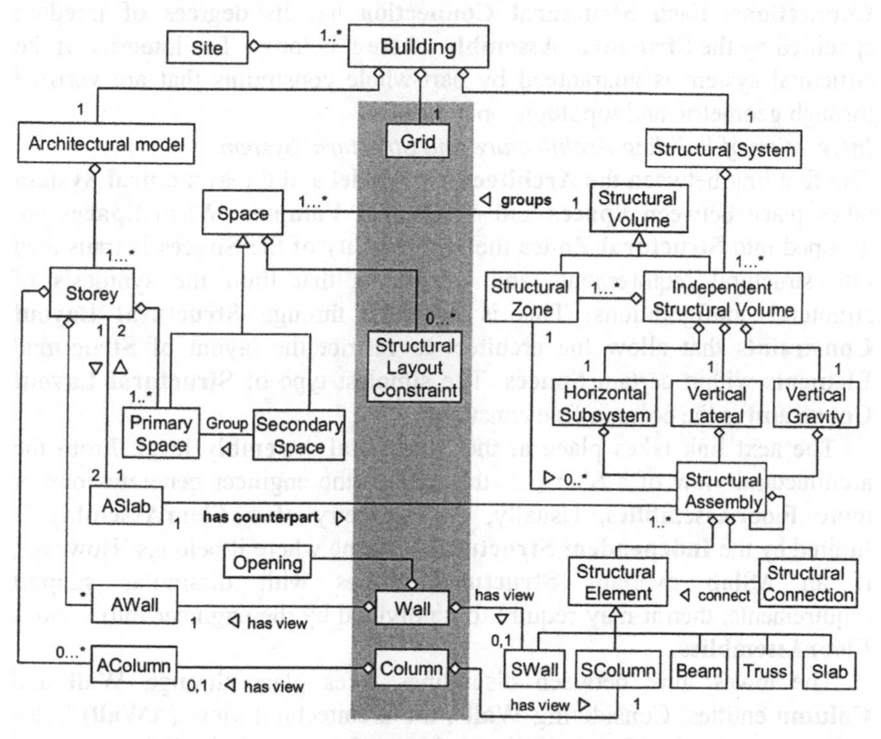


*Figure 2.* Integrated representation

*Structural_System Domain Representation:*

The **Structural_System** is subdivided into one or several **Independent Structural Volumes**. Each **Independent Structural Volumes** acts as a place-holder for the entities of its self-contained structural skeleton. **Independent Structural Volumes** aggregate **Structural Zones. Structural Zones** include additional structural requirements, such as prescribed loads, and column spacing. **Independent Structural Volumes** and **Structural Zones** are volumetric entities and therefore are classified as **Structural Volumes. Independent Structural Volumes** are also subdivided into three types of subsystems: **Horizontal Subsystem, Vertical Lateral subsystems,** and **Vertical Gravity subsystem.** Structural subsystems do not have geometry of their own, but their geometry is defined through aggregation of their constituent **Structural Assemblies. Structural Assemblies** are in turn specialized into more specific types such as **PlanarFrame, FloorAssembly,**

**Wall-Stack** and **Column-Stack**; which are not shown in Figure 2. **Structural Assemblies** are composed of **Structural Elements** (e.g. **Beams**, **Columns**, etc.), and may be composed of other **Structural Assemblies** (sub-assemblies). **Structural Elements** are joined together through **Structural Connections**. Each **Structural Connection** has its degrees of freedom specified by the **Structural Assembly** where it belongs. The integrity of the structural system is guaranteed by part-whole constraints that are verified through geometric and topologic operations.

*Integration of Building Architecture and Structural System:*

The first link between the **Architectural Model** and the **Structural System** takes place between **Spaces** and **Structural Volumes**. When **Spaces** are grouped into **Structural Zones** the functionality of the **Spaces** is translated into structural requirements and constraints that limit the synthesis of structural configurations. This is achieved through **Structural Layout Constraints** that allow the architect to restrict the layout of **Structural Elements** within certain **Spaces**. The simplest type of **Structural Layout Constraint** is the column-free constraint.

The next link takes place at the **Structural Assembly** level. From the architectural view of a Slab (i.e. the **ASlab)** the engineer generates one or more **FloorAssemblies**. Usually, the geometry of a **FloorAssembly** is limited by the **Independent Structural Volume** where it belongs. However, if an **ASlab** overlaps **Structural Zones** with dissimilar support requirements, then it may require to be divided by the engineer into various **FloorAssemblies.**

The lowest link between disciplines takes place through **Wall** and **Column** entities. Considering **Walls**, the architectural view (**AWall**) holds information that is relevant to the architect, for example its link to a given **Storey**, while the structural view **(SWall)** holds structural information only, as well as its relationship to a **Wall-Stack**.

## 5. Synthesis Algorithms

Synthesis algorithms generate structural configurations and give feedback from the emergent model to the engineer, thus allowing him/her to reason based on geometry and topology of the model being created. This feature is mostly relevant during the synthesis stage of design since it is at this stage where most decisions result from geometric and topologic concerns. Synthesis algorithms rely on the following features to assist engineers in creating feasible structural configurations:

(1) Implicit and explicit relationships among entities within the model and entities being generated. As described by Rivard et al. (2000a) the relationships between entities can be categorized in three groups: a) spatial relationships (e.g. adjacent, overlap, etc), b) class specific relationships (e.g.

aggregation, association) including "part-whole" relationships, c) domain-specific relationships (e.g. supports, attached-to, connects, etc).
(2) Knowledge encapsulated in the entities either already configured within the model or to be integrated into the model that constrains their configuration as well as the definition of related entities. For example, according to their type, **FloorAssemblies** impose limits on their spans and therefore on the dimensions of structural bays.
(3) The geometry of the model being created. Geometric constraints such as planarity, alignment and parallelism restrict the configuration of all entities within the model.

Examples of synthesis algorithms that have been implemented in a software prototype are presented in the remainder of this section, the name of the algorithms is given in *Italic* typeface:

- *findBeamSupports:* finds supports for a **Beam** that is not properly supported. It first searches for supporting **Columns** and **Walls** within the **Storey** below. If the **Beam** is still not properly supported after inspecting the vertical support system below, then the algorithm looks for primary support **Beams** within the current **FloorAssembly** that could support the **Beam**. Once a support is found, the algorithm creates a **Structural Connection** for joining the **Beam** to its support.
- *verifyWallContinuity:* verifies the vertical continuity of a given **Wall**. The algorithm considers the **Storeys** above and below and searches for **Walls** whose geometry coincides in plan view (x,y) with the **Wall** geometry.
- *uniformDepth:* generates floor framing layouts over rectangular **Structural Zones**. The algorithm is based on the principle that long, lightly loaded secondary **Beams** delivering their loads to shorter primary **Beams**, or **Girders**, will lead to uniformity of structural depth (another approach for floor framing layout is the minimum weight strategy, which will be implemented in Phase II). Note that the algorithm also works for buildings whose floor plan is an arrangement of rectangular spaces. In such cases, the engineer is required to define few primary **Beams** that divide the floor plan into rectangular areas and possibly indicate floor framing directions. For example, Figure 4a shows a column-free L-shaped space. If the engineer places girder "G1", thus dividing the **Floor Assembly** into two rectangular regions (Figure 3b), then the algorithm can take care of the actual arrangement of **Structural Elements** within the two regions, Figure 3c. In defining girder "G1" the engineer assumes that the girder will resist loads coming from both rectangular regions. In subsequent stages of the design process, activity 4.b in Figure 1, this assumption will be verified.

In the example, the engineer makes the decision of placing girder "G1" using his/her own structural engineering knowledge and experience, as well as cost and constructability concerns since the layout of girder "G1" results in uniform girders of length "d1" in both rectangular regions. This kind of knowledge has not been incorporated in the prototype.

- *verifyGravityLoadPath:* given any physical building element the algorithm finds the **Slab Element**(s) that directly support (i.e. are in contact with) the given building element. Then, from those **Slab Elements** the algorithm propagates the search for supports down to the ground. For example, Figure 4 shows gravity load paths for an object lying on the **FloorAssembly** that supports the roof of a two-storey two-bay structure. The object is directly supported by two **Slab Elements**, which are shown in darker color. From these **Slab Elements** the gravity loads are propagated to the perimeter **Beams** and then to the **Columns** that finally transmit the loads to the ground.

  Note that the algorithm simply verifies load paths through geometric operations. Tributary areas are not considered and the actual load distribution in the supports is not considered either because it requires the use of engineering first principles governing redundant systems (unless the system is statically determinate).

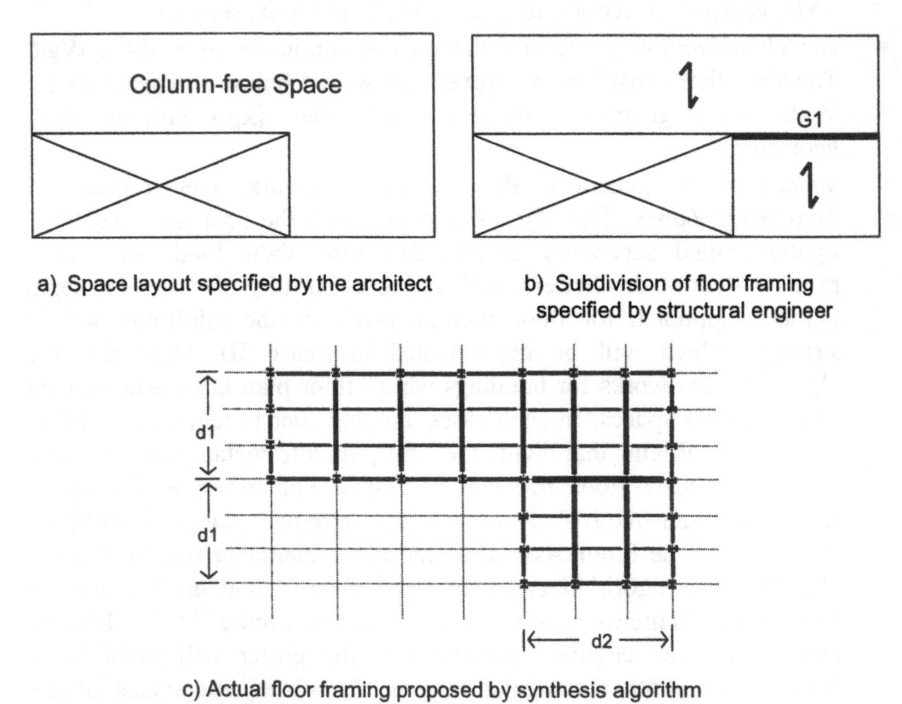


*Figure 3.* Synthesis algorithm works under instructions from the engineer

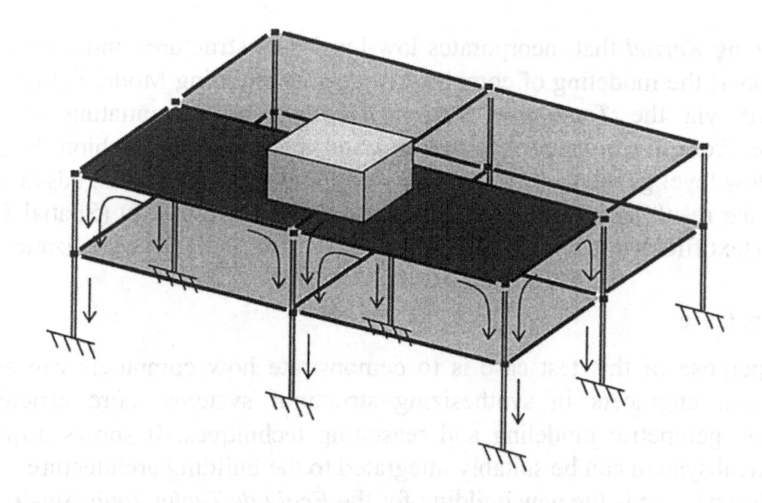

*Figure 4.* Gravity load paths

- *verifyLateralLoadPaths:* tests for the existence of a complete set of lateral load paths. It searches for **PlanarFrames** in two principal directions that are connected to floor diaphragms through **Beam** collectors and resist lateral loads either through moment connections or bracings. Thus, the number and configuration of lateral resisting **PlanarFrames** is verified to guarantee lateral support in both principal directions.

Whenever a lack of support is detected by the computer, using any of the algorithms, the engineer is notified about the problem. Then, it is up to the engineer to take corrective actions.

## 6. Software Prototype

This section describes a software prototype that aims at assisting the engineer in performing top-down design synthesis following the steps described in the proposed methodology. With reference to Figure 1, the prototype currently supports all the tasks of Phase I (i.e. excluding only activities 2.b and 4.b). The prototype has been implemented in C++. The overall system architecture is shown in Figure 5. The description that follows uses ***Bold, Italic*** typeface to identify the modules of the prototype. The user, either an architect or an engineer, enters data via an ***Alphanumeric User Interface*** (either an input text file or alphanumeric menu). The ***Top-down Design Manager*** is the module in charge of enabling the top-down synthesis process. The ***Top-down Design Manager*** interacts with the ***Synthesis Algorithms*** and with the ***Integrated Representation*** to produce the model of the structural system that is integrated to the architectural model. The entire system is built on top of ACIS®, which is an advanced ***Geometric***

***Modeling Kernel*** that incorporates low-level data structures and algorithms to support the modeling of complex artifacts. A Building Model is therefore created, via the ***Top-Down Design Manager*** by instantiating relevant entities from the ***Integrated Representation*** in a top-down fashion. Finally, only low level geometry (planes, lines, etc.) and topology (faces, edges, etc.) from the resulting building model are saved into an ACIS® (Spatial Corp 2003) text file that is read by HOOPS® (TSA Inc. 2003) for visualization.

## 7. Test Case

The purpose of this test case is to demonstrate how computers can assist structural engineers in synthesizing structural systems more efficiently through geometric modeling and reasoning techniques. It shows how the structural system can be suitably integrated to the building architecture.

The test case is the new building for the *École de Technologie Supérieure* in Montreal which is currently under construction and will open in the fall of 2004. The building has five stories above ground that occupy an area of 20,400 $m^2$ and two parking stories below ground. The building houses several activities such as: classrooms, computer rooms, cafeteria, sports facilities and various administrative services. Figure 6 shows an architectural rendering of the building and Figure 7 shows the building under construction.

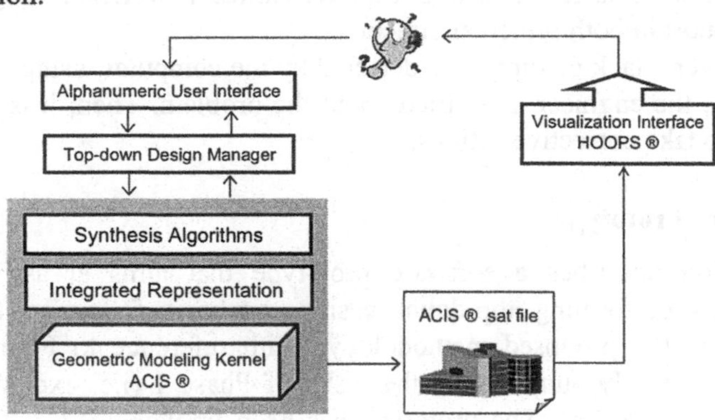


*Figure 5.* Prototype system architecture

Given the programmatic requirements, the architect performs architectural design and produces a computer model of the building. The architect also specifies modular functional dimensions in both principal directions thus defining the project **Grids**. Figure 8, shows the architectural model developed using the software prototype. Note that the circular atrium in the front was simplified into a rectilinear shape because the prototype is currently limited to these shapes.

*Figure 6.* Architectural rendering

*Figure 7.* Building under construction

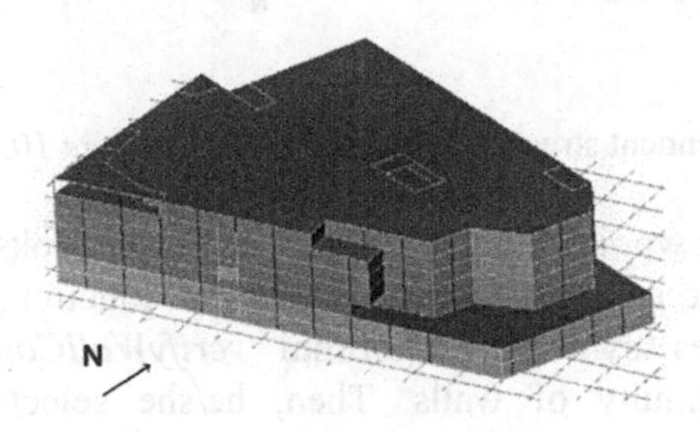


*Figure 8.* Architectural model created with the prototype

The following steps illustrate how the software prototype helps the engineer to synthesize the structural system directly from the building architecture.

- The engineer initially inspects the **Architectural_Model** in search for potential structural opportunities and difficulties (Meyer 1995). The engineer may suggest global and/or local changes to improve the structural properties of the building architecture.
- Then, the engineer begins the generation of the **Structural_System** through the definition of **Independent Structural Volumes** that group **Spaces**. As shown in Figure 9, the engineer creates two **Independent Structural Volumes**: *ISV1* and *ISV2*. For each **Independent Structural Volume** the engineer defines material(s) and structural subsystems: **Vertical Gravity subsystem, Horizontal Subsystem** and **Vertical Lateral subsystem.**
- For each **Independent Structural Volume**, the engineer groups Spaces having similar structural characteristics into **Structural Zones**. In Figure 10, *ISV1* has two **Structural Zones**: one groups spaces above ground, and the other groups parking spaces below ground. *ISV2* is in

turn divided into three **Structural Zones**: a parking zone, a zone with classrooms and administrative services, and a zone that houses the gym and a multi-functional sports space. The last **Structural Zone**, indicated lighter in Figure 10, groups Spaces that are column-free (the gym and the multi-functional sports room). Thus it carries a column-free **Structural Layout Constraint** that restricts the layout of the vertical support system.

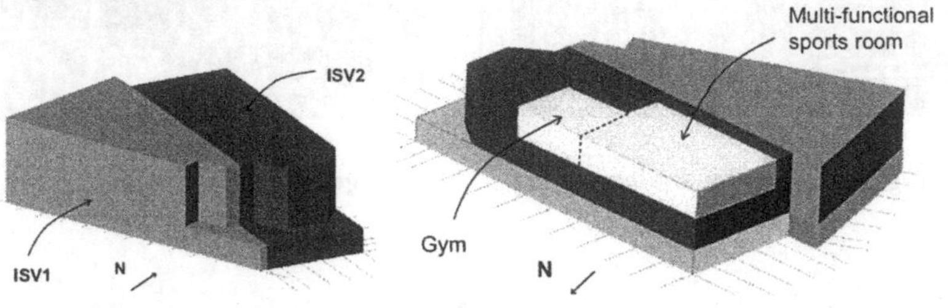


*Figure 9.* Independent structural volumes *Figure 10.* Structural zones

- Then, as shown in Figure 11, the engineer looks for architectural elements that could perform a structural function. In doing so the engineer uses synthesis algorithm *verifyWallContinuity* to verify vertical continuity of walls. Then, he/she selects core walls and retaining walls to become part of the structural system, thus defining **Wall-Stacks**. The computer automatically integrates the **Wall-Stacks** to the **Vertical Lateral subsystem** and the **Vertical Gravity subsystem**.

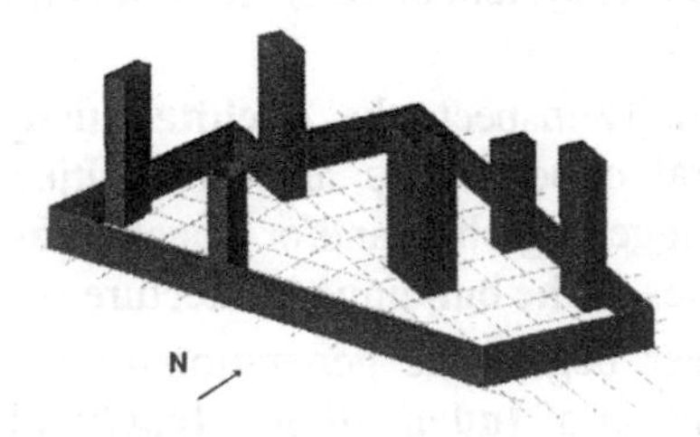


*Figure 11.* Engineer selects walls to become structural walls

- The engineer defines **PlanarFrame** types in both principal directions (Moment-Resistant Frame, Simple-Gravity Frame, etc.). Then, by following the project **Grids** and respecting space functionality the engineer lays out **PlanarFrames** in both principal directions, thus implicitly defining column lines and structural bays. Each **PlanarFrame** is automatically integrated to its corresponding **Vertical**

**Lateral** subsystem. Then, the computer generates the abstract geometry of **PlanarFrames**, i.e. a vertical plane. Figure 12 shows the abstract geometry of all **PlanarFrames**. **Wall Stacks** are automatically integrated to **PlanarFrames** to whom they are coplanar.

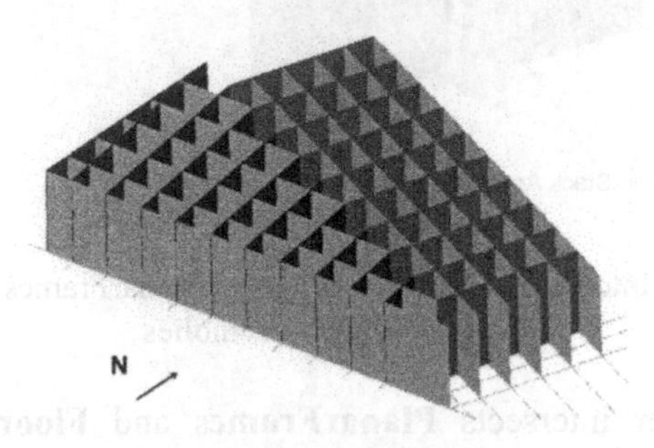


*Figure 12.* Abstract geometry of Frame2Ds

- As illustrated in Figure 13, the engineer selects **ASlabs** created by the architect and defines **FloorAssemblies.** Then the prototype computes the abstract geometry of **FloorAssemblies** (a horizontal plane), which extends throughout the **Independent Structural Volume** where they belong except where **Structural Zones** require a special type of **FloorAssembly**. This is the case of the **Structural Zone** that groups the gym and the multi-functional sports room, which is column-free thus requiring a special type of roof system. This roof system (i.e. **FloorAssembly**) is configured using algorithm *uniformDepth.* Each **FloorAssembly** is automatically integrated to the **Horizontal Subsystem.**

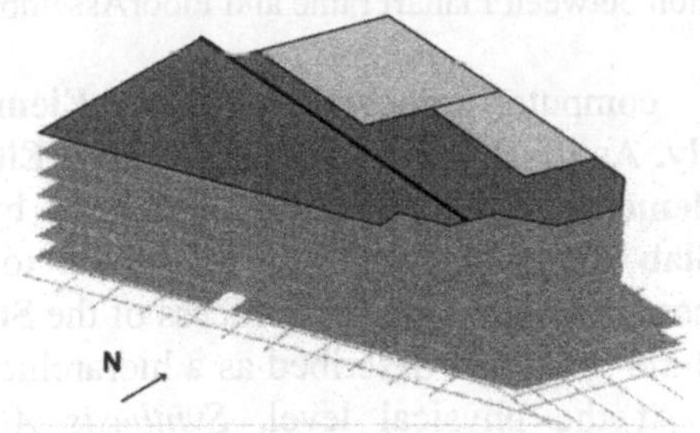


*Figure 13.* Abstract geometry of FloorAssemblys

- The computer intersects the abstract geometry of **PlanarFrames** and generates **Column-Stacks,** Figure 14. The actual definition of each **Column** within a **Column-Stack** is subject to approval by the **Structural Zone** where it belongs.

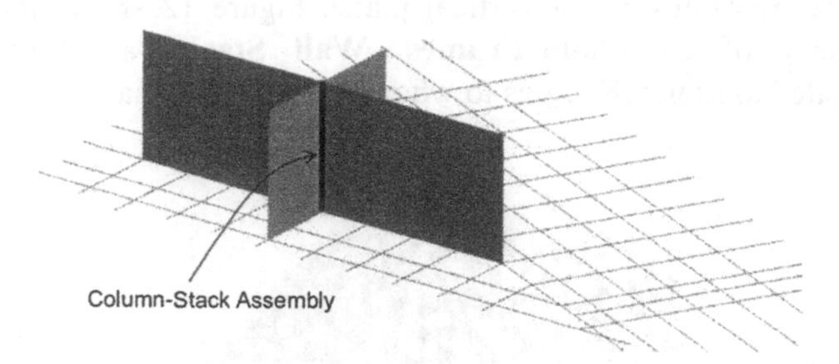


*Figure 14.* Intersection of two orthogonal PlanarFrames generates Column-Stack Assemblies

- The computer intersects **PlanarFrames** and **FloorAssemblies** thus generating primary **Beams**. Figure 15 illustrates that **Beams** are created at the intersection between the abstract geometry of a **PlanarFrame** and the abstract geometry of a **FloorAssembly**. The generation of **Beams** takes into consideration span thresholds as dictated by the **FloorAssembly** being created, as well as cantilevers and openings.

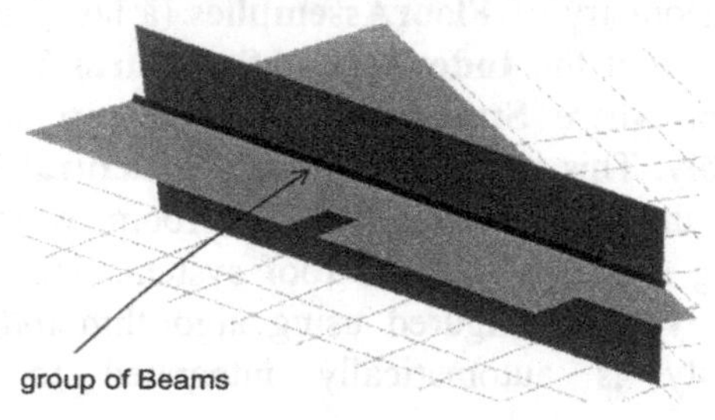


*Figure 15.* Intersection between PlanarFrame and FloorAssembly generates Beams

- Finally, the computer generates **Slab Elements** for each **FloorAssembly.** As illustrated in Figure 16, **Slab Elements** are planar **Structural Elements** whose geometry is defined by **Beams** in their perimeter. **Slab Elements** are then connected to their supporting **Beams**. This completes the synthesis process of the **Structural System.**

The resulting structural system is described as a hierarchical organization of structural entities. At the physical level, *Synthesis Algorithms* arrange structural elements and join them through connections. This completes the support at the first phase of this research project. Work is in progress however, to devise additional synthesis algorithms to anticipate potential structural problems, for example coming from asymmetric building configurations and poorly proportioned building forms. The second phase will attempt at providing support for the following activities from the proposed methodology, Figure 1:

Activity 2.b) - Determination and modeling of applied loads.

Activity 4.b) - Definition of cross-sectional properties of Structural **Elements**. The prototype considers **Structural Elements** only in a generic sense since **Beams, Girders, Joists** and **Columns** are not given any properties.

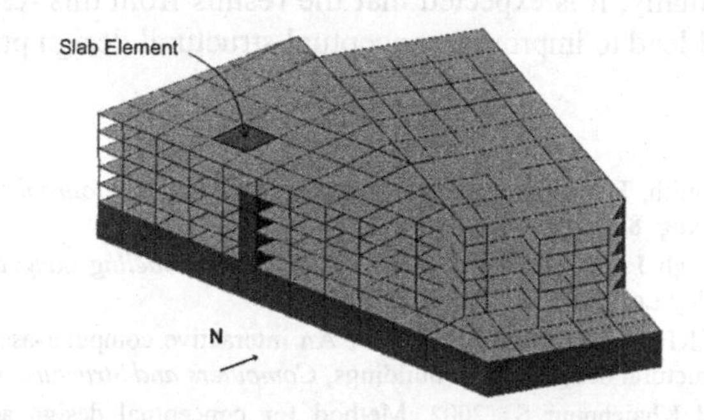


*Figure 16.* The resulting structural system

Consequently, the verification of the structural system, task 5 in Figure 1, will also be improved to consider loads and materials. For the second phase, a formal mechanism is therefore required to incorporate structural engineering knowledge regarding assembly types, loads, materials and element cross-sections, as well as constructability and cost concerns. This will lead to more knowledgeable synthesis algorithms. Finally, Phase III of this research project requires incorporating parametric capabilities for enabling changes, as well as a mechanism for generating and evaluating alternative structural configurations.

## 8. Conclusions

A design methodology has been proposed to assist engineers to synthesize structural systems more efficiently within a building architectural context. The methodology allows the engineer to make decisions at higher levels of the structural hierarchy and guide the computer to perform the most tedious and time consuming tasks at the lowest levels. This is done through a top-down synthesis approach that allows the engineer to synthesize structural schemes at different levels of the structural hierarchy and respond at each level to the requirements of the architecture. The methodology is grounded on an Integrated Representation that describes early stage concepts from the building architecture and the structural system. While, synthesis algorithms assist the engineer during the synthesis process and provide feedback from the design model. A software prototype has been implemented that demonstrates the advantages of the proposed approach. In its current state, the prototype relies mostly on geometry and topology concerns for assisting the engineer during design synthesis (Phase I). The second phase of this research project will incorporate structural engineering knowledge, as well as constructability and cost concerns; followed by the third phase that aims at incorporating changes and computer generation of feasible structural alternatives. Finally, it is expected that the results from this research project as a whole will lead to improved conceptual structural design practice.

## References


Bailey, SF and Smith, FC: 1994, Case-based preliminary design, *Journal of Computing in Civil Engineering* **8**(4): 454-468.

Booch G, Rumbaugh J and Jacobson I: 1997, *The Unified Modeling Language User Guide*, Addison-Wesley Longman Inc, Reading, MA.

Fuyama, H, Law KH and Krawlinker H: 1997, An interactive computer-assisted system for conceptual structural design of steel buildings, *Computers and Structures* **63**(4): 647-662.

Grierson DE and Khajehpour S : 2002, Method for conceptual design applied to office buildings, *Journal of Computing in Civil Engineering* **16**(2): 83-103.

Jain D, Krawinkler H and Law KH: 1991, *Logic-based Conceptual Structural Design of Steel Office Buildings*, Technical Report, Center for Integrated Facility Engineering, Stanford University, CA, USA.

Khemlani L, Timerman A, Benner B and Kalay Y: 1997, Semantically rich building representation, *ACADIA*, pp. 207-227.

Kumar, HS, and Raphael B: 1997, CADREM: A Case-based system for conceptual structural design, *Engineering with Computers* **13**(3): 153-164.

Lin TY and Stotesbury SD: 1988, *Structural Concepts and Systems for Architects and Engineers*, 2nd Ed. Van Nostrand Reinhold, New York.

Maher ML: 1985, HI-RISE: *A Knowledge-Based Expert System for the Preliminary Structural Design of High Rise Buildings*, Ph.D. Thesis, Department of Civil Engineering, Carnegie Mellon University, Pittsburgh, PA.

Maher ML and Balachandran MB: 1994, Multimedia approach to case-based structural design, *Journal of Computing in Civil Engineering* **8**(3): 359-376.

Maher ML and Zhang DM: 1993, CADSYN: A case-based design process model, *AIEDAM* **7**(2): 97-110.

Meyer S: 1995, *A Description of the Structural Design of Tall Buildings through the Grammar Paradigm,* Ph.D. Thesis, Department of Civil Engineering, Carnegie Institute of Technology, Carnegie-Mellon University, Pittsburgh, PA, USA.

Rafiq MY, Mathews JD and Bullock GN, 2003, Conceptual building design – evolutionary approach, *Journal of Computing in Civil Engineering* **17**(3): 150-158.

Ravi M and Bédard C: 1993, Approximate methods of structural analysis and design in a knowledge-based system environment, *Artificial Intelligence in Engineering* **8**: 271-275.

Rivard, H and Fenves SJ: 2000a, A representation for conceptual design of buildings, *Journal of Computing in Civil Engineering* **14**(3): 151-159.

Rivard, H and Fenves SJ: 2000b, SEED-Config: A case-based reasoning system for conceptual building design, *Artificial Intelligence for Engineering Design, Analysis and Manufacturing* **14**: 415-430.

Shen YC, Bonissone PP and Feeser LJ: 2001, Conceptual modeling for design formulation, *Engineering with Computers* **17**: 95-111.

Sisk GM, Miles JC and Moore CJ: 2003, Designer centered development of GA-based DSS for conceptual design of buildings, *Journal of Computing in Civil Engineering* **17**(3): 159-166.

Soibelman L and Peña-Mora F: 2003, Distributed Multi-Reasoning Mechanism to Support Conceptual Structural Design, *Journal of Structural Engineering* **126**(6): 733-742.

Spatial Corp.: 2003, 3D ACIS Modeler version 8.0, http://www.spatial.com.

TSA Inc.: 2003, HOOPS 3D Part Viewer version 7.0, http://www.hoops3d.com.

JS Gero (ed), *Design Computing and Cognition'04*, 57-76


# THAT ELUSIVE CONCEPT OF CONCEPT IN ARCHITECTURE

*A First Snapshot of Concepts during Design*

ANN HEYLIGHEN
*Katholieke Universiteit Leuven, Belgium*

and

GENEVIEVE MARTIN
*Université de Liège, Belgium*

**Abstract.** Design concepts in architecture typically feature in post-hoc explanations or reviews of finished design projects. By contrast, this paper tries to chase concepts during the design process, and shed more light on when, why and how they pop up in the designer's mind. To this end, the role of concepts is studied through the eyes of an architect involved in the design of a school building. The results of this pilot study amount to a tentative model of the generation and development of concepts, which tries to take into account their highly elusive character.

## 1. Introduction

Design concepts in architecture have been written about by many authors before. The majority of these writings, however, analyzes and criticizes concepts as end products of the design process. By contrast, we would like to depict them from the point of view of the design process itself. In other words, we want to look at them in a 'modus operandi' as opposed to a 'modus operatum' (Bourdieu 1977), that is how their generation, as it unfolds over time, is perceived by an architect working on a design, instead of how it looks with the hindsight of being finished. To our knowledge, hardly any empirical studies have been devoted to this issue, at least not in the domain of architecture. By way of first step, this paper therefore examines the role of design concepts from the point of view of an architect involved in the design of a school building. The protocol study reported on in this paper was originally set up to test the prototype of a design tool.

During the analysis, however, the protocol turned out to be cut out to help pin down that elusive concept of concept in architectural design.

After briefly motivating our interest in design concepts and their role during the design process, Section 2, we will describe the setting and procedure of the protocol study, Section 3. Based on the observations and results of the analysis, Section 4, we propose a tentative model of the emergence and development of concepts, which tries to take into account and explain their highly elusive character, Section 5. We discuss the lessons learned so far, Section 6 and conclude with related and future work, Section 7.

## 2. Motivation

In architecture, as in other domains, there are different viewpoints to assess the value of a design, and along with the viewpoint the judgment may vary. At one time in history considerations of form determined design quality, later functional ones. Today many people seem to assume that what makes an architectural design valuable is its underlying idea. As Lawson (1994) contends, "Good designs often seem to have only a very few major dominating ideas which structure the scheme and around which other relatively minor considerations are organized." There is nothing wrong with taking this view, provided we are clear about it being *our* view at *this* moment.

The ideas underlying architectural designs are known to architects by many names, ranging from 'image' (Alexander 1979) over 'primary generator' (Darke 1978) to 'organising principle' (Rowe 1987), but most often are called the 'parti' (Leupen et al. 1997) or 'concept' (Lawson 1994). Judging from architecture and design literature, these ideas do not necessarily require the addition of an extra ingredient. In fact, every aspect already present in the design situation, e.g. a special feature of the site or program, or a curious trait of the client, may qualify for this role. Moreover, underlying ideas do not necessarily play solo. In the Institut du Monde Arabe in Paris, for instance, Jean Nouvel combined the need for sun shading with a 'Moucharabieh'[1] pattern and the idea of a light controlling diaphragm in a camera lens (Sharp 1990). This resulted in a gigantic Islamic pierced screen, which makes this modern high-tech building a permanent reference to traditional Islamic architecture, Figure 1.

[1] In Islamic or Islamic-influenced architecture, Moucharabieh is a projecting second-story window of latticework, made up of small wooden bobbins composed into intricate geometric patterns. It is a familiar feature of urban residences in North-Africa and the Middle East, where it provides the interior with light and air while shading it from the hot sun.

Descriptions of design concepts like this one frequently pop up in lectures on or reviews of architectural projects, which suggests design concepts to play a key role in their explanation and evaluation. Since these descriptions typically derive from retrospection by the designer or critical analysis of the finished design, however, the question arises: is the role of design concepts limited to explaining or justifying post-hoc the design product, or do they feature already in the design process? If so, when, why and how do they come into play? At which point in the design process do they pop up in the consciousness of the designer? Where do they find their source? And do we need them really or are they an artifice of mediation?

*Figure 1.* Façade of Jean Nouvel's Institut du Monde Arabe (Photo: Herman Neuckermans)

This focus on process in architectural design is a relatively recent phenomenon. For many years and centuries, designing was learned implicitly by watching and working with a master. It is only recently that researchers slowly have become interested in 'how architects think in action,'[2] i.e. in the architectural design process (Akin 1979; Hamel 1990; Lawson 1980; Rowe 1987). According to Eastman (1999), there are many reasons why the present decade will – and should – be marked by a radical

[2]After the title of Schön's book *The Reflective Practitioner. How Professionals Think in Action* (1983).

focus on process in design: in order to better manage complexity, facilitate collaboration among specialists in design teams, improve design education, and enhance tool support for design. Nevertheless, there still is a great reluctance to narrow the limits of the 'black box' and make the design process explicit. To some such explicitness would remove the mysticism and art of design by reducing it to a standard procedure or method. Our focus on the architect's design process, however, is not inspired by an interest in method or systematization. Instead, it derives from the hope that a more profound understanding of design processes will ultimately improve the quality of our built environment by supporting the design thereof.

## 3. Experimental Setup

By way of first step towards such understanding, this paper tries to examine the role of design concepts from the perspective of an architect involved in the design of a school building. Let us first describe the specific conditions under which the protocol under study was generated. Originally, we have mentioned, the protocol was part of a pilot study to test a prototype design tool, more specifically to explore the potential relevance of an on-line case library for design practice (Heylighen and Neuckermans 2001).[3] To this end, four architects – two junior and two expert designers – were invited to use the prototype during a two-hour design session.

### 3.1. PROCEDURE & ASSIGNMENT

A week before the design session, the architects were given a demonstration of the prototype and a brief manual in case they wanted to 'practice' using the tool at home.

At the start of the actual design session, architects were asked for a proposal to reorganize and extend an architecture school, which is located in a 16$^{th}$ century castle, Figure 2. The task was to reorganize and optimize the West wing of the castle (design studios, lecture rooms, secretariat, photocopy room) and extend it with a reception hall, material museum and exhibition room. Small-scale plans of all floors and a number of pictures of the building exterior were provided in the room where the experiment took place.

Apart from having access to the case library, the architects could go about the design task as they preferred. Two restrictions resulted from the method used: the design session was limited to two hours and the subjects

---

[3] Because the design task used in the pilot study involved the redesign of an existing building, the protocol was also 'reused' as part of a research on redesign (Lindekens et al. 2003).

were asked to 'think aloud' as they were designing. To help them become accustomed to thinking aloud, the session was preceded by a short training exercise. Moreover, once the design session had started, the architects were encouraged to think aloud if intervals of silence lasted more than 30 seconds. During this session, all actions of the architects were audio- and videotaped.

Afterwards the drawings and notes were collected and numbered chronologically, and the tapes were transcribed.

*Figure 2.* West wing of the 16th century Castle of Arenberg (Photo: Paul Van Aerschot)

3.2. SUBJECTS

Both juniors had five years of experience as practicing architect, their senior colleagues 13 and 19 years respectively. Although not 'world-famous' architects, the latter can be considered expert designers in that their work has attracted major design awards and won important competitions. All four knew the castle quite well, since they have studied and/or are currently teaching at the school.

Only three architects finished the session. The fourth left the experiment in an early stage, because she found designing in front of a camera too stressful.[4] When analyzing the three remaining protocols, the design session of the most senior stood out as exceptionally interesting from a 'conceptual' point of view. The next section describes a qualitative analysis of this protocol. The informal quality of the analysis seems justified given its exploratory character and its aim, which is to examine whether design concepts play a role *during* the design process, and if so, to try and shed more light on this role.

Since the path of only one architect is investigated here, it is obvious that no general conclusions can be drawn. Some aspects of this protocol clearly differ from the other two protocols,[5] not to mention protocols generated under completely different conditions or real-world design processes. It is nevertheless important as preliminary work in developing a more profound understanding of the role of concepts during design, as it enables us to start addressing the questions raised above and postulate a first tentative model of concept generation.

## 4. Observations

Upon completion of the design session, we thus had at our disposal the sketches and notes produced by the architect, and the transcription of the audio and video tapes. A summary of the transcription is given in Table 1. To start with, we tried to identify those elements in the protocol data (if any) that explicitly relate to one or more design concepts.

Since the surface area available in the castle cannot accommodate all spaces listed in the program, the architect decides to extend the West wing with a new volume between the castle and the neighboring mill, Figure 3. He starts designing the new volume on line 225, and on line 638 (i.e. about 10 minutes later) he suddenly talks about a toolbox.

---

[4] While this raises serious methodological questions regarding the validity of this protocol study, if not of the think-aloud method in general, two more 'personal' factors might relate to this early resignation as well. Unlike the other subjects, this architect usually designs alone and is therefore not used to company, let alone to talking aloud during design. Moreover, although the short training exercise preceding the design session was explicitly meant to help her get used to this talking, she seemed to experience it as some sort of IQ test. The fact that she messed up the exercise completely is likely to have added to her tension.

[5] According to a comparative analysis by (Stiers and van Beuningen 2002), several of these differences might be attributed to the subjects' difference in level of expertise. Differences relate, amongst others, to the subjects' overall view of the design process (systemic or not), to how and even whether they restructure the design task, to the strategies they adopt (top-down and/or bottom-up), or to how they use analogies, first principles, the prototype design tool being tested, etc.

TABLE 1. Protocol summary

| Time | Description |
|---|---|
| 0:00:00 | Start session. Discussion about the course of the design session and the experiment. |
| 0:02:44 | The architect reads the program and looks at the plans of the existing situation. He wants to determine the scale of the plans. |
| 0:07:28 | He reads the program again and focuses on the surface areas of each of the rooms. The total net area is 770 m² so he needs roughly 1500 m² in total. |
| 0:09:20 | Subsequently he examines the area available in the castle. Since 550 m² is available in the West wing, almost twice this surface must be provided in an extension. |
| 0:13:47 | He considers different locations for the new volume, and concludes that the space in between castle and watermill can support a new building. He considers the size and characteristics of this space. |
| 0:15:51 | To determine which part of the program will be fit in the existing part and which in the extension, he discusses the identity of the castle and the extension, and determines the materials and the architectural and spatial elaboration. |
| 0:19:46 | At this point he actually starts considering the location of the different parts of the program. Two initial attempts are made: the archives could fit under the roof; the reception hall on the ground floor of the West wing, in the former stables. |
| 0:22:54 | After deciding where to put the reception hall, he focuses on the relation of this hall with the exterior neighbouring spaces: the inner court of the castle on the one hand, and the space in between mill and castle on the other hand. Also the relation between this new exterior space and the extension of the castle. He decides to put the extension next to this place because it organises the space. He decides to make a more defined qualitative space close to the river, in between mill and castle. |
| 0:26:06 | He looks at pictures of the castle to study historical aspects of the building. |
| 0:28:00 | He decides that the volume of the extension should appear autonomous vis-à-vis the castle, should have enough mass (more than one level), and could have a vertical articulation. |
| 0:28:42 | He considers relationships between inner court, reception hall, entrance, new volume and exterior space. |
| 0:31:13 | Program: argumentation pro/contra location of the archives, material museum and exhibition space. Material museum and archives can be on either ground or 1st floor, exhibition space on 1st floor? He discusses accessibility for these possibilities. |
| 0:40:00 | Character of the extension: should not be too flat, at least 2 levels. 0: glass, +1: closed, blind, with roof light. Material museum and archives on the 1st level; exhibition room and secretariat on the ground floor. |
| 0:42:34 | He discusses the relation between different parts and their relation with outside. |
| 0:51:26 | Re-evaluation of different program aspects: sizes, relations between and positions of functions. |
| 0:59:43 | The circulation in the re-designed parts is discussed. Unlike what is asked in the program, vertical circulation is placed in the extension instead of the castle. The entrance should not only provide access to the building, but also connect outside and inner court. |
| 1:03:17 | Recapitulation: at this point he redraws the choices previously made in a new drawing. In this way he reconsiders and confirms certain choices. This also brings up several new elements that have to be considered. |
| 1:10:47 | He determines the proportion of the new volume as 25 by 15 meters. |
| 1:13:00 | The position and necessity of inner separations, abutments, and beams is evaluated. |
| 1:13:48 | A pavement behind the building can make a good transition from the workshops to the grass. |
| 1:14:32 | The blind side wall of the castle justifies a blind volume in brickwork. Completely glazing the ground level underneath will make the volume look as if it is lifted. By providing a one level connection between old and new in another material, castle and extension will appear independent. The existing 'back entrance' should become larger and could have a contemporary formal articulation. |
| 1:18:20 | He discusses the possibility of a shed roof to provide light and as a way to articulate the structure. |
| 1:23:19 | Reconsidering the program and thinking up alternatives for the options taken so far. Although he rejects all of these alternatives, they make him redefine and refine his previous choice. |
| 1:34:18 | Reconsidering the proportion of the new volume with regard to the latest changes. This is dropped when he is reminded that he has only 15 minutes left. |
| 1:35:17 | He starts using the case library to find documentation about structures and daylight. |

TABLE 1 (continued). Protocol summary

| |
|---|
| 1:40:55 He considers the entrance to the archives and material museum. He decides to keep it closed at the ground level and to create a minimal perforation to go to the upper space. |
| 1:42:57 The circulation in this new part is placed in the centre, next to the exhibition space. In this way the exhibition can act as an appetizer. |
| 1:46:22 Continues using the case library without clear results. |
| 1:49:38 Stop session. |

*634 what do we get then? so we have above materials museum and archive space*
*635 that looks like that as well*
*636 that it a... a eh storing of knowledge*
*637 of knowledge*
*638 and* ***the toolbox*** *for the students and architects*
*639 and the toolbox is in ...*
*640 that is the toolbox*
*641 and the toolbox are both materials*
*642 as well as knowledge*
*643 storing of materials and knowledge as the toolbox*
*644 a ... already the start of* ***a concept***

In some sense, he explains, the materials exhibited in the material museum and the design knowledge embodied by the archived student projects can be considered as essential tools for architecture students. In the following lines, this idea is then further developed and gradually takes shape in a robust, blind, brick box lifted on top of a glazed volume, Figure 4.

Interestingly, however, the architect does not stick to this concept all the time. At a given moment, he considers giving up the idea of a toolbox all together, because the two 'tools' within the box – archive and materials museum – have contradictory daylight requirements. Therefore, he considers locating the material museum on the first floor, lit by roof light, and storing the archive away on the ground level, safely protected from the damaging daylight.

*1166 and then we have,* ***we are probably abandoning*** *the, the intention of a toolbox*
*1167* ***our concept***

Eventually, however, he decides to cover only half of the first floor with roof light, which allows him to hang on to the idea of a toolbox.

## 5. And Beyond

Upon completion of the design session, the toolbox concept seems to include in a single feature the entire design of the new volume. In one image, it reveals the intention of the creator. At that point, the design makes perfectly sense in the eyes of the observer (we understand what the designer wants to express through his construction with the concept). Regarding the precise functioning of this concept, however, several questions still remain unanswered. Where did it find its source? At which time did it pop up in the consciousness of the designer? Why did it appear?

The very difficulty of these questions lies in the inaccessibility of the architect's thoughts and operations. Only his features, gestures and words testify to his inner activity. The words he uses try to share with us his inner reasoning. But what about the unconscious operations to which the architect

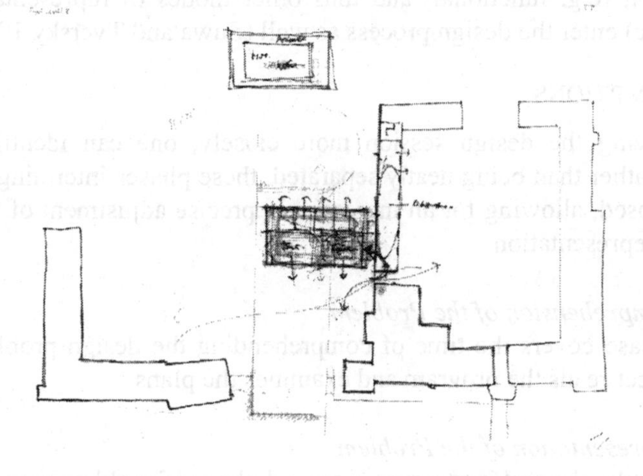

*Figure 3.* New volume between castle (right) and mill (left)

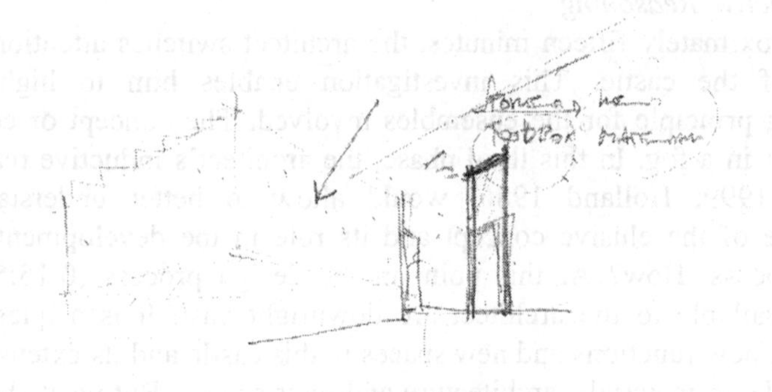

*Figure 4.* Blind, brick, robust toolbox lifted by a glazed volume

himself does not have access? Indeed, the architect's consciousness only has access to a negligible part of these processes (Damasio 2000). Thus, the concept and its development within the architect's mind are his sole property. Deeply anchored in the body, this concept surfaces through the communication tools at his disposal. We deliberately refer to 'communication tools' in the plural, because concepts in architecture may cover multiple modalities and thus require multimodal perceptual representations (Chandrasekaran 1999). Together, the words, the stories, the images, the sculptures translate the feelings which push the architect towards the creation. For even if architects, like other designers, heavily rely upon visual representations such as diagrams and sketches, other types of information (e.g. functional) and thus other modes of representation (e.g. kinesthetic) enter the design process as well (Suwa and Tversky 1997).

### 5.1. ASSUMPTIONS

By observing the design session more closely, one can identify several phases. Rather than being neatly separated, these phases intermingle and are superimposed, allowing for an increasingly precise adjustment of the design problem representation.

#### *5.1.1. Comprehension of the Problem*

A first phase covers the time of comprehending the design problem, when the architect reads the program and examines the plans.

#### *5.1.2. Representation of the Problem*

Subsequently, the architect enters a second phase of problem representation, in which he focuses on dimensions and spaces available.

#### *5.1.3. Inductive Reasoning*

After approximately fifteen minutes, the architect switches attention to the identity of the castle. This investigation enables him to highlight a globalising principle for the ensembles involved. The concept or concepts still appear in a fog. In this third phase, the architect's inductive reasoning (Lemaire 1999; Holland 1986) would allow to better understand the appearance of the elusive concept and its role in the development of the design process. How? At this point in the design process (0:15:51), the choices available to the architect are downright vast. It is a question of integrating new functions and new spaces in this castle and its extension. He must decide on materials, architecture and their spaces. But what should he hook these decisions on to? Where should he start?

#### *5.1.4. Inductive Inference*

Since the architect has little experience with re-designing historical buildings like the castle, he cannot readily recycle options or decisions taken in similar projects before. By consequence, deductive reasoning is necessarily limited: the architect is unable to build on generalizations made up beforehand to infer the particular case at stake. But how can he proceed? Starting from the elements available, he starts to seek common points and, doing so, to make inductive inferences. Out of these regularities, the concept emerges in a timid and sometimes evasive way. A word, an image thus gathers under a single label the totality of the architectural design. Each new element to be integrated by the design must have the characteristics of the whole to reinforce the 'parti'. In this manner, the concept is consolidated and emerges gradually with each new choice made. The concept thus has an integrating role in that it guides the architect through his decisions.[6]

An impasse occurs when, in front of an element or aspect to be integrated, the architect cannot answer the characteristics of the 'parti'. In this case, a reformulation of the concept is necessary. Modifications and revaluations are essential until the process of inference stabilizes.

### *5.1.5. Overall Strategy*

This inductive operation explains why the concept is so elusive in the course of the design process. Indeed, the architect does not yet possess all ingredients of the project.[7] He has to integrate the elements of the program some of which may turn out to be in complete contradiction with the initial 'parti', thus requiring a reformulation and sometimes even abandonment of the initial concept. Upon completion of a design project, however, all elements – but also some uncontrolled – are fixed. Once the building is finished, the inductive inference of the concept from the overall strategy over the characteristics is much easier, since the architect highlights only those elements that enabled him to develop the final concept. The story developed around the concept is thus much more garnished at the end of the design process because, from this moment on, all parts of the puzzle are arranged in the building.

---

[6] Note that this integrating role is not necessarily played by a concept (in the sense of 'parti'), but might as well be taken up by other elements such as a proportional system (Le Corbusier's *Modulor*, for instance), a specific building type, or a personal set of heuristics the designer has gradually developed.

[7] Indeed, many aspects or parameters of the design problem are unknown before design starts and are only discovered while designing. As Gianfranco Carrara and Yehuda Kalay (1994) contend: "The search for a solution produces insights into the design problem, that are not apparent before the search begins, revealing compromises that must be made, opportunities to be had, and trade-offs that will improve the overall quality of the solution. These insights not only produce new design parameters and define new objectives to be achieved, but may also radically alter previously stated objectives."

### 5.2. ATTEMPT AT MODELING CONCEPT EMERGENCE

In a more 'imagery' way, and by exploiting the properties of the neuronal activation mechanisms of human memory (Lemaire 1999; Churchland 1992), we propose the following model of concept emergence, Figure 5. Usually, the interactive activation and competition model is applied "to model the retrieval of general and specific information from stored knowledge of individual exemplars" (McClelland and Rumelhart 1988; 1985). We choose to apply this model in an architectural context. During his design process, the architect makes analogies with concepts in his memory (Leclercq and Heylighen 2002). He tries to find something known (a concept) to give meaning to his construction. Or, he may search to represent by a building what feeling he wants to share with the people, like an artist with his audience. The common point between these two cognitive problems is a memory system addressed by his content. We do not need all the properties about the concept unit to find it. This incomplete information about the concept's characteristics allows the designer to be guided in his choice by an "elusive concept". We have access to the concept only by its characteristics, such as a name or two, an image, a story. As such, it represents this highly personal sensation the architect feels about his work.

This connectionist model represents the concept as an active unit, linked to the others by excitatory or inhibitory connections. The excitatory connections act on units among different pools, while the inhibitory connections act on units within the same pool. The interactive processing is carried out by the bidirectional connections.

This very simplified model of the concept's emergence comprises two sets of units: on the one hand, *concept units*, the internal representations of the building, and on the other hand, accessible *characteristic units*, the mental images of shapes, structures, materials and their connections. All these aspects of the building are elements that are bound through the vehicle of a concept. This 'parti' and its relations with the various mental images are highly personal and characteristic of each individual designer.

Indeed, the connections between the units arise from the architect's own experience and memories. His knowledge is closely bricked in the weights of these connections. The activation of a unit depends on the number and the intensity of connections received by the other units it is connected to. When a concept unit is sufficiently activated, an idea of an overall principle emerges in the architect's consciousness. Its sufficient activation will interfere in the choice of other qualities of the design. The characteristic units for their part are in turn activated by the dominant concept unit and those close to it. In view of this, the design process can be described as a sort of competition between concept units that interlards the design process, whereby the architectural concept unit and its connections can be thought of as an internal representation of the architectural response or building solution. As argued above, this internal representation is necessarily multimodal (Chandrasekaran 1999), and maintains many strong relations with other characteristic units.

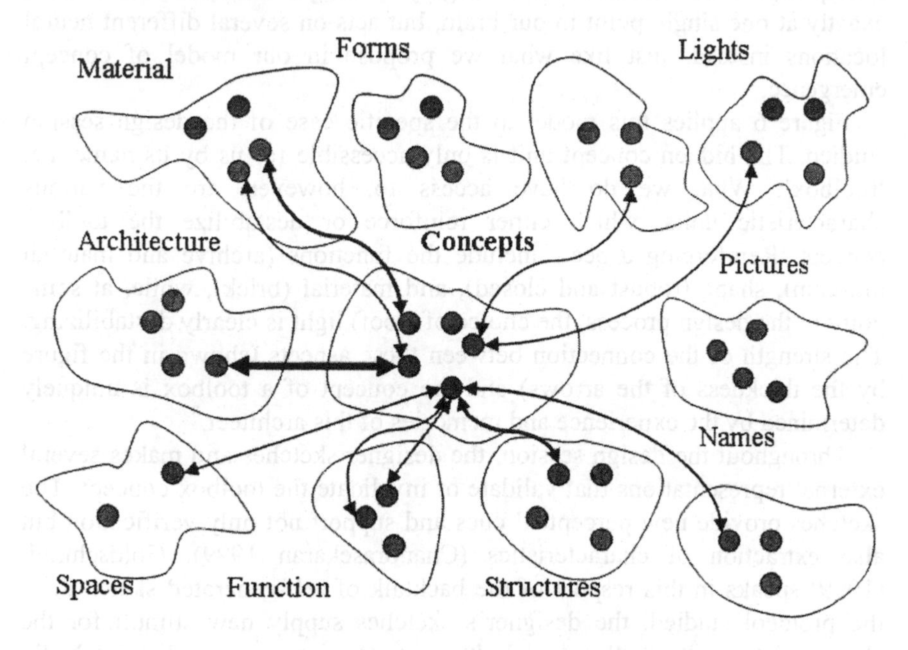


*Figure 5.* Interactive activation and competition model of the emergence of the concept

To further explain the notion of concept units, let us take the example of a door. We can give a lot of information about the characteristics of this object: the shape (a parallelepiped rectangle, but not in all cases), the material (wood, steel, ...), the weight, the name ("door"), the mechanics (the hinges, lock, handle), its function (a vehicle to close/open one space from/to another), the sounds (grating noise of old hinges, or banging and slamming), the movement to open and close it (experience from our procedural memory), our personal experience attached to the door element (experience from our episodically memory) and many other aspects. By this example, we want to show that, rather than a prototype of the door element, each person has developed an internal representation that is modelled by a network of links between the external characteristics he/she came across during his/her life. This point of view is sustained by insights from cognitive neuroscience, where researchers do not find a specific "door" neuron, but rather a distributed neural activity in the human brain. Apparently, this distributed brain activity is caused by the subsystems used in high-level vision (Kosslyn

1999). Observations reveal that the spatial properties encoding subsystem uses the dorsal pathway, while the object properties encoding subsystem uses the ventral pathway. By consequence, the stimulus of the object of perception or its visual mental imagery (Kosslyn 1999) is not located exactly at one single point in our brain, but acts on several different neural locations instead, just like what we propose in our model of concept emergence.

Figure 6 applies this model to the specific case of the design session studied. The hidden concept unit is only accessible for us by its name, i.e. 'toolbox'. What we do have access to, however, are the various characteristic units, which either reinforce or destabilize the toolbox concept. Reinforcing aspects include the functions (archive and material museum), shape (robust and closed), and material (brick), while, at some point in the design process, the choice of (roof) light is clearly destabilizing. The strength of the connection between these aspects (shown in the figure by the thickness of the arrows) and the concept of a toolbox is uniquely determined by the experience and memories of this architect.[8]

Throughout the design session, the designer sketches and makes several external representations that validate or invalidate the toolbox concept. The sketches provide new perceptual cues and support not only verification but also extraction of characteristics (Chandrasekaran 1999). Goldschmidt (1999) speaks in this respect of the backtalk of self-generated sketches. In the protocol studied, the designer's sketches supply new stimuli for the characteristic units in the model. It is an iterative process that gradually increases the activation of the concept unit, which in turn is transferred by all its connections to other units. The process stops when the building is constructed, because no new fundamental information can modify the architectural concept. The internal representation is completed, the activation of the concept unit reaches a maximum and the transfer of activation is stabilized.

## 6. Discussion

Within the domain of architecture, a design concept – also known as underlying idea or 'parti' – is typically used to explain or justify a finished design *product*. By contrast, this paper has tried to adopt a different perspective on the phenomenon by switching attention to the design *process*. To this end, we have put under the microscope a two-hour design session of

[8] By extension, the model might also account for those cases where the integrating role, here ascribed to the concept, is played by other elements (see above). In these cases, the links between the units could be thought of as belonging to the same family, such that the decisions cohere without being 'projected' explicitly on a specific concept.

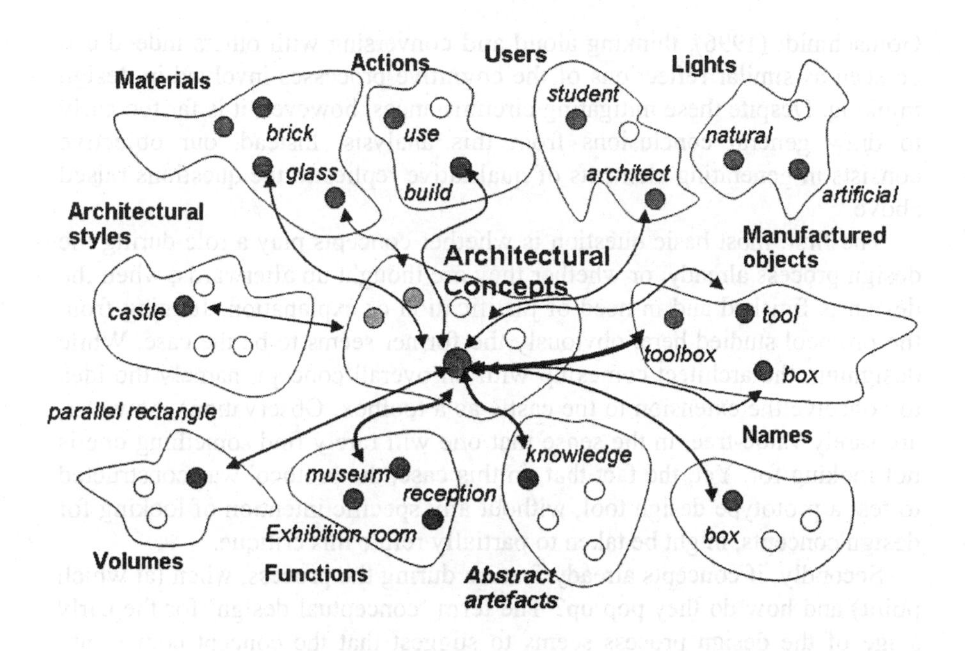


*Figure 6.* Application of the interactive activation and competition model to the emergence of the toolbox concept

an architect involved in the redesign of an architecture school. Obviously, the protocol study is seriously limited, not only in terms of time and number of subjects, but also because of the method used. Although the arguments for accepting reports of 'think aloud' exercises as a reflection of cognitive activity are well documented and substantiated (Ericsson and Simon 1984), the technique shows serious disadvantages when applied to design. Since the visual, non-verbal thinking is fundamental to how designers know and work (Cross 1982), the verbalization may produce side-effects that change the subjects' behavior and cognitive performance (Cross et al. 1996). Research in the field of product design, for instance, has illustrated that protocol analysis, and the constraint it brings both theoretically and methodologically, interferes with designing (Lloyd et al. 1996). Until we have a more coherent picture of this interference, however, we can do little more than be very explicit on how our experiment was set up (Cross et al. 1996). Perhaps worth mentioning here is that, in everyday life, our subject is used to designing in team with his design partner and thus to talking while designing. This habit might explain why he experienced the design session as less stressful than the fourth subject and, more importantly, have limited the potential side-effects of having to think aloud. According to

Goldschmidt (1996), thinking aloud and conversing with others indeed can be seen as similar reflections of the cognitive processes involved in design thinking. Despite these mitigating circumstances, however, it is far too early to draw general conclusions from this analysis. Instead, our objective consists in generating elements of qualitative replies to the questions raised above.

The first, most basic question is whether concepts play a role during the design process already, or whether they are thought up afterwards, when the design is finished and in need of justification or explanation. Judging from the protocol studied here, obviously the former seems to be the case. While designing, the architect comes up with an overall concept, namely the idea to conceive the extension to the castle as a toolbox. Observations, however, are rarely value-free, in the sense that one will rarely find something one is not looking for. Yet, the fact that, in this case, the protocol was constructed to test a prototype design tool, without any specific intention of looking for design concepts, might be taken to partially refute this critique.

Secondly, if concepts already feature during the process, when (at which point) and how do they pop up? The term 'conceptual design' for the early stage of the design process seems to suggest that the concept comes into being at the very start and henceforth can be treated as a static, invariant feature of the design project. By contrast, some authors contend that there is a continuous interaction between the design project on the one hand and the driving concept on the other: the concept would allow designers to impose an order on the design while the design would enable them to explore and develop the concept. Donald Schön (1985), for instance, describes concept generation as an experiment of which the results are only dimly apparent in the early stages of the project. In other words, often very little of the concept is understood until late in the design process (Lawson 1994). In the protocol studied here, the idea to conceive the new volume as a toolbox does not pop up at the very start of the design session, but only after a while (24 minutes to be precise). Counted in another way, the architect starts designing the new building at line 225, and only on line 638 he talks about a toolbox.

Later on, when facing an aspect that does not really support the toolbox concept (i.e. the introduction of roof light), the architect is willing to give up the idea entirely. After exploring some alternatives, he eventually decides to hang on to the idea, yet his willingness to abandon the toolbox in a relatively 'advanced' stage of the session, clearly illustrates that he does not treat the concept as an invariant of the design.

At the same time, this 'toolbox crisis' provides evidence for a possible explanation of why designing 'good' architecture (i.e. architecture that has only a very few major dominating ideas) is so difficult. Coming up with a concept is one thing, developing this concept consequentially into a built artifact is yet something different. Indeed, what makes designing good architecture extremely difficult – and at the same time extremely fascinating – is that this development is far from straightforward a procedure. Unless there is consistency and continuity from the earliest conceptual phases right through design development to detailed design, those important underlying ideas will get lost. Apparently, this 'hanging onto the big idea like grim death' is something architects tend to struggle with (Lawson 1994).

How exactly does this struggle unfold? In other words, how can a design concept get lost? And where does it come from in the first place? In order to shed more light on this complex phenomenon, we have proposed a connectionist scheme that explains the emergence of design concepts as the activation of and competition between hidden concept units, which activate and are activated by observable characteristic units. The model shows by parallel and distributed processing the emergence of the elusive concept and its instability during the design process. We do not claim that it is a model of the whole concept, as this requires a deeper study about the architectural project model representation.

## 7. Related and Future Work

Judging from the protocol studied, the connectionist model proposed seems a useful way to probe, plot and comprehend when, why and how concepts arise and develop during the design process. Moreover, some features of the model seem to chime with elements from other research, be it primarily in other design domains.

Observations of students and professionals in automotive design, for instance, confirm the interactive and iterative nature of concept development (Tovey et al. 2003). In this study, the process of moving from an initially vague concept to a detailed design proposal is likened to zooming in from an out of focus image to one that is fully detailed. According to our model, this increasing sharpness might be attributed to the growing number of characteristic units that gradually reinforce the initially instable concept unit. Just like the imagery of Kosslyn (1999) guides the attention window, the concept guides the designer's choices and plays a key role in the reasoning about "mental models". The new design is like an object you see in the fog. The external characteristics allow to recognize it, but it does not exist in your memory because it is a new creation. Since the designer has only some characteristics about the future design, arguments are based on an existing internal representation with the same properties, like the toolbox concept in the protocol studied. This concept helps to fill in missing elements and in this way guides the designer towards a fully detailed image of the design.

Another study, this time in the field of engineering design, points out the importance of each designer's unique way of paying attention to specific information (Taura et al. 2002). This highly personal selection of elements of the design situation, called the 'gazing point', seems to have a major influence on the designer's mental states during the design process. No two designers thus would go through the same mental states, even when working on the same design task. This unique 'gazing point' of every designer, and its role in the design process, is accounted for in our model by the highly individual character of the connections between the different units, which are uniquely defined by the designer's personal experience and memories.

The fact that these connections can be either excitatory or inhibitory, and their respective effect on the design process, corresponds to the characterization of the overall design process as being convergent, and yet containing episodes of divergence (Cross 2000). In the protocol studied, for instance, the overall process converges towards the concept of a toolbox, while the designer's sudden willingness to give up this concept all together can be interpreted as an illustration of (temporary) divergence.

Despite these and possibly other apparent correspondences, however, a key question remains to what extent the work of design researchers in other areas translate to architecture. Engineers, for instance, use the term 'concept' in a different way than architects. For them, the concept is a principle, often technical, or a general morphology[9] (Cross 2000). Moreover, we are the first to admit that further evidence is needed for the general usefulness and relevance of the scheme we propose, and for our protocol study in general. Future work therefore includes repeating the experiment with more and other subjects in various design contexts and at different levels of expertise, in order to test and refine the scheme proposed. In doing so, additional insights might be gained from applying Linkography (Goldschmidt 1996) to further analyze the structure of the concept generation processes recorded. The Linkography system was developed to notate design moves – an act, a step, an operation that transforms the design situation – and the links among them. Since this allows to determine for each idea direct connections or 'links' with all earlier ideas, it may help shed more light on how concepts arise and develop during design. Another option is to conduct experiments in the opposite direction, which attempt to enforce a specific concept onto the designer. Instead of following architects where they want, we might try and push them through many analogies (i.e. excitatory connections) towards a previously determined concept. Further research is also needed to comprehend and account for the interplay between short and long term memory in the cognitive processes we want to

[9] To make the translation of insights from engineering to architectural design even more confusing, engineers tend to speak about a general 'architecture'.

understand. Awaiting the results of this additional research, we hope that the present study at least has placed some markers on the way to a deeper understanding of the role of concepts in design.

## Acknowledgements

Ann Heylighen is a postdoctoral fellow of the Fund for Scientific Research – Flanders, the support of which is gratefully acknowledged.

## References

Akin, Ö: 1979, *Models of Architectural Knowledge: An Information Processing View of Architectural Design*, PhD. Dissertation, Carnegie-Mellon University, Pittsburgh, PA.

Alexander, C: 1979, *The Timeless Way of Building*, Oxford University Press, New York.

Bourdieu, P: 1977, *Outline of a Theory of Practice*, Cambridge University Press, Cambridge.

Carrara, G and Kalay, YE (eds): 1994, *Knowledge-Based Computer-Aided Architectural Design*, Elsevier, Amsterdam.

Chandrasekaran, B: 1999, Multimodal perceptual representations and problem solving, *in* JS Gero and B Tversky (eds), *Visual and Spatial Reasoning in Design*, Key Centre of Design Computing and Cognition, University of Sydney, pp. 3-14.

Churchland, PS and Sejnowski, T J: 1992, *The Computational Brain*, MIT Press, Cambridge, MA.

Cross, N: 1982, Designerly ways of knowing, *Design Studies* **3**(4): 221–227.

Cross, N: 2000, *Engineering Design Methods. Strategies for Product Design*, John Wiley & Sons, Chichester.

Cross N, Christiaans H and Dorst K: 1996, Introduction: The Delft protocols workshop, *in* N Cross, H Christiaans and K Dorst (eds), *Analysing Design Activity*, John Wiley & Sons, Chichester, pp. 1-16.

Damasio, AR: 2000, *The Feeling of What Happens: Body, Emotion and the Making of Consciousness*, Vintage, London.

Darke, J: 1978, The primary generator and the design process, *in* WE Rogers and WH Ittelson (eds), *New Directions in Environmental Design Research*, EDRA, Washington D.C., pp.325-337.

Ericsson, KA and Simon, HA: 1984, *Protocol Analysis: Verbal Reposts as Data*, MIT Press, Cambridge, MA.

Goldschmidt, G: 1996, The designer as a team of one, *in* N Cross, H Christiaans and K Dorst (eds), *Analysing Design Activity*, John Wiley & Sons, Chichester, pp. 65-91.

Goldschmidt, G: 1999, The backtalk of self-generated sketches, *in* JS Gero and B Tversky (eds), *Visual and Spatial Reasoning in Design*, Key Centre of Design Computing and Cognition, University of Sydney, pp.163-184.

Hamel, RG: 1990, *On Designing by Architects: A Cognitive Psychological Description of the Architectural Design Process*, PhD. Dissertation, Universiteit van Amsterdam, AHA books, 's Gravenhage.

Heylighen, A and Neuckermans, H: 2001, Destination: Practice. Towards a maintenance contract for the architect's degree, *in* W Jabi (ed), *Reinventing the Discourse*, ACADIA, Buffalo, NY, pp. 90-99.

Holland, JH, Holyoak, KJ, Nisbett RE and Thagard, PR: 1986, *Induction: Processes of Inference, Learning, and Discovery*, MIT Press, Cambridge, MA.

Kosslyn SM: 1999, Visual mental images as re-presentations of the world: a cognitive neuroscience approach, *in* JS Gero and B Tversky (eds), *Visual and Spatial Reasoning in Design*, Key Centre of Design Computing and Cognition, University of Sydney, pp. 83-92.

Lawson, B: 1990, *How Designers Think*, Butterworth Architecture, London.

Lawson, B: 1994, *Design in Mind*, Butterworth Architecture, London.

Leclercq, P and Heylighen, A: 2002, 5,8 analogies per hour, *in* JS Gero (ed), *Artificial Intelligence in Design '02*, Kluwer Academic, Dordrecht, pp. 285-303.

Lemaire, P: 1999, *Psychologie Cognitive*, De Boeck Université, Paris.

Leupen, B, Grafe, C, Körnig, N, Lampe, M and De Zeeuw, P: 1997, *Design and Analysis*, Van Nostrand Reinhold, New York , NY.

Lindekens, J, Heylighen, A and Neuckermans, H: 2003, Understanding architectural redesign, *in* G Aouad and L Ruddock (eds), *Proceedings of the 3rd International Postgraduate Research Conference in the Built and Human Environment*, University of Salford, Salford, pp. 671-681.

McClelland, JL and Rumelhart, DE: 1985, Distributed memory and the representation of general and specific information, *Journal of Experimental Psychology: General* **114**: 159–188.

McClelland, JL and Rumelhart, DE: 1988, *Explorations in Parallel Distributed Processing*, MIT Press, Cambridge, MA, pp.11-47.

Rowe, PG: 1987, *Design Thinking*, The MIT Press, Cambridge, MA.

Sharp, D: 1990, *Twentieth Century Architecture: A Visual History*, Facts on File, New York, NY.

Stiers, J and van Beuningen, A: 2002, Designing with DYNAMO, unpublished case study report.

Suwa M, and Tversky, B: 1997, What do architects and students perceive in their design sketches? A protocol analysis, *Design Studies* **18**(4): 385-403.

Taura, T, Yoshimi, T, and Ikai, T: 2002: Study of gazing points in design situation. A proposal and practice of an analytical method based on the explanation of design activities, *Design Studies* **23**(2): 165-185.

Tovey M, Porter S, and Newman R: 2003, Sketching, concept development and automotive design, *Design Studies* **24**(2): 135-153.

## DESIGN COGNITION

JS Gero (ed), *Design Computing and Cognition'04*, 79-96


# A COGNITIVE EVALUATION OF A COMPUTER SYSTEM FOR GENERATING MONDRIAN-LIKE ARTWORK

ANDRÉS GÓMEZ DE SILVA GARZA, ARÁM ZAMORA LORES
*Instituto Tecnológico Autónomo de México (ITAM), Mexico*

**Abstract.** In this paper we present a computer system based on the notion of evolutionary algorithms that, without human intervention, generates artwork in the style of the Dutch painter Piet Mondrian. Several implementation-related decisions that have to be made in order to program such a system are then discussed. The most important issue that has to be considered when implementing this type of system is the subroutine for evaluating the multiple potential artworks generated by the evolutionary algorithm, and our method is discussed in detail. We then present the set-up and results of a cognitive experiment that we performed in order to validate the way we implemented our evaluation subroutine, showing that it makes sense. Finally, we discuss our results in relation to other research into the computer generation of artwork that fits particular styles existing in the real world.

## 1. Introduction

Ever since computer use began to spread, people started using computers to generate artwork. At first printers and monitors could not display anything other than text, so rough figures depicting cartoon characters or everyday objects such as houses were painted by displaying sequences of ASCII characters that were designed to create an outline of the desired figure (sometimes even incorporating shading, when viewed from a distance, by using different ASCII characters for different parts of the figure to be depicted). Eventually colours and high-resolution monitors and printers became available, so computer art evolved into what it is today, able to display a multitude of abstract or realistic figures and textures in such detail that the results can now be used in movies to create animation that is virtually indistinguishable from the results of real cinematography. Another approach to computer art, rather than to try to reproduce the real world virtually, is to try to get the computer to create its own sense of style (or rather, its programmer's) and produce artwork in that style that people consider to be aesthetically pleasing. Yet another approach is to try to get the computer to create artwork that resembles (or is indistinguishable from) the creations of human artists.

It is this last approach that we have taken in the research presented in this paper. Specifically, we set out to produce a program that would generate artwork in the style of the Dutch painter Piet Mondrian, who was active mainly in the first half of the 20th century. Like many other modern painters, Mondrian started his career painting landscapes, human figures, and other realistic subjects, but eventually developed his own distinctive and abstract style (called simply de stijl, which is Dutch for "the style"). Paintings in Mondrian's style typically include vertical and horizontal black lines over a white background, with some or all of the primary colours (blue, red, and yellow), plus black, filling in some of the square or rectangular regions (or parts of the regions) separated out from the background by the black lines. It is this style that our system tries to emulate. Figure 1 shows a typical Mondrian painting in his distinctive style.

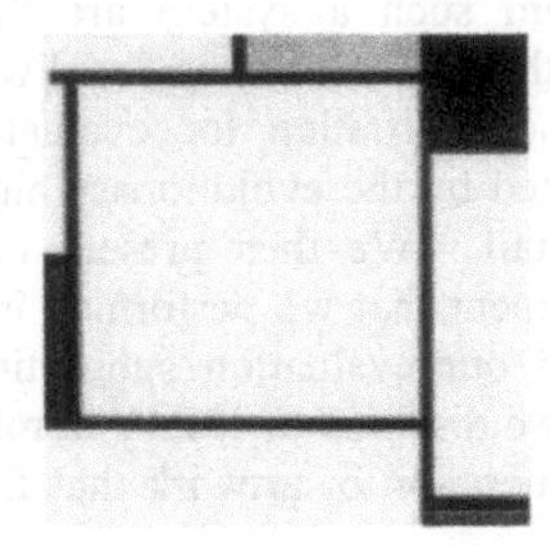

*Figure 1.* A typical Mondrian painting ("Composition 2", 1922)

The method we have used for getting the computer to generate different paintings is through an evolutionary algorithm (Mitchell 1998). This type of algorithm represents a generate-and-test, trial-and-error, brainstorming-like approach (Clark 1958): many possible paintings (populations of them) are generated quickly, by using mainly random decisions. Probably most of these paintings are of quite low quality, but after being generated they are then evaluated to determine how much they make sense. In the context of our research, making sense would imply being as close as possible to the style of Mondrian. The best paintings (according to the evaluation subroutine of the evolutionary algorithm) are kept for future evolutionary generations, and the others are discarded (so as to keep the size of the population of the algorithm constant across generations). This process ensures a monotonic increase in the average quality of the paintings in the population between generations. Depending on what is desired, either when this average quality or when the quality of just one individual painting is good enough according to the evaluation subroutine, the process is terminated. The evaluation subroutine of the evolutionary algorithm, therefore, is of critical importance to the success of the approach.

Section 2 of this paper briefly introduces evolutionary algorithms in general, and discusses some decisions we had to make in order to implement an evolutionary algorithm for the domain of Mondrian-style artwork generation in particular. Section 3 discusses a cognitive experiment we performed in order to validate the evaluation subroutine of our system, its most critical component. This is the opposite of the approach which is normally taken, in which a system is programmed based on a cognitive model resulting from observing how people (sometimes experts, sometimes not) do things. Finally, Section 4 provides a general discussion of the research topics that this project addresses, relates the project to other work in similar areas, and presents some of the results and conclusions we have obtained from the project.

## 2. Evolutionary Algorithm Implementation

Figure 2 shows the flow of subtasks performed by our evolutionary algorithm.

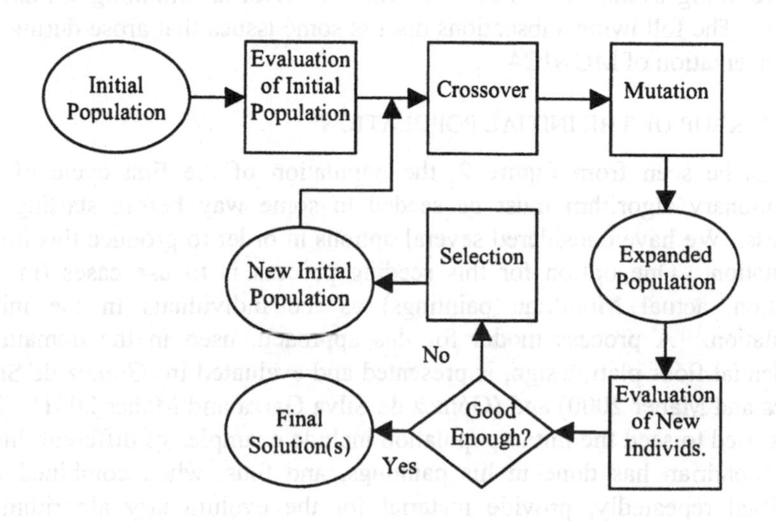


*Figure 2.* Our evolutionary algorithm's subtasks

Briefly, a population of potential solutions (in this case, paintings) is kept throughout the process. The makeup of the population changes due to the evolutionary process. New potential solutions are generated by the genetic operators of crossover and mutation, which at random combine and modify the features of old potential solutions that are already in the population. A temporarily expanded population is created by adding these new potential solutions to the original population. Each new potential solution is evaluated and a fitness value assigned to it. The fitness value in our system is a measure of how close or how far the painting is from Mondrian's style. If one or more (as desired) of the new potential solutions is already perfect (i.e., already fits the desired style) according to the evaluation procedure, the evolutionary process stops (or, alternatively, if the search has gone on for too long without successfully producing any Mondrian-like works of art). Otherwise a selection procedure sorts the potential solutions in the temporarily expanded population according to fitness value, keeps the best of them, and discards the rest. The individuals that are kept become the initial population for the next evolutionary cycle. This new population may include both old and new potential solutions (i.e., some carried over from previous generations and some newly-generated ones), and the process repeats itself.

We have implemented an evolutionary algorithm on a Windows platform in C++ (for the reasoning engine) with OpenGL (for the graphical interface) and have applied it to the domain of Mondrian-style artwork generation. The resulting system is called MONICA (MONdrian-Imitating Computer Artist). The following subsections discuss some issues that arose during the implementation of MONICA.

### 2.1. MAKEUP OF THE INITIAL POPULATION

As can be seen from Figure 2, the population of the first cycle of the evolutionary algorithm must be seeded in some way before starting the process. We have considered several options in order to produce this initial population. One option for this seeding process is to use cases (in this situation, actual Mondrian paintings) as the individuals in the initial population. A process model for this approach, used in the domain of residential floor plan design, is presented and evaluated in (Gómez de Silva Garza and Maher 2000) and (Gómez de Silva Garza and Maher 2001). The cases used to seed the initial population include examples of different things that Mondrian has done in his paintings, and thus, when combined and modified repeatedly, provide material for the evolutionary algorithm to quickly produce new potential solutions that contain Mondrian-like features. On the other hand, the use of cases can bias the algorithm too much towards creating paintings that only differ from Mondrian's in small details. We have gathered 27 cases of Mondrian paintings (pertaining to *de stijl* and not including pieces representing stylistic transitions or his rhomboidal "lozenges" as they are known) from several Internet sites, and from the books (Deicher 1999; Bax 2001). Apart from having these cases in JPG format, we have made them available to MONICA by hand-coding them into our genotype representation scheme (described below).

A different option for the seeding process is to produce an initial population entirely at random. This approach has the advantage that little prior bias can be said to exist in the initial population, and thus it is easier to claim that the computer itself really evolved Mondrian-like paintings from nothing. On the other hand, it may take too long to start producing Mondrian-like artwork, despite the bias that the evaluation procedure of the evolutionary algorithm provides in that direction, and thus may not really be a feasible approach. This method of seeding the initial population is also available to MONICA. The algorithm used in MONICA in order to produce a random initial population is shown in Figure 3 and is based on randomly deciding the sizes, coordinates, and colours of different rectangular regions to be placed in a painting. As can be seen, some slight bias that considers aspects of Mondrian's style, such as limiting the total number of coloured regions in each individual to Mondrian-type numbers, was still present in the "random" generation of the initial population, but not much. In our algorithm, coloured regions that end up getting assigned very small widths or heights end up being lines, but are not treated any differently from square or rectangular coloured regions. There is nothing in the random seeding algorithm to force coloured regions not to overlap, not to exceed the limits of the frame used for the painting, or any other things that will eventually have to happen, at the same time, for a "painting" to be accepted as Mondrian-like by the evaluation procedure.

```
For each painting to generate:
     Choose a random number N between 0 and 19.
     (N will be the total number of coloured regions to
               include in the painting)
     For each coloured region R0, R1, ..., RN:
          Randomly choose its height.
          Randomly choose its width.
          Randomly choose its position along the X axis.
          Randomly choose its position along the Y axis.
          Randomly choose its colour (between 0 and 4).

Notes: Positions are chosen to fall between 0 and 3.9999,
represent the central coordinates of a rectangular figure,
and are measured in OpenGL graphical units (not pixels).
The height and width also fall between 0 and 3.9999.
These limits ensure that the entire painting fits within
a typical computer screen.
Only five colours are considered to be valid in Mondrian's
paintings (white, black, blue, red, and yellow).
```

*Figure 3.* Algorithm for random seeding of initial population

A final option is to combine both cases and random individuals in the initial population. The question then becomes what the relative proportion of cases and random individuals should be in the initial population. For the purposes of being able to produce Mondrian-like results, it doesn't seem to matter much, though there may be an effect on convergence time. We are planning to perform some experiments with MONICA in the future to measure any possible effects. In addition, Louis (2002) uses a similar approach, combining cases and evolutionary algorithms for electronic circuit design, and has performed experiments having to do with seeding the population with different proportions of cases and random "solutions." These experiments have measured the effect of changing the relative proportions of cases and random solutions on the amount of exploration and exploitation present in an evolutionary algorithm, and have come to the conclusion that around 10% of cases is a good figure. It will be interesting to see if our future experiments validate these results or if the results are specific to the domain and representation used.

### 2.2. GENOTYPE REPRESENTATION

Figure 4 shows how we have represented in MONICA the individuals in the population of our evolutionary algorithm as genotypes at the highest level. The genotype is split into twenty regions, each of them corresponding to one of twenty possible coloured regions permitted in a painting.

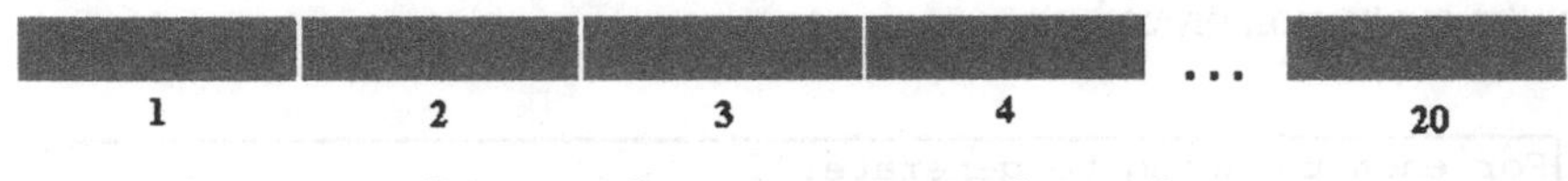


*Figure 4.* Genotype representation used

Figure 5 shows at an intermediate level how each of the twenty genotype regions is split into five sections in order to represent the colour, width, height, and x- and y-coordinates (of the center, as per OpenGL standards) of each coloured region. These last four measurements are all limited to the same range of values.

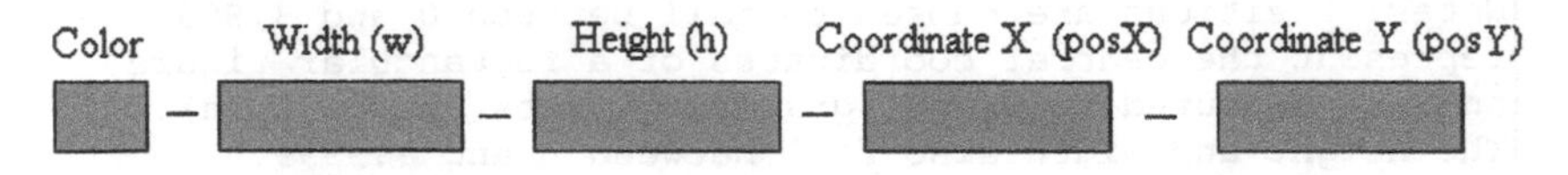


*Figure 5.* Representation of each coloured region

Figure 6 shows, at the bit level, the internal details of the representation of the colour and one of the four measurements shown in Figure 5.

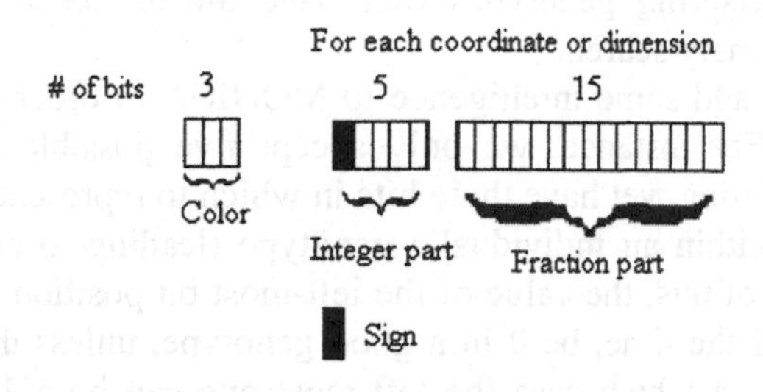


*Figure 6.* Bit-encoding of the colour and any one measurement in the representation

## 2.3. CROSSOVER AND MUTATION OPERATORS

The genetic operators of crossover and mutation, in their pure form, make all of their decisions at random. For instance, crossover combines the makeup of two original individuals by randomly choosing which two individuals to combine and randomly choosing the position(s) within their genotypes in which to cut and splice them. However, this completely blind process can produce a lot of senseless and time-wasting results, including offspring genotypes that have the exact genetic makeup as their parents. In MONICA, if a particular painting has only four coloured regions, for example, and given the fact that all genotypes are of the same length, the genes in the genotype corresponding to coloured regions 5-20 will just be filled with zeroes. If the crossover point is randomly chosen to fall in this region of the genotype for one parent, and the same occurs with the other parent, each one of the two resulting offspring genotypes will be identical to one of their parents. Thus, we would not have produced any new, and therefore potentially good, results in the new generation.

In order to avoid this problem we have added some intelligence to our implementation of both the crossover and mutation operators to ensure that any work done during evolution results in something new. This added intelligence analyses the contents of the two randomly-chosen parent genotypes, both to the left and to the right of the randomly-chosen crossover point. If it is determined that performing the crossover operation at that point will not result in offspring genotypes that are different from their parents (by carrying out the analysis described in the previous paragraph), then there is no point in performing the crossover. Instead, in these situations a different crossover point is chosen (again at random). The process is repeated until the offspring genotypes resulting from crossover are guaranteed to be different from the parent genotypes that produced them. The evaluation of offspring genotypes still needs to be performed as part of the evolutionary algorithm, but time will not have been wasted on producing and evaluating offspring genotypes that have already been encountered during the evolutionary search.

We could also add some intelligence to MONICA in order to avoid any useless results. For instance, we only accept five possible colours in a Mondrian-like painting, yet have three bits in which to represent the value of the colour gene within an individual's genotype (leading to eight possible values). Because of this, the value of the left-most bit position in the colour gene will, most of the time, be 0 in a good genotype, unless the two right-most bits are 0's, in which case the left-most one can be a 1. We could program the crossover and mutation operators to take into account these factors when making their random decisions about where to cross or mutate genotypes, in order for all the genes of offspring genotypes to get assigned values that make sense. However, we feel that this would be tampering too much with the way evolution is supposed to happen, so we have not gone so far. Again, any biasing of the population according to how much the individuals in it resemble Mondrian's style we have left to the MONICA's evaluation procedure, and we did not want to distribute its effects to other modules.

### 2.4. EVALUATION OF NEW GENOTYPES

The evaluation of new genotypes produced during the course of evolution is the most important aspect of an evolutionary algorithm, since it is where domain knowledge is used to indirectly guide the evolutionary search towards a particular objective. In MONICA we assign a fitness value between 0 and 1 to each individual in the population, where a 0 would mean that the individual doesn't satisfy any of the evaluation rules used to determine how close it is to fitting Mondrian's style, and a 1 would represent a perfect fit. We have implemented eight evaluation rules, each of which assigns a local fitness value between 0 and 1 (interpreted as above), and calculate the total fitness of an individual by adding the eight local fitness values and dividing the total by eight, thus giving each rule an equal importance or weight. The eight rules were articulated (and then programmed) by the authors after examining the 27 cases of Mondrian paintings that we had access to, and discussing the patterns that we seemed to observe in these cases. Figure 7 shows the 27 Mondrian cases. The eight evaluation rules are the following:

1. EvaluateColour: Each coloured region that is contained in a genotype must have one of the five valid colours.
2. EvaluateCoordinates: The height, width, x-coordinate, and y-coordinate of each coloured region in a genotype must all fall between 0 and 3.9999.

3. EvaluateLineThickness: Up to two black coloured regions are allowed in a genotype that are not thin, but all other black regions must be either vertically or horizontally thin (and thus represent a line rather than a rectangular region).
4. EvaluateNumberOfVerticalLines: A minimum of two and a maximum of ten vertical lines must be present in a genotype.
5. EvaluateNumberOfHorizontalLines: A minimum of two and a maximum of ten horizontal lines must be present in a genotype.
6. EvaluateLimits: Each coloured region in a genotype must be adjacent either vertically (both above and below) or horizontally (both to the left and to the right), or both, to another coloured region or to the edge of the frame (with some small tolerance).
7. EvaluateFrame: All other coloured regions in a genotype must fall within the coordinates of the frame, whose colour is white by definition and whose coordinates are represented just as any other coloured region's are.
8. EvaluateNumberOfColouredRegions: There must be at least one coloured region represented in a genotype, and at most 13, not counting lines. At most one of them can be white (and represents the frame).

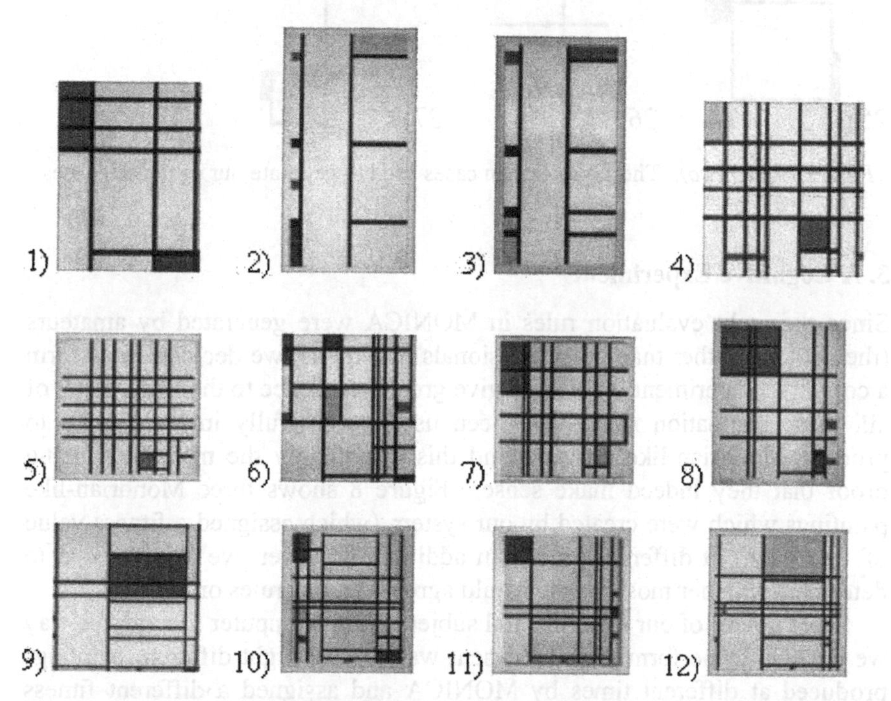


*Figure 7 (part one).* The 27 Mondrian cases used to generate our evaluation rules

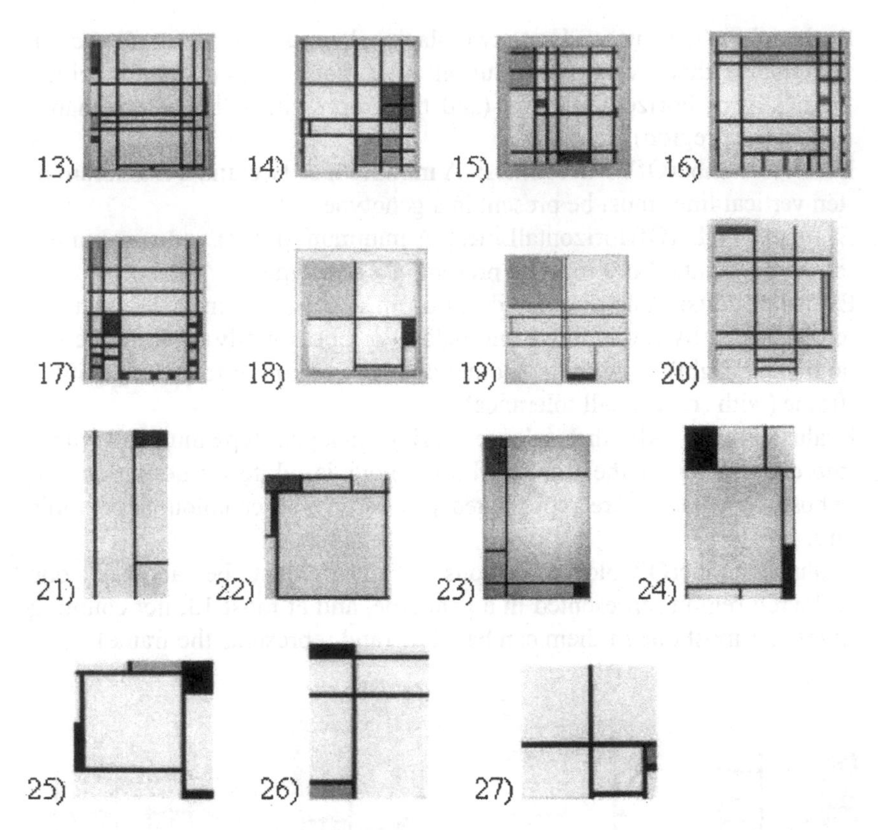


*Figure 7 (part two).* The 27 Mondrian cases used to generate our evaluation rules

## 3. A Cognitive Experiment

Since the eight evaluation rules in MONICA were generated by amateurs (the authors) rather than art professionals or experts, we decided to perform a cognitive experiment that would give greater credence to the rules. First of all, these evaluation rules have been used successfully in the system to produce Mondrian-like artwork, and this is probably the most convincing proof that they indeed make sense. Figure 8 shows three Mondrian-like paintings which were created by our system (which assigned a fitness value of 1 to them) at different times. In addition, however, we also decided to determine whether most people would agree with our rules or not.

Since not all of our experimental subjects were computer literate, the way we decided to perform the experiment was to take eight different paintings produced at different times by MONICA and assigned a different fitness value (0, 0.125, 0.250, ..., 1.0, except 0.75) by its evaluation procedure. These were rearranged randomly, labeled A-H, and placed into one file so they could be displayed together in the computer. Figure 9 shows the eight paintings used in the experiment (one of them is the right-most one shown in Figure 8). MONICA's ranking of the eight paintings shown in Figure 9 is C-G-H-A-E-F-B-D, where C is the least Mondrian-like and D is the most Mondrian-like.

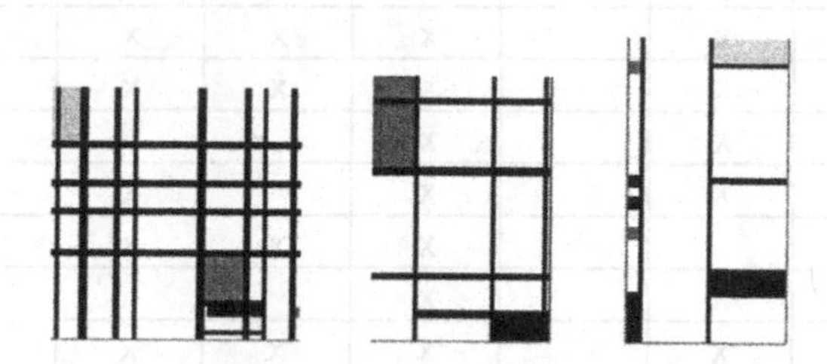

*Figure 8.* Three Mondrian-like paintings generated by our system

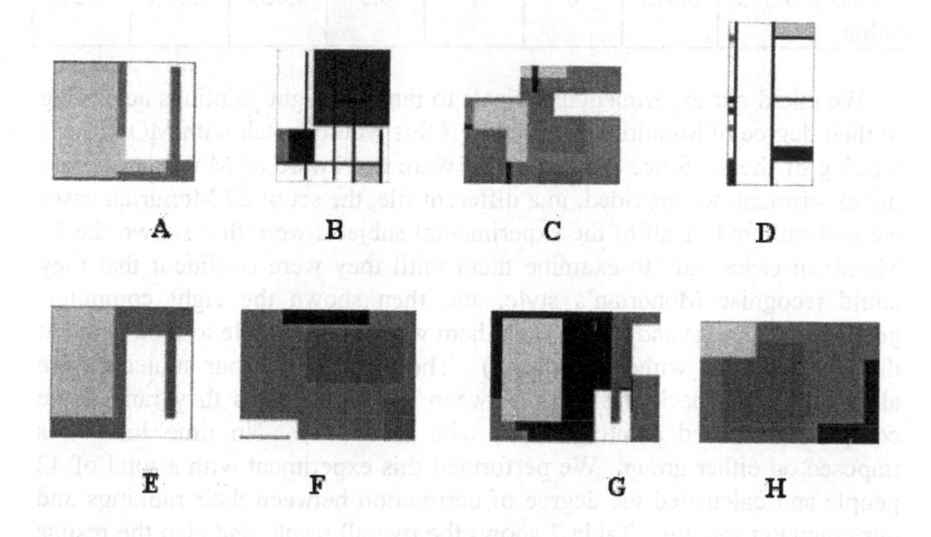


*Figure 9.* Eight paintings used in the experiment

Table 1 shows which of the paintings shown in Figure 9 satisfied which of the eight evaluation rules, and thus explains MONICA's ranking of the eight paintings. Some of the rule satisfactions shown in the table cannot be verified visually, as for example the figure does not show whether the internal genotype representation contains valid coordinate values or not for the limits of the colured regions in the genotype of a given painting (rule 2 above). This may be one of the reasons for experimental subjects not coinciding as much as they could have with the system's ranking of the eight paintings, Table 2.

TABLE 1. MONICA's evaluation of the paintings shown in Figure 9

| Satis-fied rules: | Paintings: A | B | C | D | E | F | G | H |
|---|---|---|---|---|---|---|---|---|
| **1** | | x | | x | | | | |
| **2** | x | x | | x | x | x | x | x |
| **3** | | x | | x | x | x | | |
| **4** | | x | | x | | | | |
| **5** | | x | | x | | x | | |
| **6** | x | | | x | x | x | | |
| **7** | | x | | x | | | | |
| **8** | x | x | | x | x | x | | x |
| Total fitness value: | 3/8= 0.375 | 7/8= 0.875 | 0/8= 0 | 8/8= 1 | 4/8= 0.5 | 5/8= 0.625 | 1/8= 0.125 | 2/8= 0.25 |

We asked our experimental subjects to rank the eight paintings according to their degree of Mondrianness, to see if this would match with MONICA's ranking of them. Since most subjects were not aware of Mondrian before the experiment, we provided, in a different file, the set of 27 Mondrian cases we had gathered. Half of the experimental subjects were first shown the 27 Mondrian cases, told to examine them until they were confident that they could recognise Mondrian's style, and then shown the eight computer-generated paintings and told to rank them without being able to look again at the 27 cases (i.e., without feedback). The other half of our subjects were allowed to shift back and forth between the two files as they ranked the computer-generated paintings (i.e., with feedback). No time limit was imposed on either group. We performed this experiment with a total of 42 people and calculated the degree of correlation between their rankings and our computer ranking. Table 2 shows the overall result, and also the results for subsets of our experimental subjects grouped according to gender, educational background, whether they were aware of Mondrian's paintings before the experiment or not, and whether they had feedback or not during the experiment.

From Table 2 we can see that on average our experimental subjects' rankings of the eight paintings and MONICA's rankings correlated to a degree of about 0.677. This is a strong correlation, though not exceptionally strong. Our experimental subjects, like us, were not art experts or professionals, and it would be interesting to know how much of a difference this made to the results. On the other hand, we can tell from the table that there doesn't seem to be any effect on the result due to prior knowledge (even at an amateur level) of Mondrian or his style, so repeating the experiment with experts might not lead to different results. The only results that were very different from the average was from the group of subjects that were not enrolled in, and had never obtained, a university degree of any kind. However, the sample size is too small to conclude anything definite from this observation.

In any case, the correlation found in the experimental results, plus the very Mondrian-like paintings generated by MONICA shown in Figure 8, validate and give credence to the evaluation rules we used in MONICA. We feel that this is a very valid way to program (and verify the performance of) intelligent systems, rather than the traditional method of basing the algorithms and process models used on cognitive models arising from studies and observations of humans performing the same tasks desired of the system.

TABLE 2. Experimental results

| | **Type of subject (size of sample):** | **Degree of correlation:** |
|---|---|---|
| **Gender:** | Male (21): | 0.700 |
| | Female (21): | 0.622 |
| **Educational Background:** | Less than university (2): | 0.476 |
| | Enrolled in or obtained undergraduate degree (21): | 0.655 |
| | Enrolled in or obtained postgraduate degree (19): | 0.687 |
| **Previous Awareness of Mondrian:** | Previously Unaware of Mondrian (22): | 0.620 |
| | Previously Aware of Mondrian (11): | 0.665 |
| | Unknown Awareness of Mondrian (9): | 0.757 |
| **Experimental Conditions:** | With Feedback (21): | 0.651 |
| | Without Feedback (21): | 0.671 |
| **Totals:** | All Subjects (42): | 0.677 |
| | Min. Correlation (1): | 0.190 |
| | Max. Correlation (1): | 0.976 |

## 4. Discussion and Related Work

Generating artwork by computer is not a new idea, and neither is using evolutionary algorithms for this task. Some examples of computer artwork generating systems that use ideas from evolution are presented and discussed, for example, in (Bentley 1999) and (Bentley and Corne 2002). Most of the systems described in these books that generate artwork (Todd and Latham 1999; Witbrock and Neil-Reilly 1999; Rowbottom 1999; Rooke 2002; Pagliarini and Lund 2002; Hancock and Frowd 2002; Eiben et al. 2002) do so by using the evolutionary operators of crossover and mutation, but leave it to the user(s) to decide which of the new paintings (or which of their features) to keep for future evolutionary cycles, and/or how to rank the new paintings according to their subjective (and probably unconscious) aesthetic criteria. Thus, the decisions on what is aesthetic or interesting are not made by the systems. In contrast, MONICA is designed to be a fully autonomous system requiring no user feedback as its evolutionary algorithm proceeds.

The point of this approach is to explore (and perhaps push) the limits of what computers are capable of doing by themselves. For other types of design (e.g., architectural design, mechanical design) two trends have surfaced over the past decades. One of these trends has resulted in producing interactive computer-aided design systems such as AutoCAD in which the user makes most of the important design decisions and the system simply serves as a support tool for the tedious tasks of rendering, visualisation, and simulation. The other trend has been to explore how computers can autonomously create designs without human intervention, making all design decisions themselves, and has resulted in a number of artificial intelligence projects. Perhaps most of these projects have stopped in the prototype or research stage, and have not resulted in commercially available systems, but they have produced some interesting research results and ideas. It is this second approach that we are interested in extending to artistic design in order to produce not computer-aided art systems, but rather autonomous artwork generating systems.

In order for MONICA to be fully autonomous, an evaluation procedure that captures and recognises the stylistic characteristics of Mondrian paintings had to be programmed into its evolutionary algorithm. Neither the relationship between evolutionary algorithms and Mondrian, nor the attempt to capture and automate the generation of new creations in the style of given artists or designers, is new either. However, the approach we followed in implementing MONICA is different from those that have been used in other projects that have explored these issues.

A Mondrian Evolver (as well as an Escher Evolver) is mentioned in Eiben et al. (2002). However, the Internet web-page cited does not seem to exist anymore, so we have not been able to view the program. If it works the same way as the Escher Evolver described in the book chapter cited, then anyone who accesses the Mondrian Evolver web-page can provide feedback to the program in order to influence the results of the next evolutionary cycles. As can be seen, the system is again non-autonomous, unlike MONICA.

A Mondrian Applet (Lendon 1999) and a Mondrian Machine (Lewis 1996) can be found on the Internet. The first one of these two systems is completely autonomous, and the second one semi-autonomous (because the user's clicks determine the positions of black lines in the paintings which are then automatically generated, with colours being randomly selected to fill some of the spaces between the resulting lines). Both systems have been programmed with certain rules about how to generate artwork that looks like Mondrian's according to their programmer's understanding of Mondrian's style. The difference with our work is that generating new paintings in MONICA is done completely at random through the evolutionary operators of crossover and mutation, not by following any pre-programmed generative rules. It is only in the evaluation subroutine in MONICA's evolutionary algorithm that any knowledge of Mondrian's style (according to our understanding of it) is programmed, and this knowledge in MONICA is used for style recognition rather than generation or emulation.

Returning to evolutionary algorithms, one of these has in the past been applied to generating Mondrian-like artwork, as reported in Schnier and Gero (1997; 1998). The genotype representation used in these two publications is hierarchical (much like that used in genetic programming (Koza 1992)), and relies on the observation that a large subset of Mondrian's paintings can be described by successive recursive divisions of a canvas into rectangular areas. It is the recursive subdivisions that are embodied in the hierarchical aspects of the representation. The fitness of new paintings generated by this system is determined by measuring the distance between them and the exemplars (cases) of real Mondrian paintings used to initiate the evolutionary process. The non-hierarchical genotype representation used in MONICA provides much less guidance to the system as to the structure that a Mondrian-like painting should have in comparison to the un-named system described in the two papers by Schnier and Gero. Another difference is that MONICA's evaluation subroutine explicitly captures stylistic characteristics used by Mondrian and observed by the authors, whereas the evaluation module in the system described by Schnier and Gero does not. On the other hand, the use of real cases of Mondrian paintings and the overall evolutionary framework and attempt to obtain Mondrian-like results create a close similarity between the two projects.

Work into capturing the style of particular artists or designers in the computer has often focused on shape grammars (Cha and Gero 1999) or semantic networks (Gero and Jupp 2003). However, some work has also used evolutionary algorithms to explore style. A system is described in (Ding and Gero 2001) that generates traditional Chinese architectural facades after it infers a representation of their style. The inference and subsequent learning is done with an evolutionary algorithm which in the end obtains a hierarchical genotype which represents a particular style (according to the exemplars which have been shown to it). As with the project described by Schnier and Gero, here we have exemplars and we have hierarchical genotypes. However, in this project the genotype does not describe a particular design, but rather an entire design style. This is because the genes that compose the hierarchical genotype embody syntactic and semantic primitives that combine to describe a particular style. The task of the system is to learn this hierarchical genotype given the exemplars, and the learning task is done through an evolutionary algorithm. Recognising whether a new design matches a particular style involves matching its features with the style representation embodied in the hierarchical genotype. In contrast, in MONICA style recognition is pre-programmed, not learned, and forms part of the evaluation module of the evolutionary algorithm that generates possible Mondrian-like paintings, not part of the knowledge captured in the genotype representation.

Future work in MONICA will focus on not having to pre-program any style recognition knowledge at all in order to create the evaluation module of the evolutionary algorithm. Perhaps a similar approach to the one presented in (Ding and Gero 2001) can be used, employing first a learning evolutionary algorithm to evolve a representation of Mondrian's style, and then the current evolutionary algorithm to create possible Mondrian-like paintings, but to evaluate their fitness based on the learned representation rather than using hand-coded rules for this evaluation. Or perhaps a different method (e.g., we are considering neural networks) will be able to capture Mondrian's style in such a way that it can be used by MONICA's evolutionary algorithm, but without having to program any of its evaluation or recognition knowledge explicitly. For the moment, pre-programming a particular style into the system was not too difficult, given Mondrian's geometric and relatively simple style. However, applying the same ideas used in MONICA in order to imitate other painters' styles in the computer (e.g., da Vinci) might require a different approach!

## References

Bax, M: 2001, *Complete Mondrian*, Lund Humphries (Ashgate Publishing), Aldershot, United Kingdom.

Bentley, P (ed): 1999, *Evolutionary Design by Computers*, Morgan Kaufmann Publishers, San Francisco, California.

Bentley, P and Corne, DW (eds): 2002, *Creative Evolutionary Systems*, Morgan Kaufmann Publishers, San Francisco, California.

Cha, M-Y and Gero, JS: 1999, Style learning: Inductive generalisation of architectural shape patterns, *in* A Brown, M Knight, and P Berridge (eds), *Architectural Computing from Turing to 2000, eCAADe*, University of Liverpool, England, pp. 629-644.

Clark, CH: 1958, *Brainstorming: The Dynamic Way to Create Successful Ideas*, Doubleday, Garden City, New York.

Deicher, S: 1999, *Mondrian*, Benedikt Taschen Verlag GmbH, Cologne, Germany.

Ding, L and Gero, JS: 2001, The emergence of the representation of style in design, *Environment and Planning B: Planning and Design* **28**(5): 707-731.

Eiben, AE, Nabuurs, R, and Booij, I: 2002, The escher evolver: Evolution to the people, *in* P Bentley and DW Corne (eds), *Creative Evolutionary Systems*, Morgan Kaufmann Publishers, San Francisco, California, pp. 425-439.

Gero, JS and Jupp, JR: 2003, Feature based qualitative representation of architectural plans *in* A Choutgrajank, E Charoenslip, K Keatruangkamala and W Nakapan (eds), *CAADRIA03*, Rangsit University, Bangkok, pp.117-128.

Gómez de Silva Garza, A and Maher, ML: 2000, A process model for evolutionary design case adaptation, *in* JS Gero (ed), *Artificial Intelligence in Design '00*, Kluwer Academic Publishers, Worcester, Massachusetts, pp. 393-412.

Gómez de Silva Garza, A and Maher, ML: 2001, GENCAD: A hybrid analogical/evolutionary model of creative design, *in* J Gero and ML Maher (eds), *Computational and Cognitive Models of Creative Design V*, Key Centre of Design Computing and Cognition, University of Sydney, Australia, pp. 141-171.

Hancock, PJB and Frowd, CD: 2002, Evolutionary generation of faces, *in* P Bentley and DW Corne (eds), *Creative Evolutionary Systems*, Morgan Kaufmann Publishers, San Francisco, California, pp. 410-423.

Koza, JR: 1992, *Genetic Programming: On the Programming of Computers by Means of Natural Selection*, MIT Press, Cambridge, Massachusetts.

Lendon, C: 1999, Mondrian Applet, http://www4.vc-net.ne.jp/~klivo/soft/mondrian.htm.

Lewis, M: 1996, Mondrian Machine, http://desires.com/2.1/Toys/Mondrian/mond-fr.html and http://www.ptank.com/mondrian.

Louis, SJ: 2002, Learning from experience: Case injected genetic algorithm design of combinatorial logic circuits, *in* IC Parmee (ed), *Adaptive Computing in Design and Manufacture V*, Springer-Verlag, Berlin, Germany, pp. 295-306.

Mitchell, M: 1998, *An Introduction to Genetic Algorithms (Complex Adaptive Systems Series)*, MIT Press, Cambridge, Massachusetts.

Pagliarini, L and Lund, HH: 2002, Art, robots, and evolution as a tool for creativity, *in* P Bentley and DW Corne (eds), *Creative Evolutionary Systems*, Morgan Kaufmann Publishers, San Francisco, California, pp. 367-385.

Rooke, S: 2002, Eons of genetically evolved algorithmic images, *in* P Bentley and DW Corne (eds), *Creative Evolutionary Systems*, Morgan Kaufmann Publishers, San Francisco, California, pp. 339-365.

Rowbottom, A: 1999, Evolutionary art and form, *in* P Bentley (ed), *Evolutionary Design by Computers*, Morgan Kaufmann Publishers, San Francisco, California, pp. 261-277.

Schnier, T and Gero, JS: 1997, Dominant and recessive genes in evolutionary systems applied to spatial reasoning, *in* A Sattar (ed), *Advanced Topics in Artificial Intelligence (10$^{th}$ Australian Joint Conference on Artificial Intelligence AI-97 Proceedings)*, Springer-Verlag, Heidelberg, Germany, pp. 127-136.

Schnier, T and Gero, JS: 1998, From Frank Lloyd Wright to Mondrian: Transforming evolving representations, *in* I Parmee (ed), *Adaptive Computing in Design and Manufacture III*, Springer-Verlag, London, pp. 207-219.

Todd, S and Latham, W: 1999, The mutation and growth of art by computers, *in* P Bentley (ed), *Evolutionary Design by Computers*, Morgan Kaufmann Publishers, San Francisco, California, pp. 221-250.

Witbrock, T and Neil-Reilly, S: 1999, Evolving genetic art, *in* P Bentley (ed), *Evolutionary Design by Computers*, Morgan Kaufmann Publishers, San Francisco, California, pp. 251-259.

JS Gero (ed), *Design Computing and Cognition'04*, 97-116


# COGNITIVE INVESTIGATIONS INTO KNOWLEDGE REPRESENTATION IN ENGINEERING DESIGN

JARROD MOSS, KENNETH KOTOVSKY, JONATHAN CAGAN
*Carnegie Mellon University, USA*

**Abstract.** As engineering students gain experience and become experts in their domain, the structure and content of their knowledge changes. Two studies are presented that examine differences in knowledge representation among freshman and senior engineering students. The first study uses a recall paradigm, and the second uses Latent Semantic Analysis to analyze brief descriptions written by engineering students. Both studies find that the most prominent differences between these two groups of students are their representations of the function of electromechanical components and how these components interact. The findings from these studies highlight some ways in which the structure and content of mental representations of design knowledge differ with experience.

## 1. Introduction

Engineering design is a domain in which a number of complex problem solving activities occur. As in all such tasks, cognitive processes operate upon the internal representations of the task as well as upon other relevant knowledge. These representations can change over the course of experience in order to enable a person to better respond to the problems and challenges of a domain. These representation changes are a reflection of the structure and content of a domain as well as the cognitive learning mechanisms responsible for the changes.

One motivation for studying expertise is to learn more about the general cognitive mechanisms which allow people to become experts in some domain given sufficient learning and practice. Eventually a person acquires both knowledge structures and cognitive processes that are specific to the domain of expertise and allow for efficient functioning within that domain. The knowledge structures and processes of expertise that arise are a function of both general cognitive mechanisms and the actual structure and content of the domain in which expertise is being acquired. This means that the content and to some extent the structure of the knowledge representation are determined by the particular domain of expertise. The structure of experts' representations across different domains may show some similarities either because they are constructed by the same learning mechanisms or because the domains of expertise have some common characteristics.

There are at least two types of reasons to study expertise in a particular domain. The first is to learn about the specific mental representations and cognitive processes employed by experts in that domain. This information is potentially beneficial in the design of cognitive aids that can assist experts or in improving the education of future experts in that domain. The second type of reason to study expertise in a domain is that it provides further insights into the cognitive learning mechanisms that produce the changes in representation and cognitive processes seen in expertise acquisition. One way to study these learning mechanisms is to see how they interact with the structure and content of a variety of domains in order to produce the mental structures and processes seen in experts across these domains.

A number of domains of expertise have been studied, and there have been some general findings about how the structure of domain knowledge changes with experience in these tasks. For instance, in games like chess and Go, a hierarchical database of commonly occurring piece configurations appears to exist in experts but not in novices (Chase and Simon 1973; Reitman 1976). This knowledge aids the expert in classifying the current situation and identifying good moves. Similar types of hierarchical chunking have also been identified in electronics technicians (Egan and Schwartz 1979). One of the general findings in these and other areas of expertise is that a hierarchical knowledge structure is often a component of expertise. Other general findings in the expertise literature are that experts tend to work forward from the givens in the problem rather than backwards from the desired solution, and experts tend to classify items and problems in their domain of expertise according to a deeper conceptual structure rather than surface similarities (Chi et al. 1981; Larkin et al. 1980).

Experts' knowledge representations enable them to handle problems and process information differently than novices. For example, physics experts solve physics problems in a different manner than do novices (Larkin et al. 1980). Physics experts can recognize and solve common problems in a more efficient manner than novices. In fact, in expertise it is common for the associated knowledge structures and processes to be such an integral part of the cognitive system that they affect the way the expert perceives the environment. Chase & Simon (1973) argue that what they see is perceptual chunking, and elsewhere it has been shown that while domain knowledge does not directly transfer to other domains it can still influence the way people reason about more general situations outside of their domain of expertise (Nisbett et al. 1987). So another general property of expertise is that it affects not only what is stored in memory but also how things in the world are perceived and categorized.

These effects of expertise all relate to the ability of the expert to process more domain information in a fixed period of time than can a novice. Experts have highly organized memory structures like schemas, templates, and retrieval structures (Gobet 1998; Richman et al. 1995). These memory structures allow for the easy retrieval and storage of domain information, and they affect how domain information is perceived. As information about a new problem is perceived, this information automatically activates relevant domain knowledge and processes. This allows experts to easily recognize and categorize information and solution schemas in their domain. Parts of the problem solving process are therefore more automatized in experts than they are in novices, and this enables experts to solve problems in a more efficient manner. In order to understand how experts in a domain solve problems it is necessary to examine the way domain information is represented.

Understanding the representation changes that occur as engineering students progress toward becoming professionals is essential in achieving an understanding of the cognitive processes underlying performance in engineering design. As discussed below, there are a few studies that have examined cognition and expertise in engineering design through verbal protocols and other methods, but these studies usually deal with cognition at a coarse level and do not examine mental representation. The work presented here is an initial step towards a detailed examination of the representations and processes that allow engineers to perform the complex tasks required by their profession. This paper presents two studies which begin to answer the question of what kinds of representation changes accompany the transition to expertise.

In particular, the studies presented here look at freshmen and senior engineering students in order to see what kind of representation changes accompany the early transition to expertise. The differences between freshmen and seniors may generalize to professional engineers upon further investigation, or alternatively the transition from student to professional may involve other qualitative changes in mental representation. Two different methodologies were utilized in examining representation differences in the two groups of students. The first study utilizes a recall paradigm that has been employed by a number of researchers looking at expert/novice differences (Chase and Simon 1973). This first study examines some basic differences in how components in devices are represented and chunked together. The second study uses Latent Semantic Analysis (Deerwester et al. 1990) as a methodological tool to aid in exploring and analyzing the content of students' representations. This study seeks to determine whether the seniors think about and represent devices in a more abstract functional manner than do freshmen.

## 2. Expertise in Design

There has been some relevant work on the differences between experts and novices in engineering design. For instance Atman, Chimka, Bursic, and Nachtmann (1999) have looked at differences in the design processes of freshmen and senior students by analyzing concurrent verbal protocols. They found that seniors have a better representation of what a good design process entails and can transition between steps in the design process more easily than can freshmen. Also, seniors consider more design alternatives than do freshmen which is probably one reason that seniors end up producing better quality designs in their study. Other researchers have examined the differences between the design processes of students as compared to professionals. For instance, it was found that in an artificial design task that groups of professionals exhibit more metacognitive and strategic behaviors during design (Smith and Leong 1998). Student groups rarely exhibited these behaviors, and they tended to iteratively refine their original design concept as opposed to exploring multiple alternatives as the professionals often did.

Another series of studies investigated engineering design processes using concurrent verbal protocols (Ball et al. 1994; Ball et al. 1997; Ball and Ormerod 1995). These studies found that novices use a depth-first design process while experts use more of a breadth-first approach. Both groups of designers decomposed the problem into modules, but experts tended to develop each module to a certain level of detail before moving to the next level of detail. Novices were more likely to do detailed design on one module before moving on to the next. It is proposed that the depth-first structure is advantageous for novices since it limits the amount of goal information they must store in memory. Even when experts deviated from their breadth-first structure there seemed to be principled reasons for doing so. For example, an expert may quickly follow one potential solution in depth to assess its feasibility before proceeding to other parts of the design.

There has also been some work in the domain of architectural design. For instance, Suwa and Tversky (1997) collected a set of retrospective protocols from students and a couple of professional architects. They found that the experts tended to follow certain trains of thought in more depth than the novices. Also the experts were better at reading certain types of functional information from their sketches than were novices. This same set of protocols has been the basis for other work as well (Kavakli and Gero 2001; Kavakli and Gero 2002; Suwa et al. 1998), but these have all been case studies in which either a single expert or one expert and one student have been examined so it is impossible to determine if there are any statistically reliable differences in these studies. These studies have all used retrospective protocols because it was believed that concurrent verbalization would bias the design and sketching behavior of participants. The retrospective protocols done in these studies used a video of the design session to cue verbalization. This procedure is likely to bias the results since the only recall cues to the designer are from sketching behavior. Other converging methods should be used to confirm these results to make sure the results generalize to other groups and to insure results were not overly biased by the retrospective method used. This is just one set of examples of work in the domain of architectural design that may relate to expertise in engineering design. Since it is unlikely that specific results in this domain will transfer directly to engineering design, there is a need to analyze the content and structure of the two domains in order to determine which results are relevant to expertise in engineering design. As discussed above, the cognitive learning mechanisms that enable expertise acquisition are the only thing guaranteed to hold across domains, and most studies of expertise in design do not study cognition at this level.

Overall, the work on engineering design and related areas tends to focus more on differences in the design processes of experts and novices than on representation and other cognitive issues. These studies have some things to say about the cognitive processes going on in design, but there is not much at all about the internal representations that engineers use while solving a design problem. Goel (1995) presents an analysis of design problem spaces and the results of a study which indicate that certain types of symbol systems are necessary to support design activity. His work is concerned with some general necessary properties of a representational system. However, the subjects in this study were all experts so it is unclear how these results map onto novices. As stated above, it is necessary to understand how devices and other domain knowledge are represented in memory and how these representations change with expertise. The studies presented below are a first step in understanding these important issues.

## 3. Experiment 1: Chunking of Components

This study examines how the participants chunk components into larger meaningful units. Just as experts are known to chunk elements into larger units of knowledge in other domains such as electronics (Egan and Schwartz 1979), it should be the case that the more experienced students have some way of organizing knowledge about components in a device. One hypothesis is that components will be chunked into larger meaningful units which perform a certain function in the device. Such functional units or chunks could occur across multiple devices in which the chunk performs the same function. One example would be a rack and pinion as this set of components is one common method to convert between rotation and translation and could be expected to occur in a variety of devices. However, engineers will also be able to reason about the functionality of a particular device and break it into functional units regardless of whether they are commonly occurring or not.

In order to investigate these issues, a recall paradigm was utilized that extends an approach used by others to study chunking differences in expert/novice behavior (Chase and Simon 1973; Reitman 1976). The basic method is to present a stimulus, such as a chess board in a mid-game position, for a brief period of time. The participant is then asked to recall the presented stimulus. In the original methodology both recall and perception tasks were used and chunks were identified based on inter-response times (IRTs) that were common to both tasks. However, later work examining chunks in Go (Reitman 1976) found that a common IRT could not be found for both tasks due to the fact that the chunks in Go have an overlapping structure. This was not a problem in the chess research since the chunks in chess have more of a hierarchical relationship. In our experiment, only a recall task was used, and in order to avoid problems with finding an appropriate IRT boundary, analysis of IRTs was only one of many measures used to look at representation differences. In particular, we looked at percent recall after one exposure, errors, patterns of recall, and alternate methods of identifying chunks in addition to IRTs.

## 3.1. METHOD

### *3.1.1. Participants*

Fifteen seniors majoring in mechanical engineering volunteered for the study. These students were recruited from a required senior engineering design course at Carnegie Mellon. Fifteen freshmen engineering students also participated in the study as partial fulfillment of a course requirement. All freshmen were enrolled in the engineering college at Carnegie Mellon, but students in this college do not declare a particular engineering major until after their freshman year.

### *3.1.2. Stimuli*

Three electromechanical devices were represented in schematic diagrams which indicated how components fit together in each device. The schematics were represented in an idealized fashion where only the types of components and the connections between these components were displayed. For example, all gears were represented by the same icon which includes no information about different sizes, shapes, or types of gears. Connections between components were represented by lines connecting components. An example design schematic is shown in Figure 1. The number of components in each device was 16, 13, and 14 for the drill, pressure gauge, and weighing machine respectively. The number of connections in each device is 9, 11, and 12 for the drill, pressure gauge, and weighing machine respectively. The weighing machine is similar in purpose to a bathroom scale. The number of unique components differs since some types of components were used more than once in a design. The drill had only 9 unique components, while the pressure gauge and weighing machine had 11 and 12 unique components respectively. Each diagram also had a label at the top indicating the type of device depicted as shown in Figure 1.

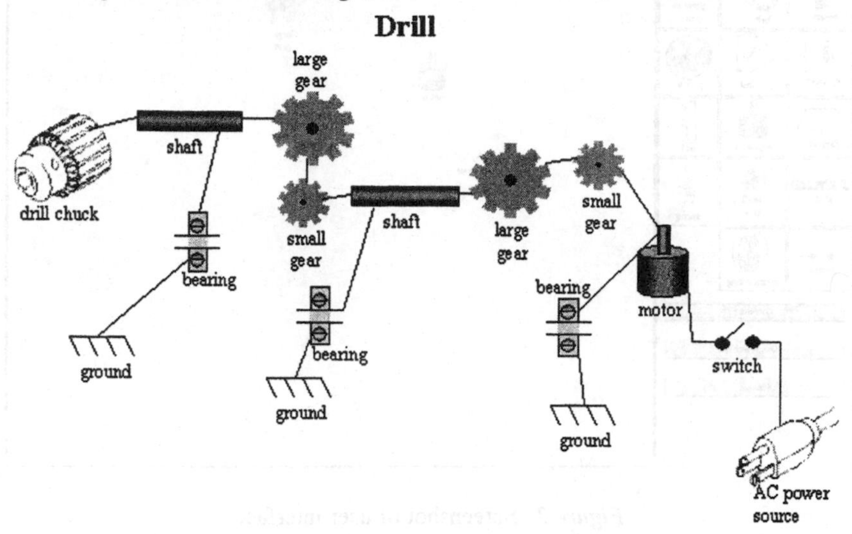


*Figure 1.* Example of diagrams seen by participants

### *3.1.3. Procedure*

Participants were asked to recall three design schematics using a graphical interface after a brief study period. Participants received instruction and were allowed to become familiar with the interface and the type of representation used in the diagrams. The user interface is depicted in Figure 2, and it consists of a set of components that can be dragged over to a drawing space where they can be moved, connected, disconnected, or removed. Participants then received a practice trial followed by three recall trials. During each trial, the initial schematic was displayed for 40 seconds, and then the display of the user interface replaced the schematic. Participants then had 3 minutes in which to recall as much of the schematic as possible. The design schematic was presented again for 40 seconds if the participants had not recalled the design completely. These periods of display and recall alternated until the participant recalled the device perfectly. The presentation order of the three design schematics was counterbalanced. The computer generated a time stamped entry in a log file for every action the participant took. The log was detailed enough so that a participant's actions could later be replayed for purposes of analysis.

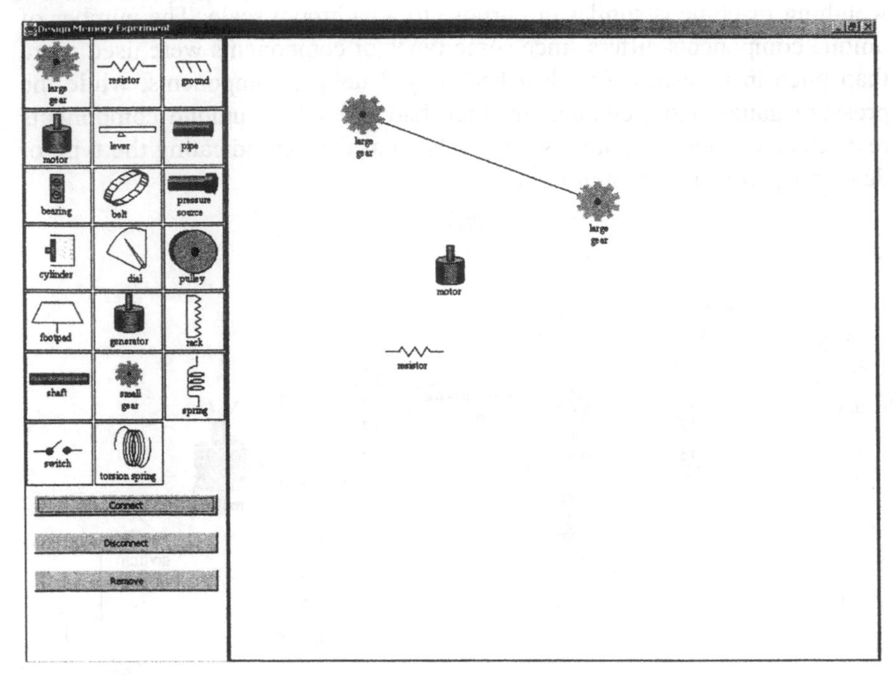


*Figure 2.* Screenshot of user interface

3.2. RESULTS

The percentage of components and connections between components recalled correctly during the first recall session of a trial was analyzed. Recall of components was almost perfect for all devices for both freshmen (M = 92.4%, SD = 9.23%) and seniors (M = 93.6%, SD = 10.2%), and experience level had no significant effect. Device type does have an effect on this measure, $F(2,56) = 3.63$, $p = .03$, and further contrasts showed that the drill components (M = 95.8%, SD = 7.58%) were recalled significantly better than both the weighing machine (M = 91.2%, SD = 10.0%), $F(1,28) = 7.81$, $p = .01$, and pressure gauge components (M = 92.3%, SD = 10.3%), $F(1,28) = 5.05$, $p = .03$. Recall of connections was lower overall than for components. There was no significant difference between freshmen (M = 77.8%, SD = 20.6%) and seniors (M = 82.8%, SD = 23.6%) on recall of connections, but again there was a significant effect of device type, $F(2,56) = 5.96$, $p = .005$. The drill's connections (M = 89.3%, SD = 14.6%) were recalled better than both the weighing machine (M = 73.8%, SD = 25.9%), $F(1,28) = 11.4$, $p = .002$, and pressure gauge (M = 78.9%, SD = 21.5%),

$F(1,28) = 6.83$, $p = .014$. These results indicate that some devices were harder to recall than others. The most difficult design to recall seems to be the weighing machine followed by the pressure gauge, and the easiest to recall is the drill. The drill has only 9 unique components with a number of repeating patterns, but the other two devices have an increasing number of unique components. It makes sense that recall difficulty would increase with the more components and locations that have to be remembered.

A number of error types were defined and analyzed. They include adding a component that is not part of the design, removing a component that is needed, connecting two components that are not connected in the original design, and disconnecting two components that should be connected. Freshmen made more errors overall than seniors, $F(1,28) = 42.1$, $p < .001$. Device type did not have an effect on overall errors, and there was no interaction between device type and experience. The component removal and disconnect errors did not occur frequently enough to be analyzed separately, but analyses were done on the add and connect errors. There were no effects of device type or experience on the add errors, but there was an effect of experience on the number of connection errors, $F(1,28) = 43.3$, $p < .001$. Connection errors can be further divided into possible connections and impossible connections depending on whether the two components could actually be connected in the real world. Connection errors are displayed in Figure 3, where it can be seen that freshmen do make more connection errors than seniors. There was no significant interaction between device type and experience with respect to connection errors.

Patterns of reconstruction were also analyzed, and one meaningful pattern was identified. A number of students started at the input of the device and reconstructed the device based on the flow of energy through the device. For example, in the drill in Figure 1, students following this pattern began with the power source and then proceeded to add the switch, motor, gear sets, and drill chuck in that order. 14 out of the 15 seniors used this pattern at least once, while only 8 of the 15 freshmen did. In addition, two of the three devices were presented so that their input was on the left and power moved through the device from left to right, but the drill was presented so that its input was on the right side of the screen. For the drill, 6 seniors and only 2 freshmen went from input to output. One of these seniors and one of these freshmen actually saw the drill as the first design, but in the other cases the participant had already reconstructed another device from left to right (input to output) and reversed for the drill. This provides some evidence that seniors prefer to reconstruct the device based on the flow of energy through the components of the device, and there was some preference for moving from input to output even when the direction of input to output was reversed on the display.

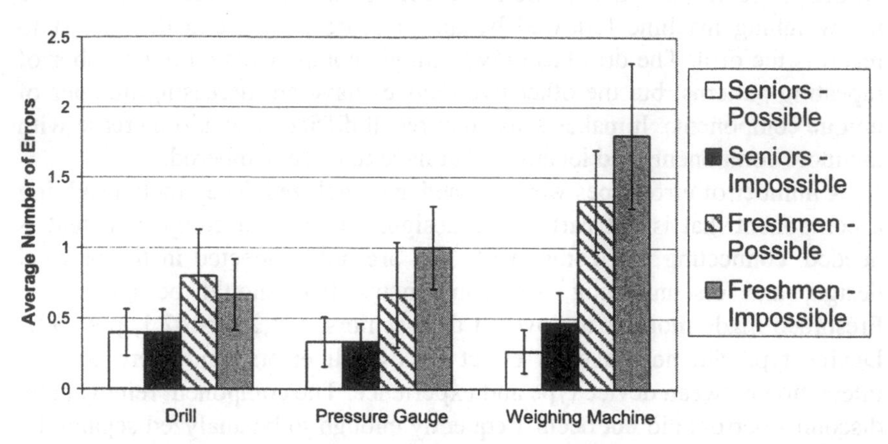


*Figure 3.* Average number of connection errors

The data can also be divided into chunks, but these divisions were not based solely on IRTs. First, an IRT criterion was set to distinguish between chunk transitions from within chunk transitions. A cutoff of 4 seconds was used. This value may be conservative as other studies have used boundaries of around 2 seconds. However, without an additional task such as the perception task used by Chase & Simon (1973) it is difficult to come up with a definite boundary. For example, in the perception task participants had to reconstruct a mid-game chess board, but they were able to glance back and forth from the board to be reconstructed and the board on which they were reconstructing the game position. It was assumed that participants encoded one chunk during each glance. Between chunk IRTs can then be distinguished from within chunk IRTs by looking at the difference in IRTs for when a participant recalled two pieces in succession without a glance at the original board and when they recalled two pieces separated by a glance at the original board.

The 4 second cutoff only includes 20% of the chunk transitions in our study. So the vast majority of transitions are still classified as within chunk transitions. However, the structure of the task also defined additional chunk boundaries. In the process of reconstructing a device, many participants would add a set of components and then proceed to connect those components before adding another set of components. This seems to provide a natural boundary whereby a participant adds a chunk, connects the components in it, and then adds another chunk. Using these two types of chunk boundaries, the data was segmented into individual chunks. Participants also drew chunk boundaries as mentioned before so their drawn chunks could be compared to the chunks generated from the recall data.

As a first step, the number of times a specific set of components were chunked together by participants was calculated. This allowed assessment of the chunks that individuals agreed on to some extent. The chunks identified by analysis of the recall data were similar to the chunks identified by the participants. This type of group level analysis indicates that both freshmen and seniors agree on the same types of chunks with the seniors being somewhat more consistent in the chunks they identify. Freshmen and seniors both produce chunks having between 2-3 components per chunk. While there seems to be agreement between freshmen and seniors when it comes to what should be chunked, the chunks identified do explain the error results mentioned earlier as will be explained below.

As mentioned before one of the most common types of errors was the connection errors, and this was the only type of error where the frequency of the error differed for freshmen and seniors. For both freshmen and seniors, 95% of their errors occurred when connecting two components that were not in the same chunk. When making a connection error, both groups are likely to make the error when connecting two different chunks, but freshmen make 2-3 times more of these errors than seniors depending on the particular problem. The difference in the frequency of connection errors then reflects the ability of seniors but not freshmen to remember how chunks of components were connected together.

### 3.3. DISCUSSION

In general it appears that seniors differ from freshmen on their understanding and ability to remember information about the connections and interactions between components. Seniors make fewer errors than freshmen, and the analyses indicate that this is mostly due to increased connection errors for freshmen. There is also some indication that freshmen may make more connection errors as problem difficulty increases (Figure 3), but this interaction failed to reach significance probably due to lack of statistical power. Seniors tend to rely more on recall methods that utilize the natural flow of power from one component to the next than do freshmen. From the connection error and chunking results, it is apparent that freshmen have more difficulty remembering how chunks of components connect together. Therefore, one of the main differences between the groups appears to be in their ability to remember how chunks of components connect together to form the overall device. This implies that freshmen are able to chunk components but have more difficulty connecting these functional units together to produce overall device behavior. The representation of chunks of components is weaker in freshmen than in seniors. While seniors are able to remember the chunks and how they interact to produce overall device behavior, freshmen are not as able to represent such interactions.

## 4. Experiment II: Functional Reasoning

In order to determine if a kind of abstract functional understanding was more prevalent in the mental representations of seniors than freshman a second study was run. Since it is apparent from the first study that freshmen lack a strong representation of how components interact, this should mean that freshmen are not as able to reason about devices in an abstract functional manner since the abstract functional level would entail reasoning about the functions of chunks of components and how those functions interact in producing device behavior.

Other work has demonstrated that people high in self-rated mechanical ability appear to reason better about the functioning of a device than do people low in self-rated mechanical ability (Heiser and Tversky 2002). In the current study it is assumed that the participants are all high in mechanical ability since they are all majoring in or intending to major in mechanical engineering. Based on these results and those of Experiment I it is hypothesized that more experienced participants will demonstrate better utilization of functional knowledge than less experienced participants. In order to investigate these issues, a new method of data analysis will be introduced. In the Heiser and Tversky (2002) work, each proposition that was written by a participant was coded as either structural or functional. This allowed them to show that high mechanical ability participants used more functional propositions. In this study, latent semantic analysis (LSA) will be used to test for higher functional content in written text. This method does not require someone to decide whether each proposition contains functional information or not. The data could also be analyzed using the proposition coding system as well just to show that the two methods are consistent, and this is part of planned future work.

In order to investigate these issues, students were asked to write brief descriptions of devices that were presented in diagrams. One assumption underlying this study is that the information students choose to include in a brief description is what they find important about the device, and that this importance is related to their mental representation of the device. In addition to the issue of functional information discussed above, another hypothesis that will be investigated is that seniors may be more mutually consistent in their descriptions than are freshmen. The reasoning behind this idea is that seniors have gone through years of formal education which may lead them all to think about the devices in a similar manner.

### 4.1. LATENT SEMANTIC ANALYSIS

The participants' descriptions were analyzed using Latent Semantic Analysis (LSA). LSA was originally developed as an information retrieval technique designed to overcome synonymy problems (Deerwester et al. 1990). It has also been used for a number of other purposes including as a model of text comprehension (Landauer and Dumais 1997). More recently it has also been used to develop similarity metrics to be utilized in the analysis of data from complex problem solving trials (Quesada et al. 2002). LSA begins with a term-by-document frequency matrix, and produces a reduced dimensionality space in which each document or term can be seen as a point or vector in that space. Similarities can then be computed between any two terms or documents by computing the cosine between the appropriate vectors. These properties make LSA an excellent tool for exploring similarities and differences between documents written by participants, thus shedding light on the content of their representations.

One set of researchers has already utilized LSA to try to identify shared design understanding among a set of designers (Hill et al. 2001). That study was concerned with building information management tools that retrieved relevant information based on a certain shared understanding of the design problem. This shared understanding was determined by using LSA to analyze documentation from design projects. This use of LSA allows for a particular- representation of the desired design project to be created, but these researchers were not concerned with identifying properties of this representation.

## 4.2. METHOD

### *4.2.1. Participants*

In this study, there were 44 volunteers from a senior mechanical engineering design class. There were also 24 freshmen volunteers from a freshman mechanical engineering class, and the study was run during their first mechanical engineering course.

### *4.2.2. Stimuli*

Three electromechanical device diagrams were used in this study. These diagrams were taken from patents for a power screwdriver, Figure 4, a cordless weed trimmer, and a drum brake system. The diagrams were mostly cross-sections of the devices and had lines labeling key components. The diagrams were used exactly as they appeared in the patent except that some labels were removed in order to ensure a similar number of labeled components for each diagram. Each diagram had 9-11 labeled components and had the name of the device printed in large bold letters at the top.

### *4.2.3. Procedure*

Participants were told that they would see diagrams of three electromechanical devices that had been taken from patents. They were told that their task was to write a description of each device but were not told what kind of information to include in their descriptions. If they asked what to include, they were told to include whatever they thought important as long as it pertained to the device shown. Participants viewed the device on a computer screen, and were told that they could click a button beneath the diagram that would remove the diagram and take them to a text area where they could type their description. There was also a button below the text area to take them back to the diagram, and they could alternate back and forth between description and diagram as often as they wanted. Each time they switched between the two views an entry was added to a log file and the current state of their description was saved to a time stamped file. Participants were instructed to spend about five minutes describing each device. They were not forced to spend exactly five minutes on a device, but they had to pace themselves to finish all three descriptions in 18 minutes. There was a clock displayed in the lower right corner of the screen to help them pace themselves.

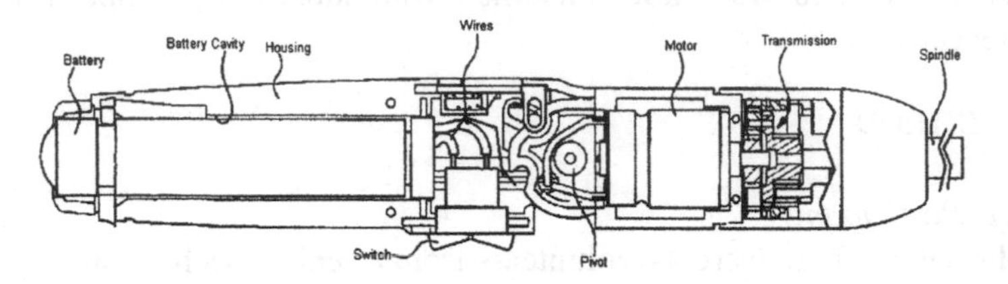


*Figure 4.* Example diagram seen by participants

The participants were then asked to rate their prior knowledge of each device (1=poor, 7=good). The freshmen participants then completed an additional set of eight true/false questions for each device before they gave their ratings. The questions consisted of a mix of questions emphasizing either the structure/composition of the device or the function of components within the device. They were not allowed to view any of the diagrams during these questions. The freshmen all participated during the second half of their first semester. The seniors were divided into two groups of 20. The first group completed the study in the first three weeks of the semester, and the second group participated after half of the semester had passed. This timing was used because it allowed us to test the hypothesized that the senior design course may be instrumental in changing the way students thought about designs since it specifically included a lecture on function structures about four weeks into the course.

4.3. RESULTS

Four seniors and two freshmen were excluded from all analyses since they failed to finish in the allotted time. This included two seniors from the early group and two from the late group.

Freshmen rated themselves as having more knowledge about the weed trimmer than the other two devices ($\chi^2 = 6.87$, $p = .03$), but there were no significant differences between devices in the seniors' ratings. There were also no significant differences between the ratings of freshmen and seniors for a particular device. All groups were therefore similar in their prior knowledge of the devices. The true/false questions were included to assess functional and structural knowledge of a device. All freshmen performed well on these questions averaging 6.8, 7.2, and 7.2 questions out of 8 correct on the brake system, screwdriver, and weed trimmer respectively. Any differences observed in functional knowledge were therefore not due to the freshmen being unable to access this knowledge.

One parameter that can be adjusted in LSA is the number of dimensions retained in the multidimensional space. Based on judgments from the number of dimensions used in previously published work with LSA, it was estimated that a good number of dimensions would be somewhere between 50 and 300. Most other LSA work has been done with much larger text corpora and optimal dimensionality was around 300 dimensions. Since our corpus of device descriptions is much smaller, a smaller number of dimensions are needed to capture most of the important information in the semantic space. The first 100 dimensions were used for all of the LSA results reported here.

The hypothesis that seniors are more consistent as a group than are freshmen was tested by computing a similarity measure between each participant's description of a device and the average vector for that device. The average vector was found by averaging the individual vectors for documents describing a particular device. A separate average was produced for freshmen and seniors for each device. For example, all freshmen drum brake device vectors were averaged to produce the average freshman brake description. Then for each device the average freshman vector was subtracted from the average senior vector. This produces a vector for each device that points from the average freshman description to the average senior description. This vector was then treated as a line in the multidimensional space with its origin at the average freshman description. All descriptions for a particular device were then orthogonally projected to a location on this line. Their location provides a way of examining freshmen and senior differences.

Due to the way the line is constructed and the way documents are projected to points on this line, the average senior description for a device will be a certain distance away from the average freshman description. The first test is whether there is a significant difference between where the senior and freshmen descriptions fall on this line, and they are significantly different, $F(1,54) = 121$, $p < .001$. The earlier hypothesis that seniors will be more similar to each other as a group than freshmen was then tested by looking at how far freshmen and seniors are from the average freshman and senior descriptions. For instance, seniors should on average be closer to the average senior description than freshmen are to the average freshman description. This difference is also significant with seniors deviating less from their average description than freshmen, $F(1,54) = 275$, $p < .001$. This means that seniors are more consistent with each other than freshmen on the information they include in descriptions of a particular device.

The search engine qualities of LSA were utilized in order to examine the hypothesis that seniors included more information about the functioning of a device in their descriptions. The documents in the multidimensional semantic space can be compared to a query vector and their similarity to this vector can be assessed using the cosine measure. In order to formulate a query that represents function information, a set of words that are associated with describing the functioning of a device were combined into a single query. Stone and Wood (2000) have developed a vocabulary to explain the internal chain of functions that produces a device's behavior. They have shown that this vocabulary can be used to represent a variety of different devices. The function words they use and the associated list of synonyms that they define for those words totals 73 words. Three of these words were judged to deal more with the structure of devices, and they were excluded from the query. These words were "connect", "locate", and "join". The remaining 70 words were combined into a query that was submitted to the LSA space.

This process functions like a search engine, and the system ranks documents according to their similarity to this query consisting of function words. Both experience level, $F(2,61) = 3.7$, $p = .03$, and device type, $F(2,122) = 12.7$, $p < .001$, have significant effects on a document's similarity to this query, but these two factors did not interact. Further contrasts reveal that the freshmen have significantly lower cosines (i.e., are less similar to the query) than the later group of seniors, $F(1,61) = 5.05$, $p = .02$. Also the drum brake system descriptions had higher cosines than both the screwdriver, $F(1,61) = 18.4$, $p < .001$, and the weed trimmer, $F(1,61) = 16.7$, $p < .001$. Document relatedness rankings retrieved from a query are often easier to interpret than the cosine values. Rankings are determined by sorting the cosines between a document and the query in descending order. The document with the highest cosine is ranked 1 and so on. The average rank (out of 192) for freshmen descriptions was 110.2 (SD = 55.4), while the seniors on average ranked 94.6 (SD = 55.5) and 82.0 (SD = 52.7) for the earlier and later groups respectively. These results indicate that seniors included more content that is similar in meaning to the function words in the query. Furthermore, the earlier group of seniors is ranked between the freshmen and the later group of seniors indicating an increase in functional content with experience.

### 4.4. DISCUSSION

The results from this study agree with and support those of the first study in that seniors are shown to incorporate more function information into their representations. Seniors do differ from freshmen on their similarity to the prototypical or average descriptions, and they differ on the amount of functional content they include in their descriptions. This means that seniors have all adopted a similar representation of the device, and that this representation includes more functional content than do the representations that freshmen use. The fact that seniors include more function content in their description adds further support to the idea that one of the main differences between the two groups of students is the ability to represent and process the functionality of chunks of components in a device.

## 5. General Discussion

The results from both studies support the idea that more experienced engineering students represent and reason about the functionality of a device and its components better than less experienced students due to differences in the representations used by the two groups. This finding seems to be the main difference in design knowledge representation at these levels of experience. This is not to say that freshmen can not or do not represent functional content, but instead that the memory structures that support this type of reasoning are not as well developed in freshmen. This level of representation provides the seniors with additional constraints when recalling the devices presented in the first study.

Senior engineering students may have a more detailed network of design knowledge which integrates content which the freshmen hold in separate representations if at all. For instance, both a freshman and senior may have similar representations of the structure of a gear and how the gear interfaces with other components. However, the senior may also have a representation of the abstract function of a gear and a set of contexts in which a gear may perform well. This idea of context is important because some functions can be performed by a group of two or more components, but not by any one of the components by itself. For instance, transforming rotational motion into translational can be performed by a rack and gear together but not by either of them separately. It may be that building up a set of such associations between sets of components and functionality is one of the main changes that take place as an engineer gains more experience. This type of learning and representation change seems similar to the learning of chunks in chess and other domains (Chase and Simon 1973).

One explanation for the findings in this paper is that the device is represented at multiple levels. The most abstract level is the overall function of the device, and the most detailed level would be the components which make up the device. In between these two levels are one or more levels in which the function chunks of components are represented. There may be multiple levels of this type of chunking in which the chunks at one level are grouped into sets at the next level. At any particular level, a chunk of components performs a subfunction which contributes to the overall functioning of a device. A similar model called *conceptual chunking* has been proposed to deal with the expertise of electronics technicians (Egan and Schwartz 1979). Using this type of model, it is proposed that the freshman differ from seniors in their ability to represent the middle levels where chunks of components perform some function. The freshmen seem to understand what types of components should go together to perform a sensible function, but they have problems linking these functions together to achieve the overall device function. In this view, seniors have a more integrative and less fragmented representation of the relation of the device and the components of which it is composed. This framework makes it easier for the seniors to recall devices since they have more constraints from which to reason. For instance seniors can reason about which components go together to perform certain functions, and they can also reason about which functions may be necessary for overall device behavior. Freshmen on the other hand may not be as able to reason about which functions are necessary for the overall behavior of the device, and so they have more difficulty connecting together chunks of the device and are less likely to talk about this level of function when describing a device.

This type of conceptual chunking relationship also relates to earlier work on the observed structure of the engineering design process in novices and experts. It has been found that experts adopt a more breadth-first design process, while novices use a depth-first process (Ball et al. 1997; Ball et al. 1994; Ball and Ormerod 1995). The designers in these studies all decomposed their design into more manageable modules. The observed structure of the design process could relate to how well developed the mental representations of the device are for experts and novices. Depth-first design processes require maintaining less intermediate information about other parts of a device, while a breadth-first approach requires thinking about how parts of a device interact at a number of different levels of specificity. It may therefore be easier for a novice to use a depth-first approach toward design.

In this way, the work on representation presented here provides some insights into why design processes differ for novices and experts.

LSA is a potentially powerful tool for investigating the structure of knowledge representations. A number of interesting questions about representation can be answered by using this automatic technique to represent a set of documents in a multidimensional space. One of the main problems in using LSA as an exploratory tool is trying to find the correct number of dimensions. However, varying the dimensionality of the space could also vary the amount of detail incorporated in the representations being examined. Seen in this light, having a variable number of dimensions could be a positive aspect of LSA as an analysis tool since the amount of representational detail being examined could be varied with one parameter.

One limitation of this work is that it only deals with engineering students. There are plans to expand this work to professional engineers, and it should be interesting to see how even greater amounts of design experience affect a person's representation of design knowledge. However, this work does capture some of the differences in the beginning stages of the acquisition of expertise. Also, even though differences associated with design experience have been identified, there is currently no mechanism that explains how these changes come about. Generating such an explanation is a necessary step in coming up with a cognitive model of the engineering design process.

## Acknowledgments

This research effort was sponsored by a National Defense Science and Engineering Graduate Fellowship and by the Air Force Office of Scientific Research, Air Force Material Command, USAF, under grant number F49620-01-1-0050. The U.S. Government is authorized to reproduce and distribute reprints for governmental purposes notwithstanding any copyright annotation thereon. The views and conclusions contained herein are those of the authors and should not be interpreted as necessarily representing the official policies or endorsements, either expressed or implied, of AFOSR or the U.S. Government.

## References

Atman, CJ, Chimka, JR, Bursic, KM and Nachtmann, HL: 1999, A comparison of freshman and senior engineering design processes, *Design Studies* **20**: 131-152.

Ball, LJ, Evans, JST and Dennis, I: 1994, Cognitive processes in engineering design: A longitudinal study, *Ergonomics* **37**(11): 1753-1786.

Ball, LJ, Evans, JSBT, Dennis, I and Ormerod, TC: 1997, Problem-solving strategies and expertise in engineering design, *Thinking & Reasoning* **3**(4): 247-270.

Ball, LJ and Ormerod, TC: 1995, Structured and opportunistic processing in design - a critical discussion, *International Journal of Human-Computer Studies* **43**(1): 131-151.

Chase, WG and Simon, HA: 1973, Perception in chess, *Cognitive Psychology* **4**(1): 55-81.

Chi, MTH, Feltovich, PJ and Glaser, R: 1981, Categorization and representation of physics problems by experts and novices, *Cognitive Science* **5**(2): 121-152.

Deerwester, S, Dumais, ST, Furnas, GW and Landauer, TK: 1990, Indexing by latent semantic analysis, *Journal of the American Society for Information Science* **41**(6): 391-407.

Egan, DE and Schwartz, BJ: 1979, Chunking in recall of symbolic drawings, *Memory & Cognition* **7**(2): 149-158.

Gobet, F: 1998, Expert memory: A comparison of four theories, *Cognition* **66**(2): 115-152.

Goel, V: 1995, *Sketches of Thought,* MIT Press, Cambridge.

Heiser, J and Tversky, B: 2002, Diagrams and descriptions in acquiring complex systems *in* WD Gray and CD Schunn (eds), *24th Annual Meeting of the Cognitive Science Society,* Lawrence Erlbaum Associates, Mahwah, NJ, pp. 447-452.

Hill, A, Song, S, Dong, A and Agogino, A: 2001, Identifying shared understanding in design using document analysis, *13th International Conference on Design Theory and Methodology,* Pittsburgh, PA, DETC2001/DTM-21713.

Kavakli, M and Gero, J: 2001, Sketching as mental imagery processing, *Design Studies* **22**(4): 347-364.

Kavakli, M and Gero, J: 2002, The structure of concurrent cognitive actions: A case study on novice and expert designers, *Design Studies* **23**(1): 25-40.

Landauer, TK and Dumais, ST: 1997, A solution to Plato's problem: The latent semantic analysis theory of acquisition, induction, and representation of knowledge, *Psychological Review* **104**(2): 211-240.

Larkin, J, McDermott, J, Simon, DP and Simon, HA: 1980, Expert and novice performance in solving physics problems, *Science* **208**(4450): 1335-1342.

Nisbett, RE, Fong, GT, Lehman, DR and Cheng, PW: 1987, Teaching reasoning, *Science* **238**(4827): 625-631.

Quesada, JF, Kintsch, W and Gomez, E: 2002, A theory of complex problem solving using latent semantic analysis *in* WD Gray and CD Schunn (eds), *24th Annual Conference of the Cognitive Science Society,* Lawrence Erlbaum Associates, Mahwah, NJ, pp. 750-755.

Reitman, JS: 1976, Skilled perception in go: Deducing memory structures from inter-response times, *Cognitive Psychology* **8**(3): 336-356.

Richman, HB, Staszewski, JJ and Simon, HA: 1995, Simulation of expert memory using EPAM IV, *Psychological Review* **102**(2): 305-330.

Smith, RP and Leong, A: 1998, An observational study of design team process: A comparison of student and professional engineers, *Journal of Mechanical Design* **120**(4): 636-642.

Stone, RB and Wood, KL: 2000, Development of a functional basis for design, *Journal of Mechanical Design* **122**(4): 359-370.

Suwa, M, Purcell, T and Gero, J: 1998, Macroscopic analysis of design processes based on a scheme for coding designers' cognitive actions, *Design Studies* **19**(4): 455-483.

Suwa, M and Tversky, B: 1997, What do architects and students perceive in their design sketches? A protocol analysis, *Design Studies* **18**(4): 385-403.

JS Gero (ed), *Design Computing and Cognition'04,* 117-134


# AN ANALYSIS OF PROBLEM FRAMING IN MULTIPLE SETTINGS

SONG GAO, THOMAS KVAN
*The University of Hong Kong, Hong Kong*

**Abstract.** Concerns have been expressed that digital tools disrupt the design process. Schön identifies the importance of problem framing in both design practice and design education. In this paper we use teamwork protocol analysis to examine the problem framing activities of architecture students to identify differences in framing activities in three different design settings, namely online co-located, online remote and paper-based co-located. In order to encode these design activities, we first developed Schön's model of "reflective conversation with the situation" into "framing", "moving", and "reflecting". The "framing – moving – reflecting" model is adopted as first coding scheme to examine framing activities. Furthermore we use Minsky's frame system as second coding scheme to investigate the different types of problem framing activities. We find that paper based design tools can afford marginally more design communications than digital based tools; however, the proportion of framing activities in online remote setting is higher than others. Implication of these findings is discussed.

## 1. Introduction

An important goal in architectural education is the nurturing of creativity. In this context, Schön (1985) points out that the main tasks of architectural education are not only to teach students the strategies of problem solving, but also to teach them how to find and formulate design problems. The relationship between creative activities, problem solving and problem finding has been extensively studied (Dillon 1982; Getzels and Csikszentmihalyi 1976; Weisberg 1988).

The activity of design has been characterized as a conversation (Schön 1985). It has been postulated that the use of digital tools has interrupted this reflective conversation (Corona-Martinez and Quantrill 2003). In this paper we are interested in investigating how architectural students engage in framing activities as they solve a simple wicked problem in three different design environments, namely digital, non-digital co-located and non-co-located. In order to approach this we use teamwork-based protocol analysis, a research method developed from verbal protocol analysis (Cross 2001).

Design can be considered a process of problem solving in which the problems can be classified as well-defined, ill-defined, or wicked (Rowe 1987). Design knowledge and expertise provide the design reasoning to solve problems (Minsky 1977; Akin 1986; Rowe 1987). Such knowledge is brought to bear on a problem in an appropriate manner that makes it susceptible to solving with the knowledge available (Weisberg 1988). The strategy adopted in solving design problems therefore depends on the nature of those problems (Van Gundy 1988). Some design tasks, especially well-formed design problems, can be solved straightforwardly from the presentation of the problem as encountered. Solutions to other problems are less apparent. In these, the designer transforms a problem from a wicked into a well-defined problem such that it can be solved (Schön 1985).

Describing design as a "reflective conversation within the situation", Schön identifies a series of design activities in which framing is an early step. Framing is therefore an essential component of creative activity (Van Gundy 1988; Weisberg 1988) covering a wide range of particular instances. Minsky has elaborated the concept by proposing a "frame-system" (Minsky 1977) in which framing is classified into four types. This paper will examine design activities using the concept of frames. In particular, we examine whether there are fewer problem framing activities in designed digital-based settings compared to paper-based design setting.

## 2. Literature Review

Design is undertaken using a variety of tools. In recent years, we have witnessed a move of the dominant design media from paper to digital. In this section, we examine descriptions of this shift and key aspects of the design conversation.

### 2.1. DESIGNING USING DIFFERENT TOOLS

Different tools have different properties, leading users to perceive and use them in different ways (Sellen and Harper 2001). Particular tools may enable particular actions to be engaged, thus different affordances offered for particular tasks (Gibson 1979). Designers use of the activities of sketching, drawing, and modeling to generate ideas in order to address problems and identify solutions (Corona-Martinez and Quantrill 2003). Paper-based design tools have been dominant in studio teaching since the formalization of design learning in the Ecole des Beaux Arts. Since then designers began to largely use paper media to aid them to engage design problems. The importance of drawing is recognized (Robbins 1994) as have other representation in other media. For example, Tchernyknov, a Russian

Constructivist, claimed that it was only by means of engaging in a range of representational and compositional technical methods including drawing that it was feasible to create 'architectural fantasies' (Corona-Martinez and Quantrill 2003).

It has been suggested that paper seems to support more accessibility than digital tools, therefore afford better communication (Sellen and Harper 2001). Some scholars suggest that digital design tools inhibit communication between mind and hand because of the precision demanded by the system, hence interrupting the conversation of design (Corona-Martinez and Quantrill 2003; Lawson 1994; Lawson and Shee 1997). Corona-Martinez and Quantrill (2003) observe that a computer is not a drawing instrument like a pencil, but "an intermediate system of drawing according to our indications provided by the pressure on the button of mouse, which in turn responds to the feedback from our sight of what appears on a screen … something new has invaded the apparently intangible craftsmanship of drawing".

The question is raised then of whether digital tools disrupt the design conversation. Interviewing several architects, Lawson found that most of them preferred using paper based design tools to help them in design thinking, then using digital tools for documentation and presentation rather than as part of design process (Lawson 1994). Burton, for example, considers that when interacting with a computer, he feels difficult to modify a drawing directly.

> "This close interaction between himself and his drawing leaves Richard Burton personally unenthusiastic about the idea of computer-aided design, of which he makes no use himself. He considers that the directness with which he can alter a drawing is missing when mediated by a computer and thus the feeling is lost" (Lawson 1994).

Similarly, Wilford prefers paper and pencil in designing. Echoing Schön's 'conversation', he calls this process of designing an iterative and comparative process, claiming that it is impossible for designers to be detached from "this very immediate process of drawing lines on paper and tracing through" (Lawson 1994).

These responses to digital tools do not acknowledge the potential that a new medium offers. Eisenman (1996) identifies this potential and claims that if architects do not challenge conventional representations they will not move beyond the paradigm of the Renaissance world view, a world view formalized in traditional paper-based representations (Corona-Martinez and Quantrill 2003). Drawings are, of course, not the only representation of design; Lawson and Shee (1997) have argued that text is as important as diagram in design process. Kvan, Wong et al. (2003) examined the effects of the two digital representations such as text and diagrams, finding that the multiple representations are essential. In all, design conversations are to be supported; the question is whether digital tools disrupt these conversations.

### 2.2. PROBLEM FRAMING

Present as an activity in all processes of problem solving (Dillon 1982), the nature of problem framing has been extensively explored in design studies. Goldschmidt (1989) describes problem representation as a process of acquiring problems from given information and identifies this activity as an essential process while designing. Kolodner and Wills (1996) suggest that appropriate problem reformulation can prevent designers from being trapped into "default assumptions about the constraints of the problem" and thus assist designers to identify new criteria and constrains that enable different approaches to design.

In this study we adopted the term of problem framing defined by Schön (1983). Schön (1983) identified the essence of "problem framing" as "knowing in action".

> "As [inquirers] frame the problem of the situation, they determine the features to which they will attend, the order they will attempt to impose on the situation, the directions in which they will try to change it. In this process, they identify both the ends to be sought and the means to be employed" (Schön 1983).

A similar view can be found in Buchanan's definition of design which includes planning and conceiving (Buchanan 1995). Articulating Simon's (1996) description of design as a search of a problem space, Minsky suggests that solutions are not found by addressing an initial formulation of the problem directly but rather that potential solutions can be sought through a reformulation of the problem space (Minsky 1977). Similarly, Gero and Kannengiesser (2002) suggest that the process of reformulating a design problem "addresses changes in the structure state space during designing" and have formalized this step using the FBS model.

We note that a wide range of activities and outcomes are described in all these descriptions of framing. If framing activities are to be observed in a variety of design contexts, it maybe necessary to describe the activity more finely. Minsky (1977) sets out a finer description in his 'frame-system' theory, articulating particular characteristics of 'problem framing'. He defines four levels of frames, namely 'syntactic frames', 'semantic frames', 'thematic frames', and 'narrative frames'. In these terms, he articulates a distinction between framing activities that propositional (syntactic frames), action centered (semantic frames), descriptive ('thematic') and evocative ('narrative'). Using these distinctions in our encoding of protocols, we can distinguish not only if there are more or fewer framing activities in each media but also identify whether the nature of the framing activity changes.

## 2.3. DESIGN KNOWLEDGE

Framing and knowledge are intertwined; framing can only occur if the designer has a knowledge context in which to devise frames. Akin (1986) points out there are two types of design knowledge (declarative and procedural knowledge). Declarative knowledge can be described as general and specific knowledge, while procedural knowledge is 'knowing-how', which is designers use network-liked and cross-disciplines domain knowledge to solve problems. He summarizes three major steps in solving ill-defined or wicked problems: first is to represent the problem and its variables; second is a body of knowledge that facilitates the transformation of problem states; the last is the search techniques enable the matching of the resources of problem-solvers with the task at hand.

Similarly, Rowe (1987) identifies two kinds of design knowledge: 'procedural knowledge' and 'substantive knowledge' and points out both of them are intertwined and related during designing.

> "The two facets of the problem are clearly intertwined. Without social purpose it is difficult to imagine how a broad understanding of and meaning for architecture can be established. On the other hand, without engagement it is difficult to see how architecture can be conveyed in the cultural mainstream, even with a strong sense of purpose" (Rowe 1987).

In his description of Simon's 'knowing how', Rowe distinguishes such design activities as 'purposefully planning', while Schön's 'know that' is the design activities of 'engagedly conveying'.

In both these discussion of design knowledge, the ability to recognize and formulate a problem is an important contributor to success. Mitchell (1994) describes this activity as a kind of intelligent effort which designers adopted to find solution. The common activity is the application of knowledge to the framing process. Knowledge itself is insufficient; framing is an essential activity of design. Minsky (1977) explains this activity in more detail and defines a structure in which to understand the activity. He points out that a frame is a tool in the armoury of an expert.

> "Here is the essence of the frame theory: when one encounters a new situation (or make a substantial change in one's view of a problem), one selects from memory a structure called a frame. This is a remembered framework to be adapted to fit reality by changing details as necessary. It is a data-structure for representing a stereotyped situation" (Minsky 1977).

Kavakli and Gero (2001) have found that there are obvious differences between the design activities of experts and novices, largely in 'problem framing' activities. Experts have a greater range of data structures with which to examine new situations and hence to identify fruitful 'stereotypes' with which to frame the new. The education process needs to support framing so that students can gain experience and exposure to the framing process.

Oxman (2002) has identified that a similar activity to framing occurs when designers work with drawings, finding new images emerging from their work as they draw. Oxman observes that designers adapt their domain knowledge to deal with visual images to develop or create new knowledge.

> "It is domain knowledge that supports our innate cognitive abilities to deal with visual images in design, to classify them and to create generic knowledge, such as generic classes of design operations" (Oxman 2002).

Minsky (1977) identifies that the information retrieval network consists of the expertise and knowledge, which links the frame systems, and in turn these frames could represent knowledge of fact, analogies, and other information. The choice of frames can act as triggers to particular chunks of knowledge; by reframing, new chunks can be accessed.

### 2.4. SUMMARY

To sum up the above literature we may conclude that the problem framing is a process by which designers use their expertise and knowledge to explore the problem space to translate a problem from the realm of wicked problems into that of the well defined, such that a solution can be found after iterations of reformulation and searching. Framing depends upon knowledge yet is essential to the formulation of new knowledge. As such, framing is a central activity in design education and an indicator of the extent to which designing is engaged.

Rephrasing the speculation of designers on the impact of digital tools in design, we can state a hypothesis that we expect to find fewer framing activities when designers use digital tools to design if they compared to when working with paper media.

## 3. Methodology

In this study we carried out protocol analysis to compare the designers' activities in paper-based and digital-based studio settings. Protocol analysis was first used by Eastman as a means to study design cognition (Eastman 1968). Since then protocol analysis has mainly been divided into two types such as process-oriented approach and content-oriented approach (Dorst and Dijkhuis 1995). Content-oriented approach focuses on information contents of what designers think (Suwa and Tversky 1996; Kavakli and Gero 2001). Process-oriented approach looks into the procedure of designing, and is mainly used in verbal protocol analysis. Compared to content-oriented approach, scholars can isolate different design behaviors that they want to concern by using process-oriented approach. Being aware of the weakness of concurrent protocol analysis (Lloyd et al. 1995), teamwork based protocol analysis has been developed (Dorst 1995).

Goldschmidt (1995) compared the activities between teamwork design with individual design and found the issues concerned by team members were similar with those of the individual. Design is not individual and private. Communication is essential for designers to exchange and develop their ideas with colleagues and clients. Furthermore team work is of importance in design process in order to address increasing complex design problems that are hard to solve individually (Stempfle and Badke-Schaub 2002; Goldschmidt 1995; Cross and Cross 1995). Through communicating design issues with each other, subjects could naturally verbalize their design thinking. The method of teamwork based protocol analysis therefore was adopted in this study.

### 3.1. EXPERIMENTAL DESIGN

The experiment has been used in previous studies to examine the efficiency of design learning in different computer supported collaborative settings (Kvan et al. 1997; Kvan et al. 1999). This time we adopted one studio setting from the previous studies, the online remote setting. For the purposes of this experiment, we developed other two co-located settings, online co-located and paper based co-located settings. Activities in these co-located settings were recorded by digital-cameras set up to capture the verbal data for further analysis; communications in the remote settings were recorded by capturing digital data.

#### *3.1.1. Three Design Settings*

To make these settings comparable we chose Microsoft NetMeeting including whiteboard and chat line as digital design tools and chose paper and pencils, etc. as paper-based design tools.

The program of Microsoft NetMeeting has been adopted in previous studies to explore the design communications in computer supported environments (Kvan et al. 1999; Kvan et al. 1997). The digital-based design tools are computer hardware and computer software. The final external representation is digital based (Kvan et al. 1999; Sellen and Harper 2001). In this study, the digital design environment is separated into two settings. In the online remote setting, subjects are located remote from one another and communicate by chat line while drawing on a shared white board. In the online collocate setting, subjects drew on a shared white board when communicating face-to-face.

Paper-based design tools have been variously used in other studies, typically paper, pens, and rulers, etc. and the final representation of the design is produced on paper media (Schön 1985; Sellen and Harper 2001).

Many researchers have studied design activities by using paper -based tools (Goldschmidt 1991; Schön and Wiggins 1992; Lawson 1994; Corona-Martinez and Quantrill 2003). In our paper-based co-located setting, subjects worked collaboratively on a shared drawing table sitting face-to-face by using paper, pens, and rulers, etc.

Table 1 summarizes the instruments adopted in the three settings; Figure 1 presents the layout of the three environments.

TABLE 1. Description of the instruments in the three settings

| | Paper based co-located | Online remote | Online co-located |
|---|---|---|---|
| **Design tools** | Paper; pencils; rulers; etc. | Hardware: two computers with keyboard and mouse;<br>Software: Microsoft NetMeeting | Hardware: two computers with keyboard and mouse;<br>Software: Microsoft NetMeeting |
| **Communi -cation** | Face to Face | Chat line | Face to Face |

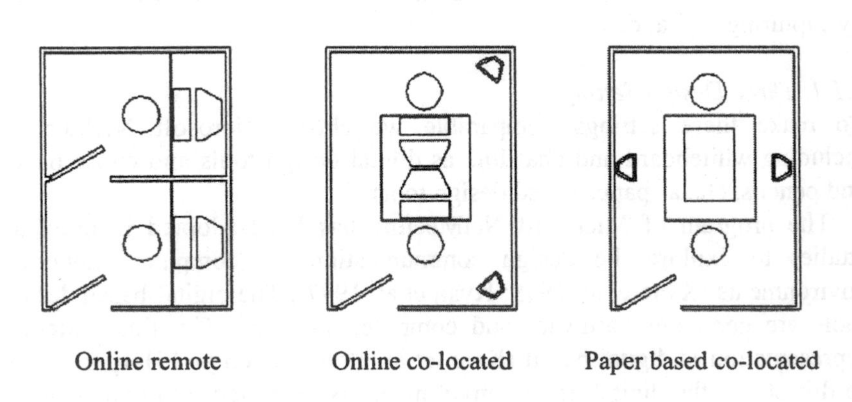


*Figure 1.* The layouts of the three settings

### *3.1.2. Design Task*

The design task is to provide access up a steeply sloping urban park from a bus stop on the lower road to the entrance of hospital on the upper road while allowing a parking area to be accessed from a side road midway up the slope. Figure 2 shows an example of final representation done in online co-located setting. During the 40 minutes' design exercises, subjects were asked to think about design issues of landscape, playground, car park, and sitting area, etc. while accommodating appropriately sloped pathways. The task is a simple open-ended, 'real-world' wicked problem, which has been used in previous study (Kvan et al. 1997; Kvan et al. 1999).

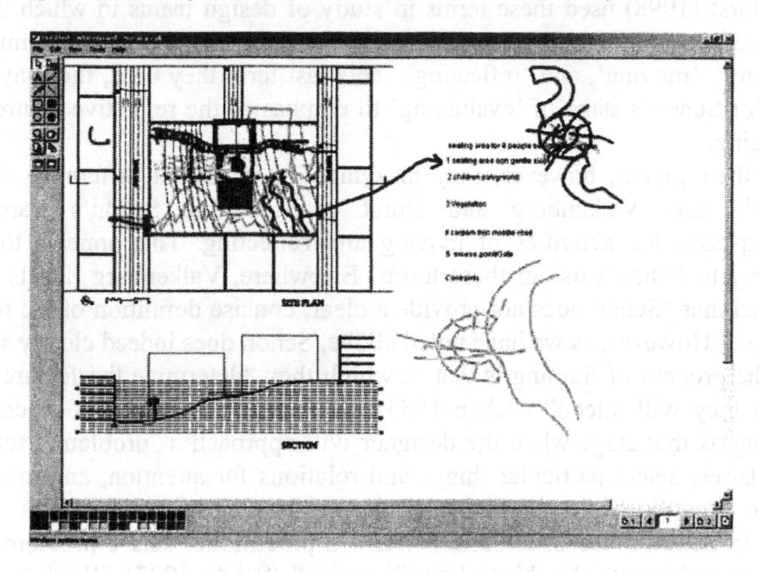

*Figure 2.* An example of final drawing in online co-located setting

### *3.1.3. Subjects*

Since this design problem is domain specific, certain design knowledge and expertise are required (Minsky 1977; Bedard and Chi 1992; Oxman 2002). Concerning on this point we chose eighteen pairs of postgraduate students to take part in this research project. The subjects have been taught architectural design for more than three years in the same educational settings, therefore they are able to deal with this simple wicked problem by using shared design knowledge domain, and can finish the design task within the given time.

## 4. Coding Scheme Development

We first encoded the protocols developed from Schön's description of the "reflective practitioner" and then re-coded those activities identified as framing using Minsky's four types of framing. Each of these coding stages is described in detail below and the results presented.

### 4.1. SCHÖN'S MODEL

Schön's theory of "reflective conversation with the situation" describes the cyclical process of design in which the designer names, frames, moves and evaluates their work as the mechanisms of design. This schema has been adopted widely to articulate the design process. For example, Valkenburg and Dorst (1998) used these terms in study of design teams in which they proposed a model for Schön's process, identifying the stages as 'naming', 'framing', 'moving', and 'reflecting'. This last term they used, they say, in place of Schön's stage of 'evaluating' to emphasize the reflective nature of designing.

In their model, however, they introduce an important difference from Schön's use. Valkenburg and Dorst propose that Schön's framing encompasses the activities of moving and reflecting. This appears to be contrary to Schön's use of these terms. Elsewhere, Valkenburg (2001) has claimed that "Schön does not provide a clear, concise definition of the term 'frame'." However, as we have noted above, Schön does indeed clearly state that the process of framing is that in which they "determine the features to which they will attend" (Schön 1983). Similarly, Schön (1988) describes framing as that stage when the designer will approach a problem, "set its boundaries, select particular things and relations for attention, and impose on the situation a coherence that guides subsequent moves". This is very close to Schön's use of naming: "When a practitioner sets a problem, he chooses and names the things he will notice" (Schön 1987). We therefore propose that naming and framing can be taken together as a distinct prior step to moving and reflecting, where framing refers to the identification of a new design problem or idea, interpreted from design brief or recognition of new design ideas as the process is engaged or introduced as new design information that has not mentioned before.

The other two terms are less problematic. Schön (1984, 1983) gives a clear definition of moving, stating that it is "to test a hypothesis ... The very invention of a move or hypothesis depends on a normative framing of the situation". Moving refers to the production of a tentative solution for the problem at hand. Thus, moving comes after framing and is dependant upon it.

Using Valkenburg and Dorst's term, we will refer to reflecting as the activity of evaluation of a solution, what Schön described as a state in which the situation "talks back", leading the designer to reframe, retest and re-evaluate. Thus, we will use the terms *framing*, *moving* and *reflecting* in the coding of protocols here. Table 2 shows the definition and example of the first coding scheme.

TABLE 2. The first coding scheme

| Coding category | | Definition | | Examples |
|---|---|---|---|---|
| **Framing** | : | Identify a new design problem; Interpret further from design brief. | : | We have to provide a sense of arrival at each site access point. |
| **Moving** | : | Proposed explanation of problem solving, a tentative solution. | : | Maybe some here can put the playground. |
| **Reflecting** | : | Evaluate or judge the explanation in 'Moving'. | : | I think it is ok. Just represent the design |

4.2. MINSKY'S MODEL OF FRAME SYSTEM

After we encoded the protocols by using the model above, we wished to understand what kind of framing activities were being engaged in. A cursory reading of the protocols indicated that the contents of activities we coded as 'framing' are similar with those illustrated in Minsky's 'frame system' (Minsky 1977). Here, he suggested four types of frames to constitute a frame-system, namely 'syntactic frames', 'semantic frames', 'thematic frames', and 'narrative frames'. Using these four types, we encoded the protocols in relevant sections. Table 3 shows the definition of these four frames and protocol examples for each category.

## 5. Results and Discussion

The eighteen protocols (six for each setting) were encoded first using the schema derived from the Schön model and subsequently using the schema from the Minsky model. These two encodings are reported separately.

5.1. FIRST CODING RESULTS

We counted the number of these design activities in terms of this "framing-moving-reflecting" model, and calculated the percentage of the three categories. Figure 3 presents the frequency of the three design activities across the three settings, and Figure 4 shows the percentage of the three design activities across the three settings.

TABLE 3. The second coding system

| Coding category | | Definition | | Examples |
|---|---|---|---|---|
| **Syntactic frames** | : | Mainly verb and noun structures. Prepositional and word-order indicator conventions. | : | We have to design the seats place inside, and the playground. |
| **Semantic frames** | : | Action-centered meanings of words. Qualifiers and relations concerning participants, instruments, trajectories and strategies, goals, consequences and side effects. | : | How many seats do we have in the playground? |
| **Thematic Frames** | : | Scenarios concerned with topics, activities, portraits, setting. Outstanding problems and strategies commonly connected with topic. | : | In fact I think the seats should be placed together to have better social interaction, maybe there is a round table, so three are bench chairs and one table. |
| **Narrative Frames** | : | Skeleton forms for typical stories, explanations, and arguments. Conventions about foci, protagonists, plot forms, development, etc., designed to help a listener construct a new, instantiated thematic Frame in his own mind. | : | So we can place many seats around here in order to have a more pleasant area. |

As can be seen in the bar charts above, there is a substantial difference in the total communications in each setting. In these, the number of framing activities is 415 in the paper based co-located setting; 391 in online co-located

setting; while in online remote this is only 181. The number of communications in the other two categories is similar and proportionate.

Thus, the results suggest that face-to-face working supports more communications than digital tools. However, when we calculated the percentage of framing in these settings, we found that the percentage of framing is higher in online remote setting than that in paper and online co-located settings (43.8% online remote, 32.5% paper base co-located versus 35.2% online co-located setting). Thus, the chat line setting appears to support framing more than other activities.

A one way ANOVA-test was performed to test the effect of these designed settings. It was found that the average distribution of framing is significantly different when comparing online remote against online collocated setting ($F(1,14)=42.83$, $p<0.001$), and also when comparing paper based collocated to online remote settings ($F(1,14)=26.4$, $p<0.001$). No significant difference is found when comparing the two collocated activities ($F(1,14)=0.22$, $p=0.64$).

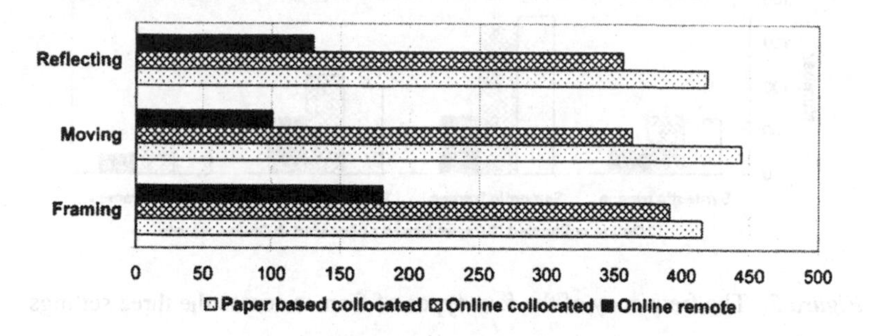


*Figure 3.* The frequency of the design activities across the three settings

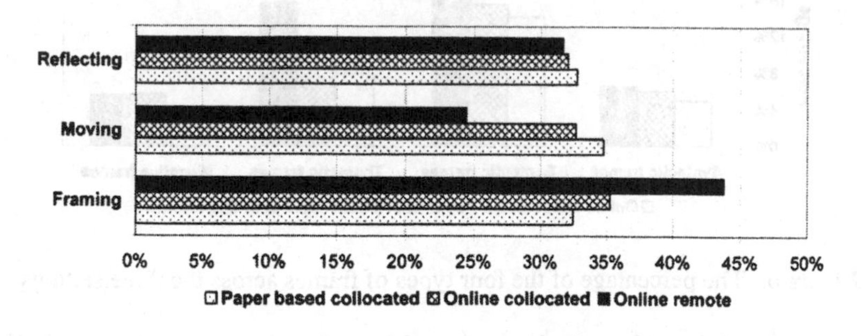


*Figure 4.* The percentage of the design activities across the three settings

5.2. SECOND CODING RESULTS

Two questions therefore were raised from the results of the first coding. First, will the significant difference between the online remote setting to other two settings be reflected by any difference in the content of the framing activities? Second if the difference of problem framing activities appears not to be significant when comparing the collocated settings, then is it also like that after we classified these activities into different levels?

Employing the schema derived from Minsky's "frame system", we counted the number of exchanges in each of the four frames types and calculate the proportion of those along the whole design sessions. Figure 5 shows the frequency of these four frames across the three settings; Figure 6 presents the percentage of those across the three settings.

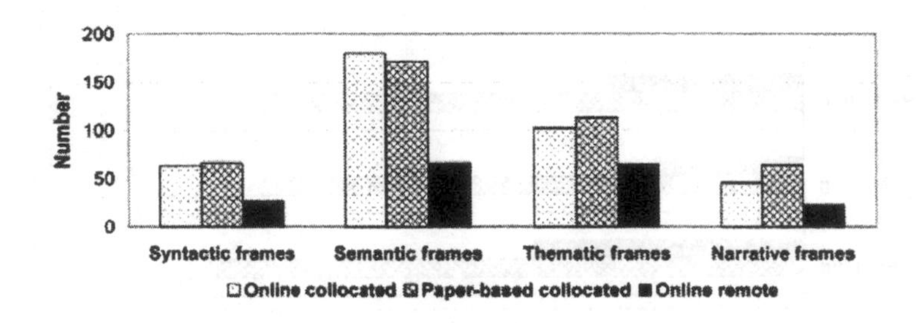


*Figure 5.* The frequency of the four types of frames across the three settings

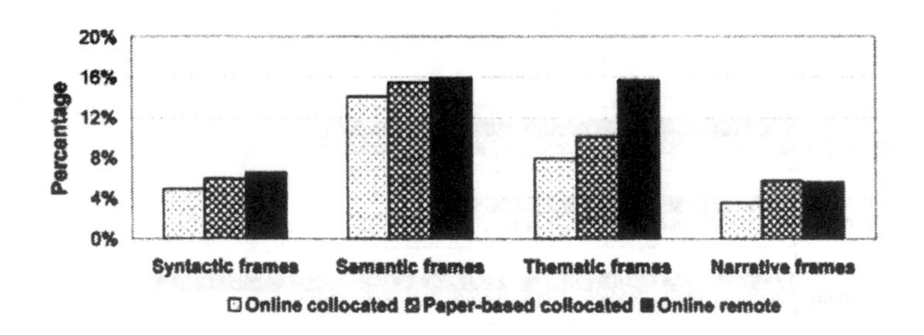


*Figure 6.* The percentage of the four types of frames across the three settings

Figure 5 shows the number of problem framing types in the online remote setting is consistently less than that in both co-located settings. Figure 6 presents the percentages of syntactic frames, semantic frames, and narrative frames among the three settings are similar, while the proportion of the semantic frames and thematic frames are proportionate in online remote setting (F (1,14)= 0.008547, p= 0.927). The data show that the proportion of thematic frames in online remote setting is higher than those in the other two co-located settings. A significant difference is found in thematic frames when comparing online remote settings to online co-located setting (F (1,14)= 12.96, p<0.01) and paper based co-located setting (F (1,14)= 16.185, p<0.01). Thus, the participants in the remote setting using chat line consistently engaged in more thematic framing than in other settings.

The results also indicate that the difference of these frame types between online co-located setting and paper-based co-located setting is found to be insignificant again. This is supported by data analysis in general (syntactic frames: F (1,14)= 0.009, p=0.926; semantic frames: F (1,14)=0.153, p=0.702; thematic frames: F (1,14)=0.377, p=0.549; narrative frames: F (1,14)=3.032, p=0.104). They also suggest that in both co-located settings the proportion of the semantic frames appear to be more dominant than the other three frames.

### 5.3. DISCUSSION

These results show that the number of problem framing activities by using chat-line to communicate is less than the number of communication in face-to-face modes but that remote communication is proportionately richer in framing activities. This repeats a pattern found in other research into chat line and face to face design communication reported in Section 2. Of course, the activities of talking in text and those of talking in face to face are different. As we have suggested, solving wicked or ill-defined problems requires problem framing, which demands designers to spend time in identifying and formulating problems. By using chat-line subjects have to write their design ideas in text, therefore this mode naturally extend the duration of subjects' design thinking. Thus this chat-line mode appears to afford more deliberation than other two co-located settings, which could be the reason that why the more framing activities happened in the online remote setting.

We can draw a parallel between Minsky's frames and the activities of design. Semantic frames can be considered to refer to the action-focused design ideas and strategies and thematic frames to descriptive or imagine-focused design problems and strategies. Minsky (1977) points out this 'frame-system' can test the effectiveness of 'imagery', and considers memory and design knowledge, consisting of a information retrieval network, are important for designers to represent novel design problems. Our data indicate that online remote communication appears to support 'imagery' more effectively than either co-located settings. Designers seem to transform and reconstruct frames more efficient in online remote setting. All of these assumptions need to be further examined.

Overall, these experimental findings show that digital design tools appear *not* to be detrimental to problem framing activities and may also enhance the efficiency of design exploration, contrary to commentary from practitioners. If digital tools are constraining inventiveness and disrupting the design conversation, it may not be in the density of framing activities but other aspects of the design process that are being disrupted. Digital design tools therefore show some potential to enable exploration beyond traditional design processes, perhaps to fulfill the wish to free designers from the traditional paradigm of designers formalized in the era of Renaissance.

## Acknowledgements

The authors would like to thank the students who participated in this research project and the members of our research group who gave invaluable comments on the development of the methodology and for reading early drafts. We also wish to thank the reviewers who gave critical comments on helping us to strengthen our paper.

## References

Akin, O: 1986, *Psychology of Architectural Design,* Pion Limited, London.

Bedard, J and Chi, M: 1992, Expertise, *Directions in Psychological Science* **1**: 135-139.

Buchanan, R: 1995, Wicked problems in design thinking, *in* V Margolin and R Buchanan (eds), *The Idea of Design,* MIT Press, Cambridge MA, pp. 3-20.

Corona-Martinez, A and Quantrill, M: 2003, *The Architectural Project,* Texas A & M University Press, College Station.

Cross, AC and Cross, N: 1995, Observations of teamwork and social processes in design, *Design Studies* **16**: 143-170.

Cross, N: 2001, Design cognition: Results from protocol and other empirical studies of design activity, *in* CM Eastman, WM McCracken and WC Newstetter (eds), *Design Knowing and Learning Cognition in Design Education,* Elsevier science Ltd, Oxford, pp. 79-103.

Dillon, JT: 1982, Problem finding and solving, *The Journal of Creative Behavior* **16**: 97-111.

Dorst, K: 1995, Analysing design activity: New directions in protocol analysis, *Design Studies* **16**: 139-142.

Dorst, K and Dijkhuis, J: 1995, Comparing paradigms for describing design activity, *Design Studies* **16**: 261-274.

Eastman, CM: 1968, Explorations of the cognitive processes in design, Department of computer science report, Carnegie Mellon University, Pittsburgh.

Eisenman, P: 1996, Visions' unfolding: Architecture in the age of electronic-media, *in* K Nesbitt (ed), *Theorizing a New Agenda for Architecture: An Anthology of Architectural Theory 1965-1995,* Princeton Architectural Press, New York, NY, pp. 554-564.

Gero, JS and Kannengiesser, U: 2002, The situated function-behaviour-structure framework, *Artificial Intelligence in Design'02, Klvwer,* Dordrecht, pp. 89-104.

Getzels, JW and Csikszentmihalyi, M: 1976, *The Creative Vision: A Longitudinal Study of Problem Finding in Art,* Wiley, New York.

Gibson, JJ: 1979, *The Ecological Approach to Visual Perception,* Houghton Mifflin, New York.

Goldschmidt, G: 1989, Problem representation versus domain of solution in architectural design teaching, *Journal of Architectural and Planning Research* **6**: 204-215.

Goldschmidt, G: 1991, The dialectics of sketching, *Creativity Research Journal* **4**: 123-134.

Goldschmidt, G: 1995, The designer as a team of one, *Design Studies* **16**: 189-209.

Kavakli, M and Gero, JS: 2001, Sketching as mental imagery processing, *Design Studies* **22**: 347-364.

Kolodner, JL and Wills, LM: 1996, Powers of observation in creative design, *Design Studies* **17**: 385-416.

Kvan, T, Vera, A and West, R: 1997, Expert and situated actions in collaborative design, *Proceedings of Second International Workshop on CSCW in Design*, Beijing, pp. 400-405.

Kvan, T, Wong, JTH and Vera, A: 2003, The contribution of structural activities to successful design, *International Journal of Computer Applications in Technology* **16**: 122-126.

Kvan, T, Yip, WH and Vera, A: 1999, Supporting design studio learning: An investigation into design communication in computer-supported collaboration, *CSCL'99*, Stanford University, pp. 328-332.

Lawson, B: 1994, *Design in Mind,* Butterworth-Heinemann Ltd., Oxford.

Lawson, B and Shee, ML: 1997, Computers, words and pictures, *Design Studies* **18**: pp. 171-183.

Lloyd, P, Lawson, B and Scott, P: 1995, Can concurrent verbalization reveal design cognition?, *Design Studies* **16**: 237-259.

Minsky, M.: 1977, Frame-system theory, *in* PN Johnson-Laird, and PC Wason (eds), *Thinking: Readings in Cognitive Science,* Cambridge University Press, Cambridge.

Mitchell, WJ: 1994, *The Logic of Architecture: Design, Computation and Cognition,* MIT Press, Cambridge, MA.

Oxman, R: 2002, The thinking eye: Visual re-cognition in design emergence, *Design Studies* **23**: 135-164.

Robbins, E: 1994, *Why Architects Draw,* MIT Press, Cambridge, MA.

Rowe, PG: 1987, *Design Thinking,* MIT Press, Cambridge MA.

Schön, DA: 1983, *The Reflective Practitioner: How Professionals Think in Action,* Basic Books Inc., New York.

Schön, DA: 1984, Problems, frames and perspectives on designing, *Design Studies* **5**: 132-136.

Schön, DA: 1985, *The Design Studio: An Exploration of Its Traditions & Potential,* RIBA Publications Limited, London.

Schön, DA: 1987, *Educating the Reflective Practioner,* Jossey-Bass, San Francisco.

Schön, DA: 1988, Designing: rules, types and words, *Design Studies* **9**: 181-190.

Schön, DA and Wiggins, G: 1992, Kinds of seeing and their functions in designing, *Design Studies* **13**: 135-156.

Sellen, AJ and Harper, R: 2001, *The Myth of The Paperless Office,* MIT Press, Cambridge, Mass.

Simon, HA: 1996, *The Sciences of The Artificial,* MIT Press, Cambridge MA.

Stempfle, J and Badke-Schaub, P: 2002, Thinking in design teams - an analysis of team communication, *Design Studies* **23**: 473-496.

Suwa, M and Tversky, B: 1996, What architects see in their sketches: Implications for design tools, *Conference on Human Factors and Computing Systems*, Vancouver, British Columbia, Canada, pp. 191-192.

Valkenburg, R: 2001, Schön revised: Describing the social nature of designing with reflection-in-Action, *Design Thinking Research Sympoisum 5: Designing in Context*, Delft.

Valkenburg, R and Dorst, K: 1998, The reflective practice of design teams, *Design Studies* **19**: 249-271.

Van Gundy, AB: 1988, *Techniques of Structured Problem Solving,* Van Nostrand Reinhold Company, New York.

Weisberg, RW: 1988, Problem solving and creativity, *in* RJ Sternberg (ed), *The Nature of Creativity: Contemporary Psychological Perspectives,* Cambridge University Press, Cambridge, pp. 148-176.

# DESIGN GENERATION

JS Gero (ed), *Design Computing and Cognition'04*, 137-156


# EXPLORATIONS IN USING AN APERIODIC SPATIAL TILING AS A DESIGN GENERATOR

KRISTINA SHEA
*Cambridge University, UK*

**Abstract.** This paper describes the computational implementation of a three-dimensional, aperiodic spatial tiling for design generation. The mathematical spatial tiling formalism is related to the shape grammar formalism to illustrate commonalties between the two approaches to spatial design generation. Computational encoding of a mathematical tiling description as a basic grammar creates a generative design tool that can be used to study the visual properties of the spatial tiling and assess its potential for interesting use in design. The generative system is then relaxed by moving to an unrestricted grammar that expands the generative power to explore alternate design viewpoints and illustrate advantages of encoding a spatial tiling as a shape grammar. Potential design contexts for using the grammar presented include architectural spaces, structures and building façade patterns.

## 1. Introduction

The development of complex three-dimensional form (3D) has long been of interest in architecture. Recently, we have seen an ever increasing amount of geometric complexity that can primarily be attributed to advances in digital technologies and the new styles they bring about. Examples include: advanced parametric CAD tools, desktop performance analysis software, an increase in using rapid prototyping for creating physical prototypes and the new geometric freedoms in fabrication and construction. Geometric complexity in architectural and structural design can arise from a number of sources including necessity imposed by challenging design sites, large scale forms and simply the desire for creating expressive designs that are non-standard, non-planar and non-orthogonal. These sources of complexity often occur when aiming to design forms using natural and organic influences, for example the curves, solid shapes and compositions found in natural macro and micro structures (Beukers and van Hinte 2001).

This paper is motivated mainly by the later two sources of complexity, visual expression and large scale forms, and explores the use of computational algorithms as a design generator capable of proposing new design styles and stimulating imaginations. The potential for computational creativity has been debated in circles of artificial intelligence and computer science for many decades and more recently in digital design circles. In a symposium on digital design that included leading architects, CAD developers, engineers and fabricators, the "imaginative" role of the computer was an important topic of discussion (Kolarevic 2003). Further, in Balmond (2002), past high profile projects in practice, both architectural and engineering, are described where mathematical and computer algorithms have played a key role in design conception and development

## 2. Background

From a theoretical viewpoint, generative computational representations and processes have been under development mainly in computer science and linguistics research areas since the arrival of the computer. The theoretical foundation behind generative design methods generally stems from two triggers; natural analogy and logical basis (Knight 2002; Knight and Stiny 2001), although they are not mutually exclusive. Nature has been widely used as a source of inspiration for creating generative and optimizing computational processes including well known work in neural networks, cellular automata, genetic algorithms, genetic programming, artificial life and more recently swarm optimization and self-organizing systems (Lipson et al. 2003). For example, recent applications of genetic algorithms and genetic programming in architecture to generate new complex forms illustrates promise for the computer as an active creative partner in design (Testa 2001).

Taking a logical basis for design generation requires a study of the underlying logic of objects and systems of objects to form rules that are used to generate a language of valid objects. Post developed the first generative system for producing well formed strings of parentheses and it was later expanded into generative grammars by Chomsky, which are now integral to computer languages and compilers (Stiny and Gips 1980). However, in architecture and design it would be more natural to compute with shape directly rather than symbols. This idea prompted the creation of shape grammars (Stiny 1980) and they have since been applied in architecture to produce languages of architectural form often to re-create and extend the style of a particular architect. Recently, Duarte (2002) has developed a grammar for the generation of mass customized housing designed by the architect Álvaro Siza at Malagueira in Portugal.

Rather than encoding existing styles, the work presented in this paper is closer related to shape grammars that are designed to generate new styles, for example *Shaper* described in Wang and Duarte (2002). Here, the computer is used to generate complex geometric forms by defining a library of shapes and the geometric relations for composing designs from them. The generated forms are algorithmic and difficult to create by hand due to the spatial complexity produced. The true power of shape grammars exists in the recognition and exploitation of emergent shapes, which has been noted by Mitchell (2002) to be the primary route to creating generative tools with enhanced creative possibilities.

Another approach to spatial design generation is mathematical tiling. Tiling can be described as a plane-filling arrangement of plane shapes whereas spatial tiling is a space-filling arrangement of three-dimensional shapes (Grünbaum and Shephard 1989). Spatial tilings also bear resemblance to packing problems often described in operations research literature (Reeves, 1995).

Shape grammars share many common attributes to planar and spatial tilings. Both formalisms are concerned with the fundamental logic and rules for composing designs of shape with certain spatial relations in Euclidean space, the boundary of which can also be constrained. Krishnamurti and Earl (1998) present research on spatial relations and generative properties for two and three dimensional rectangulations, or packings of rectangles and cuboids, using a shape grammar formalism. Most tilings, many examples of which can be found in Grünbaum and Shephard (1989), could be described by shape grammars. The reverse is not necessarily true if shape emergence is an integral feature. The relations between the two formalisms will be further explored later in this paper.

The use of mathematical spatial tiling and conception of organic forms has been of significant interest in architecture lately. Taking an example from a current project, the Beijing Aquatics Centre being designed by the architecture firm PTW and Arup, Sydney was generated from a fundamental cell arrangement, rather than a single shape, which was found by mathematicians to have the minimum surface area, that is the contact area between adjacent cells, for subdividing an infinite three-dimensional space. As a design stimulus, this fundamental unit encoded an organic style that the designers were trying to achieve and could then be arrayed and sliced in clever ways to create the desired building form, including walls and the roof structure, Figure 1. Here, the design goals included achieving an organic style with great spatial complexity and interest while limiting the number of unique member lengths, joint types and façade panel sizes, that the unaltered regular arrangement guaranteed and was somewhat maintained.

Used as architectural systems, regular arrangements, unaltered, can offer the advantages of reduced geometric variety that can adversely impact fabrication and construction costs as well as construction quality. However, taking advantages of current CAD/CAM techniques puts into question conventional notions about the added cost of irregularity. Recent interest has developed in aperiodic tiling, that is spatial arrangements of shapes that have no transnational symmetry (Grünbaum and Shephard 1989). A well-known aperiodic planar tiling is the Penrose tiles, discovered by Roger Penrose, that consists of arranging special shapes that look like "kites" and "darts". From a design viewpoint, aperiodic spatial systems have the potential to balance modularity through restriction of the allowable shapes used and visual intrigue due to their special geometric features that enable them to be combined in ways to produce compositions that are not regular patterns. The new buildings in Melbourne's Federation Square are a good example of this and use an aperdiodic pinwheel tiling.

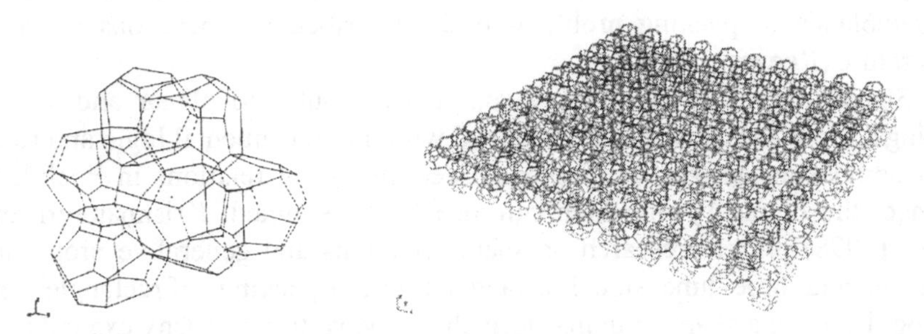

*Figure 1.* Regular spatial tiling used in a building design

## 3. Defining the ABCK shape grammar

In the search for a three-dimensional version of Penrose's aperiodic planar tiling, Danzer (1989) created a new system based on recent discoveries in quasicrystals (Nelson 1986). It was found that quasicystals exhibit long range orientational order and have no transnational symmetry, i.e. they are aperiodic arrangements. Danzer believed that there should be some sort of local conditions that cause such arrangements to be produced.

After investigation of a number of special tetrahedrons, the system Danzer created consists of four such tetrahedrons, called prototiles, named A, B, C and K, Figure 2. The family of labeled tetrahedron prototiles make up a protoset **F**. In mathematical tiling terms, the protoset **F** then admits the tiling **T** subject to matching conditions. Given a single prototile, i.e. a labeled tetrahedron, the vertices, or points, are numbered 1-2-3-4 where the 1-2-3 face sits flat on the z = 0 plane and vertex four is taken as the "apex" of the tetrahedron, for simplicity. Each vertex references a geometric point. The edge lengths that define each shape are reprinted in Table 1. It should be noted that all edge lengths are proportional to the golden ratio, $\tau^n$, which is defined as $\tau=(1+\sqrt{5})/2$, and also plays other important roles in the tiling.

**Matching conditions define the rules for composing spatial tilings from the protoset. The matching conditions here are that all mirror image faces between shapes can be matched, that is placed face to face, with one exception. Among the four tetrahedrons there are 11 unique faces where four faces are common to two tetrahedrons and one face is common to three tetrahedrons. This implies that each labeled shape can meet the mirror of itself on any face and other shapes given a common face. The exception is that a face that includes an edge length of $\tau^{-1}$, 1 or $\tau$ can only be matched with a face that is part of the mirror image of the shape itself.**

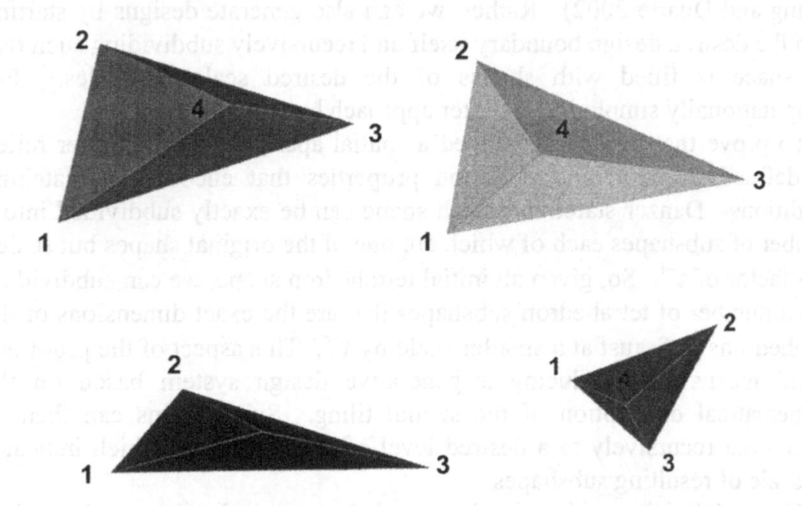


***Figure 2.*** **The set of shapes, S = {A (green), B (yellow), C (blue) and K (red)} shown at an approximate relative scale**

**TABLE 1. Labeled shape geometry specification (from Danzer 1988) where a=(√(10+2√5))/4 (cos 18°) and b=√3/2 (cos 30°).**

| ***Shape*** | ***1-2*** | ***2-3*** | ***3-1*** | ***2-4*** | ***1-4*** | ***3-4*** |
|---|---|---|---|---|---|---|
| **A** | a | $\tau$b | $\tau$a | a | 1 | b |
| **B** | a | $\tau$a | $\tau$b | b | $\tau^{-1}$a | 1 |
| **C** | $\tau^{-1}$a | $\tau$b | $\tau$ | b | a | a |
| **K** | a | b | $\tau^{-1}$a | $\tau$/2 | 1/2 | $\tau^{-1}$/2 |

**The shape grammar analogy of the description given for a spatial tiling T is that the protoset F is a set of labeled shapes, {S, L} and a set of rules R, which encode the matching conditions, generates the design language {S, L, R, I} from an initial shape in I (Stiny 1980). A shape grammar {S,L,R,I} in**

shape algebra $U_{33}$ is then defined where **S,** the set of shapes, consists of four special tetrahedrons {A,B,C,K}, **L,** the set of labels, consists of labels that distinguish the tetrahedrons and their ordered vertices, **R,** the set of rules, consists of four subdivision rules that act on each labeled shape and **I**, the initial labeled shape, is any shape in the set {**S,L**}. The design language corresponds to the spatial tiling **T**.

Computationally, spatial tiling algorithms can be implemented through composition rules, that is adding shapes one by one according to the prescribed spatial relations, i.e. matching conditions, in this case to fill a defined tetrahedron space. This is the general approach taken in *Shaper* (Wang and Duarte 2002). Rather, we can also generate designs by starting with the desired design boundary itself and recursively subdividing such that the space is filled with shapes of the desired scale or scales. For computationally simplicity, the later approach has been taken.

To prove that he had developed a spatial aperiodic tiling Danzer relied on defined inflation and deflation properties that encode the matching conditions. Danzer stated that each shape can be exactly subdivided into a number of subshapes each of which are one of the original shapes but scaled by a factor of $\tau^{-1}$. So, given an initial tetrahedron shape, we can subdivide it into a number of tetrahedron subshapes that are the exact dimensions of the tetrahedrons in **S**, just at a smaller scale by $\tau^{-n}$. This aspect of the proof is a useful means for producing a generative design system based on the mathematical description of the spatial tiling. Subdivisions can then be carried out recursively to a desired level of subdivision, *n*, which indicates the scale of resulting subshapes.

The subdivision rules implemented here encode the mathematical inflation-matrix, **M**. Starting with an initial shape in $\{A^n, B^n, C^n, K^n\}$, this matrix describes the number of each subshapes in the set $\{\tau^{n-1}A, \tau^{n-1}B, \tau^{n-1}C, \tau^{n-1}K\}$ resulting from a one level subdivision. The inflation matrix is given in rows two to five of Table 2. The total is not part of the matrix but simply lists the number of resulting subshapes from carrying out one subdivision. The nature of the spatial tiling and definition of an inflation matrix create an exponential generative algorithm, i.e. each subdivision level results in $\mathbf{M}^n$–$\mathbf{M}^{n-1}$ new subshapes where *n* is the subdivision level just completed. For example one subdivision of shape K, *n*=1, results in two subshapes, subdivision of these subshapes, i.e. *n*=2, yields nine subshapes and so on.

In relation to the inflation matrix and prescribed matching conditions, the geometry resulting from each shape "inflation", or subdivision, is shown in Figures 3 to 6. The rules are "re-write" rules described in the form that the left-hand side (LHS) is replaced by the right-hand side (RHS). The figures are sized for clarity and a relative scale among them can not be assumed.

The subdivision of each shape results in a number of mirror images that are not indicated differently, as it is not necessary for the implementation used. While the rules essentially erase the shape on the left-hand side, if emergence is enabled, the shapes at higher scales can still be recognized. Computationally encoding the mathematical tiling described by Danzer creates a generative system that can be used to consider the visual properties of the tiling and assess the potential for interesting use in design.

TABLE 2. Inflation matrix, **M**, and subshape count for one subdivision level (Danzer 1988)

| | *A* | *B* | *C* | *K* |
|---|---|---|---|---|
| $\tau^{-1}$A | 0 | 0 | 1 | 0 |
| $\tau^{-1}$B | 3 | 2 | 0 | 1 |
| $\tau^{-1}$C | 2 | 1 | 2 | 0 |
| $\tau^{-1}$K | 6 | 4 | 2 | 1 |
| **total** | 11 | 7 | 5 | 2 |

## 4. Generating Designs with the ABCK Shape Grammar

The implementation of the grammar was written in C and uses a solid modeling boundary representation. The shape grammar rules (Figures 3-6) manipulate the representation directly to generate a spatial tiling. Here, rules are applied in a specified sequence to generate the tiling encoded. A maximal line representation was not used, for simplicity, as the original intention in this work was just to be able to visualize Danzer's spatial tiling. This representation was chosen for inherent parametric capabilities and ease of encoding adjacency information. For example, mating vertices and faces between adjacent shapes results in the algorithm not suffering from geometric precision difficulties that might occur if each solid was represented individually. Parametric capabilities will be explored later in this paper. As always, a solid representation allows for ease of computing shape properties such as volume and centroid. After design generation models are exported to both VRML and DXF file formats for ease of viewing in both internet browsers and common CAD packages.

Computationally, each tetrahedron shape is represented by a graph containing a list of four faces each referencing a cyclical list of its six edge halves all of which contain pointers to four labeled vertices that each reference a geometric point. Adjacent edges and mating faces in a composition are also linked. The rules illustrated in Figures 3 to 6 are carried out by successive cuts of the edges in the shape on the left-hand side of the rule to create the new smaller tetrahedron subshapes. Due to the spatial graph representation it is not necessary to distinguish between the shapes and their mirrors. Since all subdivisions are made based on proportional lengths of edges, only the labeling of the vertices, as shown in Figure 2, is important since the mirror image shapes maintain the same topology and edge lengths between labeled vertices, Table 1.

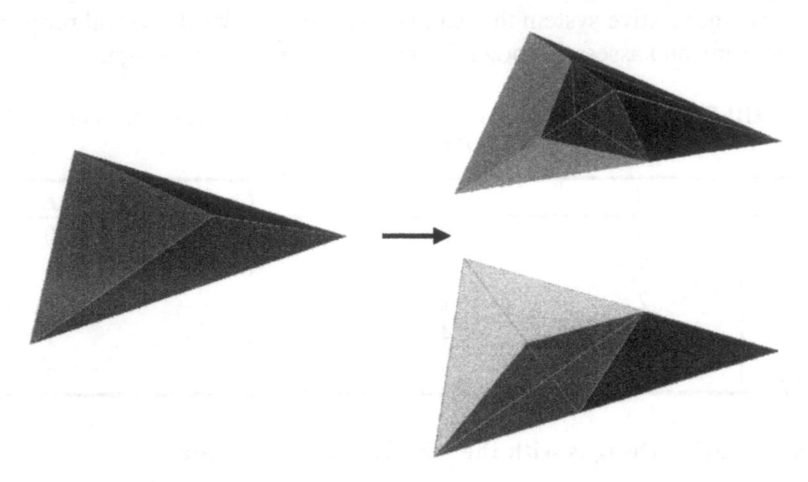

*Figure 3.* A-rule: A→ (3B + 2C + 6K) $\tau^{-1}$ ( top and bottom views shown)

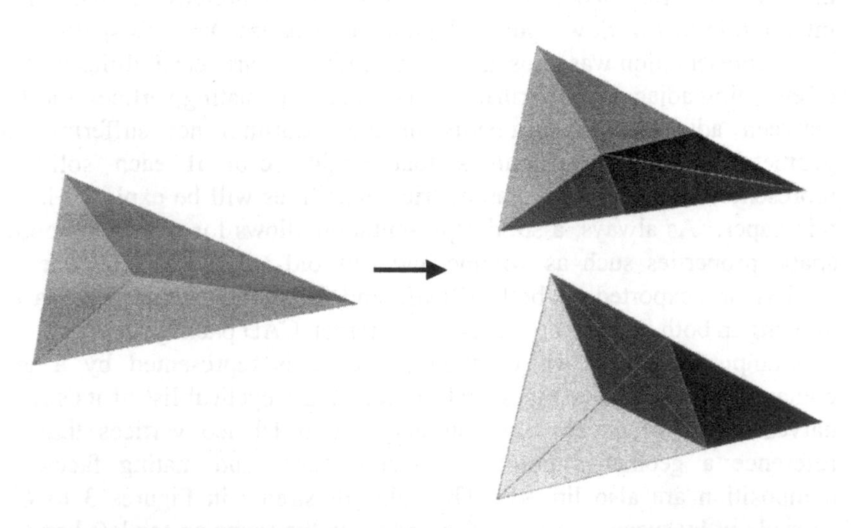

*Figure 4.* B-rule: B→ (2B + 1C + 4K) $\tau^{-1}$ ( top and bottom views shown)

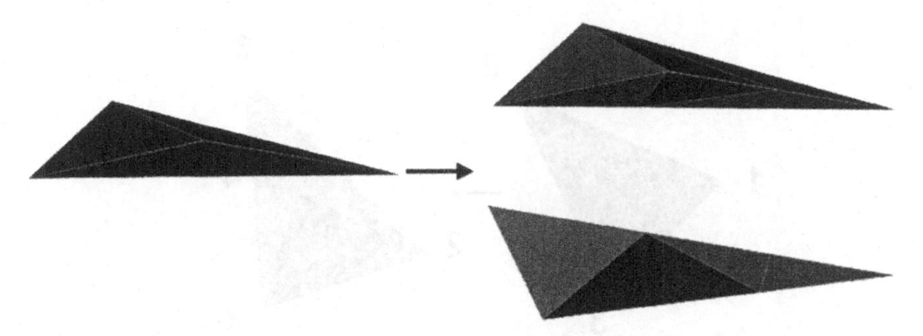

*Figure 5.* C-rule: C→ (A + 2C + 2K) $\tau^{-1}$ ( top and bottom views shown)

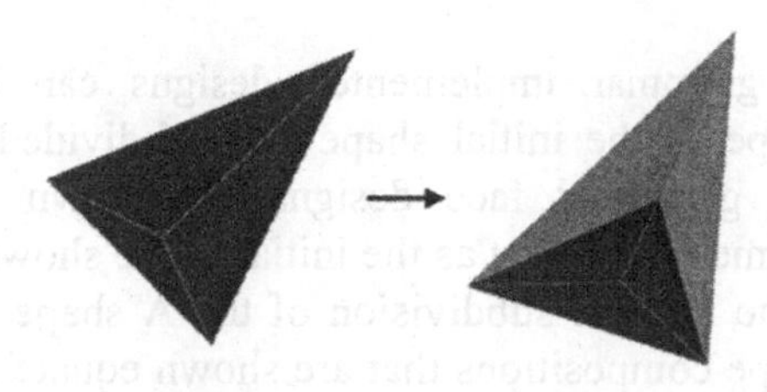

*Figure 6.* K-rule: K→ (B + K) $\tau^{-1}$

Only one edge subdivision rule is then needed to make proportional cuts of edges. Given an indexed line, *a-b*, it is cut in two parts by the proportion $\alpha$ such that a new edge with length $\alpha$ is created from vertex *a* to a new vertex *c*, Figure 7. The remaining edge from vertex *c* to vertex *b* has a length 1-$\alpha$. For example, to carry out the K subdivision rule, Figure 8, we cut the original 2-4 edge, $K_{2\text{-}4}$, into two parts by a factor $\alpha=\tau^{-3}$ as:

$K_{2\text{-}4} = \tau/2 = \tau^{-1}(B_{3\text{-}4} + K_{3\text{-}4}) = \tau^{-1}(1 + \tau^{-1}/2) = \tau/2.$

This relation holds due to the inherent property of the golden ratio, $\tau$, that is $\tau^2 = \tau + 1$. Each subdivision rule consists of cutting the starting shape sequentially into smaller tetrahedrons, enough to make the required number and type of subshapes, and reordering the vertices such as to maintain the correct ordering in each new subshape created. The cuts are always made in proportions given in the set $\{\tau^{-2}, \tau^{-1}, \tau^{-2}/2, 1/2, 1/(1-\tau^2), \tau^{-3}\}$.

*Figure 7.* Edge subdivision rule for forming new tetrahedron subshapes

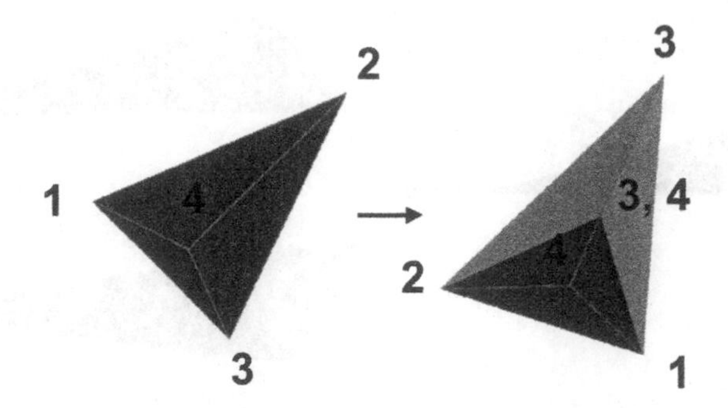


*Figure 8.* K-rule: K→ $\tau^{-1}$(B + K). The new point created is labeled vertex 3 for the new $\tau^{-1}$K shape and vertex 4 for the new $\tau^{-1}$B shape.

Using the ABCK grammar implemented designs can be generated starting from each shape as the initial shape and subdivided to a desired level, say five. The generated face designs are shown in Figure 9, approximately in the same orientation as the initial shape shown in Figure 2. Sample 3D views of the internal subdivision of the A shape are shown in Figure 10. The subshape compositions that are shown equate to calculating the inflation matrix to the power five, i.e $\mathbf{M}^5$=

| | *A* | *B* | *C* | *K* |
|---|---|---|---|---|
| $\tau^{-5}$A | 126 | 77 | 82 | 23 |
| $\tau^{-5}$B | 938 | 582 | 569 | 182 |
| $\tau^{-5}$C | 533 | 328 | 336 | 100 |
| $\tau^{-5}$K | 1722 | 1066 | 1056 | 331 |
| **total** | 3319 | 2053 | 2043 | 636 |

Through generation and visualization, potential design properties of the mathematical tiling can now be assessed. Analyzing the 3D models and images, it was observed that each composition includes slicing planes that cut right through the outer shape envelope and do not intersect any subshapes internally. This could provide for greater variation in possible design envelopes. Also, relations to Penrose's planar "kites" and "darts" are evident, for example find four red K shape faces arranged in a kite shape. In the context of facade design, the patterns exhibit unique properties of maintaining topology at the boundary where all edges intersect only at endpoints, i.e. no edge meets another edge along the edge length. Implemented computationally this particular system can also be termed a fractal since it exhibits self-similarity on all scales.

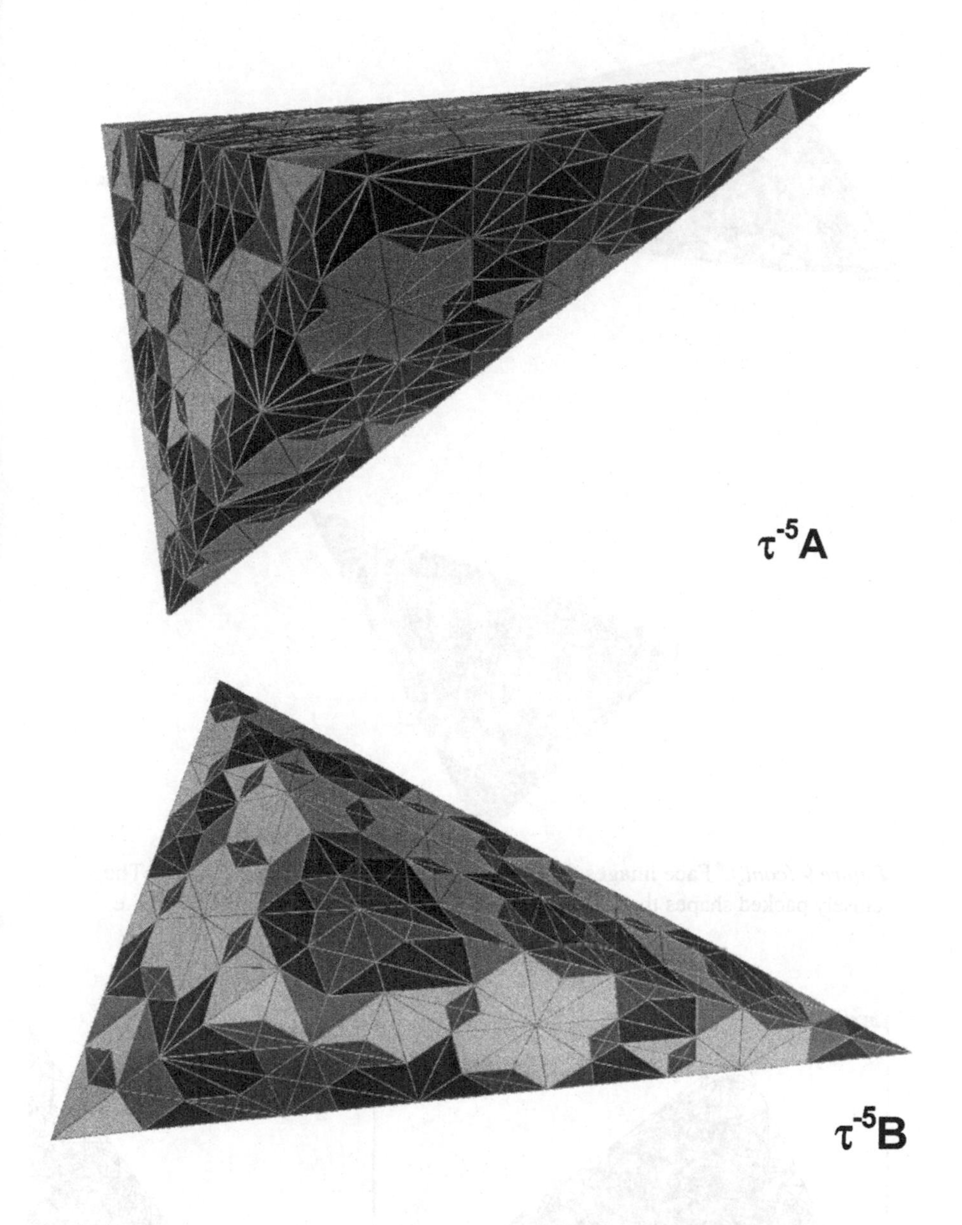


*Figure 9.* Face images for each shape subdivided to the 5th level. The closely packed shapes that compose each design are all at the same scale, i.e. $\tau^{-5}${A,B,C,K}.

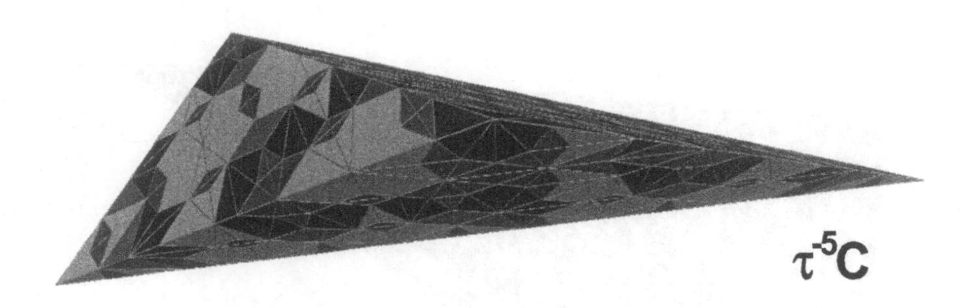


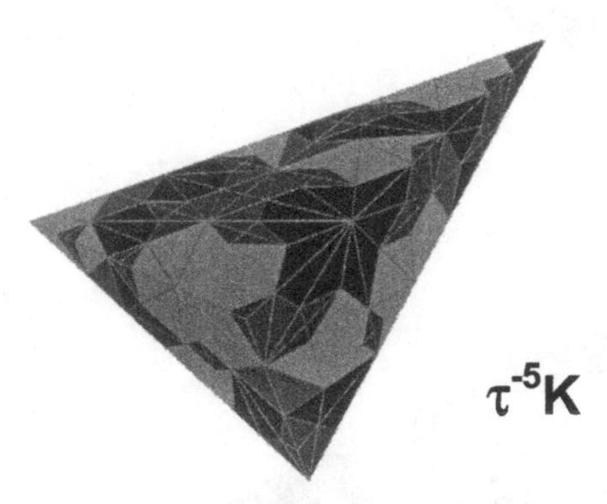


*Figure 9 (cont.).* Face images for each shape subdivided to the 5th level. The closely packed shapes that compose each design are all at the same scale, i.e. $\tau^{-5}$\{A,B,C,K\}.

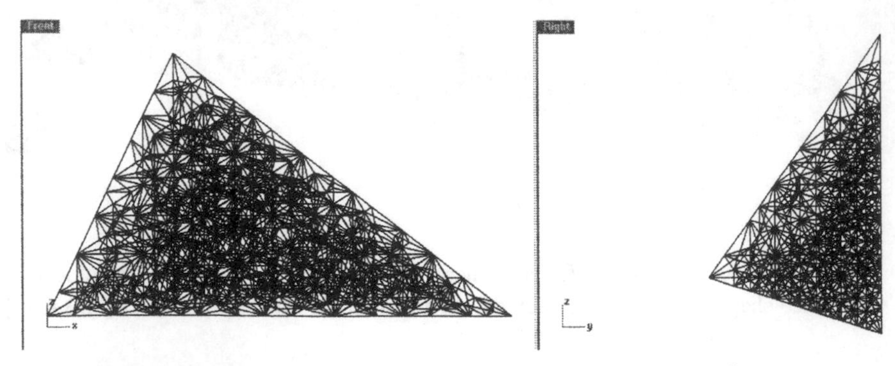


*Figure 10.* Sample 3D image of internal subdivision of shape $\tau^{-5}$A

The ABCK grammar described is exponential, as illustrated by the inflation matrix. If we start with the compositions shown in Figure 9 and apply subdivision rules for a further five levels, a 10th level inflation is generated, i.e. $\mathbf{M}^{10}$=

| | *A* | *B* | *C* | *K* |
|---|---|---|---|---|
| $\tau^{-10}$A | 171,414 | 105,930 | 105,985 | 32,725 |
| $\tau^{-10}$B | 1,280,785 | 791,594 | 791,450 | 244,640 |
| $\tau^{-10}$C | 726,110 | 448,745 | 448,834 | 138,655 |
| $\tau^{-10}$K | 2,349,710 | 1,452,220 | 1,452,110 | 448,779 |
| **total** | 4,528,019 | 2,798,489 | 2,798,379 | 864,799 |

We can observe that using an initial shape A, the largest shape, results in the largest amount of shapes being produced while shape K, the smallest, has significantly fewer. Still in just 10 subdivisions, designs are generated that consist of millions of subshapes. With exponential spatial generation the limits of some VRML viewers and common CAD tools are quickly approached. A sample of the 1-2-3 face of the K tile is shown in Figure 11 to illustrate the fineness of spatial tilling that is quickly generated. Also, note the horizontal lines that emerge, which could be useful to place floor plates. In design, compositions of different scales of shapes in the same pattern could be useful in creating a hierarchical façade and structural system.

## 5. Relaxing the ABCK Grammar

The ABCK grammar presented is a straightforward translation of the mathematical tiling and can be classified as a basic grammar according to the grammar types described in Knight (1998). The rules are in the form D → E+F where F is the set of new subshapes produced. The one similarity transformation used is the scale factor $\tau^{-n}$. While it was mentioned that the rules in their current implementation are re-write rules, the starting shape always still exists and can be recognized. The rules are applied in the same sequential order at each level to subdivide the shapes A to K that were generated in previous level, $\tau^{-n-1}$. The generative grammar is completely predictable and produces known aperiodic spatial arrangements with exact geometry.

To explore the expressive capabilities of the grammar and potential interest for design, the geometric constraints of the original tiling can be relaxed to transform the grammar into one with increased generative power. First, the grammar can be transformed into a nondeterministic basic grammar by providing means to generate different paths in the subdivisions.

The individual shape subdivision rules are unchanged; rather their application sequence is changed by allowing the algorithm to produce designs with subshapes of different scales. For example, starting with an initial shape A and stopping the subdivision of the B shape after the 3rd level, that is any occurrence of $\tau^{-3}$B, produced the composition shown in Figure 12. Allowing designs to be composed of shapes on different scales, the fractal nature of the algorithm now becomes more visible. The design language has changed. The choice of which rule(s) to make inactive at which level(s) can also be selected randomly to generate many different alternatives.

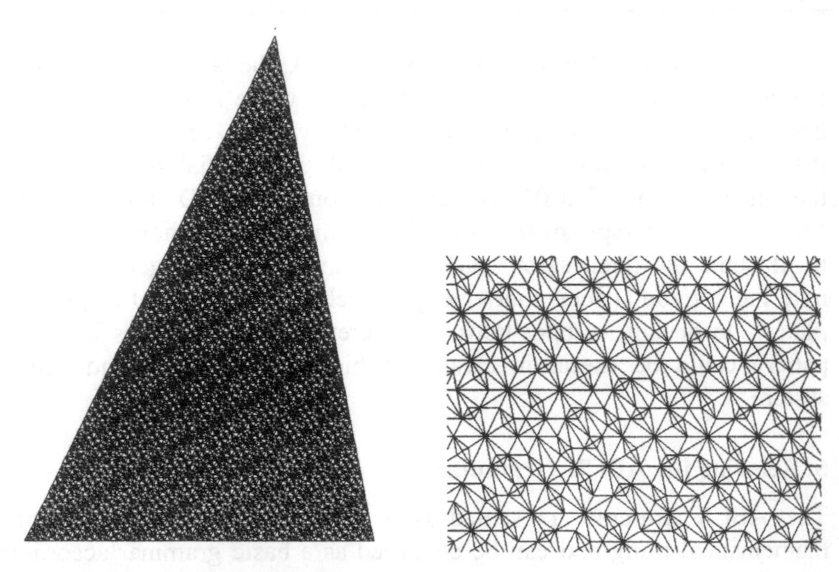

*Figure 11.* Face image and close-up view for shape K subdivided to level 10

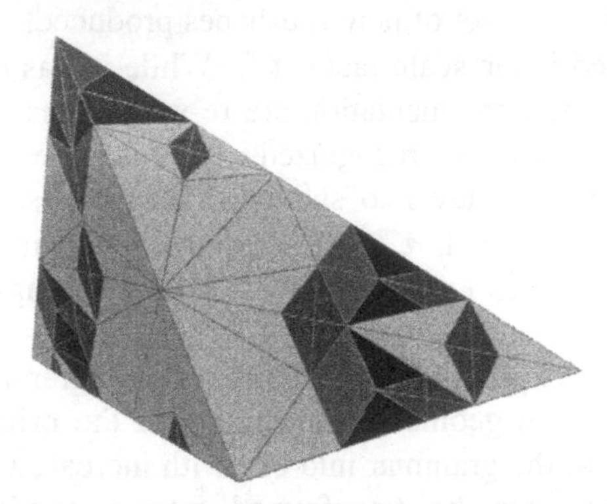

*Figure 12.* Initial A shape subdivided to level four with no subdivision of any occurrence of $\tau^{-3}$B resulting in a design with subshapes at different scales

Further increasing the generative power, the grammar can be extended to a parametric shape grammar by relaxing the constraints on the tetrahedron geometry allowed to enable parametric variations. The rules defined previously now become four general, parametric rules for subdividing tetrahedrons into a defined arrangement of parametric subshapes using proportional edge cuts. Several options exist to explore the parametric nature of the new grammar. An example is given in Figure 13 where two designs were generated for a random selection of subdivision rules at each level. The initial shape can also be changed to any tetrahedron that reflects the design scenario. For example, starting with a K mirror image shape that was elongated by moving vertex two away from the apex and moving vertices one, three and four to create a flat base resulted in the tower design shown in Figure 14. The resulting design can be compared to the original K envelope shown in Figure 9.

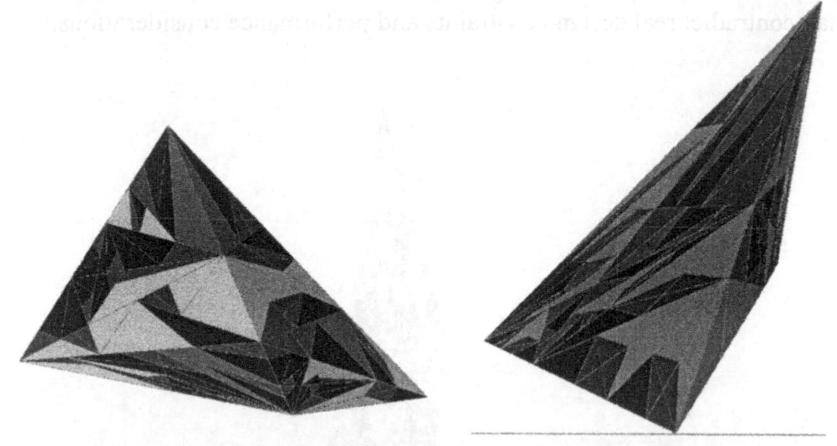

*Figure 13.* Initial A shape subdivided by random rule application to level four. The design on the left has 432 subshapes and the design on the right has 584.

As the tiling has been implemented using a shape grammar formalism, subshapes should be considered further. In the basic grammar implemented each subdivision rule acts only on the shapes with the correct shape type label and a label indicating the current subdivision level. Subshape detection of the shapes in the entire design, i.e. at any scale, is not carried out and would only result in re-subdivision of a higher scale shape. No new geometry would be produced. Rather, in the parametric shape grammar allowing rules to act on shapes of different scales is another alternative to explore that would result in producing designs with that include intersecting tetrahedrons and thus new shapes would emerge.

As the grammar becomes more relaxed, especially when also automated, an infinite number of designs with a range of subshapes can be generated for each subdivision level. The inflation matrix no longer applies. While parametric variation allows for initial shapes that deviate from the strict geometry prescribed, their subdivision no longer creates compositions of a fixed number of subshapes and faces all at a uniform scale. However, to generate designs for scenarios that fall outside the allowable shapes, or their combination, the parametric grammar still maintains defined topologic relations producing clean lines along the edges of the initial shape. Performance feedback and an optimizing process could be used to generate designs with the fundamental visual nature of the system while balancing this with design requirements. Further, generating new forms while having instantaneous feedback on their performance from different perspectives (geometric properties, space usage, structural, thermal, lighting, fabrication, etc.) would not only spark the imagination in deriving interesting spatial compositions, but also guide generation towards forms that reflect rather than contradict real design constraints and performance considerations.

*Figure 14.* Morphed initial K shape subdivided to level five

The last issue to be discussed on the ABCK grammar is that of design interpretation. Ambiguity in shape grammars is a key element to their use as creative computational partners (Knight 2003). Knight discusses ambiguity in design computation in relation to Paul Klee, "If we take into consideration the complementary relationship between two views, [a] figure may be interpreted as linear or planar. From a linear point of view, it is a square with two diagonals; from a planar point of view, it is a square divided into four triangles," (Spiller 1961).

The ABCK grammar provides a means to generate closely packed tetrahedron spaces. While the images shown so far only illustrate the packed tetrahedron shapes, other viewpoints can be taken to explore the nature of the spatial arrangements produced. For example the compositions can be visualized by different skeletal interpretations, e.g. connecting each tetrahedron centroid to its face centers, Figure 15, connecting adjacent tetrahedron centroids, connecting tetrahedron centroids to their vertices, i.e. the dual, and connecting the face centers to each other to form an internal tetrahedron, Figure 16. These are shown only for the simple grammar. Equally, another interpretation may view a certain shape, say B, to be a void space to develop alternate shape envelopes, Figure 17.

The issues and extensions discussed in this section present a starting point for exploring the ABCK grammar as a design generator itself as well as more generally the use of mathematical tilings and algorithms in design generation. The number of alternative grammar transformations and viewpoints are seemingly endless. Work is still ongoing by the author and the Advanced Geometry Unit (Arup, London) to explore design properties and potential opportunities for using the ABCK grammar in design.

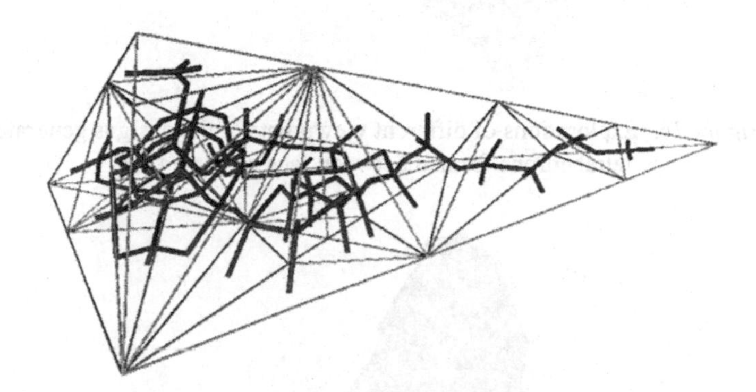

*Figure 15.* Explorations of different viewpoints of the designs generated: tetrahedron centroids to face centers($\tau^{-3}$K).

## 6. Conclusion

Synergies between mathematical tilings and shape grammar formalisms were discussed in the context of the quasicrystal spatial tiling proposed by Danzer (1989). Computationally encoding the mathematical tiling as a simple shape grammar provided a generative design tool that was used to study the visual properties of the spatial tiling and possibilities for application in building design. The mathematical spatial tiling was then relaxed to enhance the generative power by transforming the simple grammar into an unrestricted grammar. The examples presented show the range of designs that can be composed from the grammar illustrating potential for use as a design generator for architectural spaces, structures and building façade patterns. More generally, this work provides further support to the concept that many interesting geometric systems exist in the mathematical world and nature offering an interesting basis for generative design.

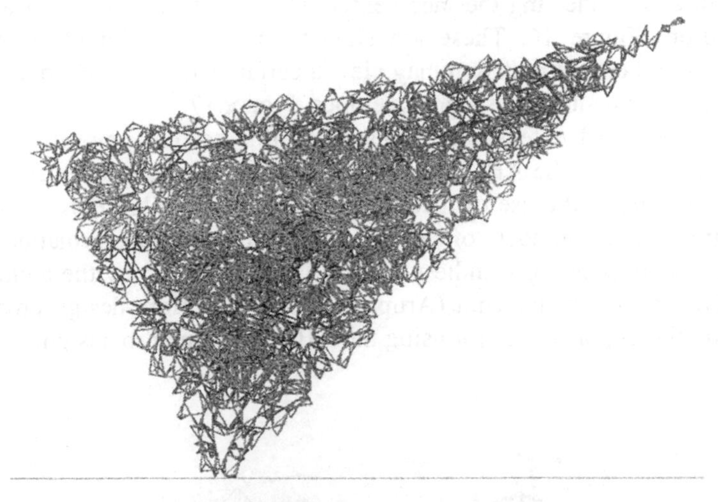

*Figure 16.* Explorations of different viewpoints of the designs generated: tetrahedrons formed between face centers ($\tau^{-5}$B)

*Figure 17.* Explorations of different viewpoints of the designs generated: interpreting all B shapes in $\tau^{-4}$A void space

## Acknowledgements

This work was primarily carried out while on a Royal Academy of Engineering Industrial Secondment to Arup R+D and Arup Advanced Geometry Unit (AGU). Special thanks are given to Charles Walker and Francis Archer as well as others at Arup for their discussions and input to this work, the Beijing Acquatics Centre team, Arup, Sydney, for their discussions on that project, and Marina Gourtovaia (EDC). The Leverhulme Trust (UK) and EPSRC provide current research support.

## References

Balmond, C: 2002, *Informal*, Prestel, London.

Beukers, A and van Hinte, E: 2001, *Lightness*, Chpt. Nature as a Role Model, 010 Publishers, Rotterdam.

Danzer, L: 1989, Three-dimensional analogs of the planar penrose tilings and quasicrystals, *Discrete Mathematics* **76**: 1-7.

Duarte, JP: 2002, Customizing mass housing: The grammar of Siza's houses at Malagueira, *Environment and Planning B: Planning and Design* forthcoming.

Grünbaum , B and Shephard, GC: 1989, *Tilings and Patterns: An Introduction*, WH Freeman.

Knight, T: 2003, Computing with ambiguity, *Environment and Planning B: Planning and Design* **30**(2): 165-180.

Knight, T: 2002, personal communication.

Knight, T: 1998, Designing a shape grammar: problems of predictability, *in* JS Gero and F Sudweeks (eds), *Artificial Intelligence in Design '98*, Kluwer, pp. 499-516.

Knight, T and Stiny, G: 2001, Classical and non-classical computation, *Architectural Research Quarterly* **5**(4): 355-372.

Kolarevic, B (ed): 2003, *Architecture in the Digital Age: Design and Manufacturing*, Spon Press.

Krishnamurti, R and Earl CF: 1998, Densely packed rectangulations, *Environment and Planning B: Planning and Design* **25**: 773–787.

Lipson, H, Antonsson, EK, and Koza, K (eds): 2003, *Computational Synthesis: From Basic Building Blocks to High Level Functionality*, Stanford, USA, Technical Report SS-03-02, AAAI Press.

Mitchell, WJ: 2002, Vitruvius Redux, in: *Formal Engineering Design Synthesis*, Chpt. 1, E. K Antonsson and J Cagan (eds), Cambridge University Press, pp. 93-125.

Nelson, RD: 1986, Quasicrystals, *Scientific American* **255**: 32-41.

Reeves, CR (ed.): 1995, *Modern Heuristic Techniques for Combinatorial Problems*, McGraw-Hill.

Spiller J.: 1961, *Paul Klee: The Thinking Eye*, George Wittenborn, New York.

Stiny, G: 1980, Introduction to shape and shape grammars, *Environment and Planning B* **7**: 343-351.

Stiny, G and Gips, J: 1980, Production systems and grammars: a uniform characterization, *Environment and Planning B* **5**: 5-18.

Testa, P: 2001, Emergent design: A crosscutting research program and design curriculum integrating architecture and artificial intelligence, *Environment and Planning B* **28**(4): 481-498.

Wang, Y and Duarte, JP: 2002, Automatic generation and fabrication of designs, *Automation in Construction* 11(3): 291-302.

JS Gero (ed), *Design Computing and Cognition'04*, 157-175


# GENERATIVE MODELLING WITH TIMED L-SYSTEMS

JON McCORMACK
*Monash University, Australia*

**Abstract.** This paper describes a generative design system based on timed, parametric Lindenmayer systems (L-systems), developed for the continuous modeling of dynamic phenomena such as morphogenesis. The specification of development functions gives the system the ability to continuously control temporal aspects of development in conjunction with the discrete rewriting for which L-systems are commonly associated. Incorporating advanced modeling extensions, such as generalized cylinders, into the interpretation of derived strings gives the system the ability to model complex shapes and forms. Examples in the design and simulation of mechanical models, plant morphogenesis and the animation of animal gaits are provided.

## 1. Introduction

Generative design offers new modes of aesthetic experience based on the incorporation of system dynamics into the production of artifact and experience. In the terminology of Thomas Kuhn (Kuhn 1996) it offers a 'paradigm shift' for the process of design and the expression of that process. The traditional modes of representation become unnecessary, being replaced by a meta-design process of activities, relationships, events and processes. A key feature of generative processes is one of database amplification, where simple sets of interacting components generate information complexity, many orders of magnitude greater than the specification.

This paper looks at a developmental model suitable for generative design, based on Lindenmayer Systems (L-systems). A temporal, developmental model is described, and its application to generative design is illustrated with examples. This model is well suited to simulating the development of a wide variety of natural and artificial phenomena.

## 2. L-Systems

L-systems are string-rewriting formalisms, originally developed in 1968 by the biologist Lindenmayer for the purposes of modeling multicellular development. Since the original formulation, L-systems have been widely studied from a variety of perspectives, including: biological simulation, mathematical formalisms, theory of computation, artificial life, and visualization. A number of variations have been introduced adapting the basic rewriting process to specific applications. Some of these applications include: cellular interaction (Lindenmayer and Rozenberg 1979); visual modeling of plants, flowers and trees (Prusinkiewicz and Lindenmayer 1990); music composition (McCormack 1996); data compression (Nevill-Manning and Witten 1997); modeling of biological organ growth (Durikovic et al. 1998); design of neural networks (Kitano 1990); architectural design (Coates et al. 1999; Hornby and Pollack 2001a); construction of artificial creatures (Hornby and Pollack 2001b); procedural design of cities (Parish and Müller 2001). This diverse oeuvre confirms the flexibility and adaptability of L-systems as a general paradigm in generative design and modeling.

Fundamental to all varieties of L-systems is the concept of parallel rewriting, whereby a set of symbols are rewritten (replaced, changed) according to some set of rules. This rewriting process occurs over the entire set of symbols simultaneously, simulating parallel development of components, similar to the way cells develop in parallel in an organism. Due to this parallelism, L-systems differ from other grammars such as Chomsky grammars, which are rewritten sequentially.

### 2.1. L-SYSTEM CLASSIFICATION: AN OVERVIEW

Key classification distinctions in L-systems are made between context sensitive and context free grammars. In the context sensitive case neighbor relations are considered in deciding which rewriting rule is to be applied (Salomaa 1973). Context sensitivity can be used, for example, to implement chemical signaling or hormone propagation in cellular simulations.

*Deterministic L-systems* are characterized by having only one rule or production for each symbol in the L-system alphabet. This means that all rewriting is deterministic and the strings produced by such L-systems will be identical provided they begin with the same initial string (known as the axiom). *Stochastic L-systems* permit rewriting on a probabilistic basis and allow the simulation of inter- and intra-species variation from the same L-system in the case of plant modeling. They can also be used to simulate Markov models, popular in musical applications (McCormack 1996).

In certain modeling contexts, the discrete nature of L-systems makes the simulation of irrational ratios or continuous properties difficult. *Parametric L-systems* solve this problem by associating real-valued parameters with symbols (collectively referred to as a module) and permit symbol rewriting to proceed by not only matching symbols but logical and arithmetic conditions involving parameters as well (Hanan 1992; Prusinkiewicz and Hanan 1990).

The key type of L-system that will be considered here is the context free, timed, parametric L-system (tp0L-system), which includes a temporal component suitable for simulation of continuous development. This is described in the next section.

## 3. Timed, Parametric 0L-Systems

Timed L-systems were proposed by Prusinkiewicz and Lindenmayer (Prusinkiewicz and Lindenmayer 1990) as an extension for achieving continuous development with D0L-systems[1]. The important developments described by the research presented in this paper are:

- the incorporation of parametric and stochastic components into timed modules;
- birth age parameters (defined below) may be expressions;
- a general development function is used to specify the relationship between module age and its realized properties (e.g. size, colour, shape, etc.).

### 3.1. DEFINITION OF TPD0L-SYSTEMS

We assume an *alphabet*, *V*, composed of a finite set of distinct *symbols* $s_1, s_2, \ldots, s_n$. tpD0L-systems operate on *timed, parametric modules,* which consist of timed symbols with associated parameters. For each module, the symbol also carries with it an *age* — a continuous, monotonically increasing, real variable, representing the amount of time the module has been active in the derivation string. Strings of modules form timed, parametric words, which can be interpreted to represent modeled structures (described in section 4). As with parametric L-systems, it is important to differentiate between formal modules used in production specification, and actual modules that contain real-valued parameters and a real-valued age.

Let $V$ be an alphabet as defined above, $\Re$ the set real numbers and $\Re_+$ the set of positive real numbers, including 0. The triple $(s, \lambda, \tau) \in V \times \Re^* \times \Re_+$ is referred to as a *timed parametric module* (hereafter shortened to *module*). It consists of the symbol, $s \in V$, its associated parameter vector, $\lambda = a_1, a_2, \ldots, a_n \in \Re$ and the age of $s$, $\tau \in \Re_+$. A sequence of modules, $x = (s_1, \lambda_1, \tau_1) \cdots (s_n, \lambda_n, \tau_n) \in (V \times \Re^* \times \Re_+)^*$ is called a *timed, parametric word.* A module with symbol $S \in V$, parameters

[1] The 'D' indicates deterministic, '0' represents the context level — in this case 0, which means context free.

$a_1, a_2, ..., a_n \in \Re$ and age $\tau$ is denoted by $\left(S(a_1, a_2, ...a_n), \tau\right)$. It is important to differentiate the real-valued *actual* parameters of modules, from the *formal* parameters (notated by underlined symbols in this section) specified in productions. In practice, formal parameters are given unique[2] identifier names when specifying productions.

We assume the following definitions:

- $\Sigma$ is the set of formal parameters, $C(\Sigma)$ is a logical expression using parameters from $\Sigma$, $E(\Sigma)$ is an arithmetic expression with parameters from the same set.
- $C$ and $E$ consist of formal parameters and numeric constants, combined using the standard operators +, –, /, *, ^ (exponentiation) $\sqrt[n]{\ }$ (*n*th root, defaulting to *n*=2 if *n* is not specified); relational operators $<,>,\leq,=,\neq$; logical operators ! (not), & (and), | (or); a number of trigonometric, stochastic and other functions; and parentheses '(', ')'. Rules for constructing expressions, operator precedence and associativity are the same as for the C programming language (Kernighan and Ritchie 1988).
- $\mathbb{C}(\Sigma)$ and $\mathbb{E}(\Sigma)$ are the sets of correctly constructed logical and arithmetic expressions with parameters from $\Sigma$. Logical expressions evaluate to Boolean values of TRUE or FALSE (equivalent to 1 or 0). Logical expressions evaluate to a real number in an arithmetic context.

A *timed, parametric 0L-system (tp0L-system)* is an ordered quadruplet $G = \langle V, \Sigma, \omega, P \rangle$ where:

- V is the non-empty set of symbols called the alphabet of the L-system;
- $\Sigma$ is the *set of formal parameters*;
- $\omega \in \left(V \times \Re^* \times \Re_+\right)^+$ is a nonempty timed, parametric word over V, called the *axiom*, and
- $P \subset \left(V \times \Sigma^* \times \Re_+\right) \times \mathbb{C}(\Sigma) \times \left(V \times \mathbb{E}(\Sigma)^* \times \mathbb{E}(\Sigma)\right)^*$ is a finite *set of productions*.

A production $\left(\underline{a}, C, \underline{\chi}\right)$ is denoted $\underline{a} : C \rightarrow \underline{\chi}$, where the formal module $\underline{a} \in V \times \Sigma^* \times \Re_+$ is the *predecessor*, the logical expression $C \in \mathbb{C}(\Sigma)$ is the *condition*, and the formal timed parametric word $\underline{\chi} \in \left(V \times \mathbb{E}(\Sigma)^* \times \mathbb{E}(\Sigma)\right)^*$ is called the *successor*. Let $\left(s, \underline{\lambda}, \beta\right)$ be a predecessor module in a production $p_i \in P$ and $\left(s_1, \underline{\lambda}_1, \underline{\alpha}_1\right) \cdots \left(s_n, \underline{\lambda}_n, \underline{\alpha}_n\right)$ the successor word of the same production. The parameter $\beta \in \Re_+$ of the predecessor module represents the *terminal age* of $s$. The expressions,

[2] Within the scope of the associated production.

$\underline{\alpha}_i \in E(\Sigma), i=1..n$ sets the initial or *birth age*. Birth age expressions are evaluated when the module is created in the derivation string. This nomenclature is illustrated in Figure 1. If the condition is empty, the production can be written $\underline{s} \rightarrow \underline{\chi}$. Formal and actual parameter counts must be the same for any given symbol.

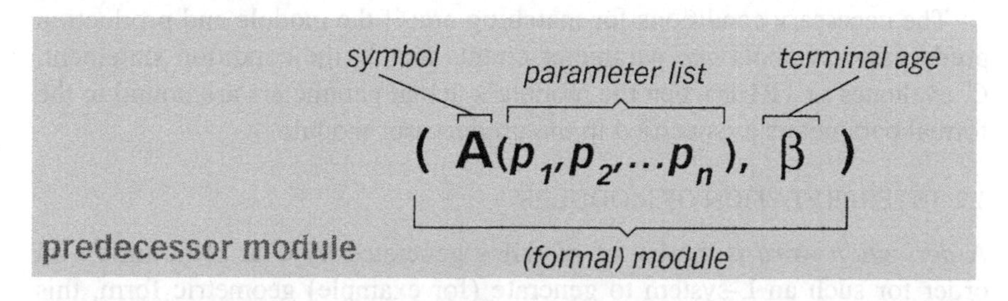


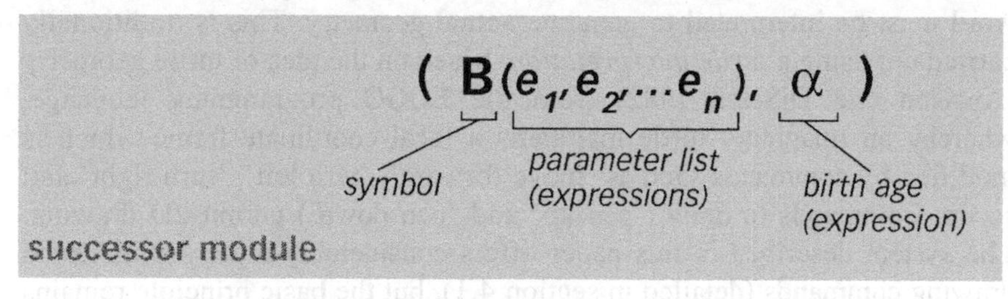


*Figure 1.* Nomenclature for predecessor and successor modules in a tpD0L-system

Here are some example productions:

$$(A(j,k)3.0) : j<k \rightarrow (B(j*k)0.0)(C(j+1,k-1)0.5) \tag{1}$$

$$(A(t)3.0) \rightarrow (A(t+1)3.0/t) \tag{2}$$

It is assumed:

- For each symbol $s \in V$ there exists at most one value of $\beta \in \Re_+$ for any production $p_i \in P$ where $(s, \underline{\lambda}, \beta)$ is the predecessor in $p_i$. If s does not appear in any production predecessor then the terminal age of $s$, $\beta_s = \infty$ is used (effectively the module never dies).
- If $(s, \underline{\lambda}, \beta)$ is a production predecessor in $p_i$ and $(s, \underline{\lambda}_i, \underline{\alpha}_i)$ any module that appears in a successor word of $P$ for $s$, then $\beta > \alpha_i$ when $\underline{\alpha}_i$ is evaluated and its value bound to $\alpha_i$ (i.e. the lifetime of the module, $\beta - \alpha_i > 0$).

Development proceeds according to some global time, *t*, common to the entire word under consideration. Local times are maintained by each module's age variable, $\tau$ (providing the relative age of the module). Rewriting begins with the axiom, $\omega$. Modules age according to *t*, and when a module reaches it's terminal age, it is replaced by a successor word, $\chi$, from the production whose predecessor module *matches* the module that has reached its terminal age. The formal parameters of the successor modules bound to the corresponding actual values when placed into the derivation string.

The necessary conditions for matching are: if the module and production predecessor symbols *and* parameter counts match; the condition statement, *C*, evaluates to TRUE when the module's actual parameters are bound to the formal parameters as specified in the predecessor module.

### 3.2. INTERPRETATION OF MODULES

A *derivation word* is the list of modules generated by *G* at some time *t*. In order for such an L-system to generate (for example) geometric form, this word must be interpreted to generate actual geometry. This is traditionally carried out using a *turtle interpretation*, based on the idea of turtle geometry (Abelson and DiSessa 1982) from the LOGO programming language, whereby an imaginary turtle maintains a local coordinate frame which is modified by commands such as 'move forward', 'turn left', 'turn right' and so on. Commands to draw ('pen up' and 'pen down') permit 2D drawing. The system described in this paper offers considerably more sophisticated drawing commands (detailed in section 4.1), but the basic principle remains the same. With a turtle interpretation, the entire derivation word is interpreted sequentially from left to right, generating the geometry. By interpreting the derivation word at different times in its development, animated sequences can be generated.

Certain symbols in the alphabet, *V*, are designated as *turtle commands*. The interpretation of these commands involves all elements of the module: the symbol determines the actual command, the module's age and parameters control various properties of the particular command. A module's parameters and age are connected to the turtle interpretation via the *development function*, specified in the next section.

### 3.3. DEVELOPMENT FUNCTIONS

Let $m = (s, \lambda, \tau) \in V \times \Re \times \Re_+$ be an actual module composed of the symbol, *s*, its actual parameters, $\lambda$ and its current relative age, $\tau$. A timed symbol $s \in V$ may optionally have associated with it a *development function*, $g_s : (V \times \Re^* \times \Re_+) \rightarrow \Re$. This function may involve any of the parameters, the current age, $\tau$, and the terminal age $\beta$ of *s* (determined by the predecessor of the production acting on *s*). Thus $g_s$ is a real valued function that can be composed of any arithmetic expression $E(\underline{\lambda_s}, \underline{\tau_s}, \underline{\beta_s})$. In addition to the formal parameters supplied, expressions can include the operators, numeric constants, and functions defined in section 3.1. The development function returns a real value, which is then used as a scaling factor for the actual parameter vector $\lambda$. That is:

$$\lambda' = g_s \cdot [\lambda] \tag{3}$$

The development function is evaluated whenever a module requires turtle interpretation, with parameter vector $\lambda'$ sent to the turtle, rather than $\lambda$ as is the case with parametric L-systems. No constraints are placed on the range or continuity of $g_s$, however if continuity is required when a production is applied (such as the accurate modeling of cellular development) $g_s$ must be monotonic and continuous. These constraints extend to the development functions for those symbols that are related as being part of the successor definition of *s*.

The development function is specified in the form:

$$g_s(\text{parameter\_list}) = \text{expression} \tag{4}$$

where `parameter_list` is drawn from the parameters, age and terminal age of *s* $\{\underline{\lambda_s}, \underline{\tau_s}, \underline{\beta_s}\}$ and `expression` is any valid arithmetic expression.

## 4. Turtle Commands

The system provides a wide variety of turtle commands. For the sake of brevity, only a subset will be detailed here. For more complete details the reader is referred to McCormack (2003).

The turtle maintains its own local coordinate reference frame, consisting of three orthogonal unit vectors representing the current Heading, Left and Up directions (denoted $\mathbf{H}$, $\mathbf{L}$, and $\mathbf{U}$ respectively). A world coordinate position, **t**, is also maintained, Figure 2.

Basic turtle commands allow movement forward ('f'), turning left ('+') and right ('–'), pitch up ('^') and down ('&') and twist clockwise ('/') and anti-clockwise ('\'). Parameters to these commands describe the amount of that particular command, i.e. f(3.2) moves forward (in the direction of $\mathbf{H}$ by 3.2 units; +(45) turns the turtle left by 45 degrees.

The turtle is a state-based system and the square brackets ('[' and ']') push and pop the turtle state (reference frame and associated drawing parameters) on and off a stack. This mechanism permits the creation of branching structures so important in plant and tree modeling applications (Prusinkiewicz and Hanan 1989).

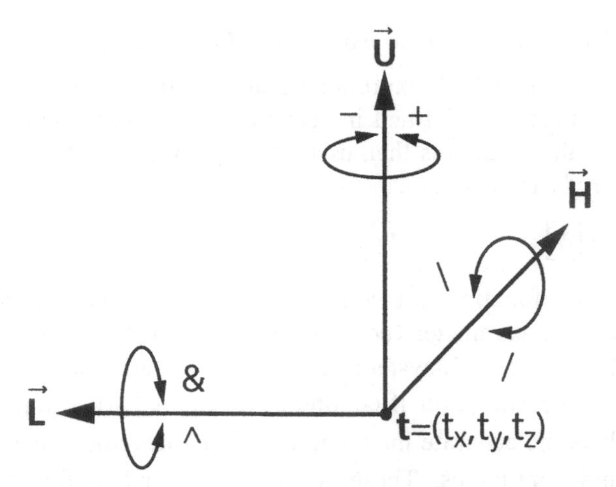


*Figure 2.* Turtle coordinate system and commands to change orientation

Other commands instance geometry, for example the 'F' command draws a cylinder of length specified by its parameter (the radius is set using the '!' command). So for example the command sequence: !(3)F(10) draws a cylinder of radius 3 units and length 10, with the principle axis in the direction of $\vec{\mathbf{H}}$. In addition, the current turtle position is updated to the end of the cylinder after drawing.

### 4.1. GENERALIZED CYLINDERS

Wainwright suggests that the cylinder has found general application as a structural element in plants and animals (Wainwright 1988). He sees the cylinder as a logical consequence of the *functional morphology* of organisms, seeing the dynamic physiological processes (function) as dependent on form and structure over time. Wainwright distinguishes the cylinder as a natural consequence of evolutionary design based on the physical and mechanical properties of cylindrical structures.

This simple cylindrical method provided by the 'F' command is not sufficient for more complex geometric modeling. Consider the case of modeling a compound segment, such as a tentacle or horn. A relatively simple L-system can be used to describe such a shape, as illustrated in Figure 3. The problem in using the cylinder symbol, F, is that compound segments exhibit discontinuities between segments.

This problem of describing more complex cylindrical shapes can be solved using *generalized cylinders*, originally developed by Agin for applications in computer vision (Agin 1972). Generalized cylinders have been used extensively by Bloomenthal to model tree limbs (Bloomenthal 1985) and a variety of natural objects (Bloomenthal 1995).

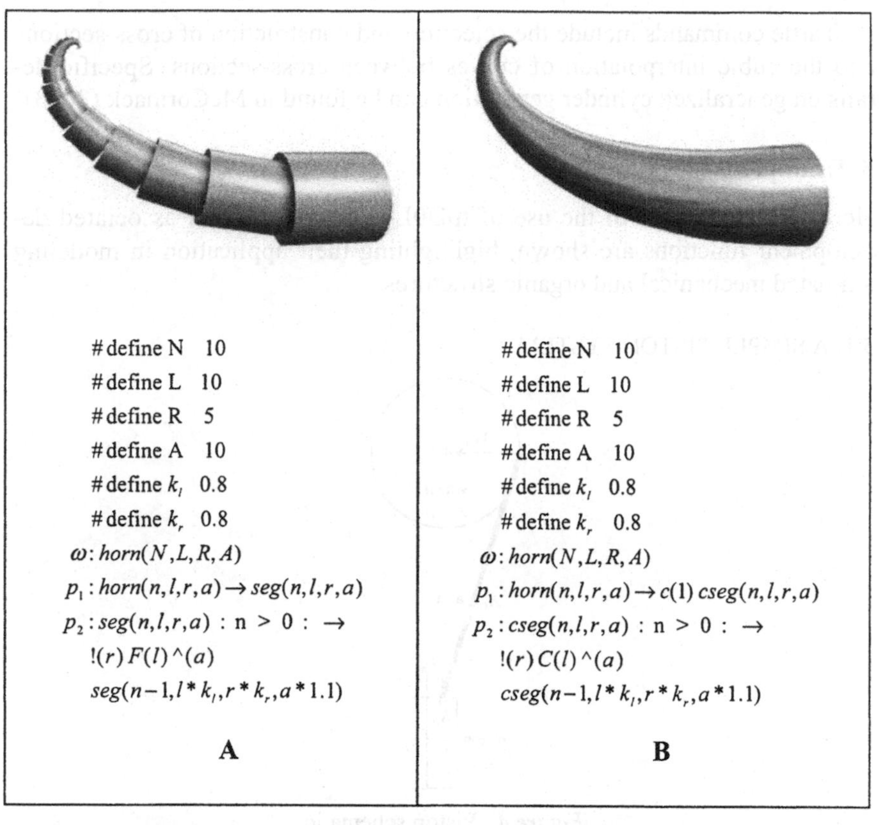

#define N 10
#define L 10
#define R 5
#define A 10
#define $k_l$ 0.8
#define $k_r$ 0.8
$\omega$: $horn(N,L,R,A)$
$p_1 : horn(n,l,r,a) \rightarrow seg(n,l,r,a)$
$p_2 : seg(n,l,r,a) : n > 0 : \rightarrow$
$!(r)\,F(l)\,\wedge(a)$
$seg(n-1, l*k_l, r*k_r, a*1.1)$
A
#define N 10
#define L 10
#define R 5
#define A 10
#define $k_l$ 0.8
#define $k_r$ 0.8
$\omega$: $horn(N,L,R,A)$
$p_1 : horn(n,l,r,a) \rightarrow c(1)\,cseg(n,l,r,a)$
$p_2 : cseg(n,l,r,a) : n > 0 : \rightarrow$
$!(r)\,C(l)\,\wedge(a)$
$cseg(n-1, l*k_l, r*k_r, a*1.1)$
B


*Figure 3.* A simple horn defined (**A**) using cylinders, which leaves noticeable gaps where the radius and angle of the cylinder changes. In **B**, this problem is fixed with the use of generalized cylinders. The parametric L-system generating each model is shown below the image (timed information is not shown)

Similar systems to the one described here also make use of generalized cylinders. The *xfrog* system of Lintermann and Deussen makes basic use of cylindrical structures in stem and branch modelling (Lintermann and Deussen 1998; Lintermann and Deussen 1999). Prusinkiewicz et. al. describe an interactive plant modeling system that makes use of generalized cylinders and is based on L-systems (Prusinkiewicz et al. 2001). Both these systems rely on external curve editing software, whereas the system described in this paper uses an extended set of turtle commands to automate generalized cylinder construction.

The basic principle for creating a generalized cylinder is to define a series of *cross-sections*, possibly of varying shape and size, distributed over some continuous curve, known as the *carrier curve*. The cross-sections are connected to form a continuous surface.

Turtle commands include the selection and construction of cross-sections and the cubic interpolation of curves between cross-sections. Specific details on generalized cylinder generation can be found in McCormack (2003).

## 5. Examples

Here some examples of the use of tpD0L-systems and their associated development functions are shown, highlighting their application in modeling animated mechanical and organic structures.

### 5.1. A SIMPLE PISTON SYSTEM

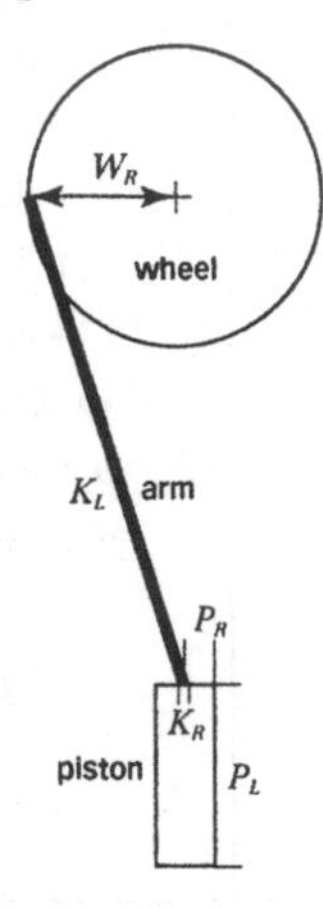


*Figure 4.* Piston schematic

This example simulates a simple piston and flywheel mechanism. The piston is connected to the flywheel via an arm of fixed length. The movement of the piston is constrained to the vertical, which drives the wheel in a circular motion. The schematic diagram, Figure 4, shows the principle features.

Figure 5 shows the L-system used to model the above system. The period of the system is determined by the constant $\rho$. In-built turtle geometry commands are sufficient to construct all the geometry. Production $p_1$ builds the system, productions $p_2 - p_4$ cycle the components in loops of period $\rho$. When a symbol reaches the end of its life, it begins again with age 0 and the cycle continues. The 'equiv' directives equate symbols with *different* names the *same* turtle function, for example the *bend* symbol is interpreted in the same way as the '+' command (counter-clockwise rotation about the **U** vector).

The development functions associated with each timed module provide the information necessary drive the animation. Thus, the productions provide structure and control, while the development functions control the animation properties of the model. The derivation word for the piston system is shown for various global time values in Table 1. The parameter values shown reflect the application of the development function.

## 5.2. MORPHOGENETIC DEVELOPMENT

L-systems are often associated with visual models of plants and plant ecosystems. This example shows the animated development of an imaginary species of plant from the interactive animation *Turbulence* developed by the author (McCormack 1994).

| | | | | | | | | |
|---|---|---|---|---|---|---|---|---|
| *# define* | $W_R$ | 50 | *# define* | $P_L$ | 60 | *equiv* | $f$ | *mov* |
| *# define* | $K_L$ | 200 | *# define* | $\rho$ | 4 | *equiv* | $+$ | *bend* |
| *# define* | $K_R$ | 2 | *# define* | E | 0.04 | *equiv* | $-$ | *turn* |
| *# define* | $P_R$ | 10 | | | | | | |

$$\omega: \ (piston, 0)$$

$$p_1: \ (piston, \mathrm{E}) \rightarrow (mov(W_R)0)!(P_R)cC(P_L, TRUE)(bend(1)0)!(K_R)cC(K_L, TRUE)(turn(1)0)f(W_R)+(\pi/2)[\ disc(0, W_R)]$$

$$p_2: \ (mov(x)\rho) \rightarrow (mov(x)0)$$

$$p_3: \ (bend(x)\rho) \rightarrow (bend(x+1)0)$$

$$p_4: \ (turn(x)\rho) \rightarrow (turn(x)0)$$

$$g_{mov}(\tau_{mov}, \beta_{mov}) = 1 - \cos\left(\frac{2\pi\tau_{mov}}{\beta_{mov}}\right)$$

$$g_{bend}(\tau_{bend}, \beta_{bend}) = \sin^{-1}\left(\frac{W_R \sin\left(\frac{2\pi\tau_{bend}}{\beta_{bend}}\right)}{K_L}\right)$$

$$g_{turn}(\tau_{turn}, \beta_{turn}) = g_{bend}(\tau_{turn}, \beta_{turn}) + \frac{2\pi\tau_{turn}}{\beta_{turn}}$$

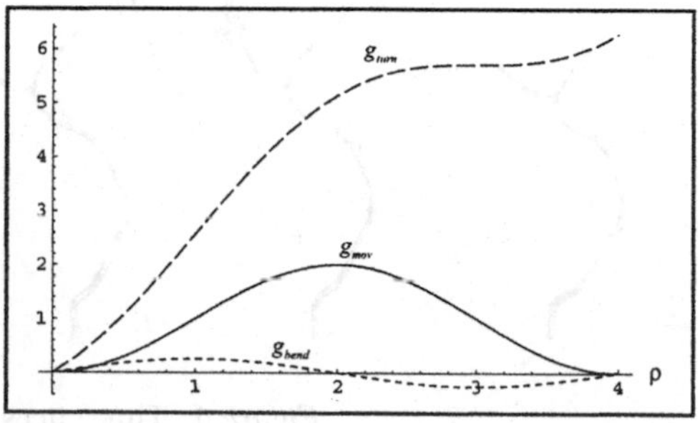

*Figure 5.* tp0L-system for the piston system and graph showing development functions over a single cycle

The full L-system describing the development is quite complex (approximately 35 productions), so in the interests of space and clarity an overview will be presented here. There are two main stages in the development of this sequence — *(i)* development of the stem, which after a certain time bifurcates into two segments; and *(ii)* the development of the flower head. The animated development of an individual model is shown in Figure 6.

TABLE 1. Derivation word at specific times for the L-system of Figure 5

| *t* (global time) | **Derivation String** (after application of development function) |
|---|---|
| 0.0 | $(piston, 0)$ |
| 0.04 | $(mov(0)0)!(10)cC(60, TRUE)(bend(0)0)!(2)cC(200, TRUE)$ $(turn(0)0)f(50)+(1.57)[\ disc(0,50)]$ |
| 1.04 | $(mov(50)1)!(10)cC(60, TRUE)(bend(0.253)1)!(2)cC(200, TRUE)$ $(turn(1.82)1)f(50)+(1.57)[\ disc(0,50)]$ |
| 4.04 | $(mov(0)0)!(10)cC(60, TRUE)(bend(0)0)!(2)cC(200, TRUE)$ $(turn(6.28)0)f(50)+(1.57)[\ disc(0,50)]$ |

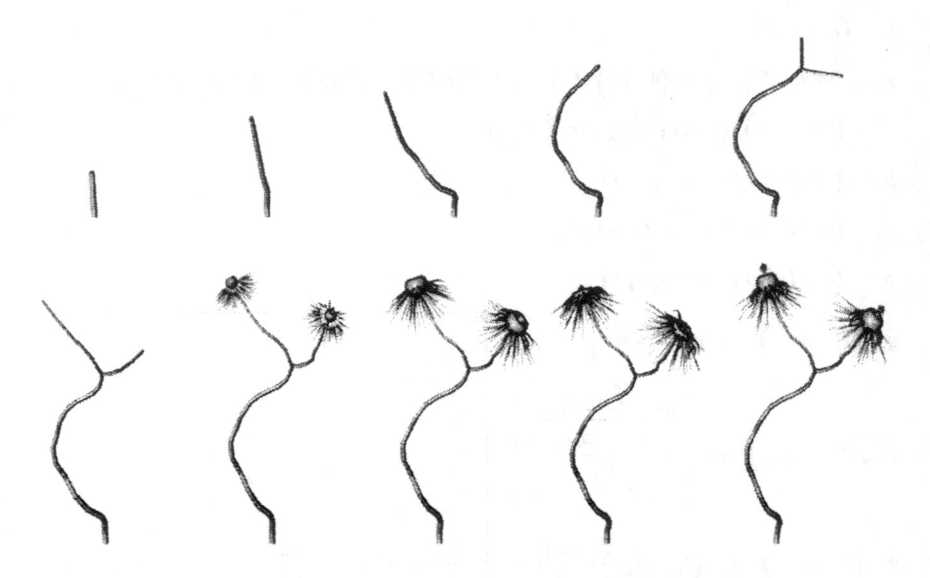

*Figure 6.* Time slices of flower growth

The randomness and variation of structure is achieved using stochastic functions that effect growth. The flower head consists of a number of animated components. The main head uses a set of three pre-defined surfaces, which are interpolated by development functions. Generalized cylinders are used to model the thorn and spike components. Note how these features animate in both shape and size.

A Bessel function of the first kind is used to model the scaling and animation of the flower head, Figure 7. Bessel functions are often used to solve motion equations for physical systems, and here the use of the function gives the animation a damped-spring-like quality (which the equation represents), as the head 'puffs' up rapidly in size and then pulsates in an oscillating rhythm, slowly damping down as the element ages. Visually similar behavior is observed in time-lapse sequences of real plants, as they respond to the rhythms of the day/night cycle.

### 5.3. LEGGED GAITS OF ANIMALS

Timed L-systems may also be used as controllers in the simulation of animal gait cycles. Here the advantage of L-systems is that they can provide both motor control and structural definition within the same grammar.

*Figure 7.* Time sequence showing development of the flower head

For the example detailed in this section, we will consider the representation of an 'animal' with multiple rigid body segments, each connected with a 2-DOF[3] articulation as shown in Figure 8.

Each body segment has two multi-jointed legs, at opposite sides of the body segment. The leg detail is also shown in the figure. A *leg* is composed of two rigid limb segments, $S_1$ and $S_2$. $S_1$ is attached to the body segment by a 3-DOF joint, $J_1$. The joint $J_2$ between $S_1$ and $S_2$ has 1-DOF.

The key advantage of an L-system specification is in the flexibility of body and limb design and specification. Similar techniques have been used as a general system to evolve novel designs of articulated figures (Sims 1994a; Sims 1994b). Arbitrary joint, limb, and body segment configurations can be achieved by modifying the productions that generate these elements.

[3] Degree of Freedom

The generalized cylinder techniques (discussed in Section 4) are used to model the complex limb and body geometries of the creature.

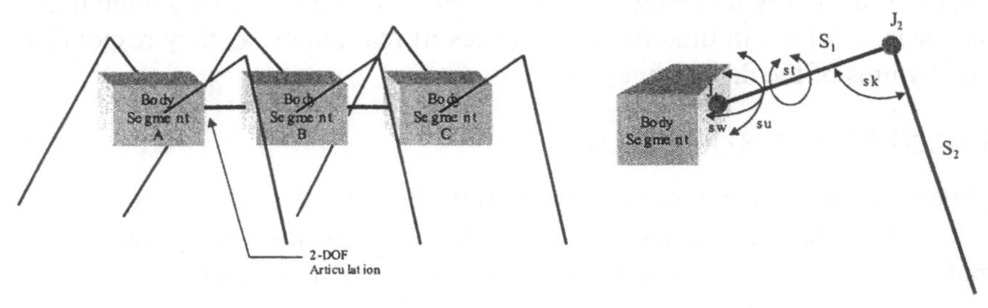


*Figure 8.* Legged animal composed of multiple articulated body segments (left) and detail of an individual segment's joint configuration (right)

The motor control structure is also specified by a tp0L-system, forming a simple finite state machine that drives the movement of the body segments and legs. Constraints on the movement of individual limbs are set within the development functions for each joint.

To illustrate how this scheme works, we will focus on the construction and animation of a single leg. The L-system shown in Figure 9 captures the essential components. The modules *sw*, *su* and *st* represent the motor control of $S_1$ (3-DOF articulation), while *sk* represents the joint angle between $S_1$ and $S_2$. The module *ls* (leg segment) calls a complex set of productions to create the geometry of the leg using generalized cylinders (these productions are not show for the sake of clarity). This module's parameter specifies the overall length of the leg segment and the turtle is placed at the position of the next joint upon completing of the geometric construction of the leg segment. The *ls* module is not timed due to the fixed structure of the leg segment itself. This constraint means that the geometric data can be cached to avoid regeneration across multiple locations and time steps. The constant $R_{S12}$ is the ratio of size between the upper ($S_1$) and lower ($S_2$) leg segments.

An individual *walk cycle* represents the movement of the leg system over a single gait. The total time for this cycle is represented by the variable $\varphi$, with individual controllers using this time or a rational ratio of it to ensure cyclic animation. For example, the twisting motion of the leg (module *st*) is separated into three distinct components which sum to the gait cycle time.

The parameter $\phi$ represents the phase of the animation cycle. As each body segment is added the phase of the walk cycle is shifted to ensure correct motion relative to the position of the segment. The parameters *n* and *l* control the magnitude of the gait and the leg respectively.

The productions of Figure 9 specify a simple state machine that encodes the geometric and temporal structure of the leg gait. The remaining information required is the associated development function for each module, detailed in Figure 10.

These functions approximate the inverse kinematical solution of the system for a walking gait. The use of periodic functions ensures that phase and perfect cycling are easily accommodated, as required for a system that must coordinate a large number of legs in a coherent fashion. In the actual generated sequences, noise perturbations based on the age of the creature are used to introduce variation and a 'natural' feel into the walk cycle, Figure 11.

$$\begin{array}{lll|llll|lll} \# define & \beta_{ls} & \varphi & equiv & + & sw & & \# define & \beta_{st} & \frac{\varphi}{5} \\ \# define & \beta_{sw} & \varphi & equiv & \wedge & su & & \# define & \beta_{sk} & \varphi \\ \# define & \beta_{su} & \varphi & equiv & / & st & & \# define & R_{S12} & \frac{3}{2} \\ & & & equiv & \& & sk & & & & \end{array}$$

$$p_1 : \left(leg\left(\phi, l\right), \beta_{ls}\right) \rightarrow \left(sw\left(1, \phi\right), 0\right)\left(su\left(1, \phi\right), \frac{3\beta_{su}}{4}\right)\left(st\left(1, \phi\right), 0\right) ls\left(l\right)\left(sk\left(1\right), 0\right) ls\left(R_{S12}l\right)$$

$$p_2 : \left(sw\left(n, \phi\right), \beta_{sw}\right) \rightarrow \left(sw\left(n, \phi\right), 0\right)$$

$$p_3 : \left(su\left(n, \phi\right), \beta_{su}\right) \rightarrow \left(su\left(n, \phi\right), 0\right)$$

$$p_4 : \left(st\left(n, \phi\right), \beta_{st}\right) \rightarrow \left(st2\left(n, \phi\right), \beta_{st}\right)$$

$$p_5 : \left(sk\left(n, \phi\right), \beta_{sk}\right) \rightarrow \left(sk\left(n, \phi\right), 0\right)$$

$$p_6 : \left(st2\left(n, \phi\right), 3\beta_{st}\right) \rightarrow \left(st\left(-n, \phi\right), 0\right)$$

*Figure 9.* tp0L-system fragment for the gait control mechanism to specify a single leg configuration. The cycle time of a single step is specified by the variable $\varphi$

$$\begin{array}{lll|lll|lll} \# define & R_{sw} & 0 & \# define & A_{su} & 0.139\pi & \# define & R_{sk} & 0.389\pi \\ \# define & A_{sw} & 0.378\pi & \# define & R_{st} & 0 & \# define & A_{sk} & 0.194\pi \\ \# define & R_{su} & 0.056\pi & \# define & A_{st} & 0.056\pi & & & \end{array}$$

$$g_{sw}\left(\tau_{sw}, \beta_{sw}, \phi\right) = R_{sw} + A_{sw} \sin\left(\frac{2\pi\tau_{sw}}{\beta_{sw}} + \phi\right)$$

$$g_{su}\left(\tau_{su}, \beta_{su}, \phi\right) = \text{if}\left(\tau_{su} < \frac{\beta_{su}}{2}, 0, R_{su} - A_{su}\left(\cos\left(2\pi\left(\frac{\tau_{su}}{\beta_{su}} - 0.5\right) + \phi\right) + 1\right)\right)$$

$$g_{st}\left(\tau_{st}, \beta_{st}, \phi, n\right) = R_{st} + \text{sign}\left(n\right) A_{st} \sin^2\left(\frac{\pi\tau_{sw}}{\beta_{sw}} + \phi\right)$$

$$g_{sk}\left(\tau_{sk}, \beta_{sk}, \phi\right) = \text{if}\left(\tau_{sk} < \frac{\beta_{sk}}{2}, 0, R_{sk} - A_{sk}\left(\cos\left(4\pi\left(\frac{\tau_{sk}}{\beta_{sk}} - 0.5\right) + \phi\right) + 1\right)\right)$$

*Figure 10.* Key development functions for the gait of a single leg

*Figure 11.* Still frame showing the legged creature running. The geometry and animation is generated using the timed L-system techniques described.

## 6. The Design Environment

This section briefly describes how, in a practical sense, the formal systems described in this paper can be used in a design environment. Design using generative methods involves the creation and modification of rules or systems that interact to generate the finished design. Hence, the designer does not directly manipulate the produced artifact, rather the rules and systems involved in the artifact's production. The design process becomes one of *meta-design* where a finished design is the result of the *emergent* properties of the interacting system (McCormack and Dorin 2001). The 'art' of designing in this mode is in mastering the relation between process specification, environment, and generated artifact. Since this is an art, there is no formalized or instruction-based method that can be used to guide this relationship. The role of the human designer remains, as with conventional design, central to the design process.

In the case of using the generative methods described in this paper, fundamentally the meta-design process must produce a tpD0L-system axiom and set of productions. To this end, the author has used three different meta-design methods: *(i)* specification of L-systems by hand, similar to the way a programmer writes a computer program; *(ii)* using a visual programming metaphor, where modules and productions are represented visually and may be manipulated topologically to create new productions; *(iii)* using artificial evolutionary techniques based on aesthetic selection (McCormack 1993).

Of these three methods, the first (direct manipulation) gives the most flexibility and control, but is the most difficult for people without a strong programming background to understand. The second gives less control, but is much more intuitive from a design perspective, and is used in commercial implementations (Lintermann and Deussen 1999). The third method, interactive evolution, is excellent for generating novel designs without needing to understand the productions involved. However, exact control is extremely difficult. The method favored by the author is a combination of explicit editing of L-system productions combined with interactive evolution. This is the methodology used to produce the examples described in Sections 5.2 and 5.3. Addressing the deficiencies of all these techniques remains an open research problem, if generative design is to achieve widespread adoption by the design community.

## 7. Conclusions

The integration of timed and parametric components to L-systems permits a new degree of flexibility and possibility for generative modeling, some of which has been illustrated in this paper. In the area of visual simulation, the use of compound structures such as generalized cylinders can be integrated into a turtle interpretation of modules generated by L-systems. The system described here enables the procedural generation of complex, time-dependent geometric structures, in ways that would be difficult or impossible to design using other methods.

Important extensions to this model are the use of hierarchical specification of grammars and the incorporation of context into developmental systems (McCormack 2003). These extensions provide even further flexibility, particularly when modeling organic structures and their morphogenesis.

## Acknowledgements

This work was produced with the assistance of a fellowship grant from the Australia Council for the Arts New Media Arts fund, and a creative development grant from the Australian Film Commission. The author acknowledges the generous support of these cultural organizations in funding this research. The paper reviewers also provided useful comments for the final version of this paper.

## References

Abelson, H and DiSessa, AA: 1982, *Turtle Geometry: The Computer as a Medium for Exploring Mathematics,* MIT Press, Cambridge, MA.

Agin, GJ: 1972, *Representation and Description of Curved Objects,* Technical Memo, No. AIM-173, October 1972, Stanford Artificial Intelligence Report, Stanford, California.

Bloomenthal, J: 1985, Modeling the mighty maple, *Computer Graphics* **19**(3): 305-311.

Bloomenthal, J: 1995, Skeletal design of natural forms, *PhD Thesis*, Department of Computer Science, University of Calgary, Calgary, Alberta.

Coates, P, Broughton, T and Jackson, H: 1999, Exploring three-dimensional design worlds using Lindenmayer systems and genetic programming, *in* PJ Bentley (ed), *Evolutionary Design by Computers*, Morgan Kaufmann, London, UK, Chapter 14.

Durikovic, R, Kaneda, K and Yamashita, H: 1998, Animation of biological organ growth based on L-systems, *Computer Graphics Forum (EUROGRAPHICS '98)* **17**(3): 1-14.

Hanan, J: 1992, Parametric L-Systems and their application to the Modelling and Visualization of Plants, *PhD Thesis*, Computer Science, University of Regina, Saskatchewan.

Hornby, GS and Pollack, JB: 2001a, The advantages of generative grammatical encodings for physical design, *Proceedings of the 2001 Congress on Evolutionary Computation*, IEEE Press, pp. 600-607.

Hornby, GS and Pollack, JB: 2001b, Evolving L-systems to generate virtual creatures, *Computers & Graphics* **26**(6): 1041-1048.

Kernighan, BW and Ritchie, DM: 1988, *The C Programming Language* (Second Edition), Prentice Hall, Englewood Cliffs, New Jersey.

Kitano, H: 1990, Designing neural networks using genetic algorithms with graph generation system, *Complex Systems* **4**(4): 461-476.

Kuhn, TS: 1996, *The Structure of Scientific Revolutions (Third Edition)*, University of Chicago Press, Chicago, Ill.

Lindenmayer, A and Rozenberg, G: 1979, Parallel generation of maps: Developmental systems for cell layers, *in* V Claus, H Ehrig, and G Rozenberg (eds), *Graph Grammars and Their Application to Computer Science; First International Workshop*, Vol. Lecture Notes in Computer Science 73, Springer-Verlag, Berlin, pp. 301-316.

Lintermann, B and Deussen, O: 1999, Interactive modeling of plants, *IEEE Computer Graphics & Applications* **19**(1): 2-11.

Lintermann, B and Deussen, O: 1998, A modelling method and interface for creating plants, *Computer Graphics Forum* **17**(1): 73-82.

McCormack, J: 2003, The application of L-systems and developmental models to computer art, animation, and music synthesis, *PhD Thesis*, School of Computer Science and Software Engineering, Monash University, Clayton.

McCormack, J: 1996, Grammar-based music composition, *in* R Stocker et al. (eds), *Complex Systems 96: from Local Interactions to Global Phenomena*, ISO Press, Amsterdam, pp. 321-336.

McCormack, J: 1993, Interactive evolution of L-system grammars for computer graphics modelling, *in* D Green and T Bossomaier (eds), *Complex Systems: From Biology to Computation*, ISO Press, Amsterdam, pp. 118-130.

McCormack, J: 1994, Turbulence: An interactive installation exploring artificial life, Visual proceedings: The art and interdisciplinary programs of SIGGRAPH 94, *Computer Graphics Annual Conference Series*, ACM SIGGRAPH, New York, pp. 182-183.

McCormack, J and Dorin, A: 2001, Art, emergence and the computational sublime, *in* A Dorin (ed), *Second Iteration: A Conference on Generative Systems in the Electronic Arts*, CEMA, Melbourne, Australia, pp. 67-81.

Nevill-Manning, CG and Witten, IH: 1997, Compression and explanation using hierarchical grammars, *The Computer Journal* **40**(2/3): 103-116.

Parish, YIH and Müller, P: 2001, Procedural modeling of cities, Proceedings of SIGGRAPH 2001 (Los Angeles, California, August 12-17), *Computer Graphics Proceedings* Annual Conference Series, ACM SIGGRAPH, pp. 301-308.

Prusinkiewicz, P and Hanan, J: 1989, *Lindenmayer Systems, Fractals and Plants*, Springer-Verlag, Berlin.

Prusinkiewicz, P and Hanan, J: 1990, Visualization of botanical structures and processes using parametric L-systems, *in* D Thalmann (ed), *Scientific Visualization and Graphics Simulation*, John Wiley & Sons, Chichester, pp. 183-201.

Prusinkiewicz, P and Lindenmayer, A: 1990, *The Algorithmic Beauty of Plants*, Springer-Verlag, New York.

Prusinkiewicz, P, et al.: 2001, The use of positional information in the modeling of plants, Proceedings of SIGGRAPH 2001, *Computer Graphics Proceedings* Annual Conference Series, ACM SIGGRAPH, pp. 289-300.

Salomaa, A: 1973, *Formal Languages,* Academic Press, New York, NY.

Sims, K: 1994a, Evolving 3D morphology and behavior by competition, *in* R Brooks and P Maes (eds), *Proceedings of Artificial Life IV*, MIT Press, pp. 28-39.

Sims, K: 1994b, Evolving virtual creatures, Proceedings of SIGGRAPH 94, *Computer Graphics Proceedings, Annual Conference Series,* ACM SIGGRAPH, pp. 15-22.

Wainwright, SA: 1988, *Axis and Circumference: The Cylindrical Shape of Plants and Animals,* Harvard University Press, Cambridge, Mass.

JS Gero (ed), *Design Computing and Cognition'04*, 177-196


# FORMALIZING GENERATION AND TRANSFORMATION IN DESIGN

*A Studio Case-Study*

OMER AKIN, HODA MOUSTAPHA
*Carnegie Mellon University, USA*

**Abstract.** This paper is an integration of two substantial endeavours. One is a general purpose 3D modelling system, ICE that introduces a new notation and an entire family of graphic design functionalities based on generative structures and manipulation handles. The other is an exhaustively annotated design studio, in which the entire graphic output of students and the annotations of their faculty have been ethnographically recorded. In this paper, we are using the ICE notation to represent the key graphic products of a selected student and the transformations between these representations. Our goal is to demonstrate that, through ICE's formal notation (1) graphic entities in complex design sequence can be unambiguously represented, (2) transformations between graphic entities in complex design sequence can be unambiguously represented, and (3) the various design sequences can be formally captured for subsequent process or cognitive analysis

## 1. Motivation

This paper is motivated with the goal of codifying design (taken both as a noun and a verb) unambiguously and formally. We believe this will lead to quantifiable representations that can help analyse designs and design generation for cognition and intent capture in design. There are two motivating ingredients of this study. One is a general purpose 3D modelling system, ICE (interactive configuration exploration) that introduces a new notation and an entire family of graphic design functionalities based on generative structures and manipulation handles, (Moustapha and Krishnamurti 2001). The other is an exhaustively annotated design studio in which the entire graphic output of students and the annotations of their faculty have been ethnographically recorded. We will briefly introduce these ingredients below. In the following sections, we will elaborate each one and conclude with a discussion of the implications of our approach for design cognition and intent capture.

### 1.1. THE ANNOTATED STUDIO

A vertical design studio in the School of Architecture, at Carnegie Mellon University was offered during the summer of 2002, by Professor Omer Akin. There were six students taking the studio, one having completed the 2nd year, two the 3rd year and three the 4th year of their college education. The entire studio work was recorded through digital photographs of student work brought to each class session and the midterm and final reviews (Akin, 2002). These graphic records were accompanied by daily diary annotations kept by the instructor for each student's progress as well as the overall progress of the studio.

Students were invited to define their own design programs or continue with a previous design problem either they experienced or experienced by their peers. Three different problems emerged, international housing prototype, dormitory housing, and a toy manufacturer's headquarters building. The studio work was complemented by visits by external faculty on a weekly basis. They gave feedback to students on their work through critics and presentations of their own work. The same faculty served on the midterm and final reviews of the studio. A typical annotation for a studio day contains segment for each student and some directed to the general issues in the studio like the one in Table 1.

TABLE 1. Typical annotation for a studio segment for Thursday, May 23, 2002

| | |
|---|---|
| Subject-W's has created a swirling shape that expresses the housing hierarchy: rooms, units, unit-clusters, wings, buildings, building-clusters. I tell her[1] to examine her ideas spatially *vis a vis* a model. I remind her of her own narrative about the new dorm lounge spaces that do not attract any social gatherings due to being out of the way. Making a conceptual diagram about social hierarchy does not always work in actual spatialization because connectivity must be solved in another medium, namely a physical model. | 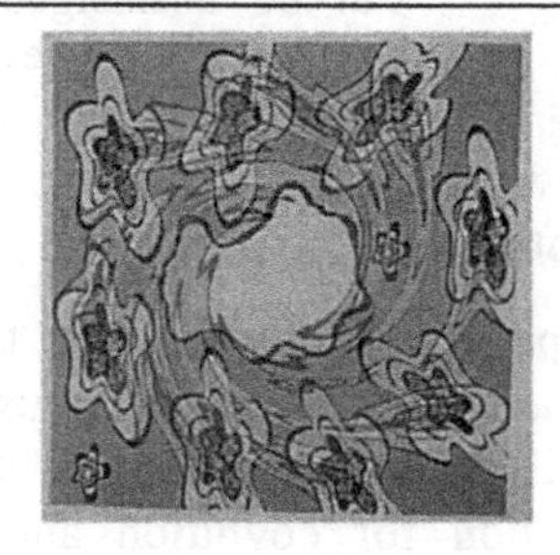 Extract from *Design The Art and Science of the Synthetic* unpublished manuscript by Ömer Akin © |

[1] We used "she" or "her" to refer to all subjects--students and critics--of the annotated studio for the purpose of anonymity. No gender implications are intended.

### 1.2. THE ICE NOTATION AND GENERATION SYSTEM

The ICE system has two major components. (1) A formal notation for representing complex design configurations based on their underlying generative and relational structures. (2) An interactive-generative computational system that supports the interactive exploration of design configurations by means of the constructs of the notation.

The formal notation summarizes any configuration into the set of minimal steps required for its generation, as well as the set of meaningful relationships required for its organization. These generative steps or relationships form the basic units of the configuration's structure, referred to as "regulators". Regulators are combined in various ways to represent the diverse types of structures observed in architectural configurations, for example symmetry, proportion, rhythm, and gradation among many others. In the ICE system, regulators control other design elements: regulators are used as handles to manipulate the design; these also control the changes happening in the design, when it is manipulated by the user. In other words, regulators are higher level entities that "regulate" the behavior of lower level design elements, hence the name regulator.

A regulator encapsulates a formula, by which it computes the positional (or other) attribute of the elements it regulates. A regulator can be a transformation, a constraint, an operation, or a variation. For instance: translations, rotations, reflections, are transformation regulators. These, along with the dilation (scale), and the curve transformations, constitute the primary regulators used for generating shapes and complex configurations. Alignments and containments are constraint regulators. These, among many others, constitute the relational regulators that further control the configuration elements. The section entitled "The ICE Notation Syntax" illustrates how shapes and designs can be generated using regulators, and their corresponding description using the ICE notation.

The ICE system is a 3D exploration tool, which uses regulators to construct design configurations, as well as to interactively manipulate those configurations. Regulators allow the structure (both generative and relational) of configurations to be redefined at any time during the exploration process.

### 1.3. THE INTEGRATION OF THE STUDIO AND THE NOTATION

We integrate these two endeavors to illustrate the potential of the ICE notation for formally describing a realistic design situation, in which the configuration is evolving such as the one presented in the annotated studio. This allows us to use the ICE notation to codify the design process in stages, which are defined by each drawing in the sequence of the design development. We also use the ICE notation to codify the transformations from each drawing in the sequence to the next one. We intend to show that significant design transformations, that would normally be quite labor intensive, can be described in a simple-straightforward way by using our ICE notation and consequently can be easily achieved using the ICE system.

## 2. The ICE Notation's Features and Syntax

The basic elements of the ICE notation are (1) the point, denoted by a lowercase letter, for instance p, and (2) the regulator, denoted by a bold uppercase letter, for instance **T**. Shapes are composite objects defined by points and regulators and are denoted as lowercase words. A prefix for the regulator, a Greek letter, indicates the type of regulator: $\Delta$ transformations, $\Phi$ constraints, $\Xi$ variations, and $\Omega$ operations. Superscripts indicate the subtype for the regulator: for example $\Delta\mathbf{C}^{p}$, $\Delta\mathbf{C}^{h}$, indicates two types of curve regulators (i.e. two distinct formulae). Subscripts (for regulators, shapes, and points) are used for indexing to differentiate elements of the same type for instance $\Delta\mathbf{T}_1$, $shape_3$.

We developed two forms for the ICE notation, a short form, which captures the regulator and the regulated object/s, for instance $\Delta\mathbf{T}(\text{shape})$ and the expanded form, which also shows the parameters of the regulator enclosed in curly brackets with vectors depicted with an overline: $\Delta\mathbf{T}^{1}[\ \{\bar{p},\bar{t},d,n\}\ (\text{shape})\ ]$. These include translation vectors and distances, rotation points and degrees, reflection and glide axes, etc. The long form is essential for system implementation. However, for the purpose of simplicity, we use the short form for all examples in this paper.

The conjunction $\wedge$ (and) is used to join two related clauses which share regulated objects. For instance: $\Delta\mathbf{T}(\text{shape}) \wedge \Delta\mathbf{A}(\text{shape})$.

The ICE notation has the following distributive property: $\Delta\mathbf{T}(\text{shape}_1, \text{shape}_2) = \Delta\mathbf{T}(\text{shape}_1) \wedge \Delta\mathbf{T}(\text{shape}_2)$.

Table 2 shows how the ICE notation is used to describe shapes.

TABLE 2. The ICE notation for generating simple shapes

| | |
|---|---|
| Straight line: $\Delta\mathbf{T}(\text{p})$<br>The translation regulator $\Delta\mathbf{T}$ sweeps the starting point p and generates a straight line. | p T |
| Curved line: $\Delta\mathbf{C}(\text{p})$<br>The curve regulator $\Delta\mathbf{C}$ sweeps p to create a curved line. Subtypes include quadratic and cubic curves. | p C |

| | |
|---|---|
| Plane: $\Delta \mathbf{T}_2(\Delta \mathbf{T}_1(\mathrm{p}))$<br>A plane is generated by the application of successive regulators. The second regulator takes as input all the generated points of the previous regulator. Note that the $\mathbf{T}_1$ regulator is applied first. | |
| Polyline: $\Delta \mathbf{C}(\Delta \mathbf{T}_1(\Delta \mathbf{T}_1(\mathrm{p})_{\#n})_{\#n})$<br>A polyline is also generated by the application of successive regulators, each one taking, as its input, only the last generated point of the previous regulator. The subscript indicates the items taken as input for the next regulator. | |
| Prism: $\Delta \mathbf{T}_3(\text{base}) \wedge \text{base} = \Delta \mathbf{T}_2(\Delta \mathbf{T}_1(\mathrm{p}))$<br>A prism is generated by sweeping a square base along the translation regulator $\Delta \mathbf{T}$. | |
| Pyramid: $\Delta \mathbf{TD}(\text{base})$<br>A pyramid is generated by sweeping a square base along a straight line while incorporating a dilation regulator D. If the scale factor is decreased, the result is a frustum. When two regulators are applied simultaneously, they are denoted as two juxtaposed regulator symbols: $\Delta \mathbf{TD}$. | |
| Sphere: $\Delta \mathbf{R}_2(\text{circle}) \wedge \text{circle} = \Delta \mathbf{R}_1(\Delta \mathbf{T}(\mathrm{p}))$<br>A sphere is generated by sweep-rotating a circle using the rotation regulator $\Delta \mathbf{R}$ | |
| Cylinder: $\Delta \mathbf{T}(\text{circle})$<br>A cylinder is generated by sweeping a circular base along the translation regulator $\Delta \mathbf{T}$. | |
| Cone: $\Delta \mathbf{TD}(\text{circle})$<br>A cone is generated by sweeping a circular base along the translation regulator $\Delta \mathbf{T}$ composed with a dilation $\Delta \mathbf{D}$. | |

The ICE notation supports several ways of generating shapes: continuous generation, discrete generation and in combination. These allows for the description of any sub part of a shape. This feature is indicated by superscripts as illustrated in Table 3.

## 3. A Sample Generative Sequence from the Studio Data

In this section, we present a series of sketches and models created by Subject-W. She is one of the students working on the dormitory housing project. The sequence we present here starts about a quarter of the way into the studio and runs through to the end, highlighting all major formal solutions produced. The data consists of annotations for each day in which the sketch or the model was created. We also include the ICE notation for each graphic display, Table 4. This is a hypothetical description of each design stage after their completion; these designs have been generated by Subject-W independent of the ICE system.

TABLE 3. The ICE notation for generating sub-shape

| | |
|---|---|
| Continuous : Cube = $\Delta\mathbf{T}_3\,(\text{square})^{<0-4>}$<br>The superscript indices enclosed in one set of brackets $^{<0-4>}$ indicate a continuous generation. The connecting line indicates that all indices between the given ranges are generated. | 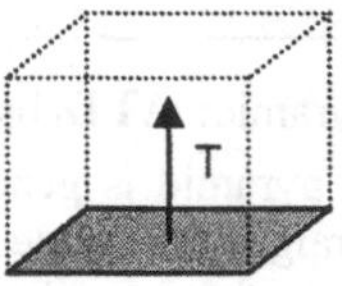 T |
| Discrete: Cube = $\Delta\mathbf{T}_3\,(\text{square})^{<0>-<4>}$<br>The superscript indices, each enclosed in a separate set of brackets $^{<0>-<4>}$ indicate a discrete generation. The discrete generation is used to define complex configurations composed of disjoint, yet related, shapes. |  T |
| Regular polygon: $\Delta\mathbf{R}(\Delta\mathbf{T}((\text{p})^{<0-n>})^{<0>-<4>})$<br>An outline polygon is generated by translating a point to construct a line, then by rotating the line, discretely, to construct the remaining sides. | 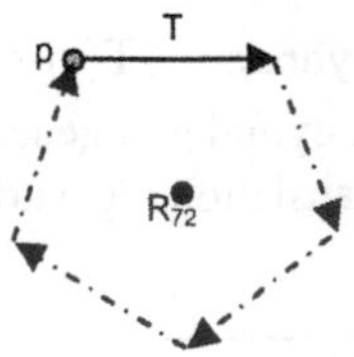 T<br>p<br>$R_{72}$ |
| Combination: Cube = $\Delta\mathbf{T}(\text{p})^{<0-1><3-4>}$<br>Here, the superscript items are grouped by brackets $^{<0-1><3-4>}$ indicating a combined generation. Each group enclosed in a bracket is a continuous group. Notice that in this example, the second index was not generated, therefore, the generated items form a subset of the possible generation of the given $^{<0-4>}$ range. | 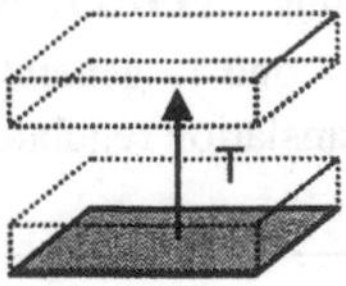 T |

We selected Subject-W's project mainly because she constantly changes directions in her development. It gives us the opportunity to illustrate how the ICE notation represents changes in design exploration paths. In Subject-W's project, the main design constructs that are repeated throughout the drawing sequence are the rooms, the dorm units, the entrances and common spaces. We will use this same relatively high level of abstraction as the main elements encoded in the ICE notation. We chose not to express further details because (i) details have not been resolved in most of Subject-W's drawings and (ii) this level of abstraction is adequate to describe the fairly complex ICE notation.

Let's begin by reviewing Subject-W drawing in Table 1. She describes her housing hierarchy of rooms, units, unit-clusters, wings, buildings, building-clusters by a seemingly unstructured swirling drawing. Nevertheless, we identified relationships between the subparts of this drawing and we describe it using the following ICE regulators: The rotation $\Delta\mathbf{R}$, the curve $\Delta\mathbf{C}$ and the dilation (scale) $\Delta\mathbf{D}$. Assuming that this drawing represents a building, and its individual flower-like objects represent dorm units, and the central form represents the common spaces, we describe it in a top down manner, as follows:

$$\text{building} = \Delta\mathbf{C}_3(\text{dormCluster}) \wedge \text{commonSpace}$$

$$\text{dormCluster} = \Delta\mathbf{RD}(\text{flowerShape})$$

$$\text{flowerShape} = \Delta\mathbf{RD}(\text{room})$$

$$\text{room} = \Delta\mathbf{C}_1(\text{p}_1)$$

$$\text{commonSpace} = \Delta\mathbf{C}_2(\text{p}_2)$$

In the next configuration, Table 4, the form of the dorm units is well defined. To derive the next configuration from the previous one, it is necessary to replace form of the dormCluster, and to replace the curve regulator $\Delta\mathbf{C}_3$ by the sequence of regulators $\Delta\mathbf{M}(\Delta\mathbf{T}(), \Delta\mathbf{R}())$.

The next configuration, Figure B in Table 5, goes back to a curvilinear theme. The main reflection axis $\Delta\mathbf{M}_2$ is maintained. To obtain the curved axis from the previous regulator sequence $\Delta\mathbf{M}_2(\Delta\mathbf{T}(), \Delta\mathbf{R}())$, $\Delta\mathbf{T}$ is deleted and $\Delta\mathbf{R}$'s rotation degree is adjusted, leading to $\Delta\mathbf{M}_2(\Delta\mathbf{R}())$. The horizontal alignment is replaced by an implicit curvilinear alignment along the curve. Within the dorm cluster, the dorm units are repositioned and re-oriented. Their reflection $\Delta\mathbf{M}_1$ axis is rotated and the common spaces become more defined.

TABLE 4. Tuesday, June 4, 2002

Subject-W is presenting a rectilinear scheme in which the modular bays of the dormitory scheme are being clustered to create a large and integrated form on the site which faces a long public edge of the campus proper as well as the service façade of the student activities building. This creates a "beads-on-a-string" type scheme.

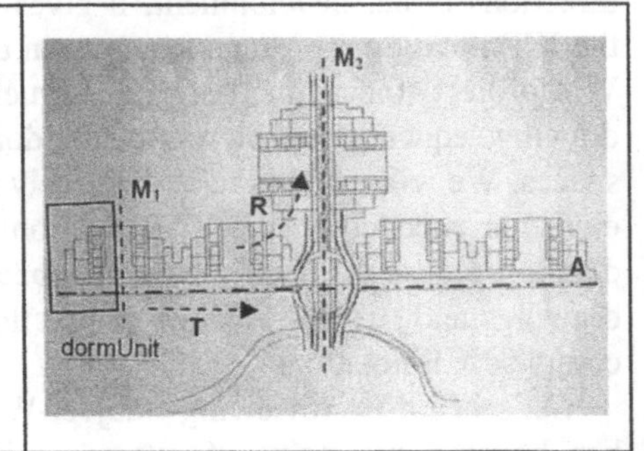


$$\text{dormCluster}_1 = \mathbf{\Delta M_1}(\text{dormUnit}_1)$$

$$\text{building} = \mathbf{\Delta M_2}(\mathbf{\Delta T}(\text{dormCluster}_1), \mathbf{\Delta R}(\text{dormCluster}_2)) \wedge \mathbf{\Phi A}(\text{dormCluster}_1\text{-dormCluster}_4)$$

$$\text{Entrance} = \mathbf{\Delta M_2}(\mathbf{\Delta C}(\text{p}))$$

The building is defined by (1) reflecting the dorm unit to form the dorm cluster, (2) translating and rotating the dorm cluster (3) reflecting the results of the previous generations and (4) aligning (regulator **A**) the dorm clusters (1 to 4). The entrance is defined by sweeping points along a curve then reflecting the curved lines.

TABLE 5. Thursday, June 6, 2002

The beads-on-a-string type arrangement has yielded to a "serpentine" form that curves with the contours, creating a concave edge for the public and a convex one for the private side of the site lot. Intuition seems to guide the form.

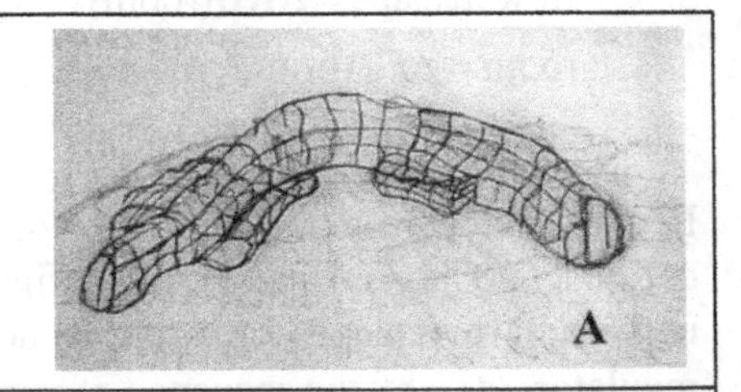


$$\text{building} = \text{serpentine} \wedge \text{commonSpace}_1 \wedge \text{commonSpace}_2$$

$$\text{serpentine} = \mathbf{\Delta C_2}(\mathbf{\Delta C_1}(\text{p}_1))$$

$$\text{commonSpace}_1 = \mathbf{\Delta C_4}(\mathbf{\Delta C_3}(\text{p}_2))$$

The serpentine building form is generated by sweeping a point along a curve, then sweeping the curve along another curve. Both common spaces are generated in the same way.

Here the serpentine form is refined. Curves turn into rotations and the central axis of symmetry from the previous is reinstituted.

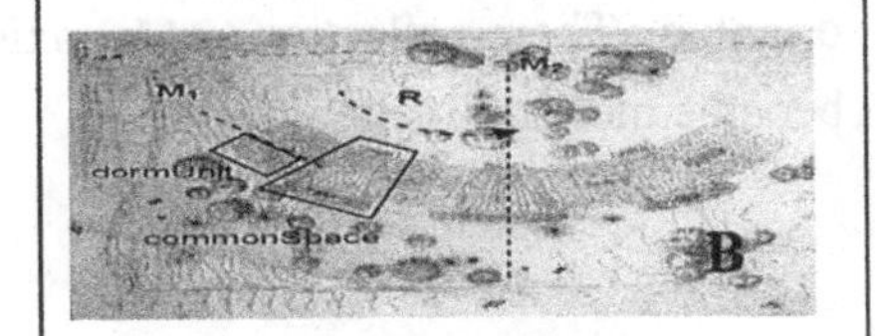

| $\text{dormCluster}_1 = \Delta\mathbf{M}_1(\text{dormUnit}_1)$<br>$\text{building} = \Delta\mathbf{M}_2(\Delta\mathbf{R}(\text{dormCluster}_1)) \wedge \Delta\mathbf{R}(\text{commonSpace})$<br>The building is generated rotating the dorm unit then reflecting it, and by rotating the common spaces (trapezoidal forms) | |
|---|---|

In the next configuration, Table 6, the curve is broken into segments, Reflection is the dominant relationship. The central axis is maintained, but slightly rotated. The $\Delta\mathbf{M}_2(\Delta\mathbf{R}())$ sequence of regulators is replaced by $\Delta\mathbf{M}_2(\Delta\mathbf{M}_3())$.

TABLE 6. Wednesday, June 12

| The next formal overhaul involves one end of the "serpentine" form bifurcating into two wings, allowing the development of a "commons" area and lobby from one of the major access edges of the site. This remains the principal *parti* for Subject-W's solution.<br> | 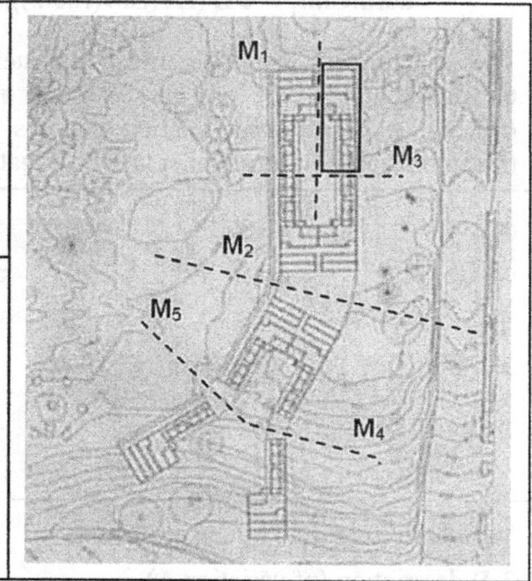 |
|---|---|

3-D description

$$\text{dormCluster}_1 = \Delta\mathbf{M}_1(\Delta\mathbf{TD}(\text{dormUnit}_1))$$
$$\text{building} = \Delta\mathbf{M}_2(\Delta\mathbf{M}_3(\text{dormCluster}_1)_{\#1}) \wedge \Delta\mathbf{M}_4(\text{dormUnit}_4) \wedge \Delta\mathbf{M}_5(\text{dormUnit}_5) \wedge \Delta\mathbf{TD}(\text{commonSpace})$$

The building is generated by reflecting the dorm cluster twice, then reflecting the individual dormUnits to achieve the bifurcations. The 3D configuration is an extrusion of the 2D configuration with a small scaling factor. The $\Delta\mathbf{TD}$ regulator copies the floor slab vertically and scales them. The subscript indicates that the reflection $\Delta\mathbf{M}_2$ only reflects the last element of $\Delta\mathbf{M}_3$ not all the dorm units.

The following two configurations , Table 7, are the variations suggested during the midterm review. Variation A is achieved by rotating the dorm cluster 180 degrees and converting $\Delta\mathbf{M}_6$ and $\Delta\mathbf{M}_5$ into rotations; while variation B is achieved by inserting another dorm cluster in the configuration, which is carried out in the notation by making $\Delta\mathbf{M}_2$ mirror both dorm units generated by $\Delta\mathbf{M}_3$.

TABLE 7. Monday, June 17, 2002, Midterm evaluation of Subject-W's work

During midterm review, Subject-W's work shows little development over the previous critic. The most significant development is the cross axis that marks the secondary entrance, along the long side of the building.

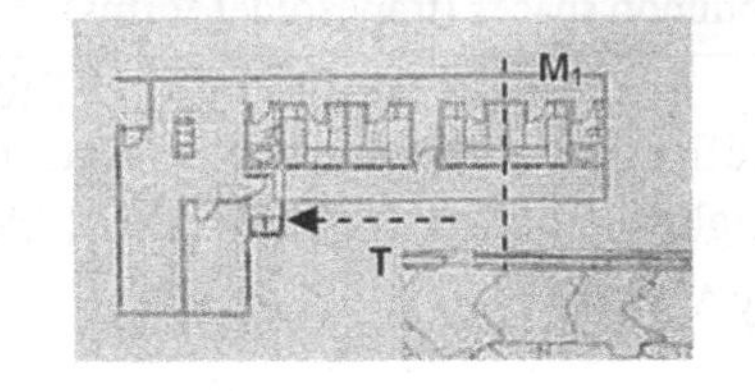


$$\text{dormUnit}_1 = \mathbf{\Phi H}\,(\Delta\mathbf{T}(\Delta\mathbf{M}(\text{room})), \text{kitchen}, \text{bathroom}, \text{balcony})$$

The containment regulator, $\mathbf{\Phi H}$, indicates that the dorm unit consists of (a translation and a reflection of the room) as well as a bathroom, a kitchen and a balcony. A containment relation imposes restrictions on constituents, such as a transformation of the container will propagate the constituents, etc.

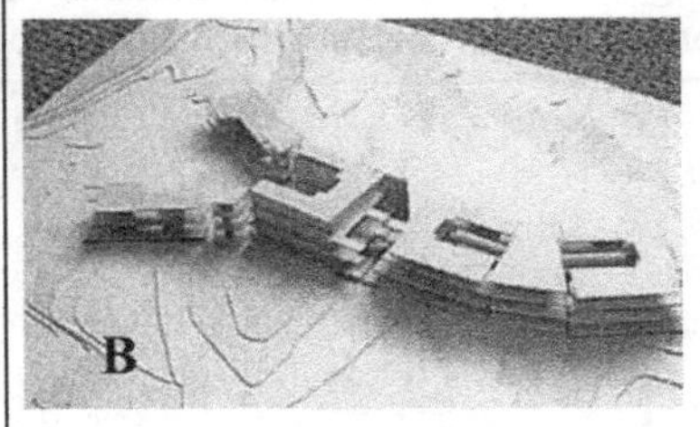


Variations suggested by the professor

3-**D** description (of variation A)

$$\text{dormCluster}_1 = \Delta\mathbf{M}_1(\Delta\mathbf{TD}(\text{dormUnit}_1))$$
$$\text{building} = \Delta\mathbf{M}_2(\Delta\mathbf{M}_3(\text{dormCluster}_1)_{\#1}) \wedge \Delta\mathbf{R}_4(\text{dormUnit}_4) \wedge \Delta\mathbf{R}_5(\text{dormUnit}_5) \wedge \text{commonSpace}$$

3-**D** description (of variation B )

$$\text{dormCluster}_1 = \Delta\mathbf{M}_1(\Delta\mathbf{TD}(\text{dormUnit}_1))$$
$$\text{building} = \Delta\mathbf{M}_2(\Delta\mathbf{M}_3(\text{dormCluster}_1)) \wedge \Delta\mathbf{M}_4(\text{dormUnit}_4) \wedge \Delta\mathbf{M}_5(\text{dormUnit}_5) \wedge \text{commonSpace}$$

The configuration in Table 8 shows a return to the curvilinear axis and the rotation of the dorm clusters. To achieve this configuration from the midterm configuration, Table 6, the bifurcation mirrors $\Delta\mathbf{M}_4$ and $\Delta\mathbf{M}_5$ are removed and $\Delta\mathbf{M}_3$ is replaced by a rotation $\Delta\mathbf{R}$. Although, this is speculation, it appears that Subject-W has created this configuration not by developing the midterm solution, but by working from the drawings in Table 5, while pair-wise integrating the common spaces from the midterm's configuration.

The configuration in Table 9 is a further development of the previous one, focusing on the redefining the lower section. Axial symmetry is still maintained.

TABLE 8. Friday, June 21, 2002

<table>
<tr><td>A new aspect of the scheme emerges. Precedent exploration based on Sant'Elias' work pushed the scheme towards "futuristic" features. The result however is not promising since the new synthesis appears to have a cartoonish resemblance to architecture. Drawings lack architectonic qualities, such as material, construction and structural specificity.</td><td>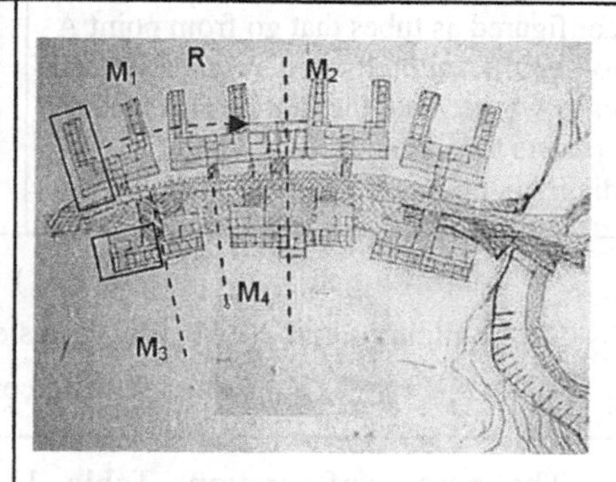
</td></tr>
<tr><td colspan="2">$\text{dormCluster}_1 = \Delta\mathbf{M}_1(\text{dormUnit}_1)$<br>$\text{dormCluster}_2 = \Delta\mathbf{M}_4(\Delta\mathbf{M}_3(\text{dormUnit}_2)_{\#1})$<br>$\text{building} = \Delta\mathbf{M}_2(\Delta\mathbf{R}(\text{dormCluster}_1), \text{dormCluster}_2, \text{commonSpace})$</td></tr>
</table>

TABLE 9. Monday, June 24, 2002

<table>
<tr><td>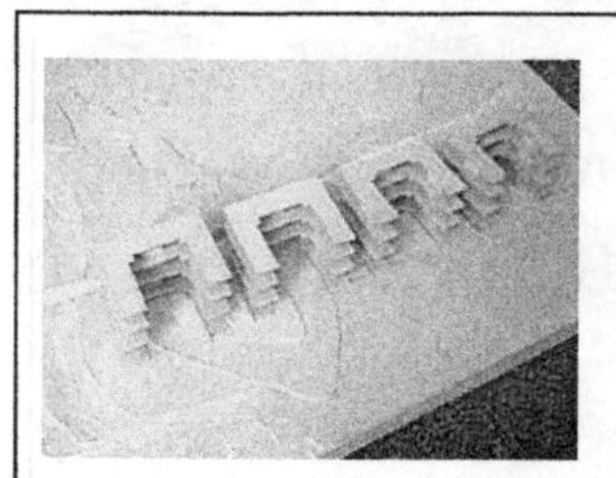<br>This is a mixed bag. While the dorm units gain architectonic clarity, the main entrance, circulation and commons areas continue to resemble "spaghetti".</td><td>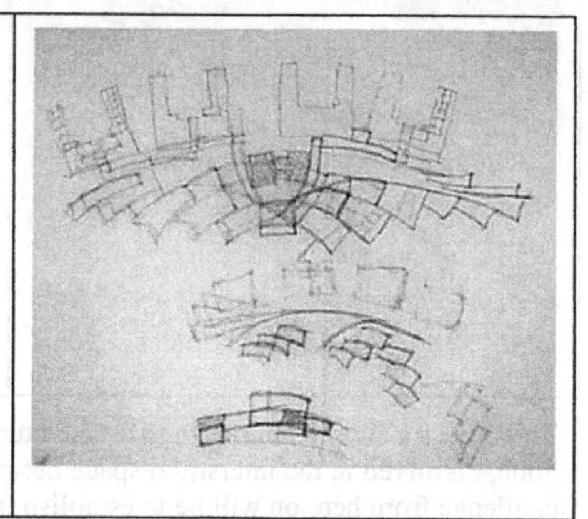</td></tr>
<tr><td colspan="2">$\text{dormCluster}_1 = \Delta\mathbf{M}_1(\text{dormUnit}_1)$<br>$\text{building} = \Delta\mathbf{M}_2(\Delta\mathbf{R}(\text{dormCluster}_1), \text{dormCluster}_2, \text{spagettiSpace})$</td></tr>
</table>

The general structure of the next configuration, Table 10, is identical to the previous configuration, but the detail of the units for the lower dorm cluster is changing. The rotation $\Delta\mathbf{R}$ is replaced by a reflection $\Delta\mathbf{M}_3$.

TABLE 10. Wednesday June 26, 2002

| | |
|---|---|
| This submission continues along the same lines as before. The "spaghetti" scheme dominates the formal development. Circulation paths are configured as tubes that go from point A to point B, without circulation and social hubs worked into the fabric. The scheme appears to be an inside-out path diagram, not architecture. | 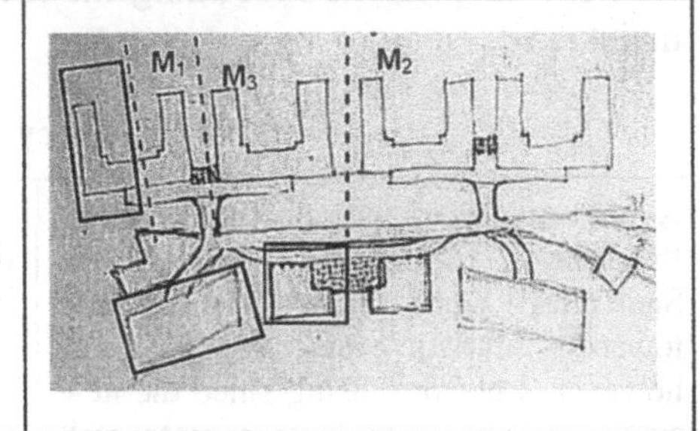 |
| $\text{dormCluster}_1 = \Delta\mathbf{M}_1(\text{dormUnit}_1)$<br>$\text{building} = \Delta\mathbf{M}_2(\Delta\mathbf{M}_3(\text{dormCluster}_1), \text{dormCluster}_2,$<br>$\text{dormCluster3}, \text{commonSpace})$ | |

The next configuration, Table 11, is a further development of the previous one, but maintains the same.

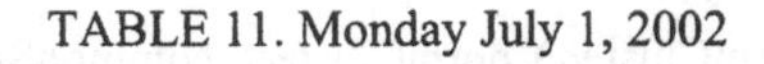

TABLE 11. Monday July 1, 2002

| | |
|---|---|
|  | 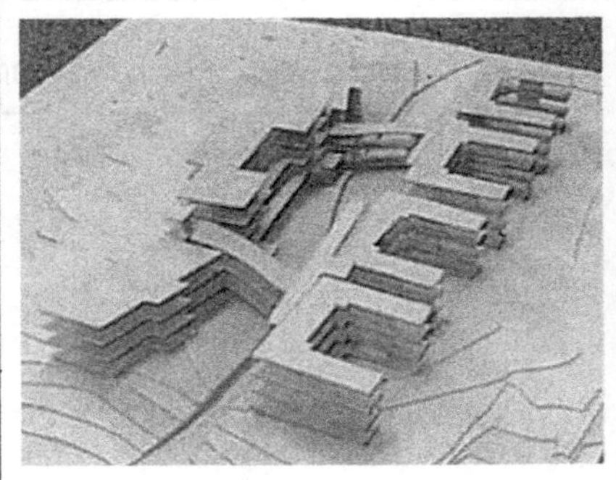 |
| This marks a significant return to architecture and architectonics. The "spaghetti" is gone, dissolved in the interstitial space between two parallel dorm wings. The challenge from here on will be to establish the dialogue between these two spines and the cladding of the open space between them. | |

$$\text{dormCluster}_1 = \Delta\mathbf{M}_1(\text{dormUnit}_1)$$
$$\text{building} = \Delta\mathbf{M}_2(\Delta\mathbf{M}_3(\text{dormCluster}_1), \text{commonSpace}) \wedge \Delta\mathbf{M}_2(\text{dormCluster}_2, \text{circulation})$$

The next configuration, Table 12, is a slight development in the dorm cluster, where a glide relationship, depicted by $\Delta\mathbf{G}$, is explored.

TABLE 12. Wednesday, July 3, 2002

This brings issues of complexity vs. simplicity to the table. The scheme does a few things well. Other building systems solutions can be layered on top of it, which remains a challenge for Subject-W.

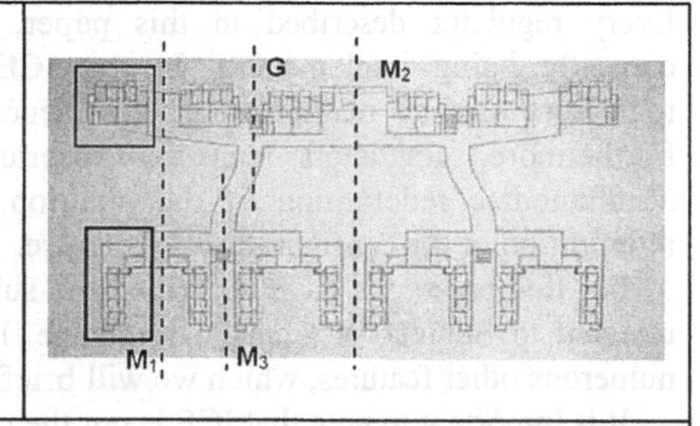


$$\text{dormCluster}_1 = \Delta\mathbf{M}_1(\text{dormUnit}_1)$$
$$\text{building} = \Delta\mathbf{M}_2(\Delta\mathbf{M}_3(\text{dormCluster}_1), \text{commonSpace}) \wedge \Delta\mathbf{M}_2(\Delta\mathbf{G}\ (\Delta\mathbf{M}_1(\text{dormCluster}_2))_{\#1}) \wedge \Delta\mathbf{M}_2(\text{circulation})$$

The final configuration, Table 13, still shows some exploration in the dorm clusters. This time the units are slightly sheared. The reflection and glide regulator sequence $\Delta\mathbf{G}\ (\Delta\mathbf{M}_1())$ is replaced by a composition of translation and shear $\Delta\mathbf{TS}$ regulators.

TABLE 13. Tuesday, July 10, 2002

Two days prior to the final review there are still basic issues of development and resolution. The final review does not bring any surprises or further development of the scheme.

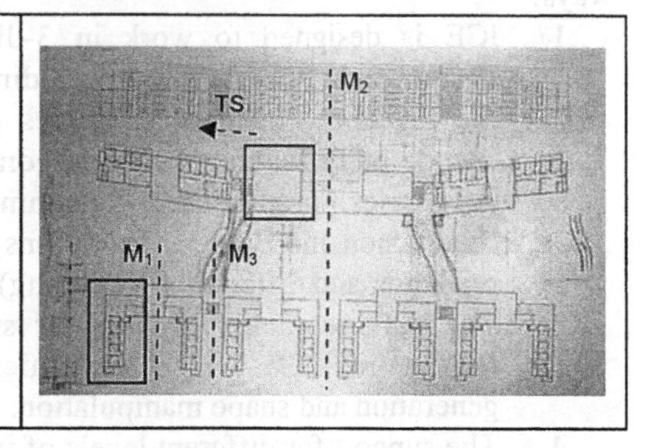

$$\text{dormCluster}_1 = \mathbf{\Delta M_1}(\text{dormUnit}_1)$$
$$\text{building} = \mathbf{\Delta M_2}(\mathbf{\Delta M_3}(\text{dormCluster}_1), \text{commonSpace}) \wedge$$
$$\mathbf{\Delta M_2}(\mathbf{\Delta TS}\ (\text{dormCluster}_2))_{\#1} \wedge$$
$$\mathbf{\Delta M_2}(\text{circulation})$$

## 4. Implications of the Notation

The ICE notation is designed parallel to a generative/manipulation system. Every regulator described in this paper, has been implemented or is currently being implemented. In the ICE system, every parameter of regulators can be manipulated, thus generating highly flexible models. Furthermore, regulators can be inserted, deleted, or replaced to accommodate redefinition of the notation string, and consequently, the redefinition of the configuration's structure.

For this paper, we have described the subset of the ICE notation that is relevant to Subject W's design sequence. However, the ICE notation has numerous other features, which we will briefly review in this section.

It is important to note that ICE is not the only notation in its class. Leyton (2001) developed a generative theory of shape, which uses the same principles of mathematics as ICE. Leyton's work focused on the mathematical theory of shape generation, while the ICE system/notation focuses on the practical aspects of implementation and usability, the most important of which the ability to manipulate the configurations generated by ICE. Leyton also addressed the issue of process-capture, which he refers to as recoverability; an issue that will be revisited it in the following section. Cha and Gero (2001) have developed a shape schema representation, based on Isometry transformations and used it to describe numerous buildings of notable architects. In addition to being part of a computational system, the ICE notation extends the aforementioned representations in the following ways:

1. ICE is designed to work in 3-dimensions. All parameters and operations in ICE are based on 3-dimentional geometry principles.
2. The regulator construct subsumes generative transformations, and encode other functions such as constraints (for instance alignment, boundaries, proportions, containments), operations (for instance subdivision and Boolean operations) and variations (such as rhythm, gradation and differential sweeping).
3. The ICE notation's support for sub-part generation is a unique feature of ICE which magnifies the possibilities for shape generation and shape manipulation.
4. The support for different levels of information greatly simplifies the representation. The short and long forms allow the ICE notation

string to be viewed at two crucial levels of abstraction (relational level, and parameterization level). The shape encapsulation feature in the ICE notation helps structure the string and avoids redundancy in descriptions.

Another distinction of our approach is that we are using a formal notation to represent an evolving design. Most attempts to formalize design representations have either encoded completed architectural design, or encoded hypothetical designs. Our paper, on the other hand, has addressed a more challenging task, a constantly changing design, which may be imperfect and often incomplete, but nevertheless illustrates a natural progress of a student at work. With the ICE notation, we codified each drawing in the development sequence as well as each transformation from one drawing to the next, thus demonstrating that the ICE notation can follow a student's exploration path.

## 5. Implications for Process and Cognitive Analysis

In the area of cognitive models of the design process one of the difficult challenges is to formally measure and compare intermediary states in a design state space and draw generalizations about human design behavior (Akin 1996). Purcell et al. (1994) have made progress in this direction. They have devised ways of unambiguously codifying and characterizing individual design activities throughout design protocols. Their codification relies on interpretations by human coders of the data and like most qualitative analysis methods on entities that emerge from the data.

The contributions of this paper are as follows: (1) we use a notation that is formal and unambiguous, (2) we use an a priori notation to code the data that is not derived from the data, and (3) we use a graphic representation that is capable of encoding both graphic entities and process transformations.

Formality of the ICE notation enables us to show quantifiable differences in the information content of both the design state representations and their transformations. The ICE notation supports of multiple descriptions for the same configuration. Therefore, it captures, for each description, different processes for generation, different applicable transformations, and consequently, different manipulation handles. For example, consider the graphic sequences in Table 14. This is subject W's midterm submission, see Table 6, as it is generated in an early version of the ICE system using two distinct generation paths, and consequently yielding different ICE notation strings. In steps 1 and 2, a dorm unit is created then reflected about $\Delta\mathbf{M}_1$. In step 3, the same arrangement is obtained (step 3A) by a reflection about $\Delta\mathbf{M}_2$, and in (step 3B) a rotation about $\Delta\mathbf{R}_1$. The generation sequence continues in distinct paths though steps 4 and 5, yielding different arrangements. In step 6 however, two different actions, reflecting about $\Delta M_5$ and reflecting about $\Delta M_6$, bring the arrangement back to equivalence. At this point the two shapes are identical, but their notation is not since the notation also captures the way in which each shape was generated. We can formally show this in the notation and unambiguously express in the final shape the difference(s) between the two generative sequences, Table 15.

TABLE 14. Generative sequence for subject W's midterm submission

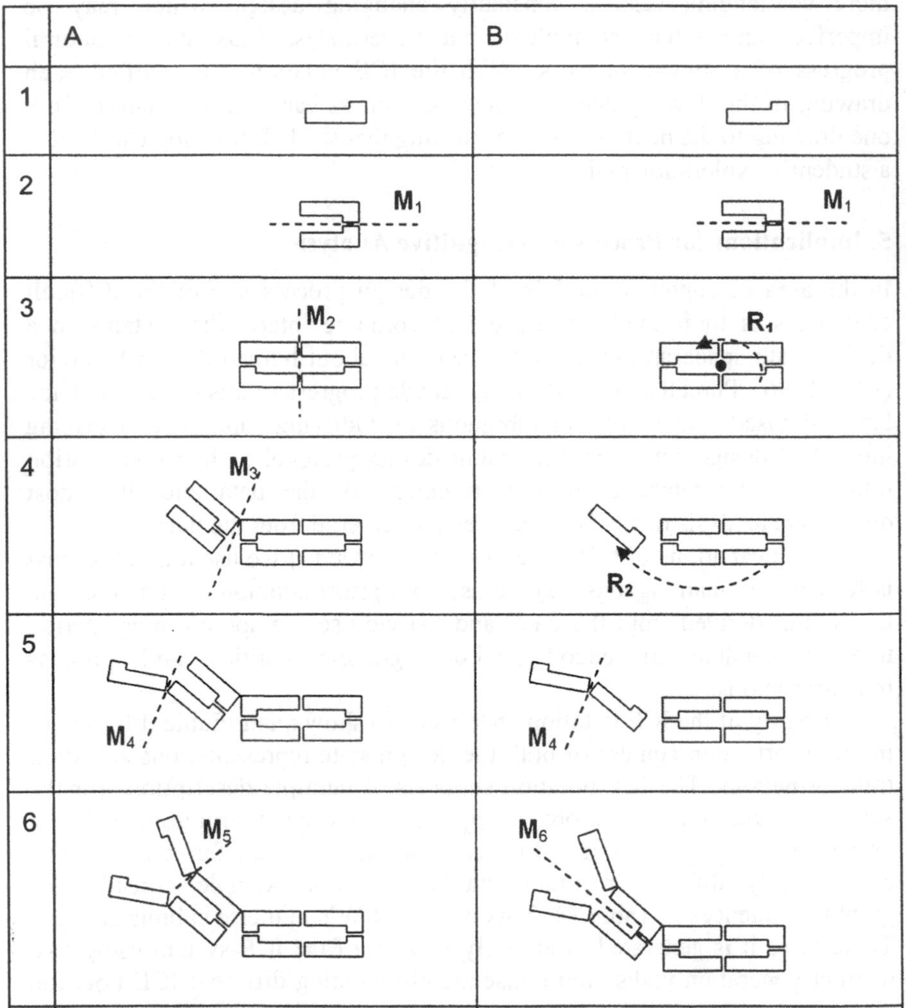

| | A | B |
|---|---|---|
| 1 | | |
| 2 | $M_1$ | $M_1$ |
| 3 | $M_2$ | $R_1$ |
| 4 | $M_3$ | $R_2$ |
| 5 | $M_4$ | $M_4$ |
| 6 | $M_5$ | $M_6$ |
| | $\Delta M_3(\Delta M_2(\Delta M_1(\text{dormUnit}_1)))$<br>$\Delta M_4(\text{dormUnit}_4)$<br>$\Delta M_5(\text{dormUnit}_5)$ | $\Delta R_1(\Delta M_1(\text{dormUnit}_1))$<br>$\Delta M_6(\Delta M_4(\Delta R_2(\text{dormUnit}_1)))$ |

| | | |
|---|---|---|
| 7 | $M_1$ | $M_1$ |
| | Move mirror $\Delta \mathbf{M}_1$ upward | Move mirror $\Delta \mathbf{M}_1$ upward |
| 9 | $M_1$ | $M_1$ |
| | Rotate mirror $\Delta \mathbf{M}_1$ counterclockwise | Rotate mirror $\Delta \mathbf{M}_1$ counterclockwise |
| | $M_3$ | $R_1$ |
| | Rotate mirror $\Delta \mathbf{M}_3$ counterclockwise | Move rotation point $\Delta \mathbf{R}$ to the right |

## 6. Implications for Intent Capture and Cognition

This capability in the ICE notation allows us to not only encode graphic and generative design sequences but also to "replay" them in the way the graphic entities were generated in the first place. This has advantages in assisting designers to visualize the genesis of a form. This can be helpful in encoding not only design histories but also the design intent. The subcomponents that make up a graphic element can help retrieve the functional requirements that go into the final form.

Secondly, we can formally and quantitatively measure the information content of each state in the state space of design representations. We can go beyond surface similarities of graphically equivalent entities and measure the steps and stages that went into creating each one. This can be used to quantify the information content of graphic designs. One goal for doing this would be to determine more parsimonious ways of producing forms.

Another purpose which is orthogonal to the first is to be able to embed handles (or structure) into shapes for further manipulation.

TABLE 15. Graphic, generative and manipulation equivalencies between sequences A and B of Table 14

| Steps | Graphics Information | Generative Information | Manipulation information |
|---|---|---|---|
| 1 | equivalent | equivalent | |
| 2 | equivalent | equivalent | |
| 3 | equivalent | NOT equivalent | |
| 4 | NOT equivalent | NOT equivalent | |
| 5 | NOT equivalent | NOT equivalent | |
| 6 | equivalent | NOT equivalent | |
| 7 | NOT equivalent | NOT equivalent | equivalent |
| 8 | NOT equivalent | NOT equivalent | equivalent |
| 9 | NOT equivalent | NOT equivalent | NOT equivalent |

For example the resulting form (step 6) in Sequence B of Table 14 has different handles than the same one in Sequence A. This has two significant results, which are illustrated in steps 7-9 of Table 14 (1) Identical manipulation-actions (for instance moving shared regulators) would result in totally different graphic configurations, step 7 and 8. (2) The different handles (non-shared regulators) allow for a different set of manipulations per graphic configuration, such as the ones shown in step 9 (moving the rotation point $\Delta\mathbf{R}_1$ or rotating the mirror line $\Delta\mathbf{M}_3$.

There are numerous possible manipulations for each sequence; those shown were just a few. Additionally, redefining the notation string by insertion, deletion, or replacement would expand the manipulation possibilities even further and redirect the exploration paths.

These types of interactions within the ICE system suggest debatable questions regarding cognition. Suppose Subject-W had a system such as ICE, would she have followed the same exploration paths as she did in the annotated studio? Would she have explored other paths and came up with different configurations? Would she have completed the exploration faster, thus giving her more time to develop details further, or would she have done many more explorations, thus sidetracking from developing the focused completed design? The more general question is whether such capabilities offer a relief in cognitive loads, or places an additional burden on the designer of understanding the structure handles and their manipulations.

Furthermore, we can codify all graphic entities by surface structure and generative structure. We believe that this has important consequences for codifying and analyzing cognitive representations of designs. Our future work in this direction will be to codify protocol (not ethnographic) data to exploit this possibility. We will also compare our approach to others in the field such as those of Purcell, *et.al*, (1994) and Suwa, *et.al.* (1998).

## 7. Potential Application Areas

The discussion provided in the above section explores the functionalities that ICE affords (or would be able to afford) us, more or less independent of the specific design applications or even specific design problems. Yet, there remain questions about how these functionalities could impact the world of computational applications. What are the potential benefits of having multiple representations of the genesis of designs? How can these descriptive techniques be helpful in prescriptive strategies in design? Are these capabilities best used in *post facto* or generative descriptions of designs?

The fact that ICE specifies *multiple ways* of creating the identical graphic entity affords us many distinct representations for each entity. We can form a rectangle by sweeping a point into a line and the line into a surface. Alternatively, we can create the same rectangle by mirroring exactly half of it along a symmetry axis. Multiple representations enable us to capture precisely the manner in which an entity is created as well as what it is. This leads to interesting design application opportunities. Can we capture different ways of making shapes that are preferred by different users? Do these correspond to drawing performance measures such as: faster, easier, consistent with the geometry of the form, and so on? While, currently, we do not have sufficient data to answer these questions, they present interesting future research avenues. Such investigations may lead to generic and customizable approaches to making graphic elements for design.

Another important goal of this approach is to develop tools that go beyond the *descriptive* accounts of design processes and assist in *prescriptive* design strategies. One of the ways this can be accomplished is through design libraries. A collection of ICE notations can constitute a library of design elements that can be used to create new design assemblies. As with a case base, this library can be used to adapt past designs and design elements to new problems. The difficulty with most case- or library-based system is the excessive overhead of populating the library or the case-base with design instances. The effort needed to build libraries is so large that most of them have impoverished instance sets. One way of overcoming this problem is to capture cases during design. ICE is perfectly suitable for this approach. It's process capture functionality can be adapted to an interactive format for the designer to store away, on the fly, instances as they design them.

Furthermore, the ICE notation can become the basis of a tool to capture *design history*. As in most complex design environments the history of designs (i.e., why some things are configured the way they are) is the most difficult requirement to satisfy with available representations. Drawings capture the "what," and the specifications capture the "how" of building designs. There are no representation systems designed to deal with the "why." We believe ICE is a natural to fill this gap. With ICE one can play back the sequence of entities created, down to the last line or point of a graphic entity, however complex they may be. We believe this will become the armature, much more effectively than any static design representation, to capture the history of design entities; to provide information about the formal genesis of each design component; to enable the modification of the design without losing information about its history; and to augment this history while preserving the design.

Finally, we believe all of these functionalities are essential for building CAD applications that capture formal design *intent* effectively and efficiently. Our future research will address some if not all of these issues.

## References

Akin, O: 2002, *Design: The Art and Science of Synthesis,* Unpublished manuscript, School of Architecture, Carnegie Mellon University, Pittsburgh, PA 15123, USA

Akin, O and Lin, CT: 1996, Design protocol data and novel design decisions, *in* N Cross and K Dorst (eds), *Analysing Design Activity*, John Wiley, pp. 35-64.

Cha, M, Gero, J: 2004, Shape pattern representation for design computation, *(http://www.arch.usyd.edu.au/%7Ejohn/publications/progress.html)*

Leyton, M: 2001, *A Generative Theory of Shape*, Springer Verlag, New York.

Moustapha, H: Forthcoming, A formal representation for generation and transformation in design, *GCAD'04 Symposium Proceedings,* Carnegie Mellon University, Pittsburgh...

Moustapha, H, Krishnamurti, R: 2001, Arabic calligraphy: A computational exploration, *Mathematics and Design 2001, Third International Conference*, Deakin University, Geelong.

Purcell, T, Gero, J, Edwards, H and McNeill, T: 1994, The data in design protocols: The issue of data coding, data analysis in the development of models of the design process, *in* JS Gero and F Sudweeks (eds), *Artificial Intelligence in Design'94*, Kluwer, pp. 225-252.

Suwa, M, Purcell, T, and Gero, JS: 1998, Macroscopic analysis of design process based on a scheme for coding designers' cognitive actions, *Design Studies* **19**(4): 455-483.

JS Gero (ed), *Design Computing and Cognition'04*, 197-215


# STYLES, GRAMMARS, AUTHORS, AND USERS

ANDREW I-KANG LI
*The Chinese University of Hong Kong, Hong Kong SAR, China*

**Abstract.** Using a grammar to understand style can be seen, for students, as involving two main tasks: cultivating a standard of stylistic correctness and converging the language defined by the grammar and the language of stylistically correct designs. This paper discusses a framework for organizing such an experience, considers how it informs the way we write grammars, presents an example (including a grammar), and reports on a classroom experience.

## 1. Introduction

Students and scholars of architecture and other visual arts often seek to understand a style based on a sample of designs in that style. What this understanding involves is explained by Stiny and Mitchell (1978b, p17, original emphasis) in the following way.

> When several buildings each create a similar impression, they are said to exemplify a particular architectural *style*. Given a finite corpus of buildings that are perceived to be alike in some sense, the problem of style consists of characterizing the basis for this likeness. Ideally this characterization has three main purposes: (1) it should clarify the underlying commonality of structure and appearance manifest for the buildings in the corpus; (2) it should supply the conventions and criteria necessary to determine whether any other building not in the original corpus is an instance of the style; and (3) it should provide the compositional machinery needed to design new buildings that are instances of the style.

These three functions can be performed by a suitable generative description of the language of designs in that style. One mechanism for articulating such a description is a shape grammar, and indeed shape grammars have been used to understand a variety of styles, from Chinese lattices (Stiny 1977) and Japanese tea houses (Knight 1981) to Wren's churches (Buelinckx 1993) and Wright's prairie houses (Koning and Eizenberg 1981).

In these and similar cases, the grammar is a hypothesis: it is based on finite empirical evidence (the corpus), accounts for that evidence (requirement 1 above), and makes predictions (requirements 2 and 3) that can be tested. And it is subject to revision, according to the correctness of its predictions.

A scholar of style tests these predictions against some standard of *stylistic correctness.* In this respect, she is like a scientist, who performs experiments which manipulate nature into evaluating her predictions, or a linguist, who asks native speakers to evaluate sentences produced by his grammar.

There is an important difference, however: stylistic correctness is almost never objective. One exception is Duarte's (2001) study of the style of a living architect. That a "native stylist" – in this case Alvaro Siza – evaluates the designs makes this study uniquely compelling.

More usually, though, no such external authority exists, and it is the author who determines stylistic correctness. She does this based on her knowledge and experience, and writes the grammar accordingly. The user studies and uses the grammar, and understands more about the style (Knight 1999–2000, "Authoring shape grammars").

But this is the style as understood by *the author.* The user does not necessarily learn how the author arrived at that understanding. Where does stylistic correctness come from? Who decides, and following what criteria? I believe that these questions are important, especially to students, who sometimes think that there is a single unambiguously correct version of a style and that their task is to learn what it is.

They should know instead that stylistic correctness can vary from person to person. As Stiny and Mitchell (1978a, p192) point out, different people "may wear different glasses." In fact, stylistic correctness can vary in the same person. It may change or develop, especially as that person sees more designs or gains more knowledge and experience.

Indeed, the most compelling aspect of stylistic correctness may be precisely that it is derived not just by observing the immediate corpus, but also by appealing to any knowledge and experience that the observer finds relevant. It matters not whether it is history, structures, function, or even a gut feeling; it is for the observer to interpret.

Thus in the end stylistic correctness is subjective. And, since stylistic correctness determines their performance, style grammars are also subjective. For any given corpus, there are as many "correct" grammars – grammars that satisfy Stiny and Mitchell's three requirements – as there are interpretations of stylistic correctness, and as many or more as there are observers.

That stylistic correctness is subjective is not obvious when connoisseurship is commonly offered as an authority; given that grammars are objective, it is even paradoxical. But to juxtapose these two tasks – composing a grammar and determining stylistic correctness – is to highlight both the limits of objective formalization and the inevitability of subjective interpretation.

We can teach students that to understand style is to interpret and apply their own knowledge and experience, and we can do it with shape grammars. In this paper, I describe a framework for just this purpose.

Briefly put, the teacher examines the corpus and formulates a working grammar, which students use to generate designs. Each student evaluates the designs, revises the grammar accordingly, and repeats until the grammar generates all and only those designs which he judges stylistically correct. In the process of evaluating the designs, the student draws on several sources of information: the corpus, the teacher's assumptions (as embodied in the grammar), and, most important, his (the student's) own knowledge and experience (for example, of history, structures, or architectural function).

Thus students have two tasks: to develop a standard of stylistic correctness, and to converge the language defined by the grammar *(the working language)* and the language of stylistically correct designs *(the target language)*. Plotting such a scenario means considering how the users – their backgrounds and agendas – affect the grammar. As we will see, there can be many points of influence.

## 2. Examining the Corpus

The object of study is building sections according to a twelfth-century Chinese building manual. This manual, the *Yingzao fashi,* was compiled by Li Jie (died 1110), court architect to the Huizong emperor (reigned 1101–1126) of the Song dynasty (960–1127), with the goal of setting standards and reducing corruption in official construction.[1]

In most cases, Li's instructions were not enumerative but generative. Liang Sicheng, the pioneering scholar of the *Yingzao fashi,* found this so remarkable that he called the manual "a grammar book of Chinese architecture" (Liang 1984, 358).

However, in some cases, Li did in fact make lists; for building sections, for example, he provided drawings and descriptions of a limited number of sections.[2] In this example we are concerned with the section of only the *ting* hall, one of two main building types. Li's information on *ting* hall sections consists of a corpus: 18 drawings, each with a written description, Figure 1.

[1] For more in English on the *Yingzao fashi,* see Glahn (1984) and Guo (1999). In Chinese, see Liang (1983) and Chen (1993).

[2] The main façade of a Chinese buildings is on the long side. Thus the section of interest is the short one, perpendicular to the main façade and parallel to the approach.

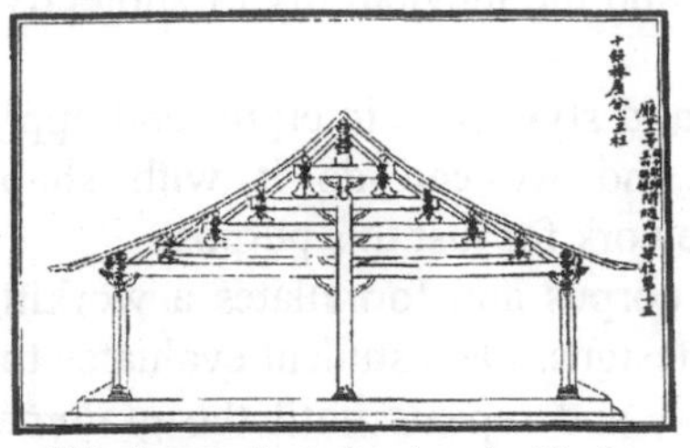

a. *Shijia chuan wu, fen xin, [yong] san zhu.* 10-rafter building, centrally divided, [with] 3 columns.

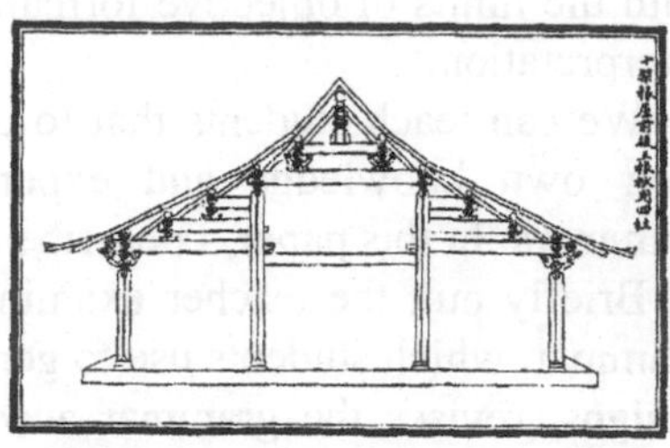

b. *Shijia chuan wu, qian hou sanchuan fu, yong si zhu.* 10-rafter building, a 3-rafter beam in front and in back, with 3 columns.

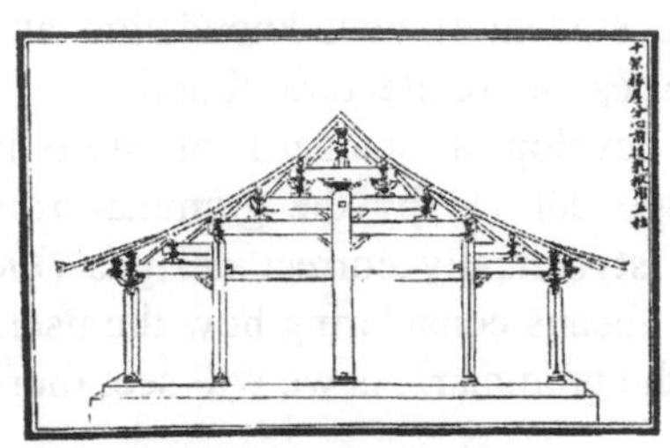

c. *Shijia chuan wu, fen xin, qian hou rufu, yong wu zhu.* 10-rafter building, centrally divided, a 2-rafter beam in front and in back, with 5 columns.

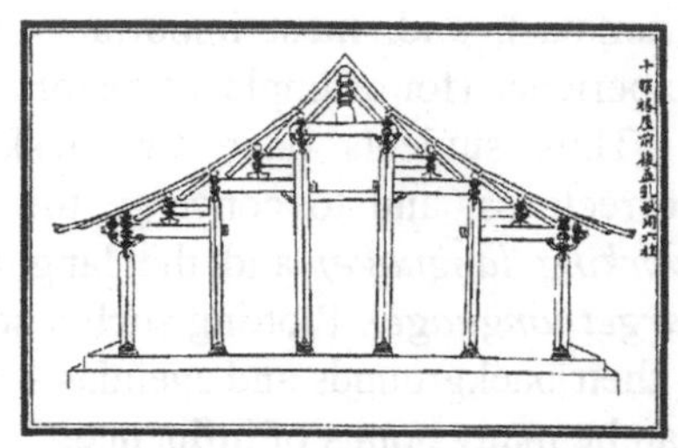

d. *Shijia chuan wu, qian hou bing rufu, yong wu zhu.* 10-rafter building, 2 2-rafter beams in front and in back, with 5 columns.

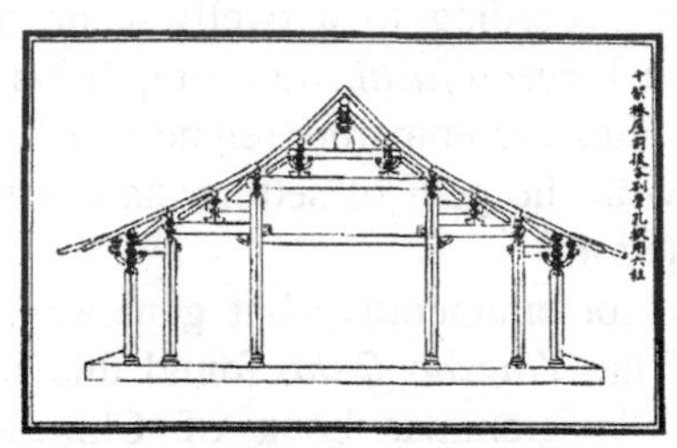

e. *Shijia chuan wu, qian hou ge zhaqian rufu, yong liu zhu.* 10-rafter building, 1- and 2-rafter beams both in front and in back, with 6 columns.

*Figure 1.* The 18 *ting* hall sections with descriptions (10-rafter buildings) (a) to (e) (Liang 1983)

Here I make some observations on the designs in the corpus. I do this in detail, because observing is the first step in moving from individual examples to thoughts about style: from observations emerge the assumptions that are embodied in the grammar. They are my answer to the question that students should ask: *Why these rules and not others?* Different assumptions are possible, and they lead to different rules.

## 2.1. STRUCTURAL ORGANIZATION

First, some principles of structural organization. Rafters *(chuan)* are not, as in the west, single components spanning from ridge to eaves. Rather, they are segmented and, in horizontal projection, equally long (except at the eaves, where they are longer). Because they are segmented, they can form a curved roof section; this section is calculated by the procedure known as *juzhe* (Liang 1983), which is the best example of Li's generative approach.

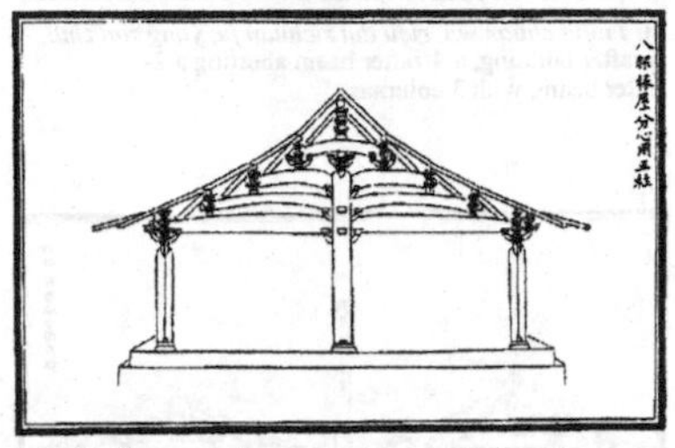

f. *Bajia chuan wu, fen xin, yong san zhu.* 8-rafter building, centrally divided, with 3 columns.

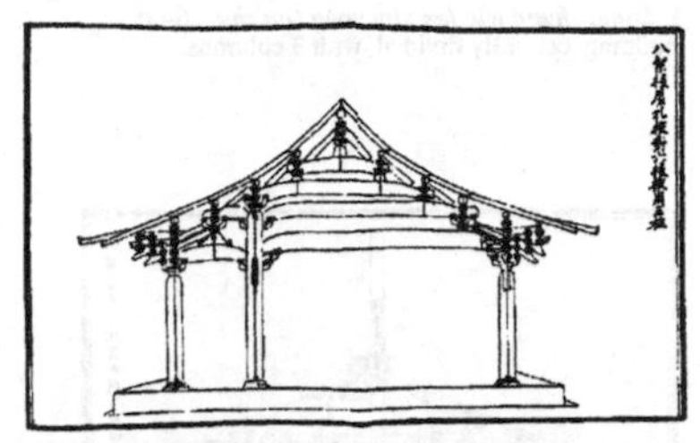

g. *Bajia chuan wu, rufu dui liuchuan fu, yong san zhu.* 8-rafter building, a 2-rafter beam abutting a 6-rafter beam, with 3 columns.

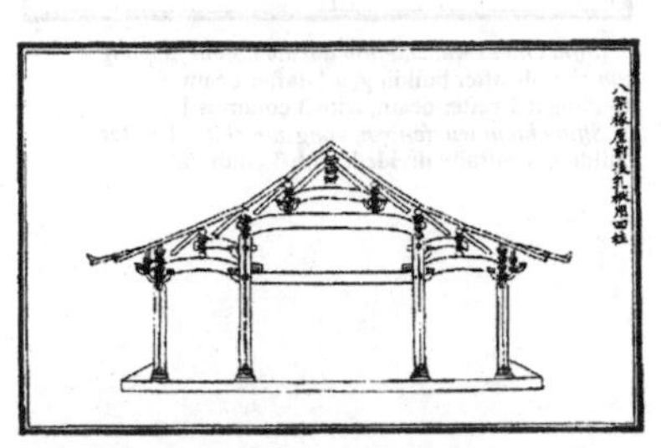

h. *Bajia chuan wu, qian hou rufu, yong si zhu.* An 8-rafter building, a 2-rafter beam in front and in back, with 4 columns.

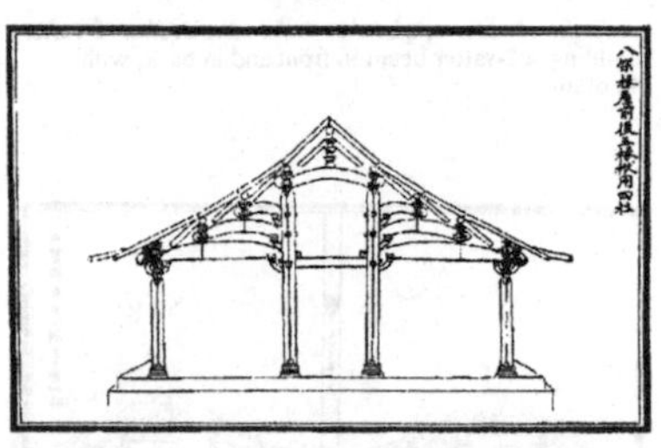

i. *Bajia chuan wu, qian hou sanchuan fu, yong si zhu.* 8-rafter building, a 3-rafter beam in front and in back, with 4 columns.

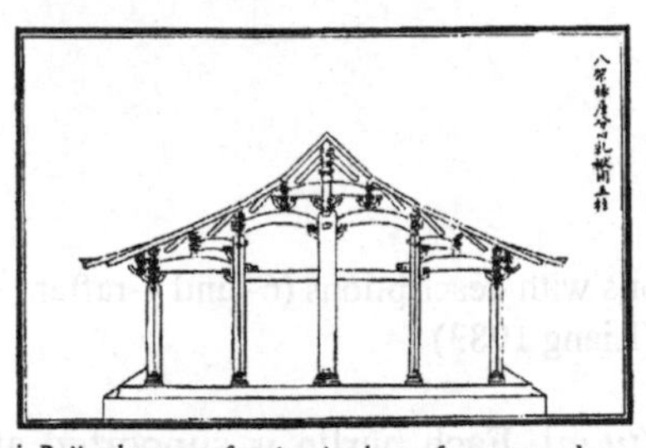

j. *Bajia chuan wu, fen xin, qian hou rufu, yong wu zhu.* 8-rafter building, centrally divided, a 2-rafter beam in front and in back, with 5 columns.

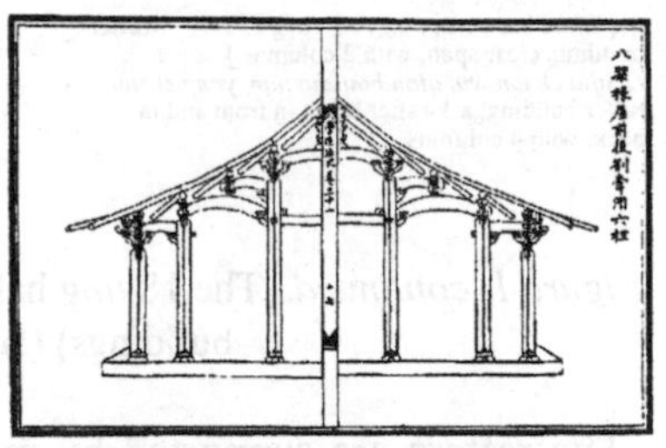

k. *Bajia chuan wu, qian hou zhaqian [rufu], yong liu zhu.* 8-rafter building, a 1- [and a 2-] rafter beam in front and in back, with 6 columns.

*Figure 1, continued.* The 18 *ting* hall sections with descriptions (8-rafter buildings) (f) to (k) (Liang 1983)

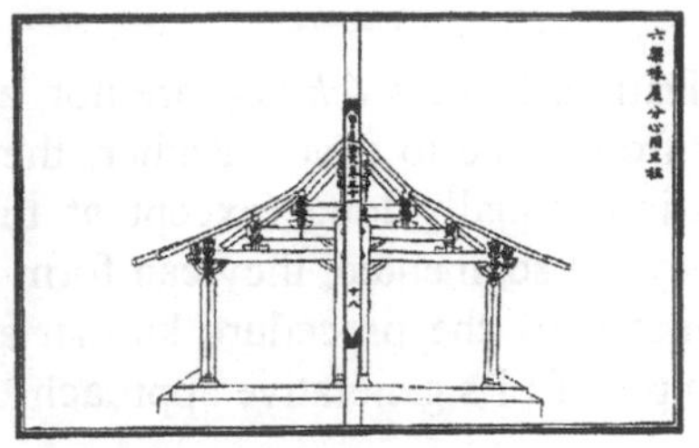

l. *Liujia chuan wu, fen xin, yong san zhu.* 6-rafter building, centrally divided, with 3 columns.

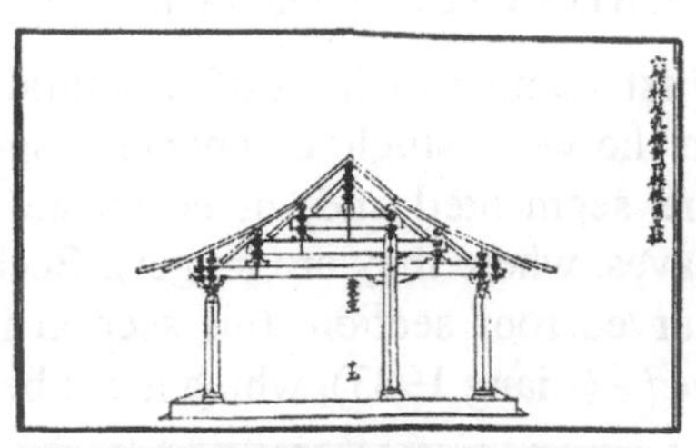

m. *Liujia chuan wu, rufu dui sichuan fu, yong san zhu.* 6-rafter building, a 4-rafter beam abutting a 2-rafter beam, with 3 columns.

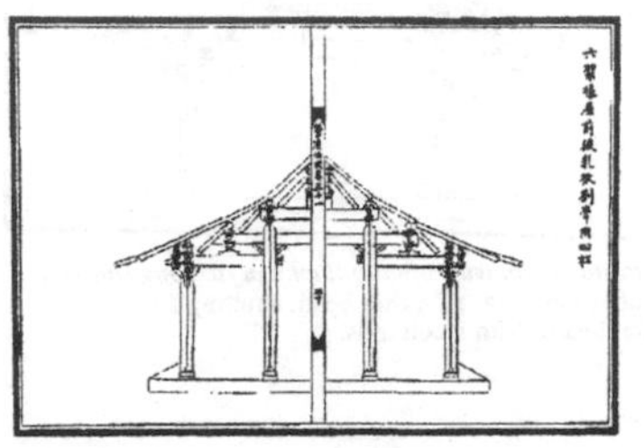

n. *Liujia chuan wu, qian hou rufu, yong si zhu.* 6-rafter building, a 2-rafter beam in front and in back, with 4 columns.

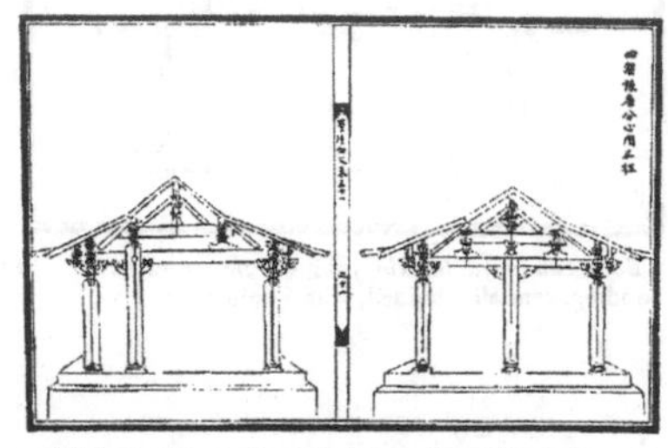

o. [*Sijia chuan wu, zhaqian dui sanchuan fu, yong san zhu.* 4-rafter building, a 1-rafter beam abutting a 3-rafter beam, with 3 columns.]
p. *Sijia chuan wu, fen xin, yong san zhu.* 4-rafter building, centrally divided, with 3 columns.

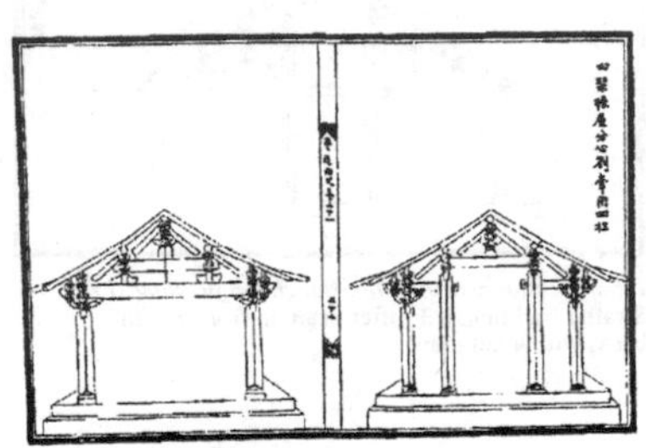

q. [*Sijia chuan wu, tongyan, yong er zhu.* 4-rafter building, clear span, with 2 columns.]
r. *Sijia chuan wu, qian hou zhaqian, yong si zhu.* 4-rafter building, a 1-rafter beam in front and in back, with 4 columns.

*Figure 1, continued.* The 18 *ting* hall sections with descriptions (6- and 4-rafter buildings) (l) to (q) (Liang 1983)

The rafters are supported by purlins *(tuan)*. Each purlin is supported at the end of a beam *(fu)*. The end of each beam is supported by a column *(zhu)*, either by resting on the top of the column or by fitting into the side of the column. Each column sits either on the floor or, truncated, on a beam above the floor; the second option increases the clear floor area. The frontmost and backmost columns are always present. As we will see below, the disposition of columns is the main distinguishing characteristic of sections.

## 2.2. DESCRIPTIONS

A second observation concerns the written descriptions on the right side of the drawings. These identify components and parameters differently than we would today; they help us see what Song builders saw.

Take as an example the section in Figure 1n, which has the description

6-rafter building, 2-rafter beams in front and back, with 4 columns
*liujia chuan wu, qian hou rufu, yong si zhu.*

A description has three parts. The first part (*6-rafter building,* in our example) specifies the depth of the building in (horizontally projected) rafters. The corpus shows buildings with depths of 4, 6, 8, and 10 rafters.

The second part of the description *(2-rafter beams in front and back)* specifies the interior disposition of columns. Three terms are used:

1. **Clear span** *(tong yan).* The lowest beam spans from the frontmost column to the backmost column; there are no interior columns (Figure 1q).
2. **Central division** *(fen xin).* There is a column below the ridge purlin (Figure 1a).
3. ***n*-rafter beams** *(n-chuan fu).* There is a beam *n* rafters in length and a column supporting it (Figure 1b).

A description contains these terms in the following combinations: *clear span* alone; *centrally divided* alone; *n-rafter beams* alone; *centrally divided* and *n-rafter beams.* The number of beams in front is equal to the number of beams in back. Interior beams are often unspecified.

The third part of the description *(with 4 columns)* specifies the number of columns. In an *m*-rafter building, there are $m + 1$ positions for columns. If we assume that the frontmost and backmost columns are always present, then there are $2^{(m-1)}$ dispositions of columns: 8, 32, 128, and 512 dispositions for 4-, 6-, 8-, and 10-rafter buildings, respectively.

Thus in our example the building is 6 rafters deep. There are a 2-rafter beam spanning from the frontmost column to an interior column, and a 2-rafter beam spanning from the backmost column to an interior column. There are 4 columns in total. Notice that the description specifies neither the central beam nor the space it defines, which we today would focus on. However, these features can all be inferred from the description.

## 2.3. OTHER OBSERVATIONS

Two final observations. First, for any section, only one description is given, even when another would appear to apply. For instance, the section in Figure 1f does not have the description

8-rafter building, a 4-rafter beam abutting a 4-rafter beam, with 3 columns
*bajia chuan wu, sichuan fu dui sichuan fu, yong san zhu.*

It would appear that a stylistically correct section can have not more than one description.

The second observation is similar to the previous one: for any description, only one section is shown. This is so even though the description does not specify all the components (such as beams, purlins, and rafters) shown in the section.

This implies that, for any description, there is only one way to complete the section. That is, the description appears to contain enough information to distinguish it from all other descriptions in the language; it appears to be *distinctive.*

Having made my observations explicitly, we are now ready to discuss the generalization of those observations: the working grammar.

## 3. Writing the Working Grammar

Why provide students with a working grammar? Why not have them write it themselves?

It is a matter of focus. Students do not have the time to acquire the expertise to write such a grammar from scratch. In the framework proposed here, the grammar is a means to the twin ends of learning about a style and learning how to understand a style. At the same time, students are quick to appreciate that grammars structure choices and present them graphically. These facts lead me to write the working grammar in a slightly unusual way.

The usual approach would be to take a labeled point with null descriptions as the initial design – to start from square one, as it were – and provide all the rules for generating from that a complete design, including those components not specified in the description. This suggests three main stages in the process:

1. Generate a *prepared design* consisting of a ground line, front and back columns, purlin positions, labels, and prepared description. Such a prepared design can have a depth of 4, 6, 8, or 10 rafters, Figure 2, top node.
2. Generate a *distinctive design,* i.e., one with just those features specified in its description, Figure 2, second row from the bottom.
3. Generate a *finished design:* clean up the labels, fill in the remaining structural components, and regularize the description, Figure 2, bottom row.

Of these three stages, only one requires the user to make any meaningful decisions: the second, creating a distinctive design. The first stage has, according to the corpus, only four possible outcomes (a prepared section of 4, 6, 8, or 10 rafters), although other outcomes (prepared sections of other sizes) can easily be imagined. And the third stage is deterministic; this assumes of course that one description maps to one complete section, but, as we have seen, this is the case in the corpus.

This leads to the unusual approach. To focus students on the decisions they make when using the grammar – the second stage – I withhold the first and third stages, and provide only the four prepared designs and the second stage schemata.

The grammar is parametric, Figure 3. I follow Stiny's (1990) formulation of a design as "an element in an *n*-ary relation among drawings, other kinds of descriptions, and correlative devices as needed." Here, a design consists of a section, a Chinese description (in *pinyin* romanization, although characters work as well), and an equivalent English description. The descriptions are handled according to Stiny (1981): simply put, the shape and its descriptions are transformed simultaneously. Here, each shape schema is associated with two description functions: one Chinese, one English.

The algorithm constructs the descriptions; the section develops accordingly. Thus the schemata are named according to the descriptions; they are as follows.

1. **Clear span** schema (CS). This leaves the "initial section" unchanged and adds *tong yan / clear span* to the descriptions.
2. **Central division** schema (CD). This instantiates a column below the ridge purlin and adds *fen xin / centrally divided* to the descriptions.
3. **Front and back beam** schemata (F1, F2, ..., F9; B1, B2, ..., B9). Each schema instantiates an *n*-rafter-long beam and its associated column in the front or the back of the section, and adds *qian (hou) n-chuan fu / n-rafter beam in front (in back)* to the descriptions. The labels ensure that the endpoint of the beam moves with each instantiation.

Students easily understand how shape schemata transform shapes. On the other hand, they are vexed by labels that constrain the applicability of schemata. For this reason, I use as few labels as possible, which makes the schemata more general, and express constraints verbally. This helps students understand both the transformations and the constraints better. In particular, when they revise the grammar, they often revise the constraints but not the schemata. The constraints are as follows.

1. Beam schemata may be applied in pairs only. That is, if a front beam schema is applied, then a back beam schema must also be applied. The lengths of the beams can be different.
2. Beam schemata may be applied as long as the new components (beam and column) do not overlap with existing components. Thus there may be more than one cycle of front-back beam instantiation.
3. A 1-rafter beam schema may be applied in the first cycle only. This means that 1-rafter beams specified in the descriptions are always in the frontmost or backmost part of the building.

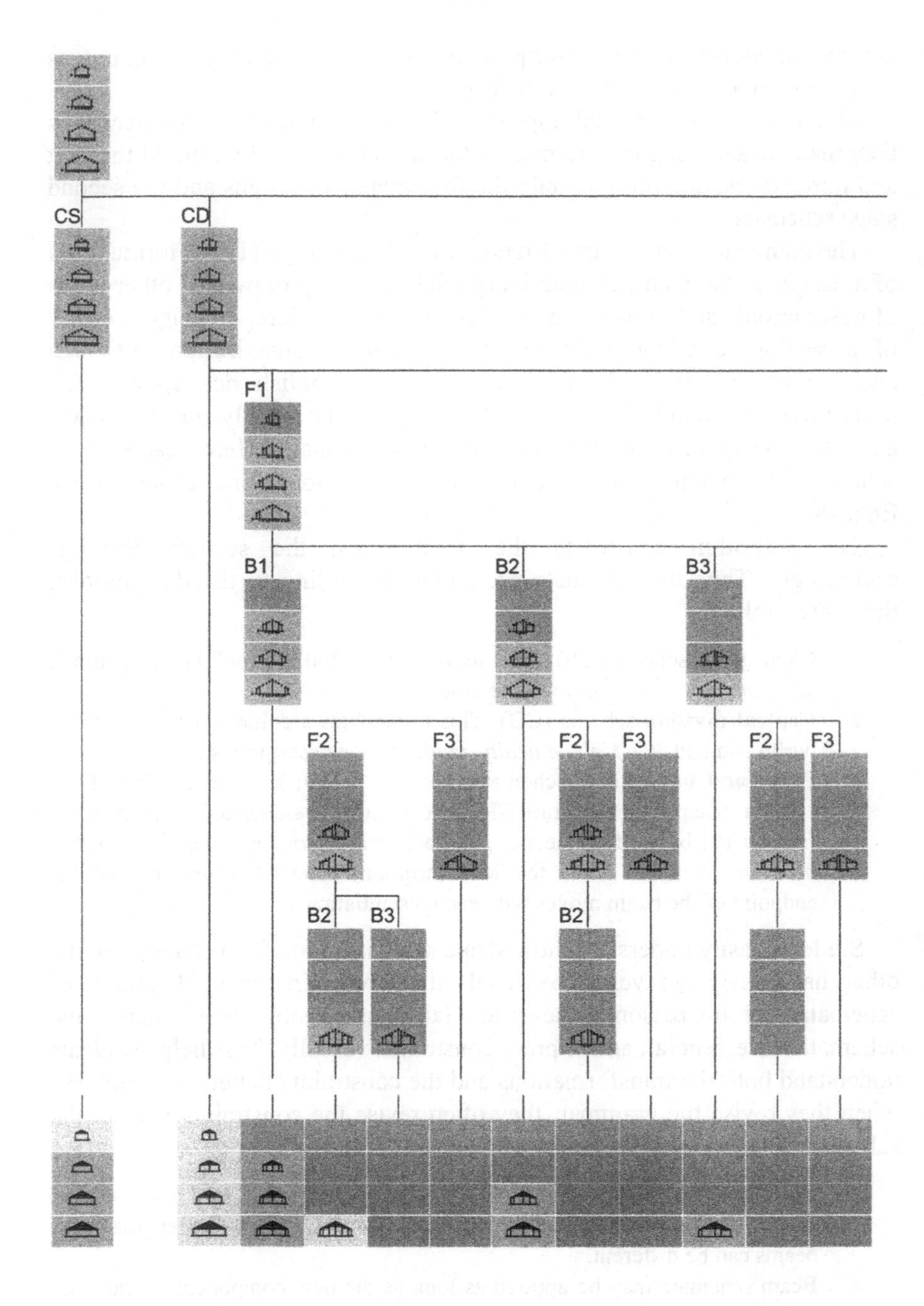


*Figure 2.* Derivation tree (leftmost 12 of 122 branches). Descriptions are omitted. Sections in the corpus have a light background. Sections that violate constraints or are found to be stylistically illegal have a dark background. All other sections have a medium background.

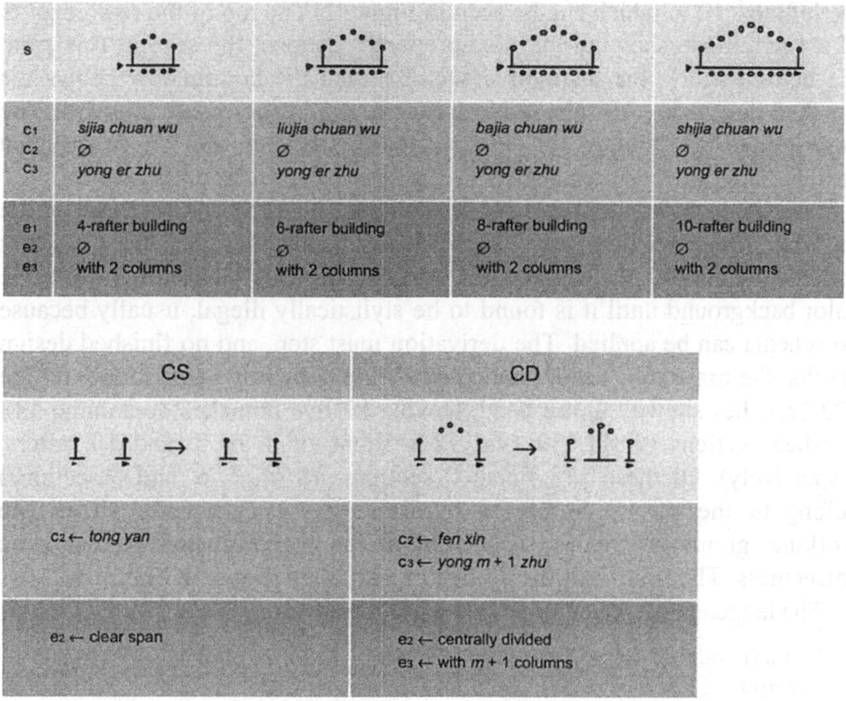

| | | | | |
|---|---|---|---|---|
| s | | | | |
| c1<br>c2<br>c3 | *sijia chuan wu*<br>∅<br>*yong er zhu* | *liujia chuan wu*<br>∅<br>*yong er zhu* | *bajia chuan wu*<br>∅<br>*yong er zhu* | *shijia chuan wu*<br>∅<br>*yong er zhu* |
| e1<br>e2<br>e3 | 4-rafter building<br>∅<br>with 2 columns | 6-rafter building<br>∅<br>with 2 columns | 8-rafter building<br>∅<br>with 2 columns | 10-rafter building<br>∅<br>with 2 columns |

| CS | CD |
|---|---|
| → | → |
| c2 ← *tong yan* | c2 ← *fen xin*<br>c3 ← *yong m + 1 zhu* |
| e2 ← clear span | e2 ← centrally divided<br>e3 ← with *m* + 1 columns |

*Figure 3.* The four prepared designs (above) and the clear span and central division schemata (below)

4. The central column may be instantiated by the central division schema only, never by beam schemata. This prevents alternate descriptions, as mentioned above, Figure 1f.
5. Schemata must be applied in one of the following sequences:
   a. The clear span schema only.
   b. The central division schema only.
   c. First the central division schema, and then the beam schemata (in pairs).
   d. The beam schemata only (in pairs).
6. Exception to constraint 2 above. A new column may overlap an existing column. In this case, the description is modified to include *dui / abutting* and the number of columns is reduced by 1, Figure 1g. This adjustment occurs in the third stage of the grammar, and is not shown.

## 4. Understanding the Working Language

The working grammar defines the students' working language. To understand it, let us consider its derivation tree. This tree has 122 branches, the leftmost 12 of which can be seen in Figure 2. The top of the row consists of a single node showing the four prepared sections; the second row from the bottom shows the distinctive sections; and the bottom row shows the finished sections. All allowable schema applications are shown. The descriptions are omitted for lack of space, but they can be constructed easily.

Each node consists of up to four sections. The prepared sections, in the top row, have medium gray backgrounds, indicating that they are not stylistically illegal. Each succeeding section in a derivation has the same color background until it is found to be stylistically illegal, usually because no schema can be applied. The derivation must stop, and no finished design results: the branch is "dead," indicated by the dark gray background. Of the 122 branches shown, 40 are dead, leaving 82 live branches containing 118 finished sections (5, 11, 30, and 72 sections of 4, 6, 8, and 10 rafters, respectively). Of these 118 finished sections, 18 (4, 3, 6, and 5 sections) belong to the corpus (indicated by light gray backgrounds). Thus the working grammar creates 100 sections to be evaluated for stylistic correctness. The first 12 of the 82 live branches are shown in Figure 4.

The language can be summarized as follows:

| Depth (in rafters) | 4 | 6 | 8 | 10 | total |
|---|---|---|---|---|---|
| In corpus | 4 | 3 | 6 | 5 | 18 |
| New | 1 | 8 | 24 | 67 | 100 |
| Total | 5 | 11 | 30 | 72 | 118 |

## 5. The Classroom Experience

By this point, I have examined the corpus, proposed the working grammar, and sketched out the working language. The stage is set for the students to begin their task. For their part, they have prepared themselves by studying relevant excerpts and drawings from the *Yingzao fashi,* reading secondary sources, and learning the fundamentals of shape grammar and generative design. They were given the following instructions.

> Investigate the language of 6-rafter *ting* [hall] sections by reconciling a grammatical definition of that language and your understanding of stylistic correctness. ... Follow these steps:
>
> 1. Using the [working] grammar, generate all the (final) designs in the language of 6-rafter sections.
> 2. You may feel that some of the designs are not in the style. Show them and explain why each is [stylistically incorrect]. Eliminate them by articulating unambiguous ... constraints to eliminate them.
> 3. You may feel that there are stylistic[ally correct] designs that are not created by the grammar (these may or may not be in the corpus). If so, show them, and modify the grammar so that it creates them.

4. Repeat steps 1, 2, and 3 until the grammar generates all and only the stylistic[ally correct] designs.

A few remarks on the students' task. First, it is presented explicitly in terms of the framework already discussed above. Second, it is reduced to 6-rafter sections only; 4-, 8-, and 10-rafter sections have been eliminated. Third, students used a Flash-based implementation (Li 2002) to generate all the designs in the working language (step 1).

The problem appears limited, but students arrived at vastly different languages, ranging from 4 to 47 designs, Figure 5. (Contrast this with the working language of 11 designs.) They revised the constraints in different ways, invoking criteria like structure and spatial sequence. For instance:

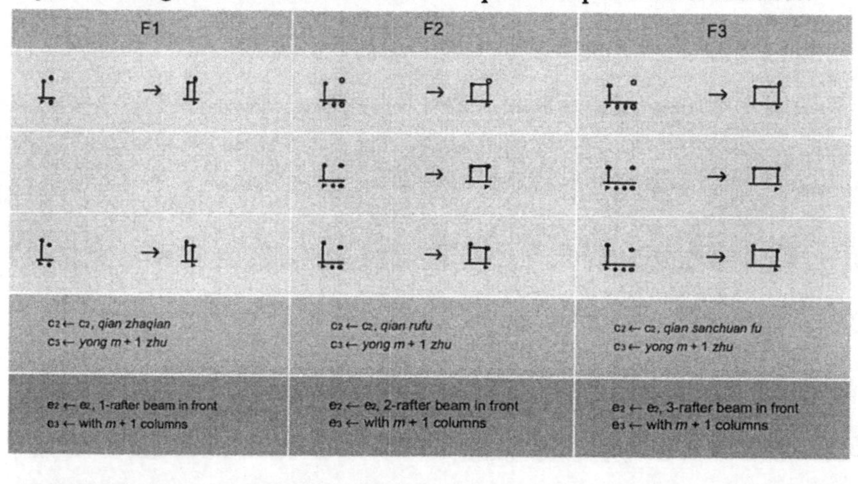


*Figure 3, continued.* The schemata for 1-, 2-, and 3-rafter beams in front (above) and in back (below)

1. A beam may be no more than 3 times as long as the abutting beam.
2. Beams in the interior of the building may not be shorter than beams on the exterior.
3. If the building has two bays, the front bay must be as deep or deeper than the back bay.
4. The building may not contain a 6-rafter beam; i.e., there is no clear span building.
5. Only a 6-rafter building may contain a 6-rafter beam. (This contradicts the corpus; see Figure 1g.)

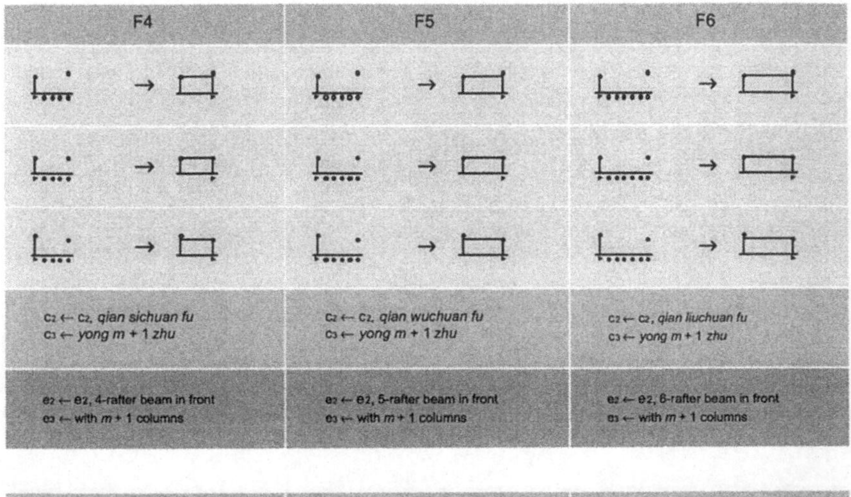


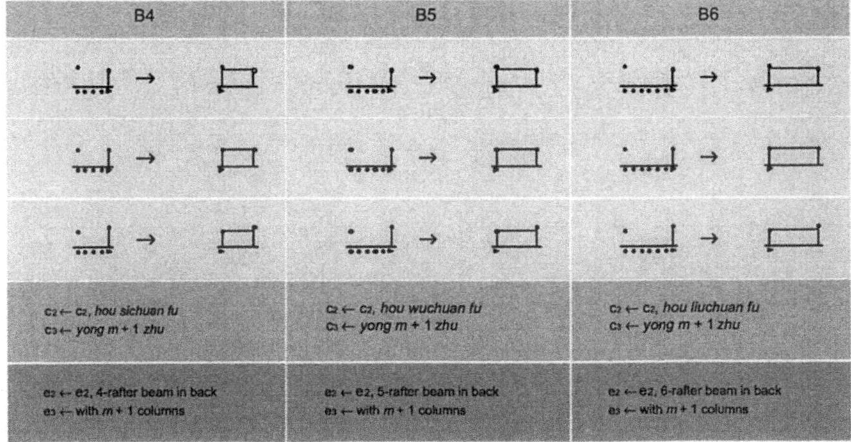


*Figure 3, continued.* The schemata for 4-, 5-, and 6-rafter beams in front (above) and in back (below)

Students manipulated both rules and constraints to refine the target language. For example, the student with only 4 designs proposed 5 constraints which eliminated most designs. As for the student with 47 designs, he added 36 that do not appear in any other student's language by introducing rules to shift columns a half-rafter's length. He reasoned that such dispositions were stylistically legal after observing that some extant Chinese buildings had columns *between* the purlins, rather than directly below them.

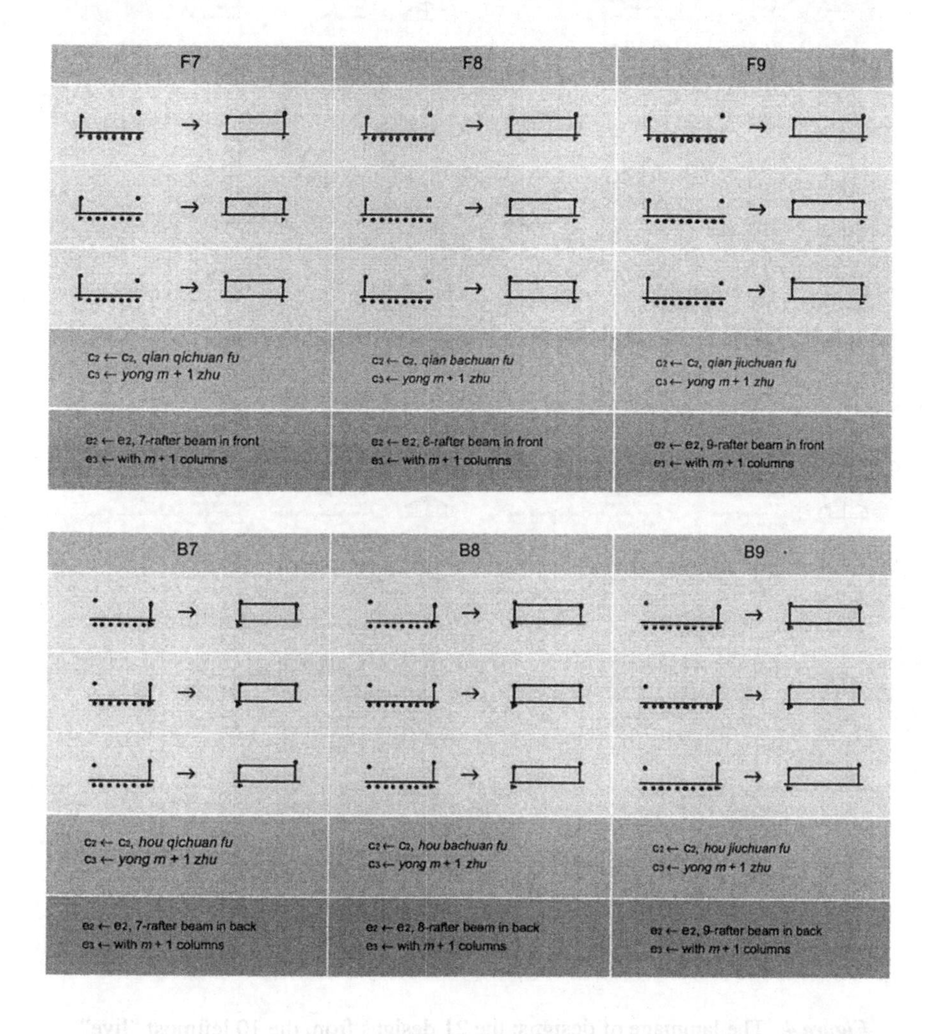


*Figure 3, continued.* The schemata for 7-, 8-, and 9-rafter beams in front (above) and in back (below)

Students understood clearly that they had to bring their own interpretation to bear on the problem of stylistic correctness. As one student wrote:

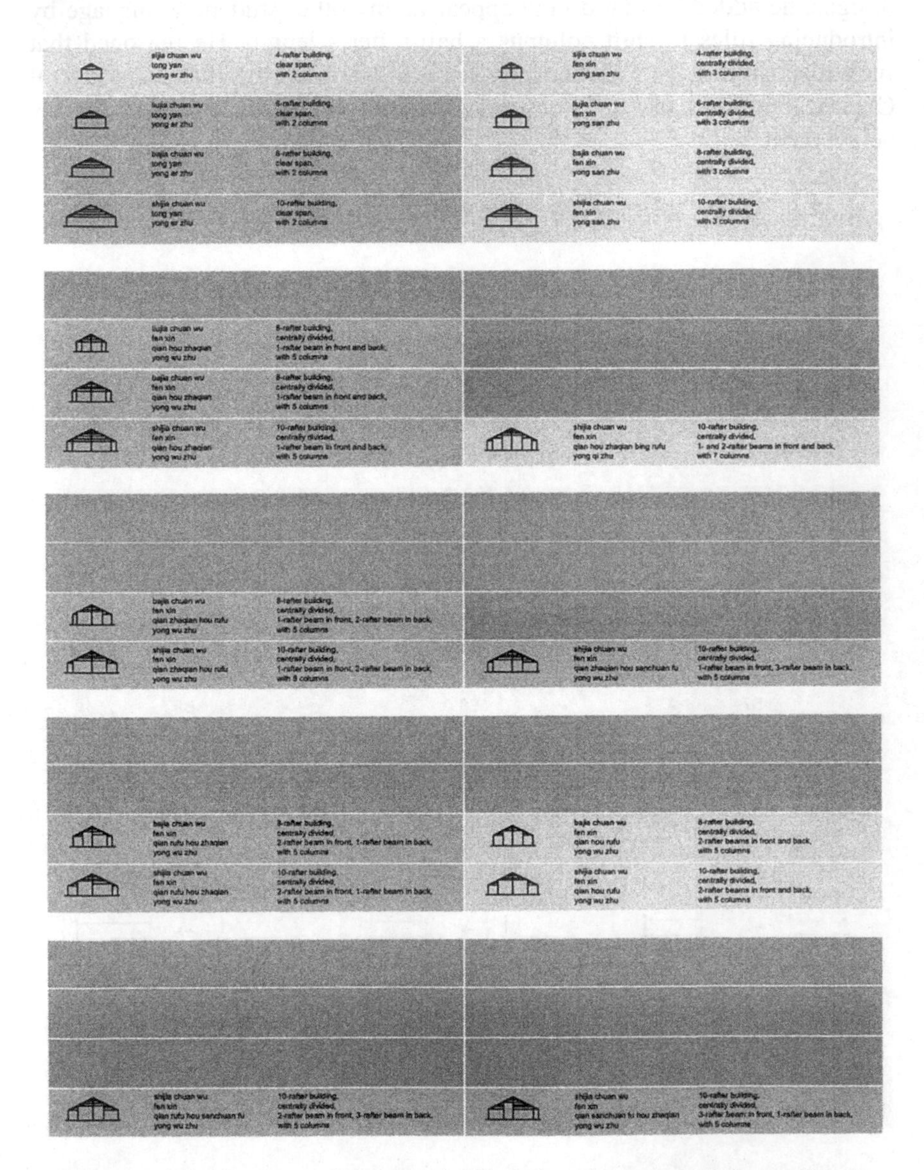


*Figure 4.* The language of designs: the 21 designs from the 10 leftmost "live" branches from the derivation tree, Figure 2. The 97 designs from the remaining branches are not shown.

Based on the original corpus, I get some rough idea about what designs are stylistic[ally correct]. The choice of this part is however base[d] on my own understanding and interpretation on the existing corpus that I have learned. So it depends on what and how I think about it. (Emily So Ching Han)

Another wrote:

Everyone has different feelings, even with the same facts. ... And that's the reason why every candidate's results of this exercise are different. (Dennis Leung Chung Hoo)

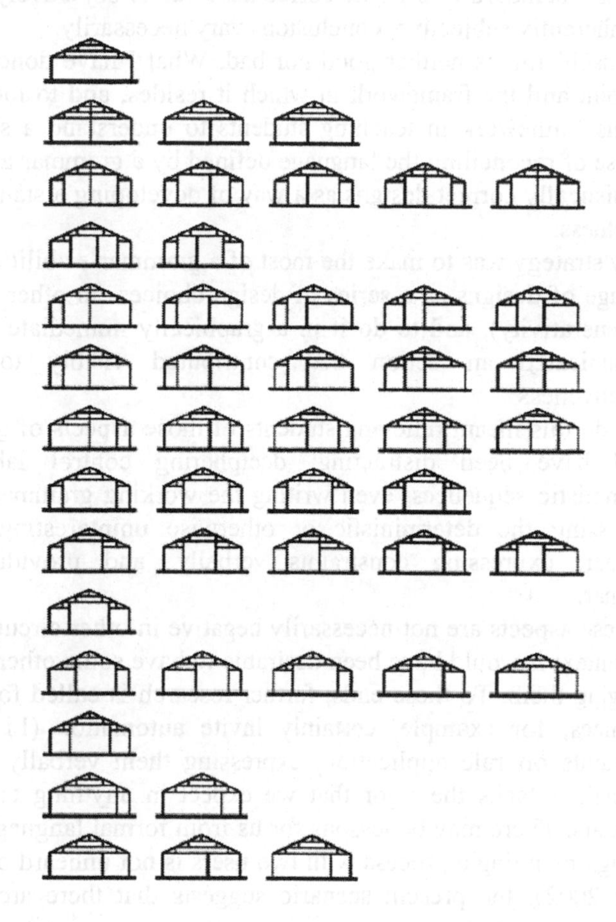

*Figure 5.* A language of sections with 47 design

A third invoked historical examples:

I think that the whole process of making and modifying Constraint no. 1 is subjective because there is not enough information to support [my] idea.

However, I think it can become more objective if we find more examples in extant buildings. (Tommy Lou Kai Chio)

## 6. Conclusion

The students began with the same corpus of designs, but ended up with markedly different conclusions about the larger language. This should not be surprising. As I have argued, extrapolating from corpus to language requires a standard of stylistic correctness that is objectively untestable and thus inherently subjective; conclusions vary necessarily.

In itself, this is neither good nor bad. What I have done is to articulate this point and the framework in which it resides, and to look for a way to use this framework in teaching students to understand a style. Hence the exercise of reconciling the language defined by a grammar and the language of stylistically correct designs as a way of developing a standard of stylistic correctness.

My strategy was to make the most of a grammar's abilities to structure a language of designs as a series of design choices (in other words, its rigor and generativity), and to do it in a graphically immediate way. Each rule communicated an action that contributed visibly to the design's distinctiveness.

To do this meant relieving students of those aspects of grammars which would have been distracting: deciphering control labels, executing deterministic sequences, even writing the working grammar. This I did by suppressing the deterministic or otherwise uninteresting stages of the grammar, expressing constraints verbally, and providing a working grammar.

These aspects are not necessarily negative in other circumstances, but in this context it would have been desirable to have some other alternatives for managing them. To these ends, further research is called for. Deterministic sequences, for example, certainly invite automation (Li 2002). As for constraints on rule application, expressing them verbally is intuitive but informal; it lacks the rigor that we expect in anything to do with shape grammars. There may be lessons for us from formal languages. And finally, although planning a process with two users is not unheard of (Knight 2000; Chase 2002), the present scenario suggests that there are many ways to involve many users.

In this context, the grammar is less a machine for production than a means of articulating, manipulating, heightening our understanding of a style; in short, it is a tool for thinking. We have long known that grammars can help us understand style. Now we see that they can respond not just to products, but to process as well. We can write grammars with subtlety and purpose to enhance our understanding of how we understand.

## References

Buelinckx, H: 1993, Wren's language of city church designs: A formal generative classification, *Environment and Planning B:Planning & Design* **20**: 645–676.

Chase, SC: 2002, A model for user interaction in grammar-based design systems, *Automation in Construction* **11**: 161–172.

Chen Mingda: 1993, *Yingzao Fashi Da Muzuo Zhidu Yanjiu* [A study of structural carpentry in the *Yingzao fashi*], 2nd ed, Wenwu, Beijing.

Duarte, JP: 2001, *Customizing Mass Housing: A Discursive Grammar for Siza's Malagueira Houses*, Ph.D. dissertation, Department of Architecture, Massachusetts Institute of Technology, Cambridge, Mass.

Glahn, E: 1984, Unfolding the Chinese building standards: Research on the *Yingzao fashi*, *in* NS Steinhardt (ed), *Chinese Traditional Architecture,* China Institute in America, New York, pp. 47–57.

Guo, Q: 1999, *The Structure of Chinese Timber Architecture,* Minerva, London.

Knight, TW: 1981, The forty-one steps, *Environment and Planning B: Planning & Design* **8**: 97–114.

Knight, TW: 2000, *Shape Grammars in Education and Practice: History and Prospects,* 1999–2000 [cited 2000], available from http://www.mit.edu/%7Etknight/IJDC/.

Koning, H and Eizenberg, J: 1981, The language of the prairie: Frank Lloyd Wright's prairie houses, *Environment and Planning B: Planning & Design* **8**: 295–323.

Li, AI: 2002, A prototype interactive simulated shape grammar, *in* K Koszewski and S Wrona (eds), *Design e-ducation: Connecting the Real and the Virtual,* eCAADe, Warsaw, pp. 314–317.

Liang Sicheng: 1983, *Yingzao Fashi Zhushi* [The annotated *Yingzao fashi*], Zhongguo jianzhu gongye, Beijing.

Liang Sicheng: 1984, Zhongguo jianzhu zhi liangbu "wenfa keben" [The two "grammar books" of Chinese architecture], *Liang Sicheng wenji* [The collected works of Liang Sicheng], Zhongguo jianzhu gongye, Beijing, pp. 357–363.

Stiny, G: 1977, Ice-ray: a note on the generation of Chinese lattice designs, *Environment and Planning B: Planning & Design* **4**: 89–98.

Stiny, G: 1981, A note on the description of designs, *Environment and Planning B: Planning & Design* **8**: 257–267.

Stiny, G: 1990, What is a design? *Environment and Planning B: Planning & Design* **17**: 97–103.

Stiny, G and Mitchell, WJ: 1978a, Counting Palladian plans, *Environment and Planning B: Planning & Design* **5**: 189–198.

Stiny, G and Mitchell, WJ: 1978b, The Palladian grammar, *Environment and Planning B: Planning & Design* **5**: 5–18.

# REPRESENTATION IN DESIGN

JS Gero (ed), *Design Computing and Cognition'04*, 219-238


# DATA VIEWS, DATA RECOGNITION, DESIGN QUERIES AND DESIGN RULES

*Representational Flexibility for Design*

RUDI STOUFFS
*Delft University of Technology, The Netherlands*

and

RAMESH KRISHNAMURTI
*Carnegie Mellon University, USA*

**Abstract.** *Sorts* present a constructive approach to representational structures and provide a uniform approach to handling various design data. In this way, *sorts* offer support for multiple, alternative data views and for data exchange between these views. The representation of *sorts* extends on a maximal element representation for shapes that supports shape recognition and shape rules. In the same way, *sorts* offer support for data recognition, for querying design information and for expressing design rules. In this paper, we present an overview of the use of *sorts* to support these functionalities. Each of these relies on the ability to alter representational structures or *sorts*, and to manipulate the composition of data forms. In this regard, we briefly consider the user interaction aspect of utilizing *sorts*.

## 1. Introduction

Computational design relies on effective information models for design, for the creation of design artifacts and for the querying of the characteristics of such artifacts. Mäntylä stated in 1988 that these (geometric) representations must adequately answer "arbitrary geometric questions algorithmically." Even without emphasis on the geometric aspects, this remains as important today. However, current computational design applications tend to focus on the representation of design artifacts, and on the tools and operations for their creation and manipulation. Techniques for querying receive less attention and are often constrained by the data representation system and methods. Nevertheless, querying a design is as much an intricate aspect of the design process as is creation and manipulation.

Design is also a multi-disciplinary process, involving participants, knowledge and information from various domains. As such, design problems require a multiplicity of viewpoints each distinguished by particular interests and emphases. For instance, an architect is concerned with aesthetic and configurational aspects of a design, a structural engineer is engaged by the structural members and their relationships, and a building performance engineer is interested in the thermal, lighting, or acoustical performance of the eventual design. Each of these views—derived from an understanding of current problem solution techniques in these respective domains—requires a different representation of the same (abstract) entity. Even within the same task and by the same person, various representations may serve different purposes defined within the problem context and the selected approach. Especially in architectural design, the exploratory nature of the design process invites a variety of approaches and representations.

Each view may rely on domain knowledge in order to provide a visualization that is particularly appropriate for the type of design object under investigation. In scientific visualizations, one can make use of the inherent dimensions of scientific data, connecting to three spatial and one temporal dimension, requiring only elementary linear algebra to lay out scientific data on a two-dimensional display (Groth and Robertson 1998). In architecture, designers commonly rely on a geometric visualization of the architectural object and its components, in both two and three dimensions, providing feedback on both aesthetic and configurational aspects of the design. A structural engineer, on the other hand, is less concerned with the geometry of the design components. Instead, a diagrammatic visualization of the design object presenting the structural characteristics of its components and their relationships is more appropriately used. Similarly, data visualizations in Geographic Information Systems (GIS) generally make use of map projections to visualize a variety of geographically related data.

Not all kinds of data structures can rely on specific domain knowledge in their visualization. For example, when exploring general information structures or databases, data may be collected from a large variety of domains and may not fit a single domain-specific visualization. In design, it can be said that there are as many design methods as there are designers. Different design methods may consider different data from different design domains and, therefore, require different visualizations. Furthermore, not all kinds of views can be envisioned a priori and specific support provided for. In such cases, the challenge is to achieve an effective mapping from data to display (Groth and Robertson 1998).

Effective visualizations enable a visual inspection of design data and information. Design queries, on the other hand, support the analysis of existing design information in order to derive new information that is not explicitly available in the information structure. Both effective visualizations, in support of alternative design views, and an expression of arbitrary design questions require flexible design information models and representations that can be modified and geared to the kinds of visualizations and queries. Supporting arbitrary design questions also requires access to information in a uniform and consistent manner, so that new queries can be easily constructed and posed based on intent, instead of on availability.

*Sorts* (Stouffs and Krishnamurti 2002; 2001a) offer a framework for representational flexibility that provides support for developing alternative representations of a same entity or design, for comparing representations with respect to scope and coverage, and for mapping data between representations, even if their scopes are not original. *Sorts* support the specification of the operational behavior of data in a uniform way, based on a partial order relationship (Stouffs 1994; Stiny 1991). *Sorts* extend on a maximal element representation for shapes (Stouffs 1994; Krishnamurti 1992) that supports shape recognition and shape rules. Data views, data recognition, design queries and design rules all relate to the concept of emergence, i.e., the recognition of information components and structures that are not explicitly present in the information and its representation, and on the restructuring of information. The concept of emergence, in turn, supports creativity and novelty (Krishnamurti and Stouffs 1997; Stiny 1993).

In a previous paper (Stouffs and Krishnamurti 2002) we explored the mathematical properties of a constructive approach to *sorts* through an abstraction of representational structures to model *sorts*. We applied this approach to representational structures defined as compositions of primitive data types, and explored a comparison of representational structures with respect to scope and coverage. We considered a behavioral specification for *sorts* in order to empower these representational structures to support design activities effectively, and provided an example of the use of *sorts* to represent alternative views to a design problem. In this paper, we consider the application of *sorts* in a broader context and present an overview of the use of *sorts* to support data recognition, design queries and design rules, next to multiple data views. Each of these functionalities relies on the ability to alter representational structures or *sorts*, and to manipulate the composition of data forms. In this regard, we briefly consider the user interaction aspect of utilizing *sorts*.

## 2. Alternative Data Views

Integrated data models are under development that span multiple disciplines and support different views. Such models allow for various representations in support of different disciplines or methodologies and enable information exchange between representations and collaboration across disciplines. Examples are, among others, the ISO STEP standard for the definition of product models (ISO 1994) and the Industry Foundation Classes (IFCs) of the International Alliance for Interoperability (IAI), an object-oriented data model for product information sharing (Bazjanac 1998). These efforts characterize an a priori and top-down approach: an attempt is made at establishing an agreement on the concepts and relationships which offer a complete and uniform description of the project data, independent of any project specifics (Stouffs and Krishnamurti 2001a).

Alternative modeling techniques that consider a bottom-up, constructive approach are also under investigation. These provide a more extensive degree of flexibility that allows for the development of information models that are context, and thus project, specific. We consider a few examples related to architectural design. Concept modeling (van Leeuwen and Fridqvist 2003; van Leeuwen 1999) allows for the extensibility of conceptual schemas and for flexibility in modeling information structures that differ from the conceptual schemas these derive from. The SPROUT modeling language (Snyder and Flemming 1999; Snyder 1998) allows for the specification of schematic descriptions that can be used to generate computer programs that provably map data between different applications. Woodbury et al. (1999) adopt typed feature structures in order to represent partial information models and use unification-based algorithms to support an incremental modeling approach.

*Sorts* (Stouffs and Krishnamurti 2002) offer a constructive approach to defining representational structures that enables these to be compared with respect to scope and coverage and that presents a uniform approach to dealing with and manipulating data constructs. Briefly, a *sort* is defined as a complex structure of elementary data types and compositional operators, and is typically a composition of other *sorts*. Comparing different *sorts*, therefore, requires a comparison of the respective data types, their mutual relationships, and the overall construction.

### 2.1. EXAMPLE A: A HIERARCHICAL STRUCTURE OF KEYWORDS

Figure 1 presents a simple example of a *sort* that represents a hierarchical structure of architectural concepts or keywords. The representation is conceived as a tree structure in which each keyword can have zero, one or more subordinate keywords. The *sort concepts*, a *sort* of labels, represents the individual keywords:

*concepts* : [Label] (1)

The subordinate relationship between keywords is expressed by the attribute operator on *sorts* ('^'). The resulting *sort*, named *conceptstree*, is defined recursively:

*conceptstree* : *concepts* + *concepts* ^ *conceptstree* (2)

The attribute operator relates to each individual keyword (*concepts*) a non-empty data form of subordinate keywords (*conceptstree*). The disjunctive composition operator ('+') allows the combination of keywords with (*concepts* ^ *conceptstree*) and without (*concepts*) attribute keywords. Thus, individual keywords are assigned either to the *sort concepts*, or with an attribute data form to the *sort concepts* ^ *conceptstree*.

```
sort concepts : [Label];
sort conceptstree : concepts + concepts ^ conceptstree;

form $concepts = conceptstree:
{ (concepts ^ conceptstree):
   { "theater"
      { concepts:
         { "infrastructure" },
        (concepts ^ conceptstree):
         { "construction"
            { concepts:
               { "load bearing structure",
                 "material" },
              (concepts ^ conceptstree):
               { "enclosure"
                  { concepts:
                     { "roof",
                       "facades" } } } },
     "format"
            { concepts:
               { "photo",
                 "scale model",
                 "text" },
              (concepts ^ conceptstree):
               { "view"
                  { concepts:
                     { "elevation",
                       "axonometric view",
                       "diagram",
                       "section",
                       "perspective",
                       "plan",
                       "site plan" } } } },
     ... } } };
```

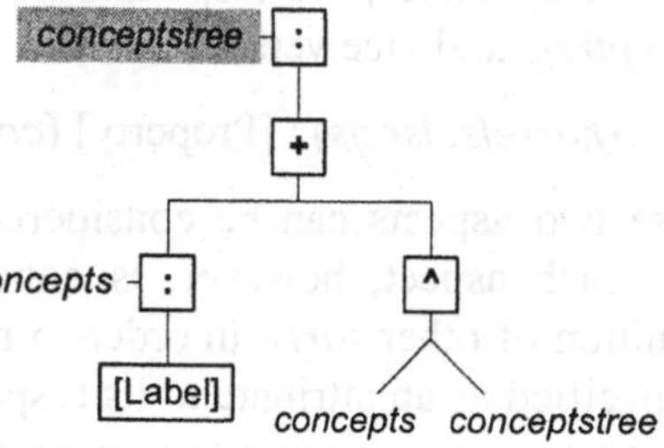

conceptstree :
+
concepts :
^
[Label]
concepts conceptstree

*Figure 1.* Textual and graphical definition of a recursive *sort* representing a hierarchical structure of architectural concepts, and the (partial) description of an exemplar data form (*Sorts* Description Language). In the definition of a *sort*, '+' and '^' denote the operations of disjunctive composition and attribute, respectively; ':' denotes the naming of a *sort*; '[Label]' is a primitive *sort* of labels.

## 2.2. EXAMPLE B: A NETWORK STRUCTURE OF KEYWORDS

An alternative view of a semantic structure (or an architectural typology) is in the form of a network or (semantic) map. A network structure distinguishes itself from a simple hierarchical structure in that a subordinate keyword may be shared by more than one keyword. Such a structure can be extended from the structure in Figure 1 by allowing references to be specified to keywords that are already defined elsewhere in the structure. Such references can be represented using a property relationship *sort* that is defined over the *sort concepts* and an equivalent *sort conceptrefs*:

*conceptrefs* : *concepts* (3)

The property relationship *sort* distinguishes two named *aspects*, *hasrefs* and *isrefs*, respectively corresponding to the relationship from *concepts* to *conceptrefs* and vice versa:

(*hasrefs*, *isrefs*) : [Property] (*concepts*, *conceptrefs*) (4)

These two aspects can be considered as two different views of the same *sort*. Each aspect, however, is considered a distinct *sort* if used in the definition of other *sorts*. In order to maintain consistency, each aspect must be specified as an attribute to its respective *sort* of origin under the property relationship, e.g., *concepts* ^ *hasrefs* and *conceptrefs* ^ *isrefs*. The first attribute *sort*, *concepts* ^ *hasrefs*, allows for the specification of keywords with one or more references to (subordinate) keywords that are elsewhere defined. The second attribute *sort*, *conceptrefs* ^ *isrefs*, allows for the retrieval of all keywords this subordinate keyword is referenced from. Both attribute *sorts*, together with the *sorts concepts* and *concepts* ^ *conceptsmap*, recursively define the *sort conceptsmap* under the disjunctive composition operator, Figure 2:

*conceptsmap* : *concepts* + *concepts* ^ *conceptsmap* + *concepts* ^ *hasrefs* + *conceptrefs* ^ *isrefs* (5)

Thus, individual keywords are assigned to the *sort concepts*, with an attribute data form (that is recursively defined) to the *sort concepts* ^ *conceptsmap*, or with an attribute data form of references to the *sort concepts* ^ *hasrefs*. If a keyword has subordinate keywords of which some but not all are defined elsewhere (and thus referenced here), then, this keyword will be assigned to both the *sorts concepts* ^ *conceptsmap* and *concepts* ^ *hasrefs*.

Figure 2 also presents an exemplar data form considering an architectural typology for Ottoman mosques (Tunçer et al. 2002). Note that the data form does not specify any data to the *sort conceptrefs* ^ *isrefs*, these are automatically derived from the data to the *sort concepts* ^ *hasrefs*.

```
sort conceptrefs : (concepts : [Label]);
sort (hasrefs, isrefs) : [Property] (concepts, conceptrefs);
sort conceptsmap : concepts + concepts ^ conceptsmap + concepts ^ hasrefs +
        conceptrefs ^ isrefs;
```

(hasrefs, isrefs) : [Property] concepts : [Label] conceptrefs : concepts

conceptsmap : + concepts ^ concepts conceptsmap ^ concepts hasrefs ^ conceptrefs isrefs

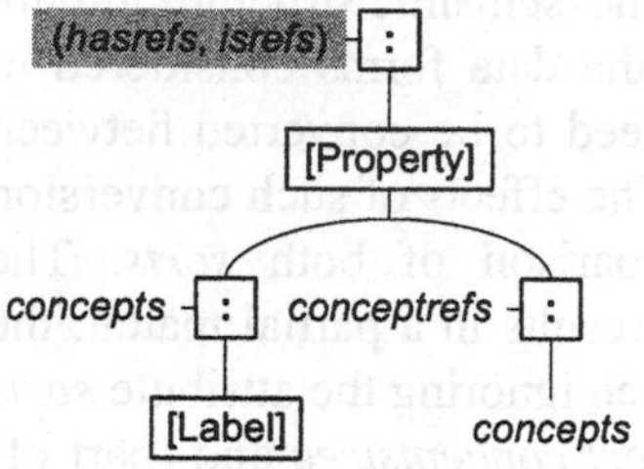

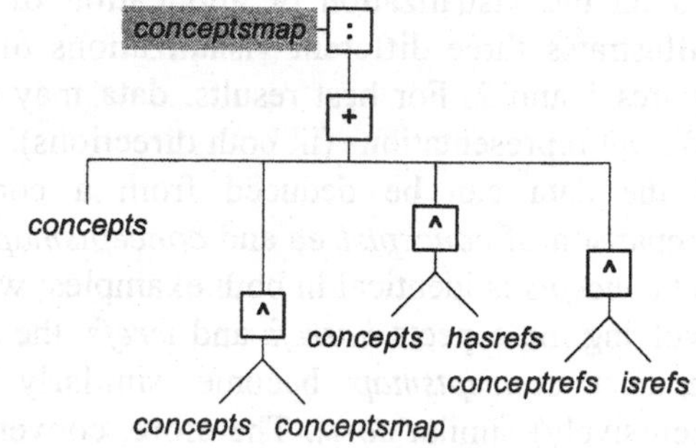

```
form $concepts = conceptsmap:
{ (concepts ^ conceptsmap):
  { "physical"
    { (concepts ^ conceptsmap):
      { "mosque"
        { (concepts ^ conceptsmap):
          { "structural"
            { (concepts ^ hasrefs):
              { #om-concepts-26 "arcade"
                { om-conceptrefs-5, om-conceptrefs-14, om-conceptrefs-19 } },
            (concepts ^ conceptsmap):
              { "arcade"
                { concepts:
                  { "spandrel" },
                (concepts ^ hasrefs):
                  { #om-concepts-11 "arch"
                    { om-conceptrefs-2, om-conceptrefs-6,
                      om-conceptrefs-16, om-conceptrefs-23 },
                    #om-concepts-13 "dome"
                    { om-conceptrefs-3, om-conceptrefs-7 } },
                (concepts ^ conceptsmap):
                  { "arch"
                    { concepts:
                      { "tympanum" } },
                    "column"
                    { concepts:
                      { "column base",
                        "column capital" } },
                    "dome"
                    { (concepts ^ hasrefs):
                      { #om-concepts-5 "crescent"
                        { om-conceptrefs-1, om-conceptrefs-4,
                          om-conceptrefs-29 } } } } },
          ... } } } } } } } };
```

*Figure 2.* Textual and graphical definition of a recursive *sort* representing a (semantic) map of architectural concepts, and the (partial) description of an exemplar data form (*Sorts* Description Language). In the definition of a *sort*, '+' and '^' denote the operations of disjunctive composition and attribute, respectively; ':' denotes the naming of a *sort*; '[Label]' and '[Property]' are primitive *sorts*, the latter defines a property relationship *sort* between two given *sorts*.

Both *sorts conceptstree* and *conceptsmap* present a possible representation of a semantic structure. The selection of any particular representation is dependent on the type of structure or semantic data, and also on the visualization or application of the semantic structure. Figure 3 illustrates three different visualizations of the data forms considered in Figures 1 and 2. For best results, data may need to be converted between different representations (in both directions). The effects of such conversion on the data can be deduced from a comparison of both *sorts*. The comparison of *conceptstree* and *conceptsmap* results in a partial match: the *sort concepts* is identical in both examples; when ignoring the attribute *sorts* involving the aspects *hasrefs* and *isrefs*, the *sort conceptstree* and (part of) the *sort conceptsmap* become similarly composed of identical and (recursively) similar *sorts*. Therefore, converting data from *conceptstree* to *conceptsmap* involves no data loss; obviously, a tree structure is a special instance of a network or map structure. However, converting data in the other direction may involve data loss; the data lost in this case is the identification of shared keywords.

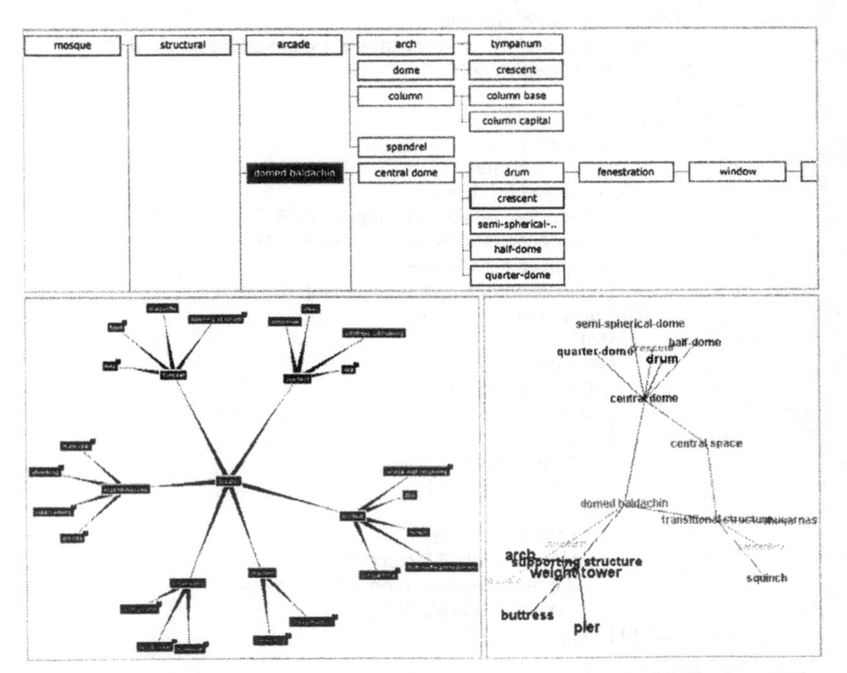


*Figure 3.* Three different visualizations of the data forms from Figures 1 and 2. The top-drawing shows a straightforward depth-first enumeration, albeit graphically enhanced; references to keywords already defined are marked with a different color border (e.g., "crescent"). The bottom left and right drawings show a 2D/3D graphical presentation of a hierarchical structure and a network structure, respectively. Image by Bige Tunçer.

## 2.3. EXAMPLE C: A COLLECTION OF KEYWORDS

Comparing *sorts* can become far more complex if one of the *sorts* is defined recursively, while the other is not. Consider another alternative representation of keywords without any (hierarchical or other) relationships. The *sort concepts* represents such a simple collection of keywords. Converting data from *concepts* to *conceptstree* (or *conceptsmap*) is fairly straightforward. Both *sorts conceptstree* and *conceptsmap* are defined in such a way that they allow for the representation of keywords without any subordinate relationships: the *sort concepts* is a part (or component) of the *sort conceptstree* under the disjunctive composition operator. The result is a partial match of identical *sorts*.

Converting data in the other direction is less obvious: the only keywords that can be converted are those that have no subordinate relationships. In the example of Figure 1 that leaves not a single keyword. However, this is not the only way that the *sort concepts* can be mapped onto a component of the *sort conceptstree*. First, the *sort concepts* also appears as a (parent) component under the attribute operator. Given such a partial match under the attribute relationship, the *sorts concepts* and *concepts ^ conceptstree* are said to be partially convertible. In the example of Figure 1, only the root keyword of the hierarchical structure, "theater," will be retained upon conversion. Second, through the recursive definition of *conceptstree*, the *sort concepts* could also be mapped onto an attribute component of the *sort concepts ^ conceptstree*. However, such a mapping cannot be considered in the context of a comparison of the *sorts concepts* and *conceptstree* as mapping *concepts* onto the *conceptstree* component of *conceptstree* would create an infinitely recursive mapping.

The conversion of a hierarchical structure of keywords (as represented by a *sort* that is recursively defined) into a simple collection of keywords as represented by the *sort concepts*, such that all or most keywords are retained in the conversion, can only be achieved through the construction of an intermediate representation. Consider a *sort concepts2* that is a composition under the attribute operator of the *sort concepts* twice:

$$concepts2 : concepts \wedge concepts \qquad (6)$$

Comparing the *sorts concepts2* and *conceptstree* at best results in a partial match of similar *sorts*, allowing for the conversion of those keywords that are directly subordinate to the root keyword in the hierarchical structure of Figure 1 and do not have any subordinate keywords themselves. Comparing the *sorts concepts* and *concepts2* results in a partially convertible match where either component under the attribute operator can be mapped onto the *sort concepts*. In this way, the keywords that resulted from the conversion above can be further converted, in two steps, into a collection of keywords (without hierarchical relationships). The intermediate representation can be extended in order to include more keywords in the conversion.

## 3. Shape and Data Recognition

Creative design activities rely on a restructuring of information uncaptured in the current information structure, as when looking at a design provides new insights that lead to a new interpretation of the design elements. It can be proven that continuity of computational change requires an anticipation of the structures that are to be changed (Krishnamurti and Stouffs 1997). Creativity, on the other hand, is devoid of anticipation.

Computationally recognizing emergent shapes requires determining a geometric, commonly Euclidean, transformation under which a specified similar shape is a part of the original shape. For example, a square must be computationally recognized as a square irrespective of its scale, orientation or location. The same approach applies to other kinds of data. For example, search-and-replace functionalities in text editors generally consider case transformations of the constituent letters. Clearly, this matching problem depends on the representational structure adopted. The maximal element representation for shapes is a particularly appropriate representation as each element type specifies its own part or match relationship (Krishnamurti and Stouffs 1997; Krishnamurti and Earl 1992).

*Sorts* can be considered as an extension of the maximal element representation to other, non-geometric, data, without necessarily considering non-spatial information as attributes to shapes. The concept of *sorts* distinguishes various behaviors data can adhere to, all based on a part relationship (Stouffs and Krishnamurti 2002). Examples are a discrete behavior, corresponding to a mathematical set, for labels or points, an ordinal behavior for numeric weights, line thicknesses or shades of gray, an interval behavior for line segments, and similar behaviors for plane segments and volumes. Consider the *sort conceptstree* and the corresponding data form in Figure 1. Keywords are assigned to the *sort concepts*, a *sort* of labels, with discrete behavior. The recognition of keywords therefore requires the full keyword to be provided, though case transformations may apply such that the word "Infrastructure" can match the keyword "infrastructure".

The behavior of a composite *sort* is derived from the behaviors of the component *sorts*, in a manner that depends on the compositional relationship (Stouffs and Krishnamurti 2002). That means that the keyword "Infrastructure" as instance of the *sort concepts* cannot be matched to the keyword "infrastructure" in the exemplar data form in Figure 1 without automatic conversion of data forms. The latter keyword, namely, is a subordinate keyword to the keyword "theater" and together form an instance of the *sort conceptstree*.

However, transformations may apply not only to the data, corresponding to the individual data types, but also to the composite structure of the data. When looking for a yellow square, one does not necessarily need to be concerned with the fact that yellow is represented as an attribute to the square, as in a graphical visualization, or instead that the square is represented as an attribute to the color yellow, as in a sorting of the geometries by color. Thus, the matching problem may involve different data views and the conversion of data between these views, all of which can be supported using *sorts*. In this case, if an instance of the *sort concepts* ^ *concepts* (or *concepts2*) is constructed using "Theater" and "Infrastructure," then the recognition of this instance in the exemplar data form will be successful because the *sorts concepts2* and *conceptstree* result in a partial match of similar sorts (see section 2.3). Similarly, the keyword "Theater" as instance of the *sort concepts* can be matched to the keyword "theater" as partial instance of the *sort conceptstree*. However, the problem still remains how the keyword "Infrastructure" as instance of the *sort concepts* could be recognized in the data form as mapping *concepts* onto the *conceptstree* component of *conceptstree* creates an infinitely recursive mapping.

The part relationship underlying the various behaviors enables the matching problem to be implemented for each primitive *sort* or data type. Since composite *sorts* inherit their behavior, and part relationship, from their component *sorts*, the technical difficulties of implementing the matching problem apply only once for each primitive *sort* or data type. As the part relationship can be applied to all kinds of data types, recognition algorithms can easily be extended to deal with arbitrary data forms, even if a proper definition of what constitutes a transformation is still necessary.

## 4. Design Queries

Querying design information, as distinguished from visual inspection, generally requires the analysis of existing information in order to derive new information that is not explicitly available in the information structure. A viable query language has to be based on a model for representing different kinds of information that adheres to a consistent logic providing access to information in a uniform and consistent manner.

Stouffs and Krishnamurti (1996) indicate how a query language for querying graphical design information can be built from basic operations and geometric relations that are defined as part of a maximal element representation for weighted geometries. These operations and relations are augmented with operations that are derived from techniques of counting and pattern matching for the purpose of composing more complex and versatile geometric and non-geometric queries. For example, by augmenting networks of lines that are represented as volumes (or plane segments) with labels as attributes, and by combining these augmented geometries under the operation of sum, as defined for the representational model, colliding lines specifically result in geometries that have more than one label as attribute. These collisions can easily be counted, while the labels on each geometry identify the colliding lines, and the geometry itself specifies the location of the collision (Stouffs and Krishnamurti 1996).

In order to consider counting and other functional behavior as part of the representational approach, *sorts* consider data functions as a data kind, offering functional behavior integrated into data constructs. Data functions are assigned to apply to a selected property attribute of a specific *sort*, which itself may be a data function. Then, the result value of the data function is computed from the values of the property attribute of the data entities of this *sort*. This result value is automatically recomputed each time the data structure is traversed, e.g., when visualizing the structure. For this purpose, this target *sort* must be related to the data function's *sort* within the representational structure under a sequence of one or more attribute relationships, with restrictions. As a data kind, data functions specify both a functional description, a result value, and a *sort* and its property attribute.

Data functions can introduce specific behaviors and functionalities into representational structures, for the purpose of counting or other numerical operations. Consider, for example, a data structure corresponding to a composition of two *sorts* where one *sort* specifies a cost to the other *sort*. Then, by augmenting the data structure with a sum function, applied to the numeric value attribute property of the cost *sort*, the value of this function is automatically computed as the sum of all cost values. Figure 4 illustrates a similar example in the context of lighting design for a stage or TV or movie studio. Consider a *sort lights*, of labels denoting spot lights or other movable lights, a *sort intensityvalues*, of numeric values representing light intensities or wattage values, and a *sort intensity,* of numeric functions:

*lights* : [Label]
*intensityvalues* : [Numeric]
*intensity* : [NumericFunction] (7)

Both labels and numeric functions adhere to a discrete behavior, while numeric values adhere to an ordinal behavior. Consider a composition of these three *sorts* under the attribute relationship, such that each intensity function has as attribute a collection of lights and each light has as attribute a single intensity value:

*lights_intensity* : *intensity* ^ *lights* ^ *intensityvalues* (8)

By instantiating the *sort intensity* with a sum function applied to the numeric value property attribute of the *sort intensityvalues*, the value of this function is automatically computed as the sum of all intensity values of the lights that are assigned as attribute to this sum function.

Next, consider an extension of this composition of *sorts* with another *sort* providing type or clustering information, e.g., a *sort beams*, of labels denoting the beams that serve to hold the lights above the stage:

$$beams : [\text{Label}] \tag{9}$$

Then, the relative position of this new *sort* with respect to the *sort* of functions has important consequences considering the number of instances of the selected function and, therefore, the result of each of these functions. Figure 5 presents two alternative data views of the same data. In the first data view (left side of Figure 5), the *sort beams* is considered as an attribute to the *sort intensity*, such that the intensity function has as attribute a collection of beams, each of which has as attribute a collection of lights and each light has as attribute a single intensity value:

$$lights\_intensity1 : intensity \wedge beams \wedge lights \wedge intensityvalues \tag{10}$$

In this case, the result of the sum function is still the total intensity of all lights, irrespective of the beams these lights are assigned to.

In the second data view (right side of Figure 5), the *sort intensity* is instead considered as an attribute to the *sort beams*, such that each beam has as attribute an intensity function, which itself has as attribute a collection of lights and each light has as attribute a single intensity value:

$$lights\_intensity2 : beams \wedge intensity \wedge lights \wedge intensityvalues \tag{11}$$

In this case, each beam specifies its own sum function and the respective results are the total intensity of only the lights on this beam.

Thus, moving data functions within the data structure by altering the compositional structure of the representation, automatically alters the scope of the function and thus the result. In this way, data functions can be used as a technique for querying design information, and moving the data function alters the query. Functions that apply simultaneously to two property attributes of two different *sorts*, or compositions thereof, can be used to compute more complex derivations. Consider cost values for linear building elements such as beams, with the cost expressed per meter. If the beam element has a property attribute specifying the length of the element, in the case of a line segment representing the beam, a function might be applied that sums the product of the length of each beam element with the respective cost per unit length. A similar approach could be considered for non-numeric functions, for example, applying to strings or vectors.

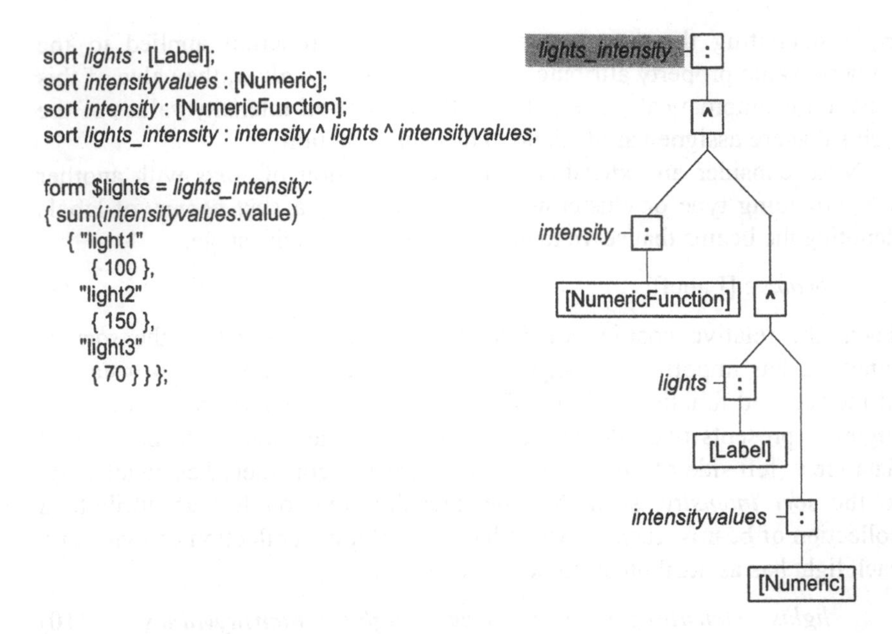

sort *lights* : [Label];
sort *intensityvalues* : [Numeric];
sort *intensity* : [NumericFunction];
sort *lights_intensity* : *intensity* ^ *lights* ^ *intensityvalues*;
form $lights = *lights_intensity*:
{ sum(*intensityvalues*.value)
{ "light1"
{ 100 },
"light2"
{ 150 },
"light3"
{ 70 } } };
lights_intensity
:
^
intensity
:
[NumericFunction]
^
lights
:
[Label]
intensityvalues
:
[Numeric]

*Figure 4.* Textual and graphical definition of a *sort* representing the intensity values of lights and including a numeric function, and the description of an exemplar data form with a sum function applied to the numeric *value* attribute property of the *sort intensityvalues* (*Sorts* Description Language). In the definition of a *sort*, '^' denotes the operation of attribute; ':' denotes the naming of a *sort*; '[Label]', '[Numeric]' and '[NumericFunction]' are primitive *sorts*, the latter defines a *sort* of numeric functions applied to a single attribute property of another *sort*.

## 5. Design Rules and Grammars

Spatial change can be viewed as a computation $s - f(a) + f(b)$, where $s$ is a shape, and $f(a)$ is a representation of the emergent part (shape) that is altered by replacing it with the shape $f(b)$ (Krishnamurti and Stouffs 1997). This computation subsumes both spatial recognition and subsequent manipulation. It can also be expressed in the form of a spatial rule $a \rightarrow b$. Rule application, then, consists of replacing the emergent shape corresponding to $a$, under some allowable transformation, by $b$, under the same transformation.

Rules can further be grouped into grammars. A grammar is a formal device for the specification of a language; it defines a language as the set of all structures generated by the grammar, where each generation starts with an initial structure and uses rules to achieve a structure that contains only elements from a terminal vocabulary. The specification of spatial rules and grammars leads naturally to the generation and exploration of possible

sort *lights* : [Label];
sort *intensityvalues* : [Numeric];
sort *beams* : [Label]
sort *intensity* : [NumericFunction];
sort *lights_intensity1* : *intensity* ^ *beams* ^ *lights* ^ *intensityvalues*;
sort *lights_intensity2* : *beams* ^ *intensity* ^ *lights* ^ *intensityvalues*;

*lights_intensity1* : ^
*intensity* : [NumericFunction] ^
*beams* : [Label] ^
*lights* : [Label]
*intensityvalues* : [Numeric]

*lights_intensity2* : ^
*beams* : [Label] ^
*intensity* : [NumericFunction] ^
*lights* : [Label]
*intensityvalues* : [Numeric]

```
form $lights = lights_intensity1:
{ sum(intensityvalues.value)
   { "beam1"
      { "light1"
         { 100 },
        "light2"
         { 150 } },
     "beam2"
      { "light3"
         { 70 } } } };
```

```
form $lights = lights_intensity2:
{ "beam1"
   { sum(intensityvalues.value)
      { "light1"
         { 100 },
        "light2"
         { 150 } } },
  "beam2"
   { sum(intensityvalues.value)
      { "light3"
         { 70 } } } };
```

*Figure 5.* Textual and graphical definition of two alternative *sorts* representing the intensity values of lights (attached to beams) and including a numeric function, and the description of exemplar data forms with the sum function applied to the numeric *value* attribute property of the *sort intensityvalues* (*Sorts* Description Language). In the definition of a *sort*, '^' denotes the operation of attribute; ':' denotes the naming of a *sort*; '[Label]', '[Numeric]' and '[NumericFunction]' are primitive *sorts*.

spatial designs; the concept of spatial elements or shapes emerging under a part relation is highly enticing to design search (Mitchell 1993; Stiny 1993).

The concept of search is more fundamental to design than its generational form alone might imply. Furthermore, there is no need to restrict it to spatial structures. In fact, any mutation of an information structure into another one, or parts thereof, can constitute an action of search. As such, a design rule may be considered to specify a particular composition of design operations and/or transformations that is recognized as a new, single, operation and applied as such. Design rules can serve to facilitate common operations, e.g., for changing one design element into another or for creating new design information based on existing information in combination with a rule. Similarly, a grammar is more than a framework for generation; it is a tool that permits a structuring of a collection of rules or operations that have proven their applicability to the creation of a certain set (or language) of designs.

Applied to *sorts*, rules and grammars can be considered as a means to contain and facilitate the flexibility and dynamism that *sorts* provides. The specification of design queries through data functions, and the transformation of *sorts* to support alternative data views, can also play a role in the application of a design rule. The central problem in implementing design rules and grammars is the *matching problem*, that of determining the transformation under which the emergent part is recognized in the data. The implementation of the matching problem for *sorts* relies on the part relationship underlying the behavioral specification of *sorts* and only applies to each primitive *sort* or data type (section 3). Rule application then results in a subtraction operation followed by an addition operation. Both operations are also defined as part of the behavioral specification of *sorts*. Stouffs and Krishnamurti (2001b) present a few examples of grammar formalisms that can be expressed with *sorts*.

## 6. User Interaction

Exploring alternative design representations requires the ability to alter representational structures or *sorts*, e.g., by adding or removing components, or by modifying the compositional relationships. Integrating data functions into design data forms similarly necessitates the ability to intervene into the data form and manipulate its composition of data entities and constructive relationships. Utilizing data recognition and design rules also benefits from the same ability to alter and build *sorts* and corresponding data forms. This, furthermore, necessitates a degree of understanding of the representational and data structures that can only be achieved using visual (graphical) means. Practical representational structures for design however may become very large and achieving a visual understanding of the representational structure may be hard to achieve. Furthermore, manipulating large representational structures by editing the individual components and relationships is far from straightforward. Achieving a desired result may require detailed knowledge or investigation of the structures and painstakingly specific manipulations. Therefore, we argue for an incremental modeling approach (see also Woodbury et al. 1999) and a user interaction to support this.

We consider *sorts* as a complex adaptive system; such systems possess the distinguishing characteristics of robustness and flexibility (Dooley 1997; Kooistra 2002). In the context of building representational structures, robustness can be considered to mean that the system offers the possibility for "correspondence" (communication) leading to an agreement on the representation that prevails in the system. At the same time, the system must offer the possibility for representations to change and in such a way that, in principle, claims on this representation generate quality improvement. With respect to *sorts*, assigning a name to a *sort* can be considered as laying claim to this *sort* with the purpose of improving quality. Correspondence on *sorts* can be achieved through incremental changes on *sorts* and by agreement on the naming of *sorts*. This implies that the incremental modeling of *sorts* in the form of defining *sorts* in terms of other *sorts* can play an important role in achieving agreement and thus in containing the "chaos" to which the construction of *sorts* can lead.

While *sorts* consider a finite vocabulary of primitive *sorts* or data types and compositional operators, practical representational structures for design can be very large and, therefore, the variety in *sorts* that can be constructed is by any practical means immeasurable. Constructing *sorts* could therefore result in a seemingly infinite series of questions or choices on which component to add when and where in the representational structure and, thus, result in "chaos." The complexity paradigm implies "systemic inquiry to build fuzzy, multivalent, multilevel and multidisciplinary representations of reality" (Dooley 1997). *Sorts* can be considered, to a certain extent, as a means to build such representations. "Order arises from complexity through self-organization" (Prigogine and Stengers 1984). In the context of building representational structures or *sorts*, the process of self-organization can take on the form of human communication or correspondence.

Correspondence on *sorts* must be facilitated through the user interaction with *sorts*. We have already referred to the ability to model *sorts* incrementally. We are also investigating kinds of actions that can be perceived purposeful in an exploratory process. These are, for example, the specification of a focus onto the structure expressing a particular interest and the selection of a part of the structure as extent of our interest. Each of these actions results in a transformation of the structure.

The expression of a focus onto a representational or data structure can be directly related to the (hierarchical) composition of the structure's entities under the compositional relationships. Entities that are considered more important are commonly found at a higher level in the structure's composition. The attribute relationship serves as a prime example, leading the focus onto the object of the relationship, while the attribute expresses a qualifier with respect to this object. For example, in an architectural design description, spatial information is commonly considered more important such that other information entities are assigned as attributes to the relevant spatial entities. Similarly, object-oriented models often adopt a hierarchical structure of functional objects at various levels of detail, reflecting upon an increasingly narrower information focus. For example, architectural design models are commonly organized by a hierarchical classification of functional areas, such as buildings, floors and zones, in that order.

Thus, expressing a focus onto the representational or data structure can result in a transformation of the hierarchical structure that raises the entity under focus towards the top of the structure. Such a transformation can be achieved automatically by reversing attribute relationships and by modifying other compositional relationships. This transformation may take place under the objective to maximize compatibility with the original representation and minimize data loss. Selecting a part of the structure can similarly lead to the breakup of compositional relationships attempting to maintain maximal compatibility with respect to the selection.

When considering that every change to a representational structure or *sort* constitutes a different data view, it can be argued that advanced support for exploring different data views at the same time facilitates the investigation and manipulation of representational structures. We are currently developing a prototype interface to build and edit definitions of *sorts*, to compare and match *sorts* and to construct corresponding data forms.

## 7. Conclusion

Representational flexibility for design cannot be simply realized by providing the user access to the representational and data structures and enabling the modification of these structures through the addition of attributes or the manipulation of the structures' entities and compositional relationships. It also has to facilitate the exploration of these structures through searching and querying the structures. Furthermore, it can be desirable to be offered the ability to identify and store common actions and manipulations for later reuse. Through support for data views, data recognition, design queries and design rules, the theory of *sorts* is a more than viable candidate for achieving representational flexibility for design. The success of this or other approach is as much dependent on the accessibility of the approach and its techniques to the user. For example, powerful query languages do not as such serve the end user (or designer) who is only interested in having easy access to the information, not in learning a new language. A visual approach can offer a solution. "Visual query languages [...] allow the user to express arbitrary queries without having to master the syntax of a rigid textual query language" (Erwig 2002). Further research and developments into *sorts* will focus onto the user interaction aspect of utilizing *sorts* for exploring alternative data views, data recognition, design queries and design rules. This should also enable us to consider and investigate more complex and practical examples.

## Acknowledgements

This work is partly funded by the Netherlands Organization for Scientific Research (NWO), grant nr. 016.007.007. The second author is funded by a grant from the National Science Foundation, CMS #0121549, support for which is gratefully acknowledged. Any opinions, findings, conclusions or recommendations presented in this paper are those of the authors and do not necessarily reflect the views of the Netherlands Organization for Scientific Research or the National Science Foundation. The authors would like to thank Bige Tunçer for the development of the semantic structures presented in Figures 1 and 2, and Michael Cumming for his work on the development of a prototype interface to build and manipulate *sorts*. The first author benefited from communication with Jan Kooistra concerning complex adaptive systems.

## References

Bazjanac, V: 1998, Industry foundation classes: Bringing software interoperability to the building industry, *The Construction Specifier* **6/98**: 47–54.

Dooley, KJ: 1997, A complex adaptive systems model of organization change, *Nonlinear Dynamics, Psychology, and Life Sciences* **1**(1): 69–97.

Erwig, M: 2002, Design of spatio-temporal query languages, position paper presented at the *Workshop on Spatio-temporal Data Models for Biogeophysical Fields*, San Diego Supercomputer Center, La Jolla, California, <www.calmit.unl.edu/BDEI/papers/erwig_position.pdf>(12 February 2004).

Groth, DP and Robertson, EL: 1998, Architectural support for database visualization, *Proceedings of the 1998 Workshop on New Paradigms in Information Visualization and Manipulation*, ACM Press, New York, NY, pp. 53–55.

ISO: 1994, *ISO 10303-1, Overview and Fundamental Principles*, International Standardization Organization, Geneva.

Kooistra, J: 2002, Flowing, *Systems Research and Behavioral Science* **19**(2): 123–127.

Krishnamurti, R: 1992, The maximal representation of a shape, *Environment and Planning B: Planning and Design* **19**: 267–288.

Krishnamurti, R and Earl, CF: 1992, Shape recognition in three dimensions, *Environment and Planning B: Planning and Design* **19**: 585–603.

Krishnamurti, R and Stouffs, R: 1997, Spatial change: Continuity, reversibility and emergent shapes, *Environment and Planning B: Planning and Design* **24**: 359–384.

Mäntylä, M: 1988, *An Introduction to Solid Modeling*, Computer Science Press, Rockville, MD.

Mitchell, WJ: 1993, A computational view of design creativity, *in* JS Gero and ML Maher (eds), *Modeling Creativity and Knowledge-Based Creative Design*, Lawrence Erlbaum Associates, Hillsdale, NJ, pp. 25-42.

Prigogine, I and Stengers, I: 1984, *Order Out of Chaos*, Bantam Books, New York.

Snyder, JD: 1998, *Conceptual Modeling and Application Integration in CAD: The Essential Elements*, PhD dissertation, School of Architecture, Carnegie Mellon University, Pittsburgh, PA.

Snyder, J and Flemming, U: 1999, Information sharing in building design, *in* G Augenbroe and C Eastman (eds), *Computers in Building*, Kluwer Academic, Boston, pp. 165–183.

Stiny, G: 1991, The algebras of design, *Research in Engineering Design* **2**: 171–181.

Stiny, G: 1993, Emergence and continuity in shape grammars, *in* U Flemming and S Van Wyk (eds), *CAAD Futures '93*, North-Holland, Amsterdam, pp. 37–54.

Stouffs, R: 1994, *The Algebra of Shapes*, PhD dissertation, Department of Architecture, Carnegie Mellon University, Pittsburgh, PA.

Stouffs, R and Krishnamurti, R: 1996, On a query language for weighted geometries, *in* O Moselhi, C Bedard and S Alkass (eds), *Third Canadian Conference on Computing in Civil and Building Engineering*, Canadian Society for Civil Engineering, Montreal, pp. 783–793.

Stouffs, R and Krishnamurti, R: 2001a, On the road to standardization, *in* B de Vries, J van Leeuwen and H Achten (eds), *Computer Aided Architectural Design Futures 2001*, Kluwer Academic, Dordrecht, The Netherlands, pp. 75–88.

Stouffs, R and Krishnamurti, R: 2001b, Sortal grammars as a framework for exploring grammar formalisms, *in* M Burry, S Datta, A Dawson and J. Rollo (eds), *Mathematics and Design 2001*, The School of Architecture & Building, Deakin University, Geelong, Australia, pp. 261–269.

Stouffs, R and Krishnamurti, R: 2002, Representational flexibility for design, *in* JS Gero (ed), *Artificial Intelligence in Design '02*, Kluwer Academic, Dordrecht, The Netherlands, pp. 105–128.

Tunçer, B, Stouffs, R and Sariyildiz S: 2002, Document decomposition by content as a means for structuring building project information, *Construction Innovation* **2**(4): 229–248.

van Leeuwen, JP: 1999, *Modelling Architectural Design Information by Features*, PhD dissertation, Eindhoven University of Technology, The Netherlands.

van Leeuwen, JP and Fridqvist, S: 2003, Object version control for collaborative design, *in* B Tunçer, S Özsariyildiz and S Sariyildiz (eds), *E-Activities in Building Design and Construction*, Europia Productions, Paris, pp. 129–139.

Woodbury, R, Burrow, A, Datta, S and Chang, T, 1999, Typed feature structures and design space exploration, *Artificial Intelligence in Design, Engineering and Manufacturing* **13**(4): 287-302.

JS Gero (ed), *Design Computing and Cognition'04*, 239-258


# PROVIDING AN OVERVIEW DURING THE DESIGN OF COMPLEX PRODUCTS

*The Development of a Product Linkage Modelling Method*

TIMOTHY AW JARRATT, CLAUDIA M ECKERT, P JOHN CLARKSON
*University of Cambridge, UK*

and

MARTIN K STACEY
*De Montfort University, UK*

**Abstract.** The lack of overview is a major problem in the design of any complex product; designers' mental processes lead to a partial and biased understanding of it. With conventional representations it is difficult to gain an even overview over the product. This paper discusses a product linkage modelling method, which was developed from earlier work on the prediction of change propagation, as a way of constructing and representing overview product models. Initial evaluation of the method has successfully occurred in two UK industrial companies.

## 1. Introduction

Complex engineering products are designed by multidisciplinary teams. Each designer in such a team typically only has a detailed view of a few of the issues involved in the design. Some designers understand specific systems, whilst others understand certain functional areas, but, at various points in the design process, each type needs to have an overview of the entire product to make the best decisions for the product as a whole. Each designer or team has different mental models and uses a wide range of external representations of the product and the product knowledge. This paper presents a method to capture overview product data and provide an effective summary of that data. Products are represented by a component-component matrix with the types of linkage that exist between components described in the cells. The component breakdown and the linkages are negotiated by a team appropriate to the product. The overview matrix can serve as a reminder to people about the connectivity in a product. It also facilitates the indexing of knowledge held by others. At the same time the matrix can be used as the basis of a method to predict how change to one component will spread through the rest of the product.

### 1.1. OVERVIEW OF THE PAPER

This paper describes the development of a product linkage modelling method. Based upon the literature on the psychology of design (Section 2) and our case studies in three UK engineering companies (Section 3), the paper proposes a representation based around Design Structure Matrices (DSM) and a classification of the linkages between components. This simple, yet comprehensive modelling method is outlined in Section 4. The evaluation of the method is described in Section 5.

## 2. Understanding Complex Products

One aspect of product complexity is the number of parts in the design: for example, a modern aeroplane consists of over 100,000 parts, a car has over 10,000, whilst a modern car engine has over 500. Even if we discount standard components, such as nuts, bolts and washers, an engine still has over 100 components designed specifically for it. This makes it very difficult for any single designer to both understand the product in depth and keep an overview of it.

When we questioned the deputy chief engineers of a helicopter manufacturer (Eckert et al. 2004), they admitted that, while they had the best overview of the product, they could only confidently claim to understand about half of the helicopter in detail, Figure 1.

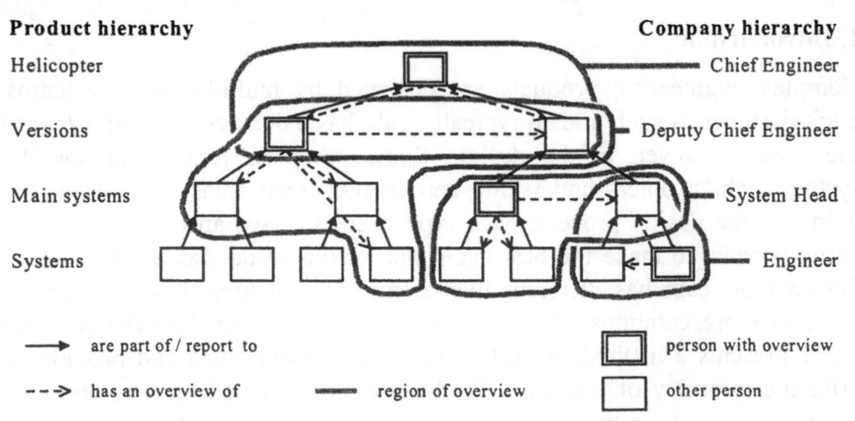


*Figure 1.* Overview over a helicopter (Eckert et al. 2004)

Any designer's understanding is biased by his or her own experience and expertise. For example, those with a mechanical engineering background did not understand the details of the avionics architecture, which was specified in-house, but developed elsewhere.

### 2.1. REQUIRING AN OVERVIEW

Nobody can understand all the aspects of a complex modern product, but even if they could, this would only be a small part of fully understanding the product. Designers need to know how the different parts are connected and what function they are intended to carry out, what requirements they fulfil and why they are designed in a particular way. In the most extreme cases, nobody in a company has an answer to these questions. For example, a brake manufacturer completely redesigned a 1930s railway wagon brake, which had an expensive copper cone in the centre. The firm wanted to replace the copper part, but did not know why it was there in the first place and struggled to predict the impact of removing it (Lindemann et al. 2001).

In most cases the consequences are more subtle, but far reaching. Without really understanding the details of a product, it is difficult to predict its exact behaviour. Design is chaotic, in that very small changes can have huge effects in very unforeseen ways. In a complex and highly integrated design like a helicopter, it is very hard to predict the spreading of vibration or noise, which result from the detailed properties of particular components. When a new product is developed the understanding of the design team emerges as the design progresses and uncertainties are gradually removed. In theory an overview is created as details are fleshed out as the design progresses from conceptual design, through embodiment design to detailed design. The reality is more complicated. Due to lead times and other constraints some parts of the design are developed earlier than others. Some components are bought in from suppliers and so the design needs to be generated around those parts, whilst others are created entirely in house requiring the collaboration of diverse teams. This process can only be co-ordinated and managed by somebody with an overview of and a vision for the product. Mistakes are often only spotted when parts are integrated and can be costly when they push a product over a contractual deadline.

Needing to have an overview of the product goes beyond being able to predict the performance of a part and the impact of a change to it. It is vital for managing the design and development process. At the beginning of a project tasks need to be planned and people allocated to these tasks. This can only be done based on an understanding of similar projects and an understanding of the expected properties of the new product. As the design progresses it is often the role of the manager to negotiate interfaces and trade-offs with other parts, which again requires an overview of the entire product to see the viewpoints of the negotiating partners (Buccarelli 1994). Individual designers have frequently complained in interviews (Eckert et al. 2001), that they do not know how their tasks fit into the bigger picture. They do not know where the design parameters and information that they are using come from and who will make use of it later on. The consequences of this are that, if they change a component or even a single value in the process, they do not know who will be affected by their decision (Flanagan et al. 2003). More fundamentally without an overview of the product, designers lose a part of their potential sense of ownership of the work, which is part of their motivation.

Industry has clearly recognised the importance of having a greater overview of a product by expressing a need for generalists to work in and manage design processes. The more complex the product, the more specialised the people working on it become. Present engineering education reinforces this by training people to a very high degree in certain specialities. To cope with the inherent complexity of a product, such as those produced by the aerospace industry, it is conceptually designed in major sub-systems with the result that these systems are optimised in themselves, but might perform sub-optimally when integrated together. Only generalists can make sense of information about the various systems and access trade-offs.

In their daily activities engineers use three-dimensional Computer Aided Design (CAD) models and two-dimensional drawings. Both are detailed descriptions of a product and designers need to know where to look on them to find information. Geometric mismatches can be shown by CAD systems, but these systems cannot give a probabilistic indication of how change could propagate. CAD systems are only just beginning to model system behaviour and at present give little indication of unwanted effects, such as vibration, temperature flows, electromagnetic interference, etc. These packages do little to give designers an overview of the product, because they do not provide an abstracted view of the product.

In many ways the most useful representation of a new design is an existing design. In fact designers often verbally describe new designs to each other with reference to existing designs. This is a very powerful and fast way to communicate, but can lead to misunderstanding due to the inherent ambiguity of verbal references (Eckert and Stacey, 2000; Eckert et al. 2003). Only the existing design allows designers to think and talk about functionality and performance easily. However considering how much the overall characteristics of a design can be affected by small changes to one part, this is an inaccurate representation.

## 2.2. THINKING ABOUT COMPLEX PRODUCTS

It is not surprising that designers struggle to have an overview over a product; with a complex product there is a great deal to remember. Designers typically either work on specific components or they carry out a certain function, for example with helicopter design a specialist team works on the undercarriage and that is all they concentrate on whilst dedicated people work on stress analysis and load calculations. These different specialists work in what Bucciarelli (1994) terms different *object worlds*. Each group has its own way of looking at design problems: shared background knowledge, concepts and terminology, problem solving procedures, and skills for creating and making sense of visual representations of various kinds of design information. Words and diagrams can mean different things to engineers with different areas of expertise, who apply different knowledge to fleshing out skeletal descriptions and inferring implications. The different groups interact by exchanging and referring to *boundary objects* (Star 1989), representations of design information that different types of engineer understand and can relate to their own concerns. In meetings designers negotiate shared (or at least compatible) understanding of design ideas (Minneman 1991); much of what they say in meetings is clarification (Maher and Simoff 2000).

Complex products cannot be described in a single representation, and some information is never described in a representation that everybody would share. Finding and employing an effective set of boundary objects is crucial for successful multidisciplinary design; developing useful boundary objects is an important goal for design researchers.

Designers interpret visual and verbal information using the concepts comprising their object world to develop mental representations of design ideas. They may have multiple representations of the same design. Some of designers' mental representations are mental models that they can use to envision how the artefact will behave (Johnson-Laird 1983). While some mental models are models of how the thing works, others map inputs to external behaviour – a user's-eye view. Designers with similar expertise will have very similar mental models, but it is easy for both designers themselves and outsiders to overestimate the similarity of their thinking. For instance we have met diesel engine designers with superficially similar backgrounds who employ radically different mental representations.

Many designers think visually and have very vivid mental imagery. Anecdotal evidence indicates that mechanical engineers are usually extremely visual and think about problems by mentally manipulating the geometry in their heads. Several mechanical engineers we have interviewed describe this as akin to a "CATIA system in their head". More analytical engineers such as stress engineers often think in terms of the correlation of parameters required to achieve a target performance. Some of them have commented to us that while they can construct mental imagery at a push, they do not naturally think in images.

Designers' mental representations of designs are limited: they may only include part of the design, and there is no guarantee that these are consistent or even coherent; people may only recognise the limitations of their mental representations when encounter questions they cannot answer. Research on mental imagery (Kosslyn 1980; 1994; Logie 1995) shows that people can have a subjective sense that their mental representations are more complete and detailed than they really are, and that details are only filled in when people focus on parts of their mental images. This is partly because the capacity of working memory is limited; Miller (1956) famously assessed its capacity as seven plus or minus two chunks. The richness of mental representations depends on the complexity of the chunks. The reliability of memory recall depends largely on the richness of the relationships between the elements to be remembered; this is increased by creating mental images of to-be-remembered information. Chunk size has been found to influence the accuracy of memory recall of, for instance, electronic circuits (Egan and Schwartz 1979) and architectural drawings (Akin 1978).

Designers frequently must think in terms of the functions that their design or parts of it need to carry out as well as its structure. They also need to think about causal processes like the transmission of noise and vibration that have only a subtle relationship to physical form. Many designers however think about new designs with reference to existing designs, using mental representations including physical embodiments as well as functions and performance factors (Schon 1988; Oxman 1990; Eckert and Stacey 2001, for discussions of the roles of types of design elements and individual examples in design thinking). However, this locks them into tacit assumptions about the structure of the new design that are very difficult to escape – a phenomenon known to psychologists as *fixation* (Purcell and Gero 1996). It requires developing mental representations of what the design should *do* that abstract away from physical embodiments. Getting designers to do this is a major purpose of many prescriptive design methods. Axiomatic design (Suh 1990) instructs engineering designers to begin with a functional breakdown and develop the concepts on a high level of abstraction, then break the function down further and then develop the form from it until the design is fully defined. Many engineers find abstract functional thinking very difficult; students who have learnt the axiomatic design method vary enormously in how easy they find it to use.

One reason why thinking in terms of abstract functional relationships is difficult is because functional properties are associated in memory with physical embodiments, which are hard to consciously ignore, and because the relationships between the components of functionally-imagined systems are sparse and more-or-less arbitrary, so they do not serve as effective cues for remembering each other. By contrast, actual machines and descriptions of physical structure have rich, non-arbitrary, mutually reinforcing spatial relationships that are relatively easy to visualise and remember, and that are effective retrieval cues for spatial information in memory. Causal relationships such as noise transmission are not salient parts of primarily geometric representations.

Hence designers, who need to get away from their assumptions about what their designs ought to look like, or who need to think about relationships between elements of their designs other than structural connections and movement, are less able to rely on their memories or diagrams showing geometry. So they have an especially great need for visual representations of design information that make explicit the aspects of the design they need to think about, and remind them of all the components and relationships they need to consider. In software development, UML is a widely used toolkit of interlocking alternative representations that make different aspects of designs explicit. In our experience, designers contributing to the development of very complex engineering products often lack a sufficiently rich understanding of the components of the product outside their specialist areas and the various relationships between them to rely on their memories and mental models. This is inevitable, but they also lack visual representations of the functional and other causal relationships between design elements, that they can search effectively for reminders of what they should know, but cannot easily recall, and pointers to their colleagues' expertise.

## 3. Visualising Product Connectivity

This paper draws on three studies into the processes of making engineering changes to existing products, Figure 2 for two examples. The needs of each company led to the development of a method to visualise linkages between components in a product and predict the way change would spread if a particular component were changed.

In 1999 we worked with a UK based helicopter manufacturer, which was interested in predicting the impact of change to aid the process of tendering for new contracts. In response we developed a tool called the Change Prediction Method (CPM), which provided a high level overview of the entire product and indicated the risk of change propagating for each component.

Since 2002 we have been working with a UK diesel engine company, which faces two key challenges: firstly large numbers of small changes being required by individual customers and secondly incremental development of their core product to meet ever more stringent legislation.

Their needs were best meet by a tool that modelled the linkages between components in the product. Our tools were evaluated in a case study with a UK jet engine manufacturer, in which the conceptual design of a simple engine was modelled. This product is being designed to be used in a variety of applications and so will have to be changed in the future. A result of this final case study was a highlighting of the need for visualisation techniques and software to support the product linkage modelling method.

*Figure 2*. Two of the products examined in this project – a helicopter and a diesel engine (pictures courtesy of Rolls-Royce plc. and Perkins Engines Company Ltd.)

The studies reported in this paper are part of an on-going interaction with industry, which aims to develop tools to improve the effectiveness and efficiency of design processes. Other studies have focussed on design processes planning and communication activity, which also require a good overview of products and processes (Eckert and Clarkson in press).

### 3.1. OUR METHODOLOGY

Both in the helicopter and the diesel engine company we interviewed over 20 engineers and engineering managers with a wide range of roles in their organisation. The interviews lasted between 45 and 90 minutes and were transcribed and analysed. In both companies we also observed meetings where engineering changes were discussed. The first author spent several weeks in the diesel engine company shadowing a senior engineer and working on the linkage modelling method with him. In all of the companies we built product models with the aid of experienced engineers and presented our findings from the interviews and our results from the model building back to the participating engineers and superiors.

### 3.2. THE CHANGE PREDICTION METHOD

Our empirical study into helicopter design (Eckert et al. 2004) has shown that the process that needs to be carried out to address a change is largely independent of the origin of the change, which could be a new requirement or an error. Due to the complexity of the product and the organisation, individuals engineers only had a partial understanding of the product, and expressed a desire for an easy way to gain an overview of it.

As mentioned above, this study resulted in the development of a method of predicting change propagation called the Change Prediction Method (CPM) (Clarkson et al. 2001). At the core of the CPM tool is a combination of the representation and analysis methods used with Design Structure Matrices (DSM) (Browning 2001), which are used to model the connectivity between the components and sub-systems that make up the product, with risk management techniques. The CPM uses a simple model of risk, where the likelihood of an event occurring is differentiated from the impact of such an occurrence. Risk is defined as the product of likelihood and impact.

The method is illustrated in Figure 3 and has three steps; it is the first of these, "Initial Analysis" that will be focused upon. Firstly a product model or breakdown must be created and put into DSM format. Once the interconnectivity between the sub-systems is represented, the change relationships can be shown. Matrices for likelihood and impact are generated with values between 0 and 1. The impact and likelihood matrices created are *direct* matrices in that they represent the risk of change propagating between linked sub-systems. *Indirect* change propagation requires the involvement of at least one intermediate sub-system and this forms a chain of change propagation. The combined impact of changing one component on another is the sum of the direct and indirect affects.

The major issue at the start is the granularity of the component breakdown of the product. Whilst a model that incorporates every single part of a product may have a certain completeness to it, there is a loss of focus to the technique along with the difficulty of handling and understanding such large arrays of information. A helicopter has over 10,000 parts. Very simple standard parts, such as nuts and bolts can be ignored. Most products are too complex to be modelled on a component level, and therefore have to be modelled as sub-systems. Finding a suitable component breakdown depends on a thorough overall understanding of the product to identify all those components that are significant for all the key parts of the organisation. The helicopter company initially suggested a breakdown of 200 components, which were grouped into 49 sub-systems. Having this number of elements in a matrix makes it difficult to capture and express the connectivity within the product. We grouped the pieces further to give a coarser granularity. This representation consisted of 19 systems, which easily fitted onto a computer screen or single piece of paper. The resulting matrix can be reordered to show those parts of the product that are most severely affected by the change.

The key point about the CPM tool is that it is probabilistic. There are a great many possible propagation paths present when a change is made, but many will not be followed because the likelihood of this happening is low.

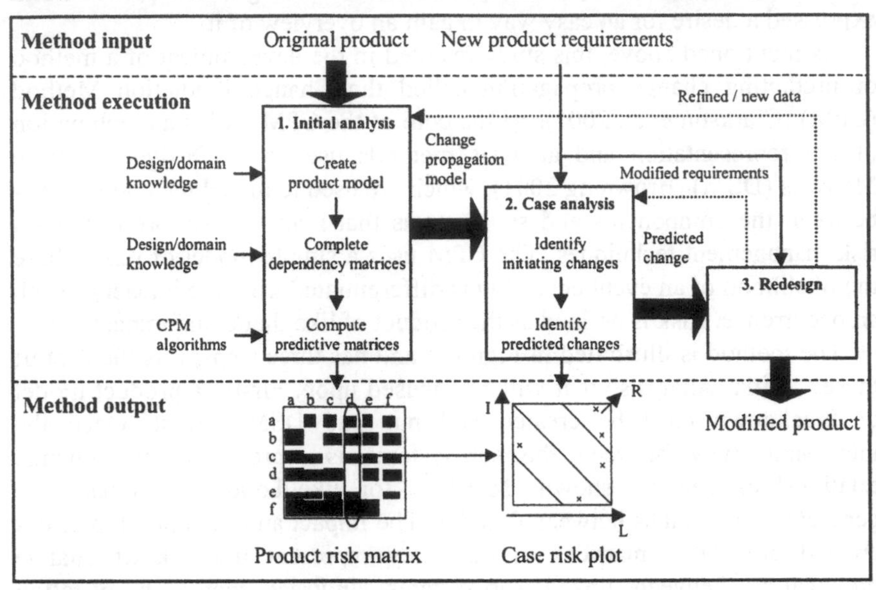


*Figure 3.* Illustration of the Change Prediction Method (Clarkson et al. 2001)

## 3.3. LINKAGE ANALYSIS

Our second case study looked at off-highway diesel engines. To meet the needs of this market, the company offers families of engines of 3-, 4- and 6-cylinders, which need to be redesigned and updated every five to six years because of the requirements of new exhaust emissions legislation (Jarratt et al. 2003). In such a mature industry the development of a new product is essentially a modification process of an existing one, where certain components, such as the gear system remain unchanged. Off-highway engines go into a very wide range of products, such as diggers, tractors, etc., but also generator sets. Each customer has slightly different requirements and requires a subtly different product. In consequence the company makes about 500 small changes to a current engine each year. While each change on its own might be simple, the overall design, logistics and manufacturing effort is considerable and engineers find it hard to appreciate fully the complexity of linkages between parts that could cause changes to propagate. A senior engineer summed up this situation well: *"We miss the other things that the components are doing because most components are doing several jobs."* One experienced design manager commented that certain connectivities were only considered because the firm had *"been burnt by them"* in the past. To support new engine development, a good overview of the linkages between components was seen as vital.

To meet the needs of this company the product modelling part of the CPM method was extended to include the nature of the linkages between components. This method has been called product linkage modelling and is described in detail by Jarratt et al. (2004).

Before discussing the elicitation process in detail, it is vital to define what is meant by a component linkage: it is a direct relationship or connection that exists between two pieces of a product. Depending upon the level of granularity chosen, the pieces can be individual parts, sub-assemblies or modules. The linkage can represent any important relationship that would connect the two pieces, from a connection, which, if broken, causes the device to cease to operate (e.g. electrical flow) to an association, which, if violated, does not affect the product's primary performance (e.g. a vibration effect). Linkages can either be symmetrical (acts in both directions i.e. A → B and B → A, e.g. when two components are bolded together) or directional (there is a flow or transfer from one component to the other i.e. A → B only, e.g. heat from the engine). Obviously change can only spread from one component or sub-system to another if they are connected in some way. Even a simple product has a complex network of linkages between its parts. As the complexity of the product examined increases, the number of potential types of linkage rises to include issues such as heat, electricity, etc.

Linkage analysis differs from functional analysis in a number of ways. Firstly, linkage analysis examines the connections between components and assemblies whereas functional analysis identifies the functionality of the individual components themselves. Added to this, linkage analysis looks at a product from a change perspective and as such has to fully embrace negative aspects (e.g. heat dissipation, electro-magnetic interference, vibration, etc.) and could include aesthetic/ form issues as well as the functional aspects. Also linkage analysis could be used to examine how the product fits into the complex web of processes that create it and support it in the market place. For example parts and sub-systems can be linked due to similar manufacturing processes or due to supply chain issues such as being sourced from the same supplier.

This approach has similarities to that of Pimmler and Eppinger (1994), who used static DSMs to reveal and examine alternative product architectures. The method investigated four different possible interactions between components/ assemblies: spatial, energy, information and material. These were related to the functional modelling concepts that were proposed by Pahl and Beitz (1996) and Suh (1990). One key difference is that this work focuses upon making product models to assist the alteration of already established products (i.e. the architecture is already established), whereas the work of Pimmler and Eppinger focused upon the development of different architecture concepts.

## 4. Building Connectivity Models

The method was developed and tried out in conjunction with the diesel engine manufacturer and evaluated in a further case study with the jet engine company. By building models with companies and observing engineers interacting with them we could also gauge the degree of overview individuals had of their product.

### 4.1. THE DIESEL ENGINE MODEL

The initial component breakdown for the diesel engines involved a long negotiation process between the first author and a senior engineer he was working with. Even when the simple, standard components have been eliminated a diesel engine still has over 100 components. In order to minimise the number of components in the matrix it was necessary to group components into systems. The final model covered 26 major assemblies, which were based upon the options offered in the company's engine reference manual; this model was put into DSM form. In negotiation with the senior engineer mechanical, spatial, thermal and electrical connections were identified as key linkages. He also provided a first version of the matrix by talking us through some of the linkages and filling the rest of the matrix in his own time.

We wanted to gauge the understanding that individuals had of the overall product. Four sessions were run with individual designers. They were provided with a blank matrix with the component breakdown and the key linkages. The subjects were given free reign to add in extra linkage types or amend the model if they felt that important connections could not be represented. The four engineers had many years of experience with diesels and a good overview of the engine; they were:

- a design analysis expert, with over 30 years experience;
- a senior design manager, who had worked as an engineer and had managed several successful new product introduction programs;
- a senior conceptual design, who had reached his position through an apprenticeship route and was responsible for the fundamental conceptual design of the engine; and
- a conceptual designer, with a degree in mechanical engineering and seven years of engine design experience.

Whilst filling in the matrix the subjects were encouraged to "think aloud" (Ericsson and Simon, 1993) and explain their reasoning behind each identified linkage. The blank matrices were printed on A1 paper and then transferred to Excel spreadsheets for analysis. The exercise was videotaped so that the subjects' responses could be reviewed later.

The approach of the subjects was different; three engineers went through the matrix column by column looking at all the connections between each pair of parts, whilst the fourth engineer looked at each type of linkage in turn. The subjects largely agreed on geometric connections (spatial and mechanical) between parts (besides some divergent interpretations of component assemblies). However, dynamic links, electrical and thermal links were interpreted quite differently. The filling in of the matrix was strongly influenced by what the engineers knew about the design and what their daily activities were. For example, the analytical designer knew most about thermal links, because they fell into his area of expertise. The exercise showed that it would be highly unlikely to gain a complete picture of the engine linkages from one person, even from an engineer with a good overview of the product and many years of experience. No single person has a sufficient overview. Therefore it was decided to assemble a group of engine experts to discuss the model and elicit a diesel engine model.

The group exercise was based around a specific engine taken from the production line: a turbocharged, electronically controlled engine was chosen with a gear driven compressor and a belt-driven alternator. An engine model made up of 41 parts and assemblies was decided upon, which was grouped into two categories 'core' and 'non-core'. The assembled team consisted of the Design Project Leader, the Concept Design Leader, the Head and Block Team Leader, a Fuel Systems Specialist, the Electronics/ Electrical Systems Manager and a Configuration Manager. This group gave comprehensive coverage of both the core and non-core areas of the engine. Two days before the exercise, each of the team was given a briefing sheet, which explained the nature of the exercise and defined the linkages. The team assembled around a production engine; the authors led the exercise and it was video taped. As with the individual exercise, the matrix was printed on A1 paper and then transferred to Excel spreadsheets for analysis.

With the engine model consisting of 41 components (1,640 possible interactions) there was a concern about the length of time required for the exercise. Thus, it was decided to 'prime' the matrix first so that the assembled team did not have to address every single possible dependency. The week before the team exercise, a leading designer went through the DSM and filled in all the links that he deemed essential. This took approximately 10 hours in total.

In the team session the engineers were not only required to provide the linkage between the components, but also the likelihood and impact of a change should it be necessary. They used a Failure Mode and Effects

Analysis (FMEA) scale that they were all well familiar with. A section of the final matrix is shown in Figure 4.

The approximate number of hours taken for each stage of the elicitation was recorded. The total number of man-hours invested in the model was just over 70. However, researchers put in about 22 hours so that input from the company was in the order of 50 man-hours.

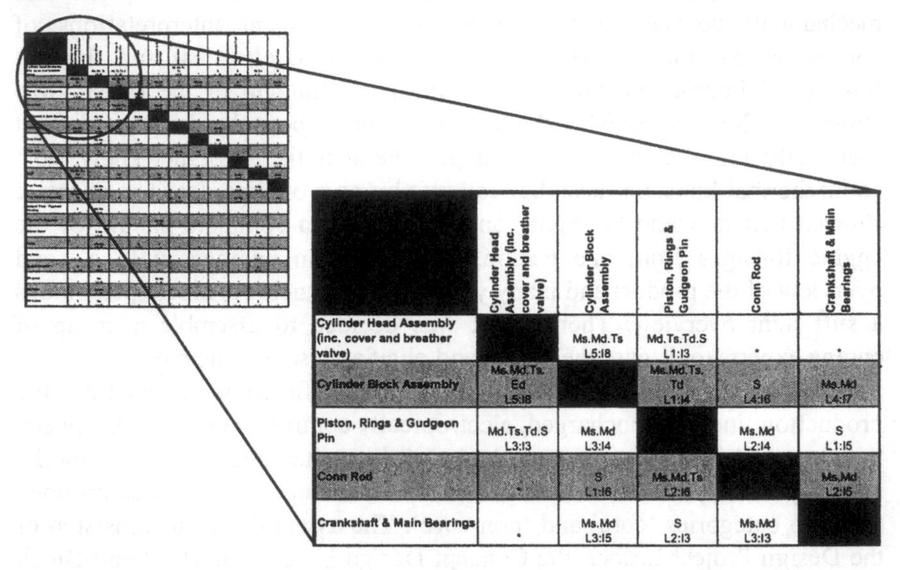

Cylinder Head Assembly (inc. cover and breather valve)
Cylinder Block Assembly
Piston, Rings & Gudgeon Pin
Conn Rod
Crankshaft & Main Bearings
Cylinder Head Assembly (inc. cover and breather valve)
Ms.Md.Ts L5:I8
Md.Ts.Td.S L1:I3
Cylinder Block Assembly
Ms.Md.Ts. Ed L5:I8
Ms.Md.Ts. Td L1:I4
S L4:I6
Ms.Md L4:I7
Piston, Rings & Gudgeon Pin
Md.Ts.Td.S L3:I3
Ms.Md L3:I4
Ms.Md L2:I4
S L1:I5
Conn Rod
S L1:I6
Ms.Md.Ts L2:I6
Ms.Md L2:I5
Crankshaft & Main Bearings
Ms.Md L3:I5
S L2:I3
Ms.Md L3:I3


*Figure 4.* Section taken from the elicited diesel engine matrix showing the linkages (each different type of linkage has a letter code e.g. Ms = Mechanical Static linkages) and change propagation values (likelihood : impact)

## 4.2. THE JET ENGINE MODEL

The change prediction and linkage method were evaluated in a case study with the jet engine company, which invited us to model the conceptual design of a small jet engine. Being a very complex product, a jet engine is usually developed by component teams that work on one part of the engine, e.g. the compressor, with little interaction with other teams. The company needs engineers with an overview of the entire product and wanted to explore using the matrix as a tool to provide an overview and act as a reminder for engineers with different backgrounds.

The jet engine could be modelled on a component level, because it had been designed to minimise the number of parts. It had only 32 parts when nuts and bolts were excluded. This could be expressed easily in a matrix. The modelling was undertaken jointly by the authors and a group of engineers in the company, who used a schematic concept diagram to assist them. A list of linkages was drawn up in about 30 minutes. After an initial two-hour session with the researchers a junior engineer filled the matrix in with his colleagues. Other parts that were not on the plan, such as the control unit were included after these discussions, because engineers in the initial meeting had not recognised their significance earlier.

The engineers liked the overall matrix method, but many found it difficult to appreciate all the linkages in the matrix; a number commented that they found it hard to visualise all the information contained within it. As a response we developed computer software for them, which allows them to explore the linkages interactively. Figure 5 shows screen shots from this software. A circle represents each component and the different types of linkages are colour coded and arrows mark their direction. Double clicking on a component causes the representation to redraw with the selected component in the centre of the screen.

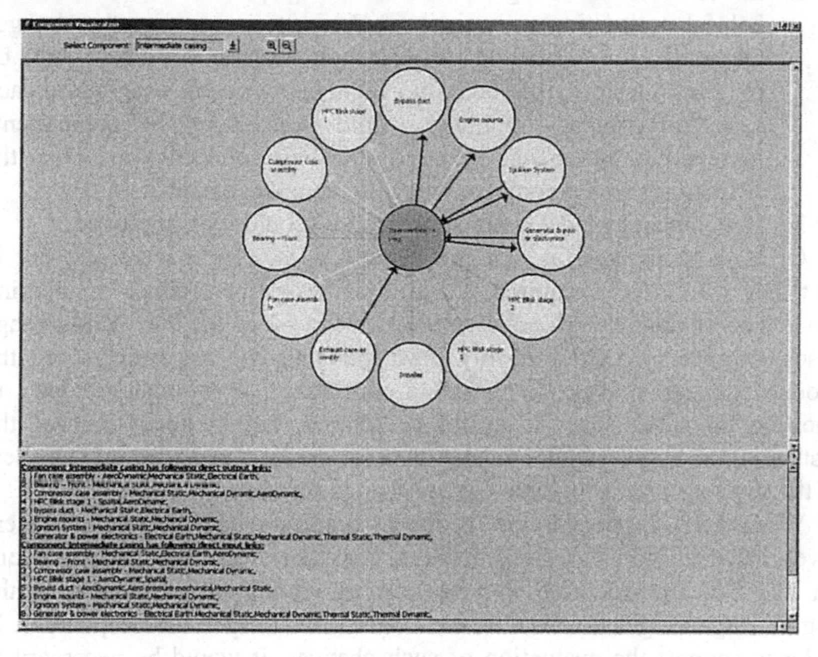


*Figure 5.* Screen shot from the visualisation software

## 5. Evaluation of the Product Linkage Models

The two models were verified in different ways. To date, it has not been possible to properly validate the improvement that such models could make to engineers' overview of products.

### 5.1. DIESEL ENGINE MODEL

With the diesel engine case study, it was possible to gather historic change data and use that to evaluate the model.

All the engineering changes made to the engine type during its first year of production were downloaded from the company's central database: in total there were 624. Each of these changes was examined in turn by the first author in conjunction with a leading diesel engine designer, who had been intimately involved with the development of the engine. Three questions were asked of each change.

1. Is this a relevant change? A great many changes are administrative (drawings requiring updating, build instructions needing to be altered, etc.) or bill-of-material issues. These do not require design resource and would not need to be supported by an overview model.
2. If it is a relevant change, would the linkage product model have been helpful in assisting an engineer or design team examine the change? Obviously the answer to this question is quite subjective and, by reviewing historic instances, hindsight becomes an issue, but in each case the company engineer would sketch out the components involved in the change scenario and their connectivities. Then the first author would compare this with the linkage model.
3. If the linkage model is not suitable, how could it be improved?

Of the 624 engineering changes examined, 94 (15%) were relevant in that they would have required a degree of design or product engineering input. The company engineer identified that in 64 of the 94 (68%) change cases, engineers would probably have gained some assistance from the product linkage model. The level of assistance is of course, open to question, in some cases it would be limited, but in about half of the instances, the engineer felt that the model accurately provided an overview of the key connections involved in the change scenario.

When the 30 instances were the model would not have been helpful were investigated further, it became apparent that there were two major reasons for this. The first of these was the level of granularity. A number of the engineering changes involved detailed alterations to specific components. In order to support the evaluation of such changes, it would be necessary to have a scalable model. The second major issue was that the model could not highlight leakage paths or potential electro-magnetic interference. These issues will be addressed in the next generation of diesel engine models.

### 5.2. JET ENGINE MODEL

The jet engine model had to be evaluated in a different way because we had modelled a concept design and therefore there was no historic change data for the product. For this model, five hypothetical change scenarios were generated based upon the company's extensive experience with other jet engine programmes. These scenarios were given to a very experienced engineer (over 25 years service with the firm) who was asked to talk through the aspects of the engine he would expect to be affected as a result of each change. The thoughts of the engineer were then compared with the model.

Overall there was a good correlation between the model and the thoughts of the experienced engineer. There were some anomalies as would be expected. Several were due to missing connections, which when the matrix was re-examined it was felt should have been present.

Evaluating a model of a concept design, which is yet to be built, is obviously an extremely difficult task, but one key aspect of the work with the jet engine company was that it highlighted the need for visualisation software tools to accompany the linkage models.

## 6. Discussion

In both case studies engineers found it very easy to fill in mechanical links and reached a consensus quickly, because they understand the spatial arrangement of the product. With a schematic or a physical product they can reason about geometry.

During the individual exercises at the diesel company, an interesting difference emerged between the conceptual designer and his analytical colleagues. The conceptual designer had a very good understanding of how the parts were connected, but when he had to think about how a change would propagate, he struggled to tell primary changes (part **A** changes, therefore part **B** changes) from secondary changes (**A** changes causing **B** to change, therefore **C** changes). He would often want to put in a linkage between **A** and **C**. His way of thinking was in terms of his experiences: e.g. *"Last time I changed A, C also changed"*. His analytical colleagues on the other hand, analysed the nature of each of the links and would provide a complex chain of connections. For them the matrices provided a structured way to express their knowledge. The difference lies between recalling linkages and actively constructing them by reasoning about the product. Both approaches are error prone, but highly complementary, with the result that the different engineers greatly respected each other.

The domain experts added many links to the matrix that had not been previously been included. For example the fuel system expert entered a further 15-20 links. More strikingly, none of the diesel engine engineers ever mentioned control linkages as an important issue during the concurrent verbalisation or other discussions. As soon as the control expert looked at the list of linkages during the group elicitation, he mentioned control linkages and everybody agreed that they were important. We told this story to the jet engine designers, when we drew up the list of linkages for their product. They chose not to include control links, but then had to revise this opinion later on, when they talked to a control expert, who pointed out how important such linkages were.

The CPM and product linkage method work together to help designers gain an overview over their products both by building the matrices and by using the change prediction methods. The method provides a simple and structured way to think and talk about connectivity, which people can pick up very quickly. The negotiations in building the matrix allow designers to see each other's viewpoint in a non-controversial way. They get exposed to the views of other engineers and see how they think about change problems. Everybody who took part in the exercises commented that they found it interesting and worthwhile. For example, the young jet engine designer gained enormous confidence through the exercise, both because he learned new facts, but, more importantly had his understanding confirmed by the domain experts.

When designers filled in the matrix they talked very naturally about changes that had happened in the past and the logical connection between components. This is a very structured way to capture both anecdotal and structured design knowledge. We are currently exploring using the output of the knowledge elicitation sessions as a knowledge capture tool. The linkage matrix and diagram is a very concise visual representation of product data, that otherwise would have to be teased out laboriously from different representations. It allows designers to assess potential connectivity quickly by focusing their attention to those areas, where a link does exist. If they do not know themselves about the connectivity in detail, they ask the owners of the relevant component or the functional expert responsible for the linkage. This enables them to pull together information much faster when planning design processes and assessing the impact of design decisions.

The change predication method adds a further level of insight into the product. By looking at the propagated change, designers have access to the insights that would usually be held by different experts. They gain a feeling for the criticality of individual components and are encouraged to think through the impact of their design decisions.

## 7. Conclusion

The lack of overview is a major problem in the design of any complex product. Designers' mental processes lead to a partial and biased understanding of the product. With conventional representations it is difficult to gain an even overview over the product. This paper discusses a product linkage modelling method as a way of constructing and representing design overview. The method has been developed in close conjunction with industry and tested with a company that was not involved in its development. It is now starting to be used in structuring design review meetings in the diesel engine company and as a knowledge capture tool in the jet engine company. Further work on the method focuses on including change logic, which allows the user to exclude components from change propagation while combining other components with logical operators.

## Acknowledgements

The authors would like to thank Westland Helicopters Limited, Perkins Engines Company Limited and Rolls-Royce plc. for all their support during this ongoing research. This project is funded by a UK Engineering and Physical Sciences Research Council Innovative Manufacturing Research Centre grant.

## References

Akin, Ö: 1978, How do architects design?, *in* JC Latombe (ed), *Artificial Intelligence and Pattern Recognition in Computer-Aided Design*, North-Holland, pp. 65-104.

Browning, TR: 2001, Applying the design structure matrix to system decomposition and integration problems: A review and new directions, *IEEE Transactions on Engineering Management* **48**(3): 292-306.

Bucciarelli, LL: 1994, *Designing Engineers*, MIT Press, Cambridge MA.

Clarkson, PJ, Simons, C and Eckert CM: 2001, Predicting change propagation in complex design, *Proceedings of ASME Design Engineering Technical Conferences*, Pittsburgh, USA, CD-ROM.

Eckert, CM and Clarkson, PJ: In press, If only I knew what you were going to do: Communication and planning in large organisations, *in* S Tichkiewitch and D Brissaud (eds) *Methods and Tools for Co-operative and Integrated Design* Kluwer Academic Publishers (in press).

Eckert, CM and Stacey, MK: 2000, Sources of inspiration: A language of design, *Design Studies* **21**(5): 523-538.

Eckert, CM and Stacey, MK: 2001, Designing in the context of fashion designing the fashion context, *in* P Lloyd and HHCM Christiaans (eds), *Designing in Context: Proceedings of the 5th Design Thinking Research Symposium*, pp. 113-129

Eckert, CM, Clarkson, PJ and Zanker, W: 2004, Change and customisation in complex engineering domains, *Research in Engineering Design*, accepted - awaiting publication.

Eckert, CM, Clarkson, PJ and. Stacey MK: 2001, Information flow in engineering companies: Problems and their causes, *Proceedings of the 13th International Conference on Engineering Design: Design Management – Process and Information Issues*, Glasgow, UK.

Eckert, CM, Stacey MK and Earl CF: 2003, Ambiguity is a double-edged sword: Similarity references in communication, *Proceedings of 14th International Conference on Engineering Design*, Stockholm, Sweden.

Egan, DE, and Schwartz, BJ: 1979, Chunking in recall of symbolic drawings, *Memory and Cognition* 7: 149-158.

Ericsson, KA and Simon, HA: 1993, *Protocol Analysis - Verbal Reports as Data*, MIT Press, MA.

Flanagan T, Eckert CM and Clarkson PJ: 2003, Parameter trails, *Proceedings of 14th International Conference on Engineering Design*, Stockholm, Sweden.

Goel, V: 1995, *Sketches of Thought*, MIT Press, Cambridge MA.

Jansson, DG and Smith, SM: 1991, Design fixation, *Design Studies* **12**: 3-11.

Jarratt, TAW, Eckert, CM and Clarkson, PJ: 2004, Development of a product model to support engineering change management, *Proceedings of the TMCE 2004*, Lausanne, Switzerland, CD-ROM.

Jarratt, TAW, Eckert, CM, Weeks, R and Clarkson, PJ: 2003, Environmental legislation as a driver of design, *Proceedings of 14th International Conference on Engineering Design*, Stockholm, Sweden.

Johnson-Laird, PN: 1983, *Mental Models,* Harvard University Press, Cambridge, MA.

Kosslyn, S: 1980, *Image and Mind*, Harvard University Press, Cambridge, MA.

Kosslyn, S: 1994, *Image and Brain*, MIT Press, Cambridge, MA.

Lindemann, U, Jung, C and Schwankl, L: 2001, Montagegerechte gestalungs eines niederhubsicherheitsventils, *ZWF* **96**(7 & 8): 373-377.

Logie, RH: 1995, *Visuo-spatial Working Memory*, Psychology Press, Hove.

Maher, ML and Simoff SJ: 2000, Collaboratively designing within the design, *in* LJ Ball (ed) *Collaborative Design: Proceedings of CoDesigning 2000, Coventry University, UK.*

Miller GA: 1956, The magical number seven, plus or minus two: Some limits on our capacity for processing information, *The Psychological Review* **63**: 81-97.

Minneman SL: 1991, *The Social Construction of a Technical Reality: Empirical Studies of Group Engineering Design Practice,* PhD Thesis, Department of Mechanical Engineering, Stanford University, Stanford, CA.

Oxman, R: 1990, Prior knowledge in design: A dynamic knowledge-based model of design and creativity. *Design Studies* **11**: 17-28.

Pahl, G and Beitz, W: 1996, *Engineering Design: A Systematic Approach*, 2nd edition, Wallace, K. (ed), Springer-Verlag, London.

Pimmler, TU and Eppinger, SD: 1994, Integration analysis of product decompositions, *Proceedings of ASME Design Theory and Methodology - DTM '94*, Minneapolis, USA, pp 343-351.

Purcell, AT and Gero, JS: 1996, Design and other types of fixation, *Design Studies* **17**: 262-383.

Schank, RC and Abelson, R: 1977, *Scripts, Plans, Goals and Understanding*, Lawrence Erlbaum Associates, Hillsdale, NJ.

Schank, RC: 1982, *Dynamic Memory: A Theory of Reminding and Learning in Computers and People*, Cambridge University Press, Cambridge.

Schön, DA: 1988, Designing: Rules, types and worlds, *Design Studies* **9**: 181-190.

Star, SL: 1989, The structure of ill-structured solutions: Heterogeneous problem-solving, boundary objects, and distributed artificial intelligence, *in* L Gasser and MN Huhns (eds), *Distributed Artificial Intelligence 2*, Morgan Kaufman

Suh, NP: 1990, *The Principles of Design*, Oxford University Press, New York.

JS Gero (ed), *Design Computing and Cognition'04*, 259-274


# AN ECOLOGICAL APPROACH TO GENERATIVE DESIGN

## *Ecoconfiguration through Agent-Environment Visual Coupling*

ALASDAIR TURNER, CHIRON MOTTRAM, ALAN PENN
*University College London, UK*

**Abstract.** In this paper we explore the use of an animat model to construct geometry. The model uses agents guided by direct (or active) visual perception of their environment to replicate human behaviour within a notional plan of an open space. The environment reacts to these agents by placing walls in order to affect their usage of the space, and thus the structure may be evolved to fit the social function of the agents within it. Here, we start with the most basic social function, to design a building that disperses agents programmed with an exploratory task across its floorplan by using an evolutionary algorithm. We investigate the effect of evolution on the generated configuration using space syntax tools. We show how the introduction of a simple rule, the desire to leave, can result in the evolution of commonly observed features — first a central axis and then a `foyer'. We discover that 'intelligibility' of the space, which might imply reduced cognitive load, may increase as the system is allowed to evolve. Finally, we consider the implications of the ecological approach for the design process.

## 1. Introduction

The aim of this paper is to show how Gibson's (1979) ecological theory of visual perception may be applied to generative design. Gibson's theory stems from his introduction of affordances, relationships conjoining actors and objects. The actor perceives her or his environment directly, without any representational model of it, and is drawn towards activities afforded by what she or he can see. For example, a chair affords sitting on it, or a corridor affords a path to walk through. He suggests that we might begin with easy-to-perceive components of the environment consisting of surfaces and surface layouts (Reed and Jones 1982). Thus, this paper investigates this primary interaction, the layout of a notional building.

We posit the possibility that the environment itself might react to its inhabitation (as though the dérive of the Situationists is actually acted upon). In response to the movement within the environment, the environment changes its configuration to attain some goal of the actors within it. Goal is perhaps too strong a word; the actors are merely engaged in an activity, and the environment adapts itself to suit the activity rather than any particular goal. The activity we investigate herein is 'exploration' (or possibly again, 'dérive'). Agents move in the direction of open space visible to them, which affords them onward movement to another space; the response of the environment is that it should be explored as evenly as possible. In order to allow the environment to change, we use a simple evolutionary algorithm (EA). The EA adapts a two-dimensional plan configuration, evaluating potential environments according to the distribution of actor numbers that visit each part of it.

The action of the actors and environment may be understood as a process of structural coupling (Maturana and Varela 1980). The context of structural coupling is cell biology, although others (e.g. Luhmann 1995) have since applied the idea to social systems. In Maturana and Varela's model of cell operation, each `unity' maintains itself as a distinct entity through a recurrent process called *autopoiesis*, which they regard as the defining attribute of living beings. If the unity is engaged with the environment, then the engagement occurs between their respective structures, which they call *structural coupling*. They propose an apposite analogy in view of our task:

"Thus, for example, in the history of structural coupling between lineages of automobiles and cities there are dramatic changes on both sides, which have taken place in each one as an expression of its own structural dynamics under selective interaction with the other." (Maturana and Varela 1987, p. 99)

In this paper, we propose to enact a kind of asymmetric structural coupling. We will evolve lineages of buildings in order to evolve to a natural coupled state between the occupier and the environment, that is an agent representing the human placed within a building structure. In terms of a whole human, its legs, visual tendency, and ability to walk are structurally coupled to an environment that is walkable and seeable. Maturana and Varela regard the final structurally coupled ecology as cognitive, although with simple reactive units as our own, it is impossible to suggest that they themselves are capable of cognition. However, the result of the process, seen in terms of an evolved, and naturally coupled, environment is in one sense a map of that actor's engrained knowledge (or rather, that internal process of the unity), and therefore some, measurable, cognitive representation.

In the next section, we will start by looking in more detail at how the system we propose relates to cognition, and whether or not the outcome can be regarded in some sense as cognitive. We look at how techniques developed from space syntax principles (Hillier and Hanson 1984) can measure this outcome as being intelligible (or unintelligible), that is, as having some cognitive value. Next, we describe in detail a system that combines reactive animats, within an evolving configuration. We demonstrate how the activity of exploration leads, sometimes, to 'intelligible' layouts emerging, while adding the necessity to leave causes the environment to relate to its entrance, often through the addition of a central axis, and in some cases, later in evolution, what we might call a 'foyer'. We conclude by looking at the implications of the system for designers.

## 2. Background

Gibson's direct perception has become known as active perception to researchers in the field of robotics. The classic implementation of an active perception is Brooks's (1991) subsumption architecture, which leads Brooks to suggest that there may be knowledge without representation. He makes a robot that is engaged directly in the environment for the task of centering itself within a room. It is important to realise that the organism, in this case a robot, or more generally an animat (i.e, either hardware or software robot), relates to the environment in the modes available to it. If the modes do not exist then the robot cannot conceive of them, indeed, the modes themselves are meaningless — as Gibson points out, to walk, or to see, requires both the thing that 'walks' or 'sees' *and* the thing to be walked on or to be seen. It is this realisation that leads Clancy (1997) to propose a *situated* cognitive process. Consideration of the situated process has further led researchers at Sussex to examine the possibility of *evolving* embodied animats to fit tasks within the world Harvey et al. (1997). Within this programme of research, Dale and Collett (2001) have produced a model in which animats are given visual sensors and one of two motor capabilities, either to fly or to walk. They show that evolution of the control mechanism to approach a post results in 'bee-like' and 'ant-like' behaviours. To Dale and Collett, the process of evolution thus always adapts the animat to the task, regardless of the internal configuration of the neural control units. That is, the situated task leads to the structural coupling that occurs, directly, without evolution affecting the final coupled structure. As already elaborated, the fact of conceiving the world is regulated by what is conceivable, which is thereby regulated by the structure of the animat and environment.

However, to 'conceive' mentally for an animat with a simple architecture of motors and sensors with the most basic layer of control might seem a little extreme. The animat is simply a reactive unit. For there to be cognition, it is usually suggested that the animat must form a mental model of its world. However, Cruse (2003) has recently argued that there is overlap, that cognitive structures do emerge naturally from the evolution of reactive systems. In a sense, this is of no surprise, but the level at which Cruse demonstrates `cognition' (in terms of a map of all possible configurations of an arm unit, which thereby allows planning) is at the tens of neural units. Be that as it may, the animats we will demonstrate herein are most certainly not cognitive, although they are situated, as Brooks's (1991) original robots were situated. We should be aware, though, that consideration of situated cognition has recently led Gero and Fujii (2000) to reverse the model, so that the tools that construct the environment itself form a system of embodied agents. Gero and Fujii call this the situated computing paradigm. Maher and Gero (2002) provide a detailed account of a system that follows the paradigm in order to build a three-dimensional environment in a collaborative virtual environment around its inhabitation.

As we have suggested, the system we propose does not comprise cognitive agents, merely situated, reactive agents. However, Gero and Fujii's idea of reversing the model, so that the environment reacts in accordance with the its coupled activity with its inhabitation, is appealing. Thus, in earlier papers, we have looked at the conjunction of reactive agents with agents that control the configuration of the environment Turner (2002; 2003b), that is to look at an ecological process, the evolution of structurally coupled agent-environments. Indeed, we have looked at more complex scenarios than described here, considering an ongoing process of interaction between people agents and artwork agents (Turner 2003b). Herein, however, we turn our attention to the outcome of running the system. The environment and actor may be evolved to a single solution, rather than engaged in an ongoing process, and the result is a single configuration; the configuration is of interest.

Hillier and Hanson (1984) suggest a building or city configuration itself is measurable in terms of axial lines — an algorithmically extractable set of 'longest' lines through the space of a system, forming, through their crossing points, a connected topological graph of the configuration, an axial map. Graph measurements of the relationships between lines in axial maps have been shown to correlate with pedestrian movement (Hillier et al. 1993), which has led both Penn (2001) and Hillier (2003) to investigate why this should be, and how the axial map might relate to a human cognitive map (Tolman 1948). They suggest that the lines may form a minimal set to engage with the environment, and thus form the basis for the most elementary map. Kuipers et al. (2003) supports this case with evidence from robotics, showing that a 'skeletal cognitive map' provides a shorthand of the environment in terms of relationships such as 'left of' and 'right of' paths through it, and thus a good representation of the environment for robot recall. For our purposes, it is important to notice that the action of the task within the environment leads to a specific skeletal or axial map. In effect, the task can be seen as a generator of relationships that are inscribed on the environment. The nature of the interaction, we suppose, will lead to an environment attuned to its usage. That is, the process retrieves and enacts a cognitive map related to its context.

## 3. The Ecoconfigurational System

An ecoconfigurational system will be one where the inhabitants and environment coevolve to form an inhabited configuration suited to their common activities. We can regard the system from an autopoietic standpoint, with each actor-architecture comprising a structurally coupled unit, as shown in Figure 1. Our ecoconfigurational system will be asymmetric, and rather than coevolving, comprise two elements: animats (which will not evolve) and environment (which will evolve), described in turn below.

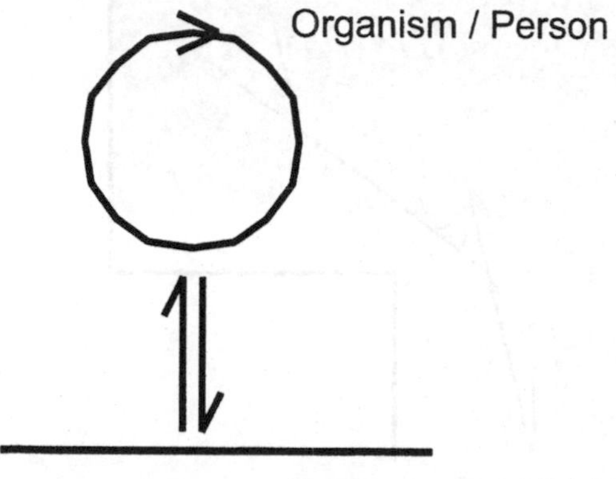


*Figure 1.* Autopoietic relationship between person and architecture, after Maturana and Varela (1987)

### 3.1. THE ANIMAT INHABITANTS

In previous papers, we have looked at the use of reactive animats, or software agents to model pedestrian behaviour (Turner and Penn 2002; Penn and Turner 2002). The agents are considered from the Gibsonian of view, as being situated within the environment. Thus, rather than have the agents look out onto the environment (which is computationally costly for many agents), we have them sample the existent possibilities at a location within the environment. That is, we precompute what Gibson would call an 'ambient optic array', that is, an array of possibilities available to the agent at any point within the environment. This is implemented as a grid of cells overlaid on the environment, where each cell contains information about the locations that can be seen directly from it, that is, a list of locations that afford onward movement within the environment. As the visual possibilities lie outside the agent, in the environment, we call the visual architecture *exosomatic* (outside the body). The agent wonders freely, and samples the ambient optic array for the cell to which it is closest in order to see. The set of visible locations is subdivided into angular 'bins', so that the agent takes a group of bins to form an angular field of view, Figure 2.

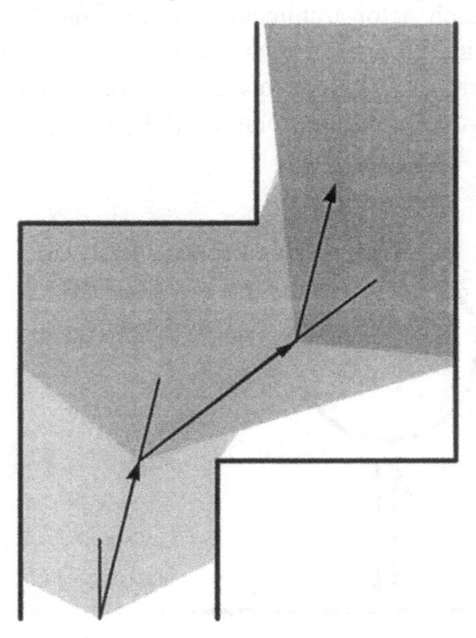

*Figure 2.* An agent that iteratively chooses a visible location at random from a set of all visible location is guided configuration, after Turner and Penn (2002)

We originally used a 0.75m grid spacing of cells in order to give a dense coverage of potential visible locations, on the basis that it approximates human step length. The agents were then programmed to walk at 1.5m/s across the grid, choosing the grid point closest to their current location to form their field of view. We found that if the agents are programmed with field of view of about 170° (that is, approximating human visual field) and the agents take 3 steps between each new destination choice then the pattern of movement correlates with observed aggregate movement of people, but less so otherwise. We have applied the agents to both the Tate Britain Gallery in London (Turner and Penn 2002), and an area of the City of London (Turner 2003a), and compared the numbers of agents passing through gates (doorway thresholds, or a notional gate on a street segment) with the observed number of people walking through the gates. In the Tate Britain Gallery, for a sample of about 60 gates, the correlation is $R^2 = 0.76$ for agents released from the main entrance; in the City of London, for a sample of about 80 gates, a correlation of up to $R^2 = 0.67$ is found, depending on the way the agents are released into the system. Since the agents seem to reproduce in some way and to some extent aggregate pedestrian movement, we propose that it is not unrealistic that they may perform a similar role in proposed building plans. Therefore, in this paper, we use identical 3-step agents, albeit that move at a more leisurely notional speed of 0.75m/s although we might want to suggest instead that the scale of the building is simply double, the two are equivalent.

## 3.2. THE EVOLUTIONARY ENVIRONMENT

For the evolutionary environment, we start with a context for the explorational task to take place in. The context is an open space with an entrance in the middle of the southern end, as shown in Figure 3. If the agents described above are released into this environment, then the behaviour is quite uninteresting. They mill around in the centre of the room. Figure 3(b) shows trails of 100 agents released into the room and left to roam for 10 minutes (600 agent steps). In order to create interest for the agents programmed with these rules, and in order for them to do something more interesting, we are required to introduce configuration. To enable this in a systematic way, the context space is zoned into 49 4m x 4m areas, each of which can be divided from the others by means a wall, thus, for a 28m x 28m space there are 7 x 7 'rooms', divided by a potential 84 walls. Thus, the evolutionary environment encodes the configuration in terms of which walls exist and which do not as an 84-bit genome. A random configuration of walls generated from this encoding, is shown in Figure 3(c), once again with the corresponding agent trails marked.

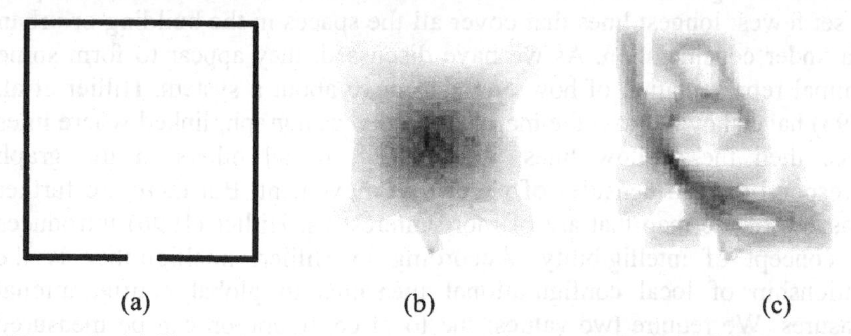


*Figure 3.* (a) The environmental context, a space notionally about 28m x 28m with an entrance (b) Agent trails from a ten-minute exploration task. The areas are darker where more agents have trodden. (c) A randomly generated configuration, again with ten-minute agent trails

The environment is given the task of dispersing movement within itself as much as possible. It does this with a simple penalty formula. 300 agents are released from the entrance (contrary to our previous implementations, the agents do not obstruct each other, so these are essentially 300 separate agent explorations, run in parallel). As the agents walk through the 49 'room' zones they trigger a counting mechanism. At the end of each 10 minute exploration, the environment calculates the difference between the average number of agents passing through all rooms and the number that have passed each room in particular, normalised according to the value of the mean. The differences are then summed for all rooms in the system. Rooms that are completed unvisited (i.e., cut off from the rest of the system) are given a penalty of two times the mean. The complete penalty function is shown in equation 1, where $P_0$ is the penalty, and $x_n$ the count of agents that have walked through 'room' *n*.

$$P_0 = \sum_{n=1}^{49} \begin{cases} 2 & \text{if } x_n = 0 \\ \left| \dfrac{x_n - \bar{x}}{\bar{x}} \right| & \text{otherwise} \end{cases} \tag{1}$$

A population of 100 environments are generated automatically, and each one assessed according to the penalty function. The system is then evolved using a standard genetic algorithm (Goldberg 1989). The operators used are uniform crossover, with a 5% mutation rate, and rank selection.

## 4. Analysis: Axial Maps and Intelligibility

Each building plan generated by the ecoconfigurational system was analysed using the 'space syntax' method of axial mapping. We have already mentioned the axial map in section 2, as a tool that may be used to compare different configurations. Hillier and Hanson (1984) describe the axial map as the set fewest longest lines that cover all the spaces in the building or urban area under consideration. As we have discussed, they appear to form some minimal representation of how one can move about a system. Hillier et al. (1993) have shown that if the map is regarded as a graph, linked where lines cross, then the shallow lines with respect to all others in the graph correspond to high densities of pedestrian movement. But there are further measures of the map that are of more interest us. Hillier (1996) introduces the concept of intelligibility. According to Hillier, intelligibility is the relationship of local configurational measures to global configurational measures. We require two values: the local configuration can be measured via the *connectivity* of each axial line (that is, the number of other lines that cross it). The global configuration can be measured via the *mean depth* to all other lines in the system, that is the average number of connections that have to be made in order to get to any other line in the system. The intelligibility is assessed by a scatterplot of connectivity against mean depth. If the two correlate then it is said to be intelligible. That is, the well-connected areas are shallow to the rest of the system, and the less densely connected areas are deeper with respect to the other lines. Conroy (2001) provides experimental evidence that the measure of 'intelligibility' does actually correspond in some way to the intelligibility of a space, by asking people to find objects in systems defined as 'intelligible' and 'unintelligible' by our definition; she shows that 'intelligible' spaces allow easier navigation to the object, and shorter discovery times, than 'unintelligible' ones. The question for us is, is there any rationale in the environments that the ecoconfigurational system generates? Hence, we will consider how intelligible the environments that are generated actually are.

## 5. Results

We provide the results for two implementations of the ecoconfigurational system, in the first, the system is implemented as described, with 300 agents release from the entrance and allowed to wander the system. The system is evolved for 5000 generations, and we look at the results at various stages. In the second implementations, the system is modified, so that the agents also exit the system, and the configuration penalised if they do not. We look at the consequences of the modification.

### 5.1. THE EXPLORATION SPACE

Sample results for a particular run of the system are in Figure 4. The systems do not seem that different initially; however, if we look at the axial maps generated, we can start to make qualitative observations about the system. After 1000 generations, what might seems an identical layout to later ones, is actually a `tangled mess' of axial lines to one side of the entrance; by 2000 generations, this tangle has been unravelled. This might seem all that is necessary to distribute movement across the system, however, this is not the case. At 5000 generations, the structure of the unravelled system has actually been reinforced: there are lines crossing straight through the system at the top, middle and down the right side, rather than a set of broken lines as there were at 2000 generations. Movement of the agents has been directed around the system in continuous loops in order to distribute it.

We should note however that all this manipulation has left us with a system that no longer bears any relationship to the entrance. In fact, the relationship starts out as quite strong as is evidenced by the sheaf axial lines extending out through the entrance at 1000 generations, but it is weakened until by 5000 it has disappeared completely. The consequence is that this may be a wonderful space to explore, but the inhabitant has little chance of finding her or his way out! Therefore, in the next section, we suggest a refinement to the algorithm, a possibility to exit.

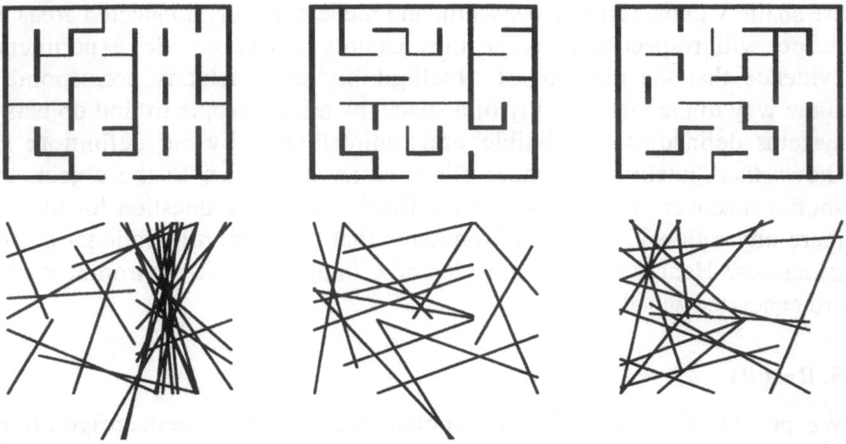

*Figure 4.* Results from running the ecoconfigurational system. Top, left to right, configurations after 1000, 2000 and 5000 generations. Bottom, axial maps for each configuration.

Configurations produced by each run do vary: the system quickly finds a set of configurations which score around 18 on the penalty function, and the results shown do seem to typify the sorts of configuration produced by the system. Figure 5 shows the average peak fitnesses at each stage for a total of three system runs[1]. The slow decrease in fitness seems to suggest that the results from generation 2000 onwards are simply drifting across a plateau of approximately stable configuration fitness, with just minor increases in fitness as the system progresses. It is assumed that the population level fitness, although not calculated in these experiments, will show rapid convergence over this period, so what we see is a fit, population level response.

5.2. FROM EXPLORATION TO EXPLORATION WITH EXEUNT

It is important at this point not to break the paradigm of direct perception. In order to ask the agents to leave we could merely invoke a rule telling them where the entrance is so as to guide them back to it. But this is to imply we have knowledge of an objective structure that resides in the agents' heads: some map of the system. Instead, we must look at what active perception of the environment allows us to do. We know that we want the agent to leave,

[1] Note that the peak fitness fluctuates because each best performing configuration is reassessed at each 500 generations, and as the agents sample the system stochastically, they may turn out a slightly varying penalty result each time.

we know that the exit affords it to leave, and so we simply have to program the agent such that after a certain time, if it sees the entrance, then it moves towards it in order to leave, and exits. So as to affect the configuration through this modification, we should also apply a penalty if the agent is unable to leave. For the purposes of the algorithm, we will simply state that for each of the 300 starting agents that has not left the system within 5 minutes of its decision to leave (after 10 minutes as before), a penalty of value one will be applied. The new penalty $P_1$ function is shown in equation 2, in terms of the original penalty function $P_0$ and the number of agents that fail to leave the system in the required time, $k$.

$$P_1 = P_0 + k \tag{2}$$

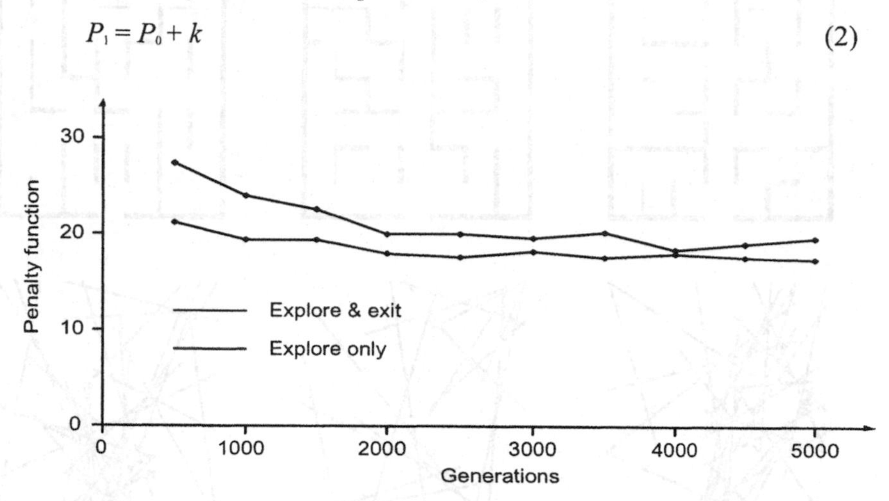


*Figure 5.* Peak fitness against number of generations of evolution for the basic ecoconfigurational system (dashed, below) and the system with the condition that agents should exit (dotted, above)

The results of running this modified algorithm are shown in Figures 5 and 6. As with the basic system, the peak fitness quickly converges after 2000 generations, although it remains slightly higher, due to the fact it has to maintain as easy exit for the agents within the system. This means that we see a clear focus of the entrance in every system produced, as demonstrated in Figure 6, which shows a typical run of the system. At first, this is apparent as a central axis through the system, but at 5000 generations, this particular run of the ecoconfigurational algorithm found a new solution. It creates a strongly bound set of axial lines around the entrance, a kind of 'foyer' area to the main system. It is interesting at this point to note how closely this follows the pattern of foyer suggested by Alexander (1977), a transitional place that bears the hallmarks of both inside and outside. The axial lines create a link to the outside, while maintaining a centrality to the internal circulation. In other runs, the foyer does not always emerge, although interestingly, the system always seems to go through a similar reunification after a few thousand generations: the basic corridor result, witnessed after 2000 generations, always seems to emerge. Then, later in the evolutionary cycle, a cross bar, separating upper and lower systems, seemingly corresponding with a slightly fitter result. However, this slightly fitter result also has a corresponding gain which we might not anticipate: it seems to be more *intelligible*.

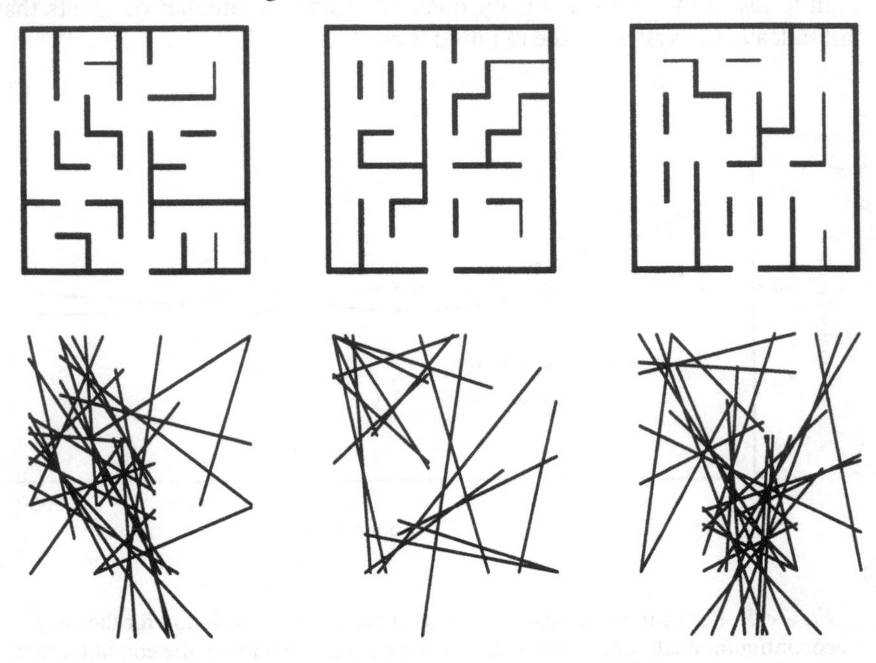

*Figure 6.* Results of running the ecoconfigurational system with the condition that the agents should exit. Top, left to right, configurations after 1000, 2000 and 5000 generations. Bottom, axial maps for each configuration.

5.3. INTELLIGIBILITY

So far we have said nothing about the measurement of intelligibility of these systems. It is difficult for such small systems to know whether changes in mean depth and connectivity are results of a genuine characteristic change, or simply due to some error introduced by having so few lines. Certainly, the measure is mostly close to its maximum of one. Again, this is a feature of small systems, where it is difficult to achieve a variation between connectivity and mean depth, and mean depth tends to follow directly with inverse connectivity. However, there does seem to be a repeated phenomenon in the values for systems with central axes and those with the foyer: the foyer systems seem to slightly more intelligible (around 0.9–1.0) than the central axis configuration (around 0.7–0.8). Whether or not this is a genuine finding for the system will have to wait until larger configurations are tested.

## 6. Conclusion and Implications

This paper has presented an ecological approach to generative design through the application of an ecoconfigurational system. The system combines agents with vision with an environment that evolves over time. The agents embody Gibson's (1979) most basic model of affordance – they see available 'walkable' surface and move to it by means of a stochastic process. These agents generate aggregate movement patterns that tend to correspond with those observed in actual buildings and cities (Turner and Penn 2002; Turner 2003a); we therefore regard the movement of the agents as an indicator of possible human movement within a configuration, and thus that the design generated may be applicable to the real world.

The agents are placed within a two-dimensional plan of a notional building environment broken by surfaces. The position of the surfaces is governed by an evolutionary algorithm. The surfaces are at first randomly placed, and the agents allowed to walk through the generated configuration. The environment then tries to adapt around the movement of the agents so that it is dispersed as evenly as possible within it. The outcome of several thousand generations of configurations is that an intelligible (in Hillier's (1996) terms) environment is generated. What is more, when the agents are programmed with a desire to leave the environment (which they accomplish through a direct perception of the exit), the system evolves features found within actual building environments (first a central axis, then, in some cases, a foyer), implying that the result of historical evolution of building may be in fact a natural response to the task of inhabitation.

The starting point for the approach was to look at the natural interaction between body and environment through what Gibson calls an ecological process. In Gibson's words:

> "When no constraints are put on the visual system, we look around, walk up to something interesting and move around it so as to see it from all sides, and go from one vista to another. That is natural vision..." (Gibson 1979 p. 1).

This led to the view that we might be able to construct a structurally coupled system of agent and environment, the natural outcome of interaction between bodies engaged in a task when combined with a malleable environment over time. We argued that the outcome might be seen in some way as a 'natural' cognitive map of the relationship taking place within the contextual boundaries of the system — that the dérive of the Situationists would become embodied as praxis (and therefore, in one sense, the system itself would have a political nature). Of course, such thinking is not new to architecture. Since Heidegger introduced the concept of Dasein (the human as seen objectively that only makes sense in terms of its environment), architects have thought about the importance of relationships between the inhabitants and the environment, rather than the two as separable entities. Norberg-Schulz (2000) perhaps states it most aptly:

> "Life, then, is understood as a series of relationships, and existence is seen as a way of openly mirroring oneself in an array of different ways of being. Thus, man stops being an observer and becomes a participant..."

The step further, then, is to suggest that the participant might actively provoke a response in the environment, that architecture may be conceived by the action of its participants. There is much room for further work, of course. The participants at the moment act only on the environment, they do not interact with each other. Although we have looked at interaction of human and artwork in previous papers (Turner 2002; 2003b), the goal of course, would be to build around a pair, or group, of structurally coupled agents, so that the building becomes a mirror of the whole activity intended for their occupation.

We should also consider, though, the practicalities of implementation. Generative design generally has the effect of pushing the architect away. However, at least within the context of this system, the architect both retains a role and the role is clearly defined – the one who understands the social function, and how it is affected by the contextual conditions of the system. The nature of the aesthetic form itself, of course, may enter into the social usage, so the architect must not be deluded that a system such as the one we have demonstrated will not later take on aspects of form design as well as configuration. Of course, we should state explicitly that our aim is not to replace the architect, but present an ideal of a symbiotic relationship between computer and architect where the design is informed but not overly controlled by the process. Nonetheless, it is apparent that this system currently appears far from real-world use; configurational design is unlikely to be handed over to a computer in the near future. However, the system as it stands is ideally suited to applications such as placement and sizing of signage — a current topic of research and one where input is required (e.g. Kichhanagari et al. 2002; Bourdeau and Chebat 2003) — and so could be put to use on evolving suitable signage for agents with goals.

Thus, we hope that an ecological approach to generative design may lead to a tool that is both of theoretical interest and practical use.

## Acknowledgements

The input of Erica Calogero, whose diploma work concentrated on agent-environment interaction in similar 7 x 7 dynamic grid structures, and her supervisor, Prof. Stephen Gage is gratefully acknowledged.

## References

Alexander, C: 1977, *A Pattern Language: Towns, Buildings, Construction*, Oxford University Press, New York.

Bourdeau, L and Chebat, JC: 2003, The effects of signage and location of works of art on recall of titles and paintings in art galleries, *Environment and Behavior* **35**(2): 203–226.

Brooks, RA: 1991, Intelligence without representation, *Artificial Intelligence* **47**: 139–159.

Clancy, WJ: 1997, *Situated Cognition*, Cambridge University Press, Cambridge, UK.

Conroy, RA: 2001, *Spatial Navigation in Immersive Virtual Environments*, PhD thesis, Bartlett School of Graduate Studies, UCL, London.

Cruse, H: 2003, The evolution of cognition — a hypothesis, *Cognitivie Science* **27**: 135–155.

Dale, K and Collett, TS: 2001, Using artificial evolution and selection to model insect navigation, *Current Biology* **11**: 1305–1316.

Gero, JS and Fujii, H: 2000, A computational framework for concept formation in situated agent design, *Knowledge-Based Systems* **13**(6): 361–368.

Gibson, JJ: 1979, *The Ecological Approach to Visual Perception*, Houghton Mifflin, Boston, MA.

Goldberg, DE: 1989, *Genetic Algorithms in Search, Optimization and Machine Learning*, Addison-Wesley, London, UK.

Harvey, I, Husbands, P, Cliff, D, Thompson, A and Jakobi, N: 1997, Evolutionary robotics: The Sussex approach, *Robotics and Autonomous Systems* **20**: 205–224.

Hillier, B: 1996, *Space is the Machine*, Cambridge University Press, Cambridge, UK.

Hillier, B: 2003, The architectures of seeing and going — is there a syntax of urban spatial cognition?, *Proceedings of the 4th International Symposium on Space Syntax*, UCL, London, UK, p. forthcoming.

Hillier, B and Hanson, J: 1984, *The Social Logic of Space*, Cambridge University Press, Cambridge, UK.

Hillier, B, Penn, A, Hanson, J, Grajewski, T and Xu, J: 1993, Natural movement: Or configuration and attraction in urban pedestrian movement, *Environment and Planning B: Planning and Design* **20**: 29–66.

Kichhanagari, R, Motley, R, Duffy, SA and Fisher, DL: 2002, Airport terminal signs—use of advance guide signs to speed search times, *Transportation Research Record* (1788), 26–32.

Kuipers, B, Tecuci, D and Stankiewicz, B: 2003, The skeleton in the cognitive map: A computational and empirical exploration, *Environment and Behavior* **35**(1): 80–106.

Luhmann, N: 1995, *Social Systems*, Stanford University Press, Stanford, CA.

Maher, ML and Gero, JS: 2002, Agent models of 3D virtual worlds, *ACADIA 2002: Thresholds*, California State Polytechnic University, Pomona, pp. 127–183.

Maturana, HR and Varela, FJ: 1980, *Autopoiesis and Cognition: The Realization of the Living*, D. Reidel, London, UK.

Maturana, HR and Varela, FJ: 1987, *The Tree of Knowledge: The Biological Roots of Human Understanding*, Shambhala Publications, Boston, MA.

Norberg-Schulz, C: 2000, *Architecture: Presence, Language and Place*, Skira editore, Milan.

Penn, A: 2001, Space syntax and spatial cognition, or why the axial line?, *Proceedings of the 3rd International Symposium on Space Syntax*, Georgia Institute of Technology, Atlanta, Georgia, pp. 11.1–11.16.

Penn, A and Turner, A: 2002, Space syntax based agent models, *in* M Schreckenberg and S Sharma (eds), *Pedestrian and Evacuation Dynamics*, Springer-Verlag, Heidelberg, Germany, pp. 99–114.

Reed, ES and Jones, R (eds): 1982, *Reasons for Realism*, Lawrence Erlbaum Associates, Hillsdale, NJ.

Tolman, EC: 1948, Cognitive maps in rats and men, *The Psychological Review* **55**(4): 189–208.

Turner, A: 2002, Ecomorphic dialogues, *in* C. Soddu (ed), *Proceedings of Generative Art 2002*, Politecnico di Milano, Milan, Italy, pp. 38.1–38.8.

Turner, A: 2003a, Analysing the visual dynamics of spatial morphology, *Environment and Planning B: Planning and Design* **30**(5): 657–676.

Turner, A: 2003b, Reversing the process of living: Generating ecomorphic environments, *Proceedings of the 4th International Symposium on Space Syntax*, UCL, London, UK, pp. 15.1–15.12.

Turner, A and Penn, A: 2002, Encoding natural movement as an agent-based system: An investigation into human pedestrian behaviour in the built environment, *Environment and Planning B: Planning and Design* **29**(4): 473–490.

JS Gero (ed), *Design Computing and Cognition'04*, 275-293


# MAKING A WALL

*An Investigation into How the Choice of Material and Sequential Positioning in a Construction Process Affects the Relationship between the Design Intention and the Physical Outcome*

RACHEL CRUISE, PHILIPPE AYRES
*University College London, UK*

**Abstract.** All decisions made to create a design proposal, are based on the information the designer has about the specific criteria that influence the design: the desired function and aesthetic. The design proposal is influenced by the kind of data that is presented to the designer about the design requirements, and by the methodologies that are used to search this data for possible designs. The Dry Stone Waller is a piece of software that was written to investigate the possibilities of a given set of material in achieving certain design goals. Creating this program became an investigation into the 'strategic' nature of design tools and their relationship to the sequential 'tactical' decisions made in choosing and placing particular stones to a make a construct out of irregular components – a dry stone wall.

## 1. Introduction

Michel de Certeau (1984) in the 'Practice of Everyday Life' describes two different methods of making decisions:

A Strategic decision:

> *' The Calculus of a force-relationship which becomes possible when a subject of will and power can be isolated from an environment. A strategy assumes a place that can be described as proper and thus serve as the basis for generating relations with an exterior distinct from it.'*

A Tactical decision:

> *' In the super market, the housewife confronts heterogeneous data and mobile data - what she has in the refrigerator, the tastes and appetites and moods of her guests, the best buys and their possible combinations with what she has on hand at home: The intellectual synthesis of these given elements takes the form however not of a discourse but of the decision itself, the act and manner in which the opportunity is seized.'*

We would like to extend these definitions as set out by de Certeau (1984) to apply to decisions made in design and construction systems. Design is perceived as a linear action where it is supposed that the designer selects the critical information which informs the design, prior to drawing the proposal. Current computer design tools tend towards strategic design decisions being made, because of the methodology that is embedded in them.

Dry Stone Walling is a craft that uses an understanding of particular pieces of stone to decide how to select and place them in a particular context, to achieve certain design goals. The Dry Stone Waller software was created to investigate the possibility of making CAD tools that allow for tactical decisions to be made. This suggests a design system that recognises the potential of using the varying tolerances of standardised components and the way they are related to create a unique wall, even if they were intended to be repeated components that have identical relationships with the other components and the surrounding environment, Figure 1.

*Figure 1.* A photograph taken by the author of a dry stone wall in the Lickey Hills, Birmingham

## 2. Strategic and Tactical Design Decisions

Translating a whole design proposal into a physical construction requires the design to be broken into a number of manufacturing processes. This often means that the design depending on it's complexity is re-designed as an assembly of multiple components or parts.

A design proposal often describes purely visual goals prior to the consideration of processes needed to achieve the intention. As potential construction or manufacturing process are considered, possible sequences of fabrication methods are generated. Different fabrication methods will inevitably lead to a particular constructed result that could not be achieved using any other fabrication method. Depending on the fabrication system chosen any constructed solution will be in varying amounts different from the initial design proposal. All goals within an initial design proposal are not always achievable, or have to be modified to be feasibly made.

Once the precise nature of construction components is defined virtually and then physically it is hard to reconsider the components possibilities. When the construction system has been chosen the methods by which the construction is created are not often reconsidered during the fabrication to see if the reality fulfils the design intentions.

So designs defined by strategic goals are difficult to achieve because they are defined by ideals, that are not necessary created using the realities of the construction processes. These strategic decisions made about a design are not the act of creation, they only generate further tactical decisions themselves to be solved. Strategic decisions are not the defining decisions, those made on site, consciously or unconsciously, during the actual fabrication are. The strategic decisions set up how the proposal is perceived and the methodology required to achieve it. However if the strategic decisions are not made with an understanding of the tactical decisions they create there will be a greater degree of separation between the proposal and the reality.

Evans (1997) gives us a beautiful example of the shift that occurs between design and construction in his description of the Royal Chapel, Anet designed by Philibert de l'Orme where the tiled floor depicts what is documented by Philibert as the plan of the dome. As Evans (1997) describes the plan has been drawn so it appears more aesthetically pleasing rather than documenting the actual construction of the dome.

*'All the drawings made of the chapel from the sixteenth to the late nineteenth century are manifestly incorrect .. Simply count the number of intersections along one of the eighteen longitudinal lines of the dome, and then count the number of intersections along a corresponding radius on the floor. In the dome there are eight, on the floor six'*

The traditional act of drawing has nowadays been translated into CAD software whose tools mimic the physical tools used for drawing, for example rulers and line tools. CAD software does not mimic the physical actions of a construction process so designers cannot experiment with construction techniques through their notation. Instead experiments are conducted with the placement of lines and their visual impact. At best the lines are records of mental experiments in fabrication processes. The tools used to record the design parameterise it solely in geometrical constructs representing the form, colour and the size of an object. Whether it is described in two dimensions or three the drawing rarely has physical properties linked to it. The design is explored with a lot of missing information that is critical to the performance of the proposal in the real world. These virtual constructs have no weight, no compaction under loads, no frictional properties, no desire to fall under gravity and no solidity which prevents overlaps. Through the process of drawing a proposal designers are not constrained by physical realities. This might be seen as a benefit, but if the goal of the designer is to fabricate the proposal, like craftsmen, designers could use an investigation of the process of manufacture or construction as a tool to generate more innovative proposals, thereby removing the need to reanalyse drawn proposals to find ways of making them, and through that modifying the design proposal.

## 3. Difference in the Same

The repeated manufacture of the same product requires repeating the processes that make each type of component used to construct the product. Being able to reproduce a product to a high level of accuracy became an important idea in the industrial revolution. Often experiments in being able to reproduce a certain aesthetic outcome failed and the aesthetic result of the product ended up being one that was found during experimentation to be consistently repeatable.

Machines that create these repeated objects themselves are made of repeated parts: repeated cutting blades, dies, presses and the repetitive nature of the circular cog. Each machine part has been manufactured itself. Each interaction between a machine part and the unprocessed or partially processed component is unique, created by the precise interactions of sometimes hundreds of machine parts, levering, drawing, rotating and interlocking. The potential for the processed part to deviate from the desired is made less than in a hand-made production, but not only does the machine act on the material, the material acts on the machine, blades blunt, hot materials cause expansion, deformation occurs and friction wears parts down. Surface upon surface, these mechanical processes are reduced to interactions which are hard to keep constant. Any deviation however small will prevent the creation of an exact outcome. The problem compounds the more interactions create the process.

Even parts of one type are, in fact, different from one another, the range of difference or tolerance depending on the precision to which the object is made cannot be completely removed, only reduced. There is always a smaller scale to go to, and each smaller scale offers an increased number of potential variables. These variables added together create differences between the made components. It is the repetition in the actions of manufacture that enable us to see them as 'the same'. It is our perception of them as 'the same' that allows them to potentially replace one another in a construction system.

Manufactured parts are then simulations of the ideal intention but they are simulations that always deviate from the ideal. So the design proposal is made by stating the ideal parameters of the parts and the tolerance within which the parts are acceptable manifestations of the design intention. In any assembly of parts where the differences between parts are recognised, the choice in using any particular part rather than another from the selection of parts is critical to the outcome of the construction process. Combinations of tolerances can build up and produce perceptible deviations in the final product or whole.

This methodology of exercising great control over a produced component, which is then often placed in an un-controllable context, still leads to tolerances being generated. It is not that the components are not precise but the way that they interact with a partially completed construction is and this makes the precision of the objects somewhat redundant. Components may be rejected on site not because they are unacceptable in terms of the design requirements but because they do not work in relation to the placement of the other components and the conditions of the site.

The components are initially perceived as physical manifestations of the component design ideal. During the assembly of these components the perception of these produced objects changes. Deviations between parts becomes more apparent as they are placed in the context of other components. This tests only the relative precision to which they have been reproduced. The assembly of the components into a construction therefore varies to a degree partially due to the tolerance of the part and also to do with the accuracy of how the components are assembled. The design goal becomes eclipsed by the necessity of physically relating the components together.

The construction system of bricks is designed so that the ideal part fits everywhere. This component is to the designer an ideal rectilinear shape that tessellates perfectly. The use of this standard component seems to give the designer more control of the outcome, as in his perception every brick is the same as the next so the maker of the wall does not take any decisions that affect the outcome of the wall. However the real component always falls short of the ideal. In the world of the made object bricks are not identical, leaving the maker of the wall to assess the deviations of the bricks in order to tessellate them and decide if they should be included in the wall. The maker uses the mortar in the wall as a substance to take up the certain deviations. So whilst the deviations are undesirable most can be eliminated by using the mortar and the tactical decisions of the maker. This allows the designer to still perceive the bricks as exact replicas of an ideal.

## 4. Dry Stone Walls

At the other extreme from designed construction components is a material such as field stones which are used for dry stone walling. It is a material that is processed only very crudely, if at all, before it is used. It is in this kind of construction, Figure 2, that the tactical design decisions have a huge impact on the construction, because by nature, stones will have a large degree of variability between one another. The dry stone waller has little or no control over his construction components in comparison to a designer's control over manufactured parts. The components or stones are seen as 'given', because they exist prior to the proposal. However the waller's understanding of them is built up over time by observation and touch. It is accepted that the components may bear some relationship to one another through their materiality but they are never thought of as the same. Since they have no assigned position in a proposed wall the skill in making a dry stone wall is in using the found material to create the desired wall.

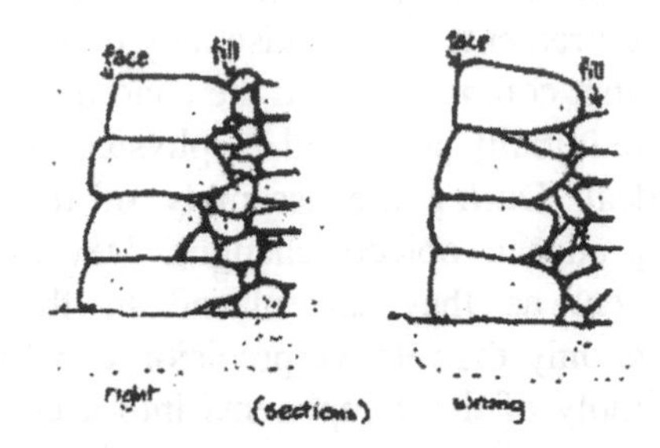


*Figure 2*. Details of the construction of a dry stone wall. Image from the British Trust for Conservation Volunteers http://www.btcv.org/skills/walls

This means that the sequential choice of stone, order of assembly and relational positioning of the material are critical to achieving the finished wall. The aesthetics and functional performance of the wall depends on the analysis of the stones, their selection and their positioning. This is a process, which makes use of the differences in size, shape and quality between stones. The dry stone wall is an example of a construction where tactical design decisions are magnified due to the magnified differences between components.

The pile of stones the waller starts with is a pile of possible outcomes and the order with which these are placed in the wall determines the wall's form, aesthetic and how the wall achieves the design goals. The waller is participating in a system where his perception of the material shifts due to the physical testing of possible ways to achieve the desired wall. This perception is based on his knowledge of the placed stones and his understanding of the stones left in the pile. The possibilities for the arrangement of the stones within a wall are so numerous it would be impossible to consider all positions available for each stone.

The wall is a construction made by the sum of the dry stone waller's design decisions, which are informed and carried out by his actions made during to the construction process.

## 5. The Dry Stone Waller

The pile of stones the waller starts with is an unknown set of data. He may have a feel for the size distribution of the pile but he cannot yet know the properties of each individual stone. The choice of picking a stone holds one set of possible outcomes that are multiplied by the almost infinite possible orientations of the stone within the chosen place. These possibilities are increased again by the way the stone interacts with another surface either the ground or another stone that holds its own possibilities.

The Chinese method for laying stones is described as finding one point of contact between the stone to be placed and the stone below and rotating the stone until a best fit is found. This is a very simple and beautiful method to search for possibilities of a stone within a particular position.

Stones could be classified as:

1. Flat stones to level the ground.
2. Foundations are made up of:
   - Ties that span the width of the foundations.
   - Runners that are smaller than the width of the foundations.
3. Courses are made up of:
   - Coursing / facing stones that are shorter than the width of the course.
   - Through stones that span the two lines of facing stones.
4. The rubble infill is made up of awkward and smaller stones.
5. The coping could be chosen to be a number of arrangements of stones and one example is shown in Figure 3.

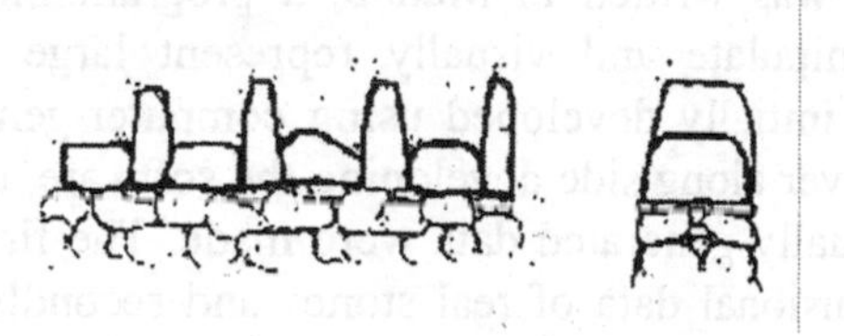

*Figure 3.* A does and bucks coping: Image from the British Trust for Conservation Volunteers: http://www.btcv.org/skills/walls

These divisions of stones may seem clear, but the classification of a particular a stone as a large stone is a relative classification. The size of this stone might be reassessed when during another search through the stone pile many larger size stones are found. The 'large' stone might then be reassessed by the waller as a medium size stone. The stones seeming reclassify themselves as more information is available on which the classification is made. It is not the size and shape of each stone considered in isolation that defines how the stone might be useful in a construction, but the relative difference between a stone's shape and size and the rest of the pile. The choice of each stone is based on an incomplete understanding of all of the stones available. Likewise the context each stone will be placed in is unknown. The waller's understanding of the context for any given gap in the wall comprises of the knowledge he has of the previous stones he has laid, the ground conditions he is working with and his strategic design objectives such as the wall's height and the wall's width.

The type of soil on which the wall is built determines how the ground will compact due to the increasing amounts of loading. The ground will react to the additional weight of every stone added to the wall as the stone's weight is transferred through the wall to the ground. Through this the act of building the wall starts to change the environment in which the wall is built. The waller, like De Certeau's house wife, is dealing with a complex set of data that continues to change after every decision is made.

## 6. The Virtual Stone Waller

The program for making a dry stone wall was written as a way of making proposals that attempt to simulate the tactical decisions made by a dry stone waller. The program was not written as an 'expert system' to find optimum design solutions but as a program that sets a certain methodology of fabrication, through which the design proposal is created. The design proposal itself becomes a simulated outcome of the fabrication steps needed to physically make the proposal.

The program was written in Matlab, a programming language that is designed to manipulate and visually represent large sets of data. The algorithms were initially developed using computer-generated sets of data for stones. However along side developing the software, data input tools that replaced the virtually generated data were made. The first was a scanner to input three dimensional data of real stones and secondly a set of scales to weigh the individual stones.

The scanner defines the stone as a set of coordinates taken from the surface of the stone. These coordinates are taken in a series of circumferences or rings that are traced at an increasing height around the stone. When these rings are stacked together they form a three dimensional representation of the stone. The stone's data is stored in two different ways: one which preserves the ring structure of the stone data in an array and the second which just lists the coordinates in order to create a visual representation of the algorithms.

These coordinates are used to calculate the geometrical center of the stone. Assuming that the stone has uniform density the 'Centre of Mass' variable for each stone is defined as the geometric centre, see Figure 4.

While a stone has not yet been chosen to be part of the wall the Centre of Mass is set as the origin (0,0,0). But once the stone is chosen and moves into the simulated construction site the origin changes as the stone moves position.

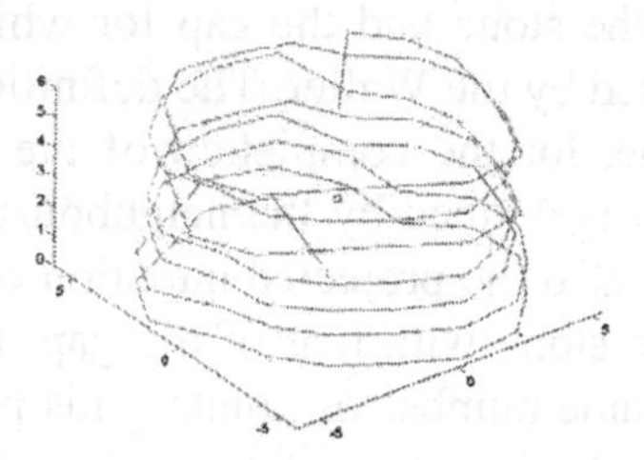

*Figure 4.* A scanned stone's data set showing the geometrical centre

Each stone is given an identity number which is initially stored in a logging variable called 'The Pile of Stones'. Selecting any of these identity numbers automatically calls up the three dimensional data, the Centre of Mass and the Weight variable of the identified stone.

The stone data is not the only input required for the program. The site in which the wall is to be constructed is defined geometrically by the size and fall of the simulated land. The type of soil the wall is to be built on is also defined. The amount of compaction of the soil, in reaction to a placed load, varies between very soft soils to bedrock which does not compact at all.

Input variables are also used to control the skill of the Waller. One variable controls how accurately the Waller defines a stone's shape. A second input variable controls the number of potential stones that he can remember. Finally the strategic decisions needed for the design were created as input variables defining the starting point of the wall within the site, the shape of the path of the wall as well as it's height and length. An example of the waller completing a row of stones to fulfil a strategic input is shown in Figure 5.

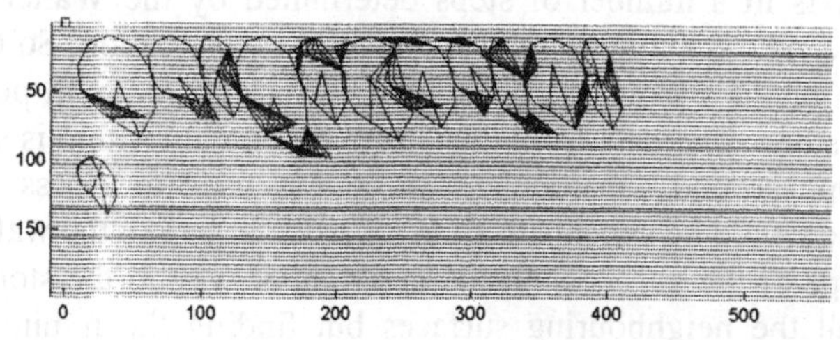


*Figure 5.* In plan, starting a new row of stones after achieving the 400cm width of the virtual wall

6.1. PICKING A STONE

A random number generator is used to simulate picking a stone from the logging variable: 'The Pile of Stones'. The identity number of a chosen stone is removed from the logging variable and the corresponding stone's data is called up. The shape of the stone and the gap for which the stone is being proposed are both analysed by the Waller. The definition of the gap or space that next requires a stone, for the completion of the wall, is defined as a complex series of data. It is defined by the neighbouring stones, sometimes the ground condition and also the projected intention of the wall. In order to compare the shape of the stone with that of the gap, the gap and the stone must be defined by the same number of points. This process was viewed as shape analysis where the Waller depending on his skill would reduce the data sets to record the most dramatic geometric changes. A unskilled Waller might define the shape of a stone by three data points and a skilled Waller might define the shape with many more points. Graphical representations of this are shown in Figure 6.

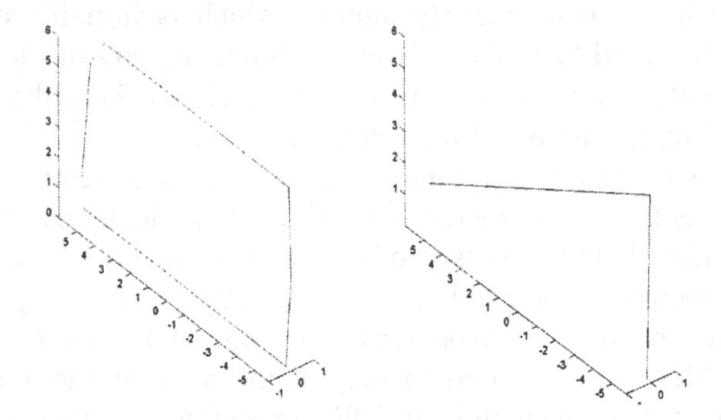

*Figure 6.* The Dry Stone Waller analysing the vertical section of a stone with varying skill. First defining it with six points, and then with three points

The shape of a stone is analysed by rotating the virtual stone about a particular axis in a number of steps determined by the Waller's ability to assess shape. The maximum point is found at each rotation. So the Waller's information about a stone is based on a series of maximum points plotted around the stone. The data from this shape finding algorithm is stored as the stone's shape around a particular axis and is used to assess whether the picked stone would be suitable to fit in a particular gap in the wall.

The shape of the gap is analysed in a similar way to the stone by taking points of all the neighbouring surfaces but finding the minimum point at each rotation, to describe the shape of the gap, Figure 7. By comparing the gap's shape data to that of the stone's determines whether the stone is acceptable or not.

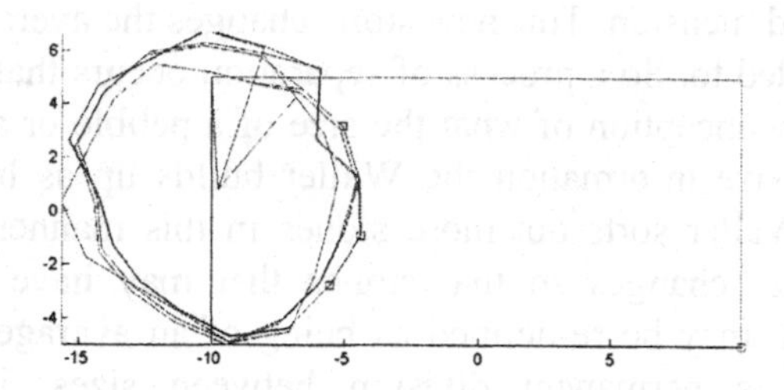

*Figure 7.* The Dry Stone Waller analysing the horizontal shape of a gap beside a scanned stone

6.2. THE KNOWLEDGE PILE

If the stone is found not to be acceptable all the information that the Waller has learnt about this particular stone is stored and the identity number of the stone is logged in the 'Knowledge Pile'. The Knowledge Pile holds the identity number of stones that have had their shape analysed but have not yet been picked for the construction. The Waller knows more information about the stones in the Knowledge Pile than any other stones and if it isn't empty it is the first pile he will search to find a suitable stone for any gap in the wall.

The Waller cannot infinitely analyse stones and hold all of the shape data. So the program was designed with a limited memory for the shape analysed stones. This limit defines how many stones can be remembered at anyone time thereby altering the skill of the Waller.

When the Knowledge Pile is searched and a stone from the pile is found to be acceptable the identity number for the stone is taken from the Knowledge Pile, reducing the number of stones held there and the stone is given a new identity number as part of the wall. If on the other hand no stone from the Knowledge Pile is found to be acceptable then the Waller will continue to search for a suitable stone by shape analysing other stones.

If the next analysed stone is also found to be unacceptable and the Knowledge Pile becomes full the first stone that was accepted into this pile will be deleted allowing the latest unaccepted stone to replace it.

6.3. THE SIZE PILES

When a stone is deleted from the Knowledge Pile not all of the information that the Waller has learnt about it is lost. An average diameter of the stone is measured and using this it is placed into one of three 'Size Piles': a 'Pebble

Pile' for smaller stones, a 'Stone Pile' for average sized stones and a 'Boulder Pile' for larger stones. The selection process looks at the average dimensions of the stones in each pile and places the stone in the pile with the closest average dimension. This new stone changes the average dimension of the pile it is added to. So a process of separation occurs that is based not on the Waller's preconception of what the size of a pebble or a boulder should be, but on the size information the Waller builds up as he analyses each stone. As the Waller sorts out more stones in this manner the Size Piles' average diameter changes so that stones that may have originally been defined as small may be re-defined as being of an average size. Therefore there is never a permanent division between sizes, it is a relative classification.

The Dry Stone Waller was programmed so that it always would always find a stone for any gap and so that no stone is ever completely rejected as impossible to use in a wall's construction. The Waller looks for an appropriate stone for a gap by firstly searching the Knowledge Pile and then the appropriate Size Pile is searched and finally the Waller will return to the original Pile of Stones and start analysing new stones. Any stone chosen is found acceptable by comparing it's shape data to that of the gap to be filled. There is unlikely to be an exact match so there must be an acceptable tolerance within which the stone is found to be acceptable. For any particular gap if Waller returns to the original Pile of Stone and picks an unacceptable stone the Waller will increase the difference that is acceptable between the gap and the stone before he researches the hierarchy of stone piles. So that the possibility of finding an acceptable stone increases until one is found. This means that the set of stones is not searched and re-searched infinitely to find a stone of a particular size and shape that does not exist.

### 6.4. PLACING A STONE

When an acceptable stone is found it is rotated into the chosen orientation. It is placed in the gap in the wall so that it is not touching any other stone and then rotated in small steps about the given orientation to find a position of best fit between the complete set of data of the stone and the complete set of data of the gap.

It is easy mathematically to find the maximum and minimum points of stones in different axis so that the placed stone does not over lap any other but the irregular shape of the both the placed stone and the stones that make up the gap in the wall makes it difficult to predict how the stones will touch.

If the stone in question formed an interaction with an ideally flat surface the relationship between the known points and this completely understood surface is predictable. Critical points can be predicted as being crucial to their relationship. However by replacing the flat surface with a complex set of data such as another stone the interactions are less easy to predict. The stones' shapes may interlock, Figure 8, and the points of intersection between these stones will not be just between known points, they will be between known points and the spaces between known points. Here the connections between known points on the surface of a stone must be considered. In the Dry Stone Waller the connections between points were considered as straight lines. In reality this is clearly not really the case. But these interactions have to be considered as approximations because however high the resolution is there are always spaces between the defined points on a stone that have to be considered during these interactions.

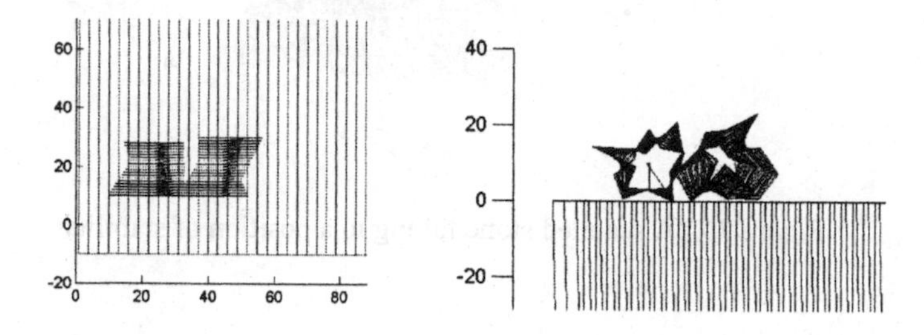


*Figure 8.* The plan and elevation of two virtual stones that are as close as possible to one another and interlocking

So to find the points of intersection between stones the placed stone is moved gradually closer to the other stones in small steps until points of intersection are found with it's neighbours in all three axis, Figure 9.

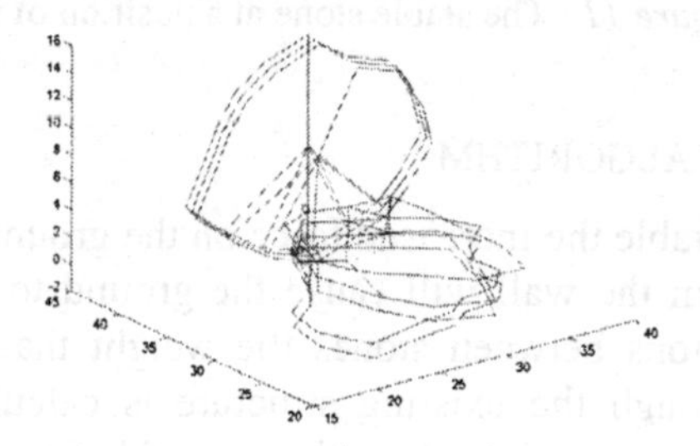

*Figure 9.* Points of intersection being found between two scanned stones

## 6.5. GRAVITY, STABILITY AND FRICTION

Wherever the stone is in the wall it is very important to establish that the placed stone is stable in it's located position. In reality gravity acts on an unstable stone casing it to fall to a position of stability.

Using the placed stone's weight, moments are taken about the intersection points of the stone with the surface below. If the moments in three dimensions do not balance and the horizontal component of the weight force overcomes the frictional force between stones, the stone is rotated in the direction of the resultant force about an intersection point until a new intersection point is found, Figure 10. Then the moments are re-taken. This is repeated until the stone is found to be stable, Figure 11.

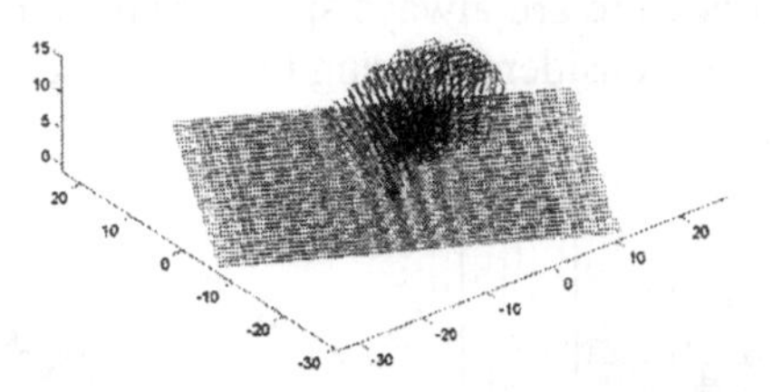

*Figure 10.* A scanned stone falling to a position of stability

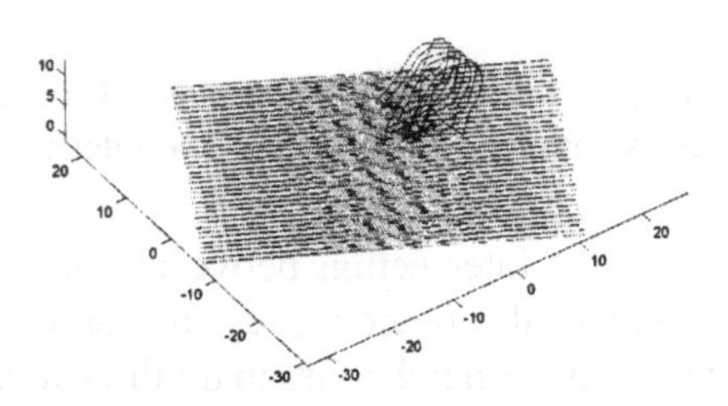

*Figure 11.* The stable stone at a position of rest

### 6.6. COMPACTION ALGORITHM

Once the stone is stable the increased force on the ground by the new stone's additional weight in the wall will cause the ground to compact. Using the points of intersections between stones the weight that is transferred from stone to stone through the existing structure is calculated. This way the increased load at various points along the ground is found, Figure 12, and the ground compacts a certain amount determined by the type of soil as shown in Figure 13. This changes the position of all the stones supported by that piece of ground.

The stone now becomes part of the wall and it is logged in variables that locate precisely where it is in the wall so that it can be used to define the next gap that is left in the wall.

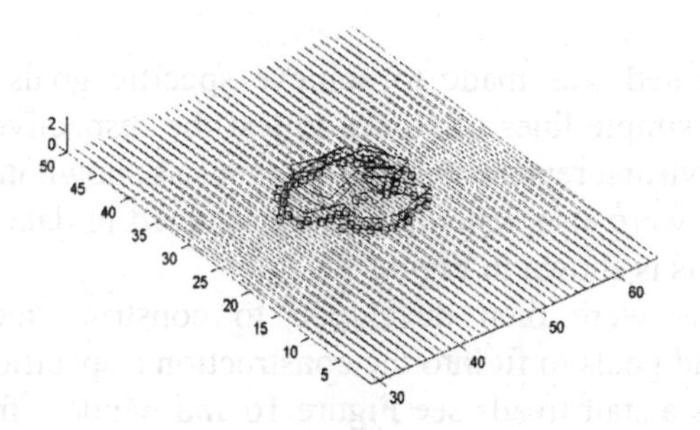

*Figure 12.* A stable stone showing it's points of intersection with the ground plane

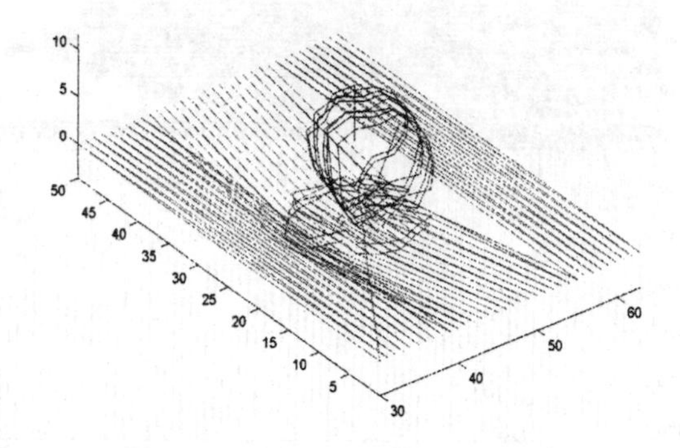

*Figure 13.* One stone on top of another causing the soft soil of the ground plane to compact

## 6.7. MONITORING ALGORITHMS

As the wall is built the choice of stone is not purely based on how well one stone fits by the side of another. The Waller needs to be aware when he has fulfilled the set goals of the walls. More than this the Waller needs to be choosing the stones in order to fulfil chosen strategic goals of the desired wall. A series of algorithms monitor how close the Waller is in achieving these goals and as the wall gets closer and closer to the set goals, the monitoring algorithms change some of the criteria that defines the data set of the next gap in the wall. This means that the stone chosen for a particular gap allows the Waller's decisions to attempt to get as close as possible to the desired dimensions of the wall. It is these programs that decide that a row of stones is completed, when the course is finished and when the wall has been successfully constructed.

## 7. The Walls

Each wall generated was made to achieve specific goals. Sometimes the walls were very simple lines of stones to test the respective algorithms and were made with virtual representations of stones as shown in Figure 14.

Similar walls were also generated using scanned in data from real stones an example of this is shown in Figure 15.

The programs were then developed to construct more complicated proposals that had goals to fit into the construction a specific size and weight of object such as a stair treads see Figure 16 and window frames see Figure 17. These additions caused the stones to be constructed into more complex geometries and affected the choice of stones so that the construction components fitted as accurately as possible.

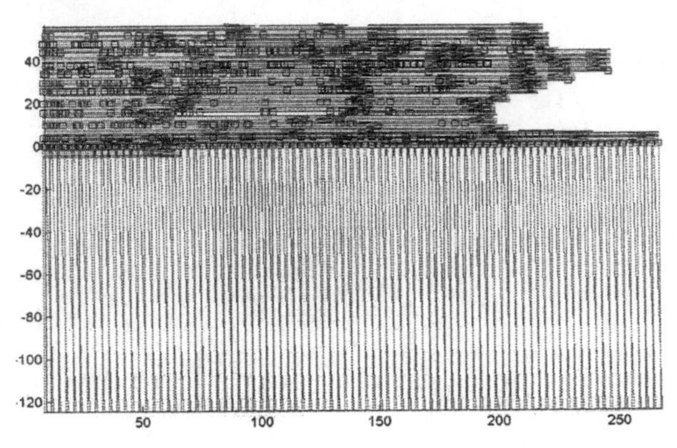


*Figure 14.* An elevation of a virtual dry stone wall aiming to be 50 cm high and 200cm long. Constructed using computer generated stones by the Dry Stone Walling program and showing the points of load transfer between stones.

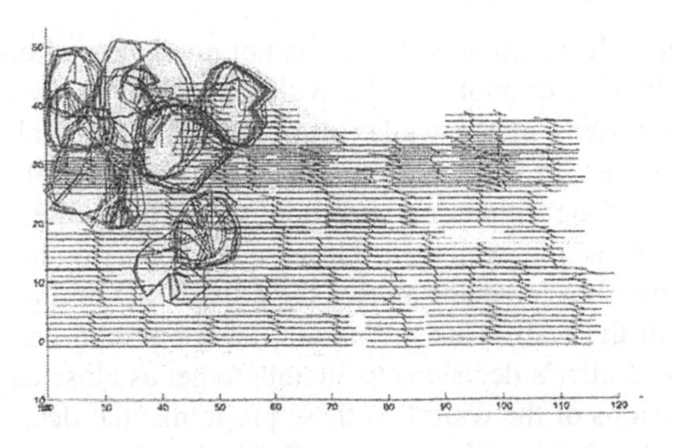

*Figure 15.* An elevation of a wall built with scanned in stones aiming to be 40 cm high and 120 cm long

*Figure 16.* In elevation, stones chosen to support stair treads

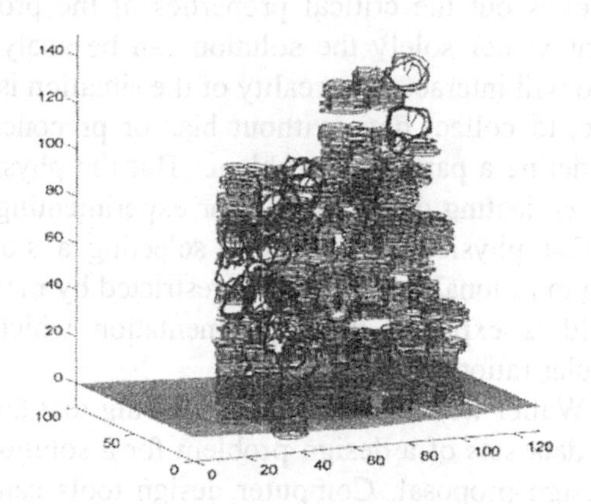


*Figure 17.* In elevation, part of a wall built to hold a window frame that slots into the wall

## 8. Conclusion

The strategic decision pre-empts or pre-rationalises the given data of the context or proposed problem instantly drawing a line between relevant and irrelevant information. This reduces the search space by using pre-defined criteria on which to make decisions. The strategic decision chooses the critical data not by looking at the realities of a problem but by following decisions made which define a boundary around the problem. The manner in which the solution is thought about affects the data chosen to represent the problem, which in turn affects the proposed solutions. The possibilities of the context or solutions are therefore limited by the very first preconceptions of the solution that were made. However this reduction of the data set enables faster solutions to be found as only part of the search space is considered.

In a tactical decision the critical data in the data sets are not treated as static, they are reconsidered at each stage because it is realised that a data set might have different critical information depending on the type of solution proposed. The relationship between the context and the solution is not defined by solely understanding the context, or solely understanding a design proposal. It is their interaction that matters. The tactical decision treats the context and the solution as data spaces that search one another for possibilities.

Only by physically relating a proposed solution to the context, can the complex data be reduced to find the critical information that is important to create that particular relationship. It is the relationship and only the relationship that draws out the critical properties of the proposal. Neither solely the data context nor solely the solution can be analysed enough to predict how the two will interact. The reality of the situation is that it is hard, perhaps impossible, to collect data without bias or preconceived ideas of how the data will define a particular problem. But the physical world is a very rich place for collecting data sets and for experimenting with possible design solutions. Our physical actions, e.g. selecting a stone, are taking inputs that are hard to rationalise and are not restricted by mental constructs. The physical world is explored by experimentation which is not often defined by a particular rational.

The Dry Stone Waller was an experiment showing that the methodology used to search the data sets of a design problem for a solution is key to the outcome of the design proposal. Computer design tools can be created to deal with complex data sets to find solutions to very physical problems. They can accept the realties of the physical world and enable designers to explore and use it to their advantage. If designers could use the potential of a group of construction components whose selection and position within a construction was used to overcome contextual irregularities, perhaps this suggests that construction components could be made to lower tolerances, thereby reducing production costs. This might bring the virtual world closer to the physical, rather than forcing our perception and the modes of fabrication to move closer to a virtual world that only deals with problems defined by a pre-selected set of data.

## Acknowledgements

I would like to acknowledge the contributions of Philippe Ayres and Nic Callicott of Sixteen Makers for their support in developing the project, Professor Stephen Gage and members of the Interactive Workshop of The Bartlett, School of Architecture, University College London for giving me the opportunity to carry out the project, and a further acknowledgement to Oliver Williams of Techniker for discussions about the skills involved in constructing a dry stone wall.

## References

De Certeau, M: 1984, *The Practice of Everyday Life*, University of California Press, California, pp. 82.

Evans, R: 1997, *Translations from Drawing to Building and other Essays*, Janet Evans & Architectural Associations Publications, London, pp. 175.

## PATTERNS IN DESIGN

JS Gero (ed), *Design Computing and Cognition'04*, 297-316


# EMERGENT ELEMENTS IN PERIODIC DESIGNS: AN ATTEMPT AT FORMALIZATION

KATARZYNA GRZESIAK-KOPEĆ
*Jagiellonian University, Poland*

**Abstract.** The exploration of emergence is one of the most important and one of the least formalized areas of research in the field of graphical design. This paper presents a formal description of emergent elements in selected periodic designs, namely rosettes, borders and two-dimensional designs. The introduction covers the basic concepts for the area of creative design and emergence. In the following chapters a formal model for graphical design and our concept of using it to define emergent elements is provided.

## 1. Introduction

### 1.1. CREATIVITY IN DESIGN

The design process is strongly connected to decision making and making choices naturally involves searching for solutions. A pathway leading to a solution a designer is looking for cannot be predicted. That is why the question of creativity arises in computer aided design. Creativity in design has many interpretations (Boden 1991; Gero and Maher 1993; Kim 1990; Sternberg 1988; Weisenberg 1986). There is a clear distinction between considering creativity as residing only in the artefact and considering that certain processes have the potential to produce artefacts which may be evaluated as creative (Gero 1996). In this paper we are going to call *creative* the kind of design which produces unexpected results for the designer.

### 1.2. EMERGENCE

The theory of emergent evolution was introduced in the years between 1920 and 1930 by English philosophers and biologists, mainly S. Alexander, Ch.D. Broad, J.Ch. Smuts and C. Lloyd Morgan (Morgan 1923). According to this theory, the world undergoes a continuous evolution, which causes changes in the quality of individuals from a lower stage and creates higher qualities which compose a hierarchical sequence. This kind of evolution is called an emergent evolution, in other words *emergence*.

The notion of emergence in philosophy is analogous to the one in design. In design, those features which are not explicitly represented but emerge from a design structure are called *emergent*. What is more, only those features which can be represented explicitly in a design structure are considered to be emergent. While representing emergent features in a design structure we, according to the emergent evolution, create a new quality different individual.

Visualization plays a crucial role in representing data. There are many ways of visualization which means a transformation of thoughts into visible forms. For example we can do this with the help of a two-dimensional drawing. It is this kind of representation we are going to use in this paper. We assume that white spaces are an empty background and black regions are actual designs.

Let us examine the example in Figure 1. A simple change of position of copies of the given shape, Figure 1A, is not enough to obtain an explicit representation of the emergent cross in Figure 1B. In order to achieve it, the designer has to add some extra elements which will represent the cross. This is the way the emergence enhances creativity in the design process - a new quality different individual has come into being.

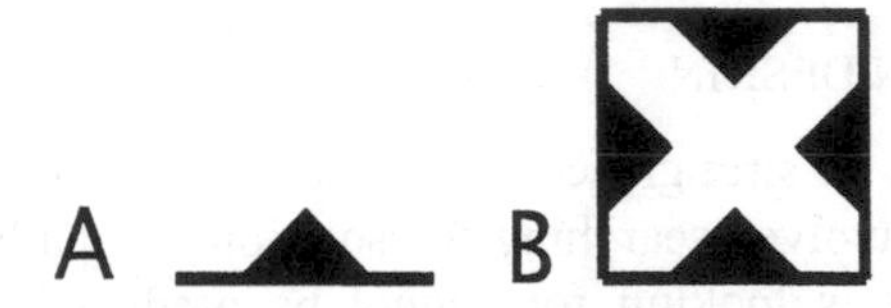


*Figure 1.* (A) Single object (B) Configuration of four copies of the given object which implies emergent cross

Seeing demands some activity from the person who is watching. It is not enough to wait passively for an image on the retina. A retina is like a movie screen on which changing sequences of pictures are displayed but a brain, which manages everything, intentionally notes only a few of them. On the other hand, only a blurry visual impression or even a very small detail is enough for us to decide that we have seen this or a quite different thing (Rasmussen 1999). The ability to recognize properties and characteristics or functions which are not originally intended in design is an important aspect of human visual perception. Cognitive theory explores this phenomenon and tries to explain the factors which cause such a variety of possible interpretations, groupings and segmentations of images. This ambiguity in perception comes from several sources. It is worth mentioning here the Gestalt nature of human perception: It has been widely recognized that local features of a scene, such as edge positions, disparities, lengths, orientations, and contrast, are perceptually ambiguous but that combinations of these features can be quickly grouped by a perceiver to generate a clear separation between figures and ground (Grossberg and Mignolla 1985).

According to the Gestalt psychologists people appear to intuitively distinguish the foreground from the background in their reading of shapes. In Figure 2 there is a famous example (Arnheim 1977) of a black vase which materializes into two white human profiles. It depends on the viewer's imagination which part of a drawing constitutes a foreground. To one viewer it would be a black vase and to another it would be two opposing profiles.

*Figure 2.* Emergent faces

Emergence is a very general concept and covers not only visual and structural aspects of artefacts but functional and behavioral aspects as well. However, in this paper we are going to focus only on emergence in the structural level of design and we are going to propose a coherent formal model of design space including emergent elements. Many different shape representations, while considering emergence, have been introduced, from infinite maximal lines and their constraints (Gero and Yan 1994), through a logic-based framework for shape representation using half-planes (Damski and Gero 1996) to shape algebras (Chase 1996). Nevertheless, all those works were focused on recognizing some emergent features and not on describing them. Also the field of our consideration is slightly different. We are interested in periodic designs in decorative art and not in architectural ones.

### 1.3. DESIGN SPACE

Computer aided design requires a formal definition of the design space. In many applications, graphical models are composed of transformed copies of shapes from a given initial set. That is why, in this paper, we have assumed a similar approach to the problem. From now on, we are going to call the initial set an *alphabet of basic shapes* (Woodbury et al.1990).

Let us assume that we are working in $R^n$ with Euclidean norm, written $(R^n, |\cdot|)$. For each set $A \subset R^n$ let $Int(A)$ denote the interior of $A$ and

$Fr(A)$ denote the frame. First of all, we have to define the basic concept for our graphical model, the concept of *a shape*.

Definition 1. The *shape* in the n-dimensional Euclidean space $R^n$ $(n \geq 1)$ is a compact and connected subset of the space.

We can examine the notion of a shape in Figure 3. There is an example of a rosette design and its elements. Only the element 3B is a shape.

*Figure 3.* (A) Design and its elements (B) a shape (C) a part which is not a shape

In our model, together with shapes, we have to consider a set of possible operations on them. We assume that there are no constraints defined on the set and a range of these operations is chosen by the designer. However, in order to achieve some non-trivial results the designer should take into account at least a set of similarities. From now on, we are going to call this set a *set of admissible transformations*.

In our model, the choice of a set of admissible transformations, denoted as $F$, is a starting point of the design process and, as we can see from the definitions below, influences the *alphabet of basic shapes*. Two shapes A and B are said to be *conjugate* in respect to $F$ if there exists $f \in F$ such that $f(A) = B$.

Definition 2. The set of shapes in $R^n$ denoted as $S$ is called the *alphabet of basic shapes in relation to* $F$, where $F$ is the set of admissible transformations in $R^n$ which contains at least similarities and their composites, if $(\forall A, B \in S)\, A$ and $B$ are not *conjugate* in respect to $F$. Each element of the alphabet is named *a basic shape*.

While transformed, basic shapes constitute a *spatial pattern* which is underlaid by a *spatial structure* defined as follows:

Definition 3. The *spatial structure* is the set of pairs $Q \subseteq S \times F$, where $S$ is the alphabet of basic shapes and $F$ is the set of admissible transformations.

Definition 4. The *spatial pattern* for $Q$ is $Z(Q) = \bigcup_{(s,f)\in Q} f(s)$. Each pair $(s, f) \in Q$ determines a *component* $f(s) \subset Z(Q)$.

In other words, a spatial pattern is a sum of components.

Let us now introduce three spatial structures $\underline{Q_1}$, $\underline{Q_2}$ and $\underline{Q_3}$ for the example rosette design we are going to consider during our investigation.

Let our alphabet of basic shapes $\underline{S}$ contains only one shape $A$ visible in Figure 4, $\underline{S} = \{A\}$. The set of admissible transformations $\underline{F}$ consists of: glide reflections, translations and rotations. According to our definition of a spatial structure $\underline{Q_i} \subset \underline{S} \times \underline{F}, i \in \{1, 2, 3\}$. In our design space we create spatial patterns $Z(\underline{Q_1})$, $Z(\underline{Q_2})$, $Z(\underline{Q_3})$ as in Figure 5, Figure 6 and Figure 7.

*Figure 4.* The shape $A$ for the rosette design example

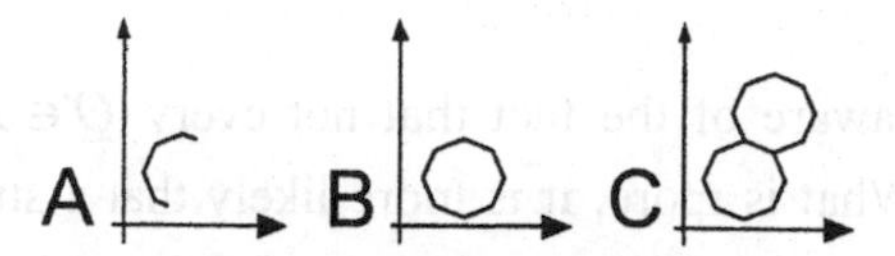


*Figure 5.* (A )-(B) some of the steps to create the final pattern (C) the final spatial pattern $Z(\underline{Q_1})$

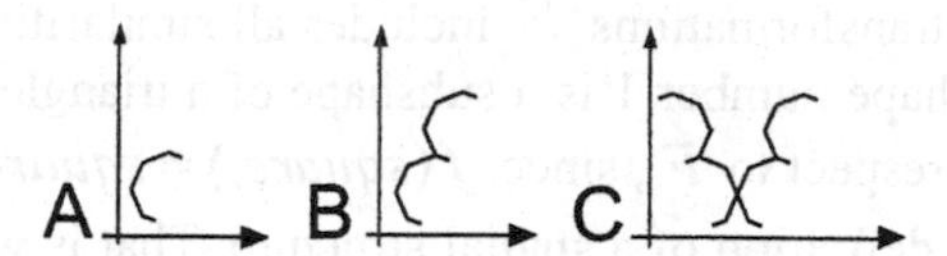


*Figure 6.* (A)-(B) some of the steps to create the final pattern (C) the final spatial pattern $Z(\underline{Q_2})$

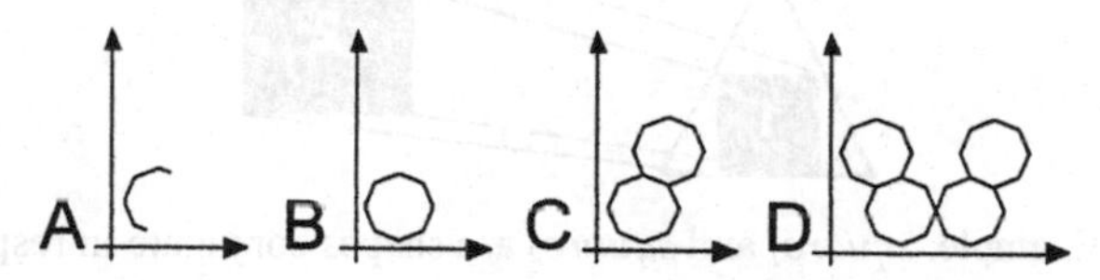


*Figure 7.* (A)-(C) some of the steps to create the final pattern (D) the final spatial pattern $Z(\underline{Q_3})$

With such a defined spatial pattern we are going to identify the graphical model, which is a subset of $R^n$ that satisfies some special conditions. Since the change of only a part of a design often leads the designer to a final solution, we cannot miss this important feature in our model. Therefore, we need the notion of a *fragment* of a shape and the notion of a *subpattern* of a spatial pattern.

Definition 5. The *fragment* of the given shape $A$ is each shape $B$ so that $B \subseteq A$.

For a given alphabet of basic shapes $S$ we define another shape alphabet containing fragments of basic shapes from $S$ in the following way:
Definition 6. Let $S$ be an alphabet of basic shapes. The *complete shape alphabet in respect to* $S$ is defined as follows $\overline{S} = \{\overline{s} : \exists s \in S$ so that $\overline{s}$ is a fragment of $s\}$.

Now, let us define a *family of substructures* for a given structure $Q$.
Definition 7. Let $Q \subseteq S \times F$ be a spatial structure. The *family of substructures* for $Q$, written $P(Q)$, is defined in the following way: $P(Q) = \{Q' : Q' \subset \overline{S} \times F \wedge \forall(\overline{s}, f) \in Q' \exists (s, f) \in Q$ so that $\overline{s}$ is a fragment of $s\}$.

We should be aware of the fact that not every $Q' \in P(Q)$ has to be a spatial structure. What is more, it is more likely that a structure $Q'$ chosen from a family is not a spatial pattern at all. That is because $\overline{S}$ is almost certainly not an alphabet of basic shapes in relation to $F$. Let us assume that our starting set $S$ consists of two shapes: a triangle and a square. The set of admissible transformations $F$ includes all similarities. As we can see in Figure 8, the shape number 1 is a subshape of a triangle and is conjugate to a square (2) in respect to $F$, since $f(square_1) = square_2$. This situation is contrary to our definition of a spatial structure. That is why, the family of substructures is not called a family of spatial substructures but just a family of substructures.

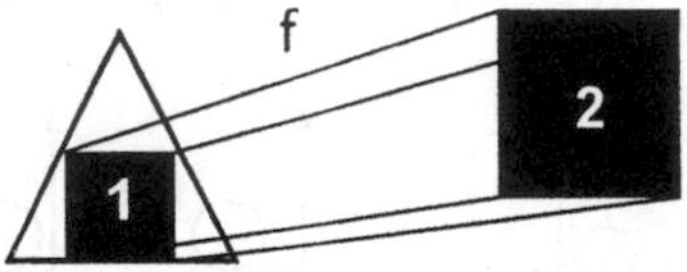

*Figure 8.* Example: $square_1$ and $square_2$ are shapes conjugate in respect to $F$

At last, we can define a *family of subpatterns* for a given pattern $Z(Q)$.
Definition 8. Let $Z(Q)$ be a spatial pattern. The *family of subpatterns for* $Z(Q)$, written $P(Z(Q))$, is defined as $P(Z(Q)) = \{Z(Q') : Q' \in P(Q)\}$.

In order to better understand this definition let us consider the spatial pattern $Z(Q)$ to be just a subset of the given space and denote it $Z$. According to our intuition each set $A$ which is a subset of $Z$ should be considered as a subpattern and this is the case. All patterns which are members of a family of subpatterns for $Z(Q)$ satisfy the expectation.

## 2. Periodic Designs

In this paper we are going to search for emergent elements only in selected types of periodic designs, namely *rosettes*, *borders* and *two-dimensional periodic designs* (patterns). These kinds of designs were also investigated by Doris Schattschneider who elaborate their classification (Schattschneider 1986). However, before considering emergent features, we have to start with a precise definition of a *periodic design* itself and its main characteristic, a *motif.*

Definition 9. Let $Z(Q)$ be a spatial pattern for $Q \subseteq S \times F$ in $R^n$ $(n \geq 1)$ and $I$ be a set of isometries on $R^n$ without the identity. Let $Q_M \subset Q$. The spatial pattern $Z(Q)$ is called *periodic* if and only if $\exists i \in I$: $Z(Q) = M \cup \bigcup_{k=1}^{m-1} i^k(M), m \geq 2, m \in N$, where $M = \bigcup_{(s,f) \in Q_M} f(s)$. $M$ is called a *motif.*

In other words, a motif is a selected spatial pattern which is to constitute a design. Every periodic design is composed of $m$ copies of a motif. We recognize as periodic only those designs where at least two copies of a motif are present. Knowing the concept of a motif, let the introduced before spatial patterns $Z(\underline{Q_1})$, $Z(\underline{Q_2})$, $Z(\underline{Q_3})$ constitute three motifs $M_1$, $M_2$, $M_3$ for our periodic design example, Figures 5, 6 and 7.

Now we are going to introduce an additional concept – the *multiplicity of a periodic spatial pattern.*

Definition 10. The number of transformations of the *minimal motif* needed to generate the whole design is called the *multiplicity of the periodic spatial pattern* and is denoted as $m$.

Following the definition, the multiplicity of a periodic spatial pattern is determined by a *minimal motif,* which has not been defined yet. A *minimal motif* is directly related to a type of the periodic design. It has a different meaning for a *rosette,* different for a *boundary* and different for a *two-dimensional periodic design.* In this case, we are going to explain this concept separately for each type of design.

As we can see from the definitions, we do not assume any constraint on a motif. We may thus construct a very complex design even though the set of admissible transformations contains, for example, only similarities.

### 2.1. ROSETTE

One of the most popular geometrical motifs frequently used in decorative art was, and still is, a circle. It appears as a separate motif, like in Korean art, Figure 9, or as a boundary for other motifs placed within, like on Japanese coats of arms (Morant 1981). That is why the *rosette* is one of the types of periodic designs we are going to consider.

*Figure 9.* Rosette in Korean art

Definition 11. The periodic spatial pattern $Z(Q)$ is called a *m-multiplicity rosette*, $2 \leq m < \infty$ if and only if there exists an isometry $i$ so that:

1. the isometry $i$ is a rotation,

2. $Z(Q) = M \cup \bigcup_{k=1}^{m-1} i^k(M)$,

3. $M = i^m(M)$.

Every rosette is a periodic spatial pattern with special limitations on transformations. First of all, the isometry $i$ has to be a rotation. Secondly, the (m+1)-copy of a motif covers the first one, $i^m(M) = M$. This condition follows our perception of a rosette as a periodic design enclosed within a circle. Note that in a rosette its multiplicity (which is the number of motif rotations during its construction) does not have to be equal to its degree of rotational symmetry. Let us illustrate this feature in Figure 10.

*Figure 10.* (A) Rosette design with 8-fold rotational symmetry (B) multiplicity is equal to 4 (C) multiplicity is equal to 8

When having a whole rosette design, like in Figure 10A, we are not able to define its multiplicity explicitly, because we do not have any knowledge of its motif. In order to remove any ambiguity we introduce the concept of a *minimal motif* which must be given as a starting point, together with a rosette, to our consideration.

Definition 12. The motif $M$ provided in the definition of a rosette is called the *minimal* motif. The minimal motif determines the multiplicity of the rosette.

Let us now consider motifs $M_1$, $M_2$, $M_3$ defined before as minimal ones for the rosette design $D$ in Figure 11. We have three pairs: $(D, M_1)$, $(D, M_2)$, $(D, M_3)$. As far as the first two pairs are concerned, our rosette design $D$ is of 8-multiplicity, Figures 12 and 13. However, if we take into account the third pair $(D, M_3)$ our design becomes of a multiplicity of 4, Figure 14.

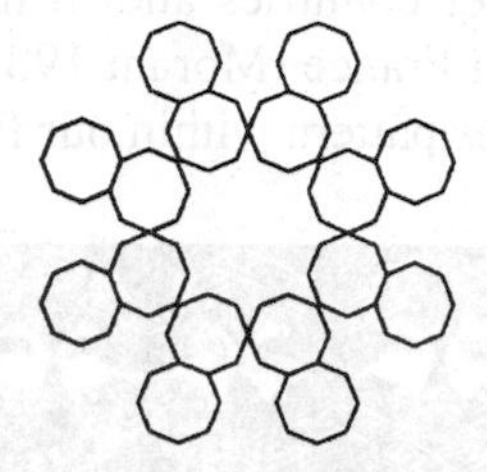

*Figure 11.* Example rosette design $D$

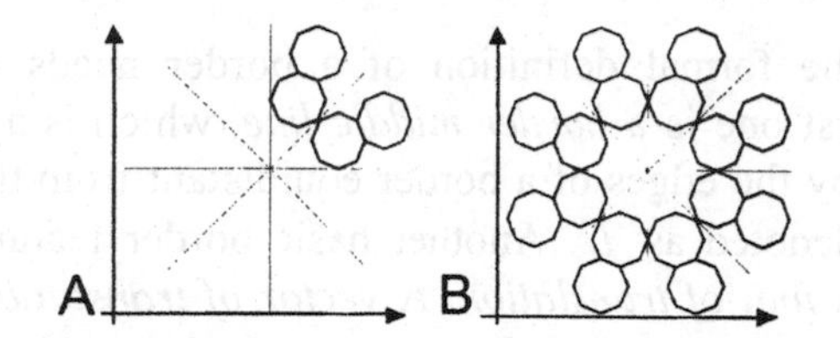


*Figure 12.* Rosette example: 8-multiplicity (I)

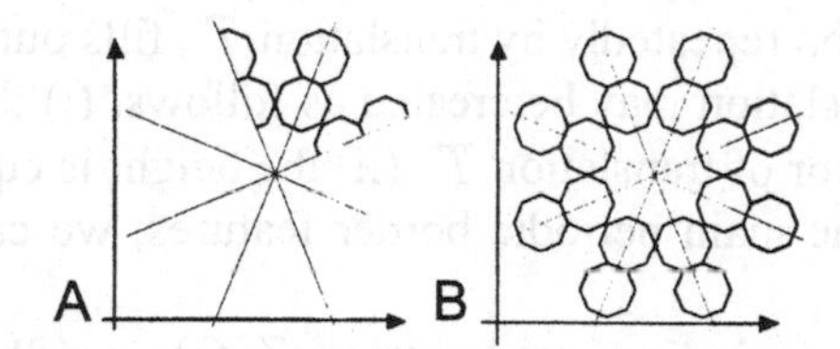


*Figure 13.* Rosette example: 8-multiplicity (II)

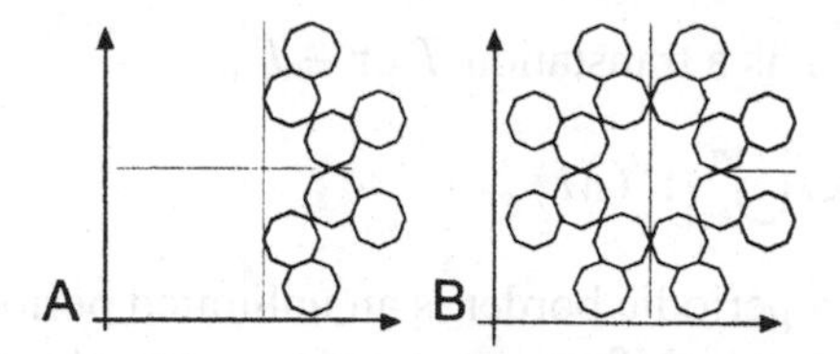


*Figure 14.* Rosette example: 4-multiplicity

We can clearly see from our example how important the choice of a minimal motif is, which we assume is selected by the designer. This choice, as we will see in following sections, determines not only multiplicity but other important features of periodic designs as well.

## 2.2. PERIODIC BORDER

Borders in the shape of a line or a ribbon where popular elements in separating other motifs in prehistoric decorations, Greek vases, and others. A border in the shape of zigzag which signified water in ancient Egypt, can also be found in many other countries and in many different epochs from China, Figure 15, to Roman France (Morant 1981). Therefore, we are going to describe a *periodic border* pattern within our framework.

*Figure 15.* Periodic border in Chinese decorative art

Introducing the formal definition of a border needs some additional concepts. The first one is a *border middle line*, which is a line placed in a strip delineated by the edges of a border equidistant from them and parallel to them and is denoted as $L$. Another basic border feature is a *vector of translation* and a *unit of translation*. A *vector of translation* is the shortest vector of translation invariable in respect to a border and is denoted as $T$. A *unit of translation* for a periodic border is the minimal area of a plane, which when acted on repeatedly by translation $T$, fills out the whole design. Such a unit of translation may be created as follows: (i) the base is equal to the length of a vector of translation $T$, (ii) the height is equal to the border's width. Knowing the main periodic border features, we can now define the border itself.

Definition 13. The periodic spatial pattern $Z(Q)$ is called a border if and only if there exists an isometry $i$ so that:

1. the isometry $i$ is a translation $T$ or $-T$,
2. $Z(Q) = M \cup \bigcup_{k=1}^{\infty} i^k(M)$.

In other words, a periodic border is an unlimited periodic design which is generated by recurrent shifting of a unit of translation by a vector of translation $T$ or $-T$. Periodic borders are enclosed between two parallel infinite lines, the edges of the borders. An example of a periodic border is in Figure 16.

*Figure 16.* Periodic border

In Figure 17 a periodic border is presented together with its middle line $L$ and the vector of translation $T$, and the unit of translation.

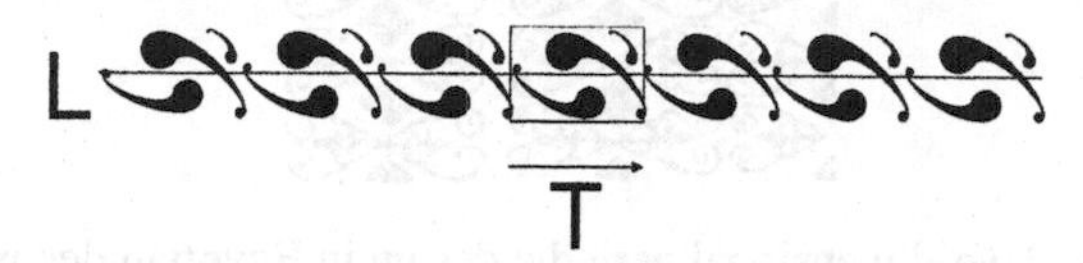


*Figure 17.* Periodic border characteristics

We have to stress here that a unit of translation may be selected in many different ways. There are many possible units of translation for a given border as we can see in Figure 18.

*Figure 18.* Two admissible units of translation for a given periodic border

Now we can define the concept of *minimal motif* in the context of a periodic border.

Definition 14. The motif enclosed in the unit of translation is called *the minimal motif.*

There is no fragment of a minimal motif which shifted by a vector shorter than a vector of translation $T$ would generate a whole border. A minimal motif may be composed of one or more *generating regions*.

Definition 15. The minimal plane which, when acted on repeatedly by the symmetries of the design, fills out the periodic border is called the *generating region*. We assume here that the set of *generators*, the set of acceptable transformations, include more symmetries than only translations.

There exists a very important taxonomy of borders and two-dimensional patterns. We distinguish two kinds of these designs, filled and free. However, this distinction does not influence emergent elements discussed here.

In this paper we assume, as we can see in the definition, that a periodic border is unlimited which allows us to build a unified characteristic of borders and *two-dimensional periodic designs*.

### 2.3. TWO-DIMENSIONAL PERIODIC DESIGN

The last kind of design we want to consider is a *two-dimensional periodic design*, which was widely used not only in Islamic decorative art but in the art of Greece and Egypt, Figure 19, as well (Morant 1981).

*Figure 19.* Two-dimensional periodic design in Egyptian decorative art

As we have already said, a unit of translation for a border is not uniquely defined. This fact was not so important while defining borders but it is essential while defining *two-dimensional designs*.

Definition 16. The periodic spatial pattern $Z(Q)$ is called a *two-dimensional periodic design* if and only if $Z(Q)=\bigcup_{k=1}^{\infty} T^k(B)$, where $B$ is a periodic border and $T$ is a translation $T'$ or $-T'$ so that $T \neq 0$ and $T$ does not lie on the middle line of the border $B$.

In other words, the unlimited periodic design generated from the periodic border $B$ when acted on repeatedly by a translation $T'$ and $-T'$ is called a two-dimensional periodic design. From a given periodic border we can often define many two-dimensional periodic designs depending on a choice of a unit of translation, as we can see in Figure 20.

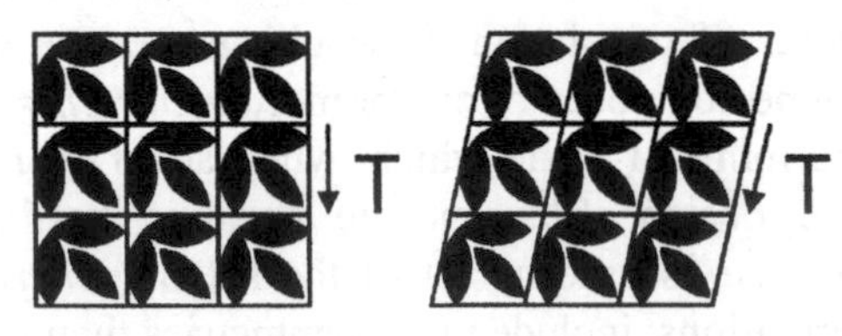


*Figure 20.* Two different two-dimensional periodic designs generated from the same border for two different units of translation

Like for other sorts of designs, we are now going to define the *minimal motif* of a two-dimensional periodic design.

Definition 17. The *minimal motif* of the two-dimensional design is the minimal motif of the periodic border it is generated from.

## 3. Formal Model of Emergent Elements in Periodic Designs

In this section we are going to describe potentially emergent elements which may come into existence in periodic designs like a *rosette*, a *border* or a *two-dimensional periodic design*. Even though we are going to consider only a few, precisely defined types of designs, the location of emergent features causes a lot of problems.

We have already described all the types of designs we are going to search for emergent elements. Now we are going to make an attempt to formalize the *emergence* itself. Let us define once again the problem we are facing. We would like to point out *potentially emergent elements* in a given periodic design. The attribute *potentially* is not here without reason. In this paper we are not exploring the field of pattern recognition and we will not describe recognized shapes. The final decision whether a selected element may be classified as the emergent one is left to the designer or to some expert system. From now on we assume that each element distinguished by us is emergent, which will not influence the result of our search but will make it clear.

Since we do not assume any constraints on a minimal motif of a periodic design, we are not going to analyze the motif in order to find emergent elements in its structure but we will try to point out all elements which emerge from the motif combinations instead. What is more, as emergent we describe only those elements which come into existence after all transformations were applied to the motif during the design process. The choice of designs we have made implies a following taxonomy of emergent elements: *emergent shapes*, *holes*, *grooves* and *negatives*. The nomenclature we took from Ryszard Jakubowski (Jakubowski 1986).

This grouping of emergent elements is not accidental. It is possible to consider a single design in two different ways. On the one hand an author's target may be to give some form to materials he works with. On the other hand he may work with a free space and acknowledge the creation of its appearance as the main meaning of his work (Rasmussen 1999). In order to recognize all emergent shapes in examined designs we have to look at these designs from both points of views. On one occasion we may arrange structural forms and we will search for *emergent shapes*. On another, we may model empty spaces to find *holes*, *grooves* and *negatives*.

### 3.1. EMERGENT SHAPES

At the beginning we will describe our concept of an *emergent shape*, which is a structural element of a design not regarded by the designer during the process of creation.

Definition 18. Let $Z(Q)$ be a spatial pattern of the given periodic design $P$. Each subpattern $Z(Q') \in P(Z(Q))$ which comes into existence after all transformations were applied to generate $P$ is called an *emergent shape*.

A very simple example of an emergent shape is in Figure 21.

*Figure 21.* Emergent shape: (A) configuration of two triangles (B) emergent rectangle

3.2. HOLES

In the following sections we will deal with those emergent elements which do not have a structural representation within the design but can be seen at first glance. The first kind of such element is *a hole*.

The concept of a *hole* is an ordinary one and quite obvious for a human being. We will put it in a formal way according to the model we have presented.

Definition 19. Let us consider the spatial structure $Q \subseteq S \times F$ and the spatial pattern $Z(Q) = \bigcup_{(s,f) \in Q} f(s)$. There exists exactly one *set of holes* $H(Z(Q))$ if the following conditions are satisfied:

1. $\exists F', S', Q'$ such that $F' \neq \varnothing$ is a set of transformations without their composites, $S' \neq \varnothing$ is a set of shapes, $Q' \subseteq S' \times F'$,
2. $\forall (s', f') \in Q'$ $f'(s')$ is a shape and $f'(s') \cap Z(Q) = Fr(f'(s'))$.

Each $f'(s')$ so that $(s', f') \in Q'$ is called a *hole* and $H(Z(Q)) = \{f'(s') : (s', f') \in Q'\}$.

According to our assumption that, as emergent, we only recognize those elements which come into existence after all transformations are applied to generate a design, we will probably not take into account a single hole but a whole set of holes. For a rosette of m-multiplicity, as we can see in Figure 22, a set of holes if not empty may have one element or many elements. However, this set is always of a finite number of elements. What is more, the number of holes for a rosette of m-multiplicity, except the case of only one hole or none, is always a multiple of m or a (multiple of m) + 1.

*Figure 22.* Rosette's holes: (A) no hole (B) single hole (C) many holes

We have to remember that we do not take into account holes in motifs, that is why in some cases a set of holes for a given rosette may be quite diverse depending on a minimal motif. Let us consider the rosette in Figure 23.

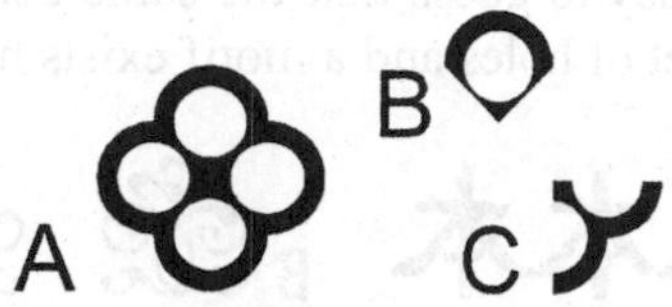


*Figure 23.* (A) Rosette and its minimal motifs (B) does not generate any holes (C) generates a set of 4 holes

As we can see from the example in Figure 23, a set of recognized holes has either 4 elements, assuming that the motif 23C is defined as a minimal motif, or is empty when the motif 23B is a minimal one. Our leading rosette example is highly relevant to the consideration. Let us first investigate a set of holes for the rosette with a given minimal motif $M_1$ and after with a minimal motif $M_2$ like in Figures 24 and 25.

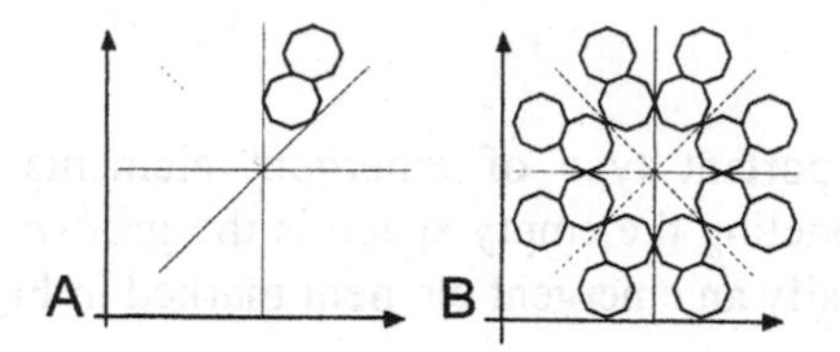


*Figure 24.* Rosette and its minimal motif which generates one hole

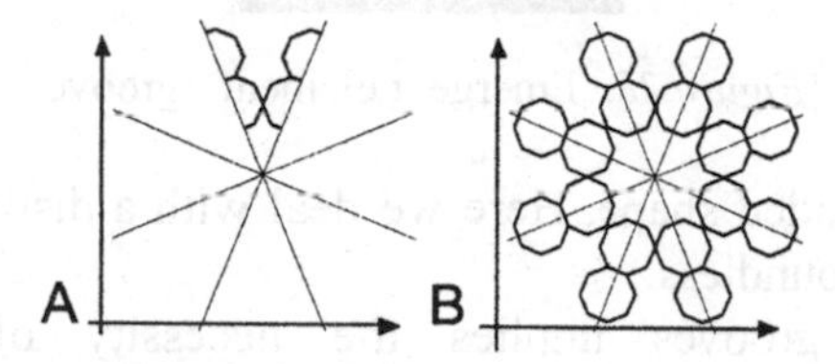


*Figure 25.* Rosette and its minimal motif which generates a set of (2 times 8) + 1 holes

In the first configuration, a set of holes is composed of only one hole, while in the second one of (2 times 8) + 1 holes.

It is the role of a designer to point to a minimal motif which will determine the exploration. Nevertheless, we have to realize that considering the importance of the designer's decisions does not disqualify the solution we propose in this paper. We do not attempt to examine a motif but try to localize emergent features in designs which are created from it in a precisely formalized way. Analysis of the motif itself is left to its creator.

When we explore periodic borders, Figure 26, or two-dimensional periodic designs, Figure 27, a set of holes is either empty or infinitely countable which is an explicit implication of infinite multiplicity of the types of designs. It is easy to guess that the same correlation, as in the case of rosettes, between a set of holes and a motif exists here as well.

*Figure 26.* Border's holes: (A) no holes (B) countable number of holes

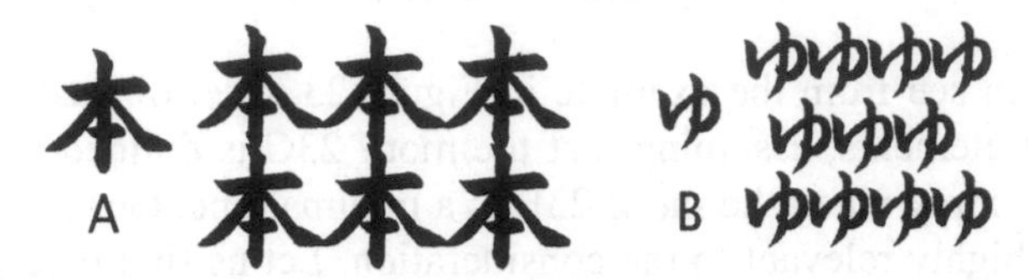


*Figure 27.* Two-dimensional periodic design's holes (A) countable number of holes (B) no holes

### 3.3. GROOVES

The next very important type of emergent elements which come into existence while modeling the empty space is the *groove*. Without a *groove*, how would we classify an emergent element marked in Figure 28?

*Figure 28.* Emergent element - groove

A hole is a bounded shape. Here we deal with a distinguished subspace where one part is boundless.

Distinguishing grooves implies the necessity of their structural representation. We have to find their representation because of the definition of emergent elements we have established - we have to be able to represent it explicitly in our design. This is quite a difficult problem because grooves are boundless. However, some features of human perception make this problem much easier. A human being usually perceives as figures these objects which are convex and concave ones identifies with backgrounds. That is why we are going to enclose grooves within minimal convex sets using rectangles.

Definition 20. Let us consider the spatial structure $Q \subseteq S \times F$ and the spatial pattern $Z(Q) = \bigcup_{(s,f)\in Q} f(s)$. A set $G(Z(Q))$ is called a *set of grooves* if the following conditions are satisfied:

1. $\exists F', S', Q'$ such that $F' \neq \varnothing$ is a set of transformations without their composites, $S' \neq \varnothing$ is a set of shapes, $Q' \subseteq S' \times F'$,
2. $\forall (s', f') \in Q'$ $f'(s')$ is a shape,
3. $\forall (s', f') \in Q'$ $f'(s') \cap Z(Q) \subset Fr(f'(s'))$,
4. $\forall (s', f') \in Q' \exists$ minimal rectangle $R$ so that:
   a. $f'(s')$ is one of the shapes from the set $R \setminus Z(Q)$,
   b. $Fr(f'(s')) \cap R$ is a single line segment,
   c. $Fr(f'(s')) \cap Z(Q) = Fr(f'(s')) \setminus R$.

Each $f'(s')$ so that $(s', f') \in Q'$ is called a *groove* and $G(Z(Q)) = \{f'(s') : (s', f') \in Q'\}$.

In Figure 29 there are some examples of periodic designs with single grooves indicated.

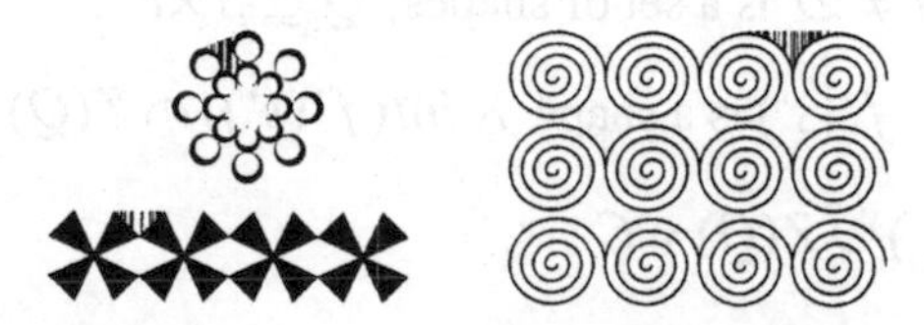

*Figure 29.* Grooves in periodic designs

While considering patterns we take into account all grooves which emerge from the two borders which constitute them. So even though it is obvious that no groove can exist unless there is a connection between motifs in a design, the case is more complicated while considering a two-dimensional periodic design where one border of those which make up this design with connected motifs is enough to imply grooves, Figure 29.

We are not interested in single grooves, as we are not in single holes, but in sets of grooves which come into being after all transformations needed to generate a final design have been applied. A number of grooves in a given design is similarly dependent on the kind of design as the number of holes is. If not empty, a number of elements in a set of grooves of a rosette of m-multiplicity is a multiple of $m$ and of borders and two-dimensional periodic designs it is countably infinite.

### 3.4. NEGATIVES

The last group of emergent elements to consider is the group of *negatives*. We come across the necessity of describing *negatives* while modelling open spaces, just like with holes or grooves. We have to face a sort of emptiness like this in Figure 30.

*Figure 30.* Emergent element - negative

There is no way to classify the emergent element from above either as a hole or as a groove. That is why we have to define another category of emergent elements, *negatives*. Such elements may appear only in those designs which are not connected. Let us begin with a general definition of a *negative*.

Definition 21. Let us consider the spatial structure and the spatial pattern $Z(Q)=\bigcup_{(s,f)\in Q} f(s)$ which is not a connected one. Let $C$ denote a *complement area*. A set $N(Z(Q))$ is called a *negative* if the following conditions are satisfied:

1. $\exists F', S', Q'$ such that $F'\neq\varnothing$ is a set of transformations without their composites, $S'\neq\varnothing$ is a set of shapes, $Q'\subseteq S'\times F'$,
2. $\forall (s',f')\in Q'$ $f'(s')$ is a shape $\wedge\, Int(f'(s'))\cap Z(Q)=\varnothing$,
3. $\bigcup_{(s',f')\in Q'} f'(s')\cup Z(Q)=C$.

$$N(Z(Q))=\bigcup_{(s',f')\in Q'} f'(s').$$

A complement area in negatives, just like minimal motifs, have to be considered for each type of design separately since it is closely connected to its structure.

While considering rosettes we complement it to a sphere, Figure 31. The choice is obvious when we look at the way this kind of design is created and all other complement shapes would be unnatural.

Definition 22. Let $Q\subseteq S\times F$ be the spatial structure and $Z(Q)=\bigcup_{(s,f)\in Q} f(s)$ be the spatial pattern in $R^n$ which is not a connected rosette. A minimal sphere $\bar{k}(x_0,r)\subset R^n$ so that $Z(Q)\subset\bar{k}(x_0,r)$ is called a *complement area* of the rosette.

*Figure 31.* Rosettes and their negatives

In order to find the negative of a border we choose a complement to a stripe with unlimited length and limited height, which is closed from its bottom and its top, Figure 32.

Definition 23. Let $Q \subseteq S \times F$ be the spatial structure and $Z(Q) = \bigcup_{(s,f) \in Q} f(s)$ be the spatial pattern in $R^n$ which is not a connected border. A minimal stripe $U \subset R^n$ so that $Z(Q) \subset U$ is called a *complement area* of the border.

*Figure 32.* Periodic borders and their negatives

Finally, we will define the complement area of a two-dimensional periodic design, Figure 33.

Definition 24. Let $Q \subseteq S \times F$ be the spatial structure and $Z(Q) = \bigcup_{(s,f) \in Q} f(s)$ be the spatial pattern in $R^n$ which is not a connected two-dimensional periodic design. A *complement area* of the two-dimensional periodic design equal to $R^n$.

*Figure 33.* Two-dimensional periodic design and its negative

In this way we have proposed a complete formal model for periodic designs like rosettes, borders and two-dimensional periodic designs and their emergent elements.

## 4. Conclusions

In this paper, we have proposed a formal way of describing the emergent elements in selected periodic designs, namely rosettes, borders and patterns, which we found very useful while looking for emergent features. We call emergent only those elements which are produced by all the transformations used to generate a given design. The study of elements which emerge after each transformation will be the subject of further research. It will also cover emergent structural features occurring in finite borders and finite two-dimensional periodic designs.

## Acknowledgements

I am very grateful to Zenon Kulpa for his valuable comments and suggestions for improvement of an earlier version of this paper. I would also like to thank my supervisor Ewa Grabska for her insights and constant encouragement.

## References

Arnheim, R: 1977, *The Dynamics of Architectural Form*, University of California Press, Berkeley.

Boden, M: 1991, *The Creative Mind, Myths and Mechanisms,* Wiedenfeld and Nicholson, London.

Chase, SC: 1996, Design modeling with shape algebras and formal logic, *Design Computation: Collaboration, Reasoning, Pedagogy,* Proceedings of ACADIA '96, Tuscon, AZ.

Damski, JC and Gero, JS: 1996, A logic-based framework for shape representation, *Computer-Aided Design* **28**(3): 169–181.

Gero, JS: 1996, Creativity, emergence and evolution in design: Concepts and framework, *Knowledge-Based Systems* **9**(7): 435–448.

Gero, JS and Maher, ML (eds): 1993, *Modeling Creativity and Knowledge-Based Creative Design,* NJ: Lawrence Erlbaum , Hillsdale.

Gero, JS and Yan, M: 1994, Shape emergence using symbolic reasoning, *Environment and Planning B: Planning and Design* **21**: 191–218.

Grabska, E: 2001, Emergent shapes in graphical design, *Working Paper*, Jagiellonian University, Cracow.

Grossberg, S and Mingolla, E: 1985, Neural dynamics of perceptual grouping, *Perception and Psychophysics* **38**: 141–171.

Jakubowski R: 1986, A structural representation of shape and its features, *Inf. Sci.* **39**: 129–151.

Kim, SH: 1990, *Essence of Creativity*, Oxford University Press, New York.

Morant, H: 1981, *Historia Sztuki Zdobniczej od Pradziejów do Współczesności* [History of decorative art from its origins until now, in Polish], Arkady , Warszawa.

Morgan, CL: 1923, *Emergent Evolution,* Williams and Norgate, London.

Rasmussen, SE: 1999, *Odczuwanie Architektury* [Feeling architecture, in Polish], Wydawnictwo MURATOR sp. z o. o.,Warszawa.

Schattschneider, D: 1986, In black and white: How to create perfectly colored symmetric patterns, *Computers & Mathematics with Applications* **12B**(3/4): 673–695.

Sternberg, R(ed.): 1988, *The Nature of Creativity,* Cambridge University Press, Cambridge.

Weisberg, RW: 1986, *Creativity: Genius and Other Myths*, W. H. Freeman, New York.

Woodbury, RF and Carlson, CN and Heisserman, JA: 1990, Geometric design space, *Intelligent CAD, II 1990*, North Holland, pp. 337–354.

JS Gero (ed), *Design Computing and Cognition'04*, 317-335


# ANIMATION IN ART DESIGN

EWA GRABSKA, GRAŻYNA ŚLUSARCZYK, MICHAŁ SZŁAPAK
*Jagiellonian University, Poland*

**Abstract.** The paper concerns designing and animation of periodic patterns. First, the characterization of rosettes, borders and periodic planar designs is given. Then, an animated generative system DARTAN, which produces a great variety of periodic designs, is proposed. The role of perceptual actions and animation in supporting perception of emergent shapes in generated patterns is also described.

## 1. Introduction

Decorative art and graphic design are an important part of every culture. Nowadays, as computers posses great computing power, they offer new possibilities to support creation of art designs. One of these possibilities is using animation to visualize design processes.

The main goal of this paper is to emphasize the importance of using animation in generative design systems with the iconic representation. The paper lays the foundation for the development of a generative system that will ultimately enable the designer to begin with basic patterns in one, two or three dimensions and let the system generate and animate much more complex patterns.

Dynamic approach to design considers a design process as a sequence of actions which change the external world. Different modes of design actions are distinguished (Suwa et al. 2000). Physical actions concern drawing and overdrawing elements. Perceptual actions refer to attending to features of depicted elements. Conceptual actions allow the designer to evaluate the obtained solutions. Computer animation allows one to illustrate the sequence of physical actions producing art designs by means of a key-frame animation. The successive key-frames show the design being created after applying single actions. Visual perception of the animation is essential for the considered categories of perceptual actions. These actions enable one to discover emergent shapes and to invent new ideas and goals and therefore often lead to creative outcomes.

This paper considers design of periodic patterns aided by animation. The art designs are generated by means of the system DARTAN (Decorative ART ANimation) (Szłapak 2003). This program, written in Flash, enables the user to choose motifs for generated designs. Then the animation of structures composed of the motifs is activated and the appearance of a design continually changes.

It seems that design of periodic patterns can be effectively supported by animation techniques. Simultaneous rotations and translations of motif's copies, which constitute periodic patterns, produce captivating and unexpected effects. The emergent shapes occurring in designs of this type are easy to discover when the animation of motifs is used. As the pattern changes in a periodical way, new configurations composed of the copies of the same motif are formed and perceived. Since the emergence is rooted in constraints that govern the constructions of spatial patterns (Grabska 2003) some emergent shapes in periodic designs can be anticipated but most of them are identified only after inspecting different visualizations.

The way in which an animation affects the appearance of decorative patterns has been studied using the system DARTAN. Periodic designs are generated by transforming motifs using transformations, which form groups of symmetry, and can be classified according to these groups. Observations of animated patterns enabled us to infer that some transformations applied in animations change symmetry groups of the patterns. Animation of an initial pattern of a given symmetry group can result in a pattern, which needs a different set of transformations to be generated, than a starting one does. Changes of this kind are described using mathematical formalism presented in the next section.

The presented method of art designs synthesis enables the designer to explore the generative potential of a set of basic patterns and to find in the process new and interesting basic patterns to use in future as well as space-filling designs of interest. From this perspective, the DARTAN provides a design computing capability that can augment the design cognition of the human.

## 2. Periodic Patterns

As it was mentioned, periodic patterns are a class of designs considered in this paper. The thorough characterization of these patterns is presented in (Schattschneider 1986). Periodic designs can be classified according to their symmetry groups. Using animation in the generation process allows one to notice interesting mathematical properties of periodic designs related to changes of symmetry groups. Some formal concepts needed to characterize these properties will be presented in this section.

All periodic designs composed of one motif can be created using only four types of transformations: translations, rotations, reflections and glide reflections. These are the only isometries, which preserve shapes while repeat motifs to generate the whole design.

We consider here rosette designs, periodic border designs and periodic planar designs, and use animation to support imagination and creative skills of the designer producing these patterns. Our approach to periodic designs differs from the one presented in (Schattschneider 1986) as we treat these designs as a special kind of spatial patterns (Woodbury et al. 1990).

We assume that periodic patterns are constructed with transformed copies of basic geometrical shapes that are defined as primitives. Let us define spatial patterns in a formal way.

Let F be a group of transformations of $R^n$ into $R^n$, where each $f \in F$ transforms bounded subsets onto bounded subsets. An element $f \in F$ is called an ***admissible transformation***.

**Definition 2.1**
Let F be a group of admissible transformations of $R^n$ into $R^n$. Let S be a given set of bounded subsets of $R^n$.
If $f(p)=q$ for $f \in F$, where $p, q \in S$, implies that $p=q$, then elements of S are called **primitives** over F and S is called an **alphabet of basic shapes**.

A spatial pattern is the union of transformed primitives called components. A ***component*** is represented as a pair (p,f), where p denotes the primitive and f is the appropriate transformation.

**Definition 2.2**
Let S be an alphabet of basic shapes over F, where F is a group of admissible transformations of $R^n$ into $R^n$. Let $Q \subset S \times F$ be a pattern structure.
The **spatial pattern** for Q is denoted by Z(Q) and is defined as follows:

$$Z(Q) = \bigcup_{(s,f) \in Q} f(s).$$

In this section we consider only periodic patterns defined in $R^2$. Each periodic pattern is composed of repeating copies of a single motif, which we propose to define as follows.

**Definition 2.3**
Let $Q \subset S \times F$ be a pattern structure in $R^2$.
The two-dimensional pattern for Q selected by a designer and denoted by $M = \bigcup_{(s,f) \in Q} f(s)$ is called a **motif**.

Using the presented notion of a motif we define a periodic pattern.

**Definition 2.4**
Let F be a group of admissible transformations of $R^2$ into $R^2$ being isometries.
A given pattern P in $R^2$ is called **periodic** iff there exists a motif M such that $P = \bigcup_{i=2,\ldots,n} f_i(M)$, where $f_i \in F$. A number $n$ is called a **multiplicity** of P.

One of the most popular periodic patterns is a rosette. Each rosette design is inscribed in a circle centred at a given point. To create a rosette with $n$-fold rotational symmetry a chosen motif is rotated $n$ times about a fixed central point. A smallest shape whose rotated copies generate a rosette is called a minimal motif. The symmetry group of such a rosette consists of the $n$ rotations and is called a ***cyclic group*** $C_n$ .

**Definition 2.5**
A periodic pattern R with an $n$-fold rotational symmetry is called a **rosette**. Let $R = \bigcup_{i=2,\ldots,n} f_i(M)$, where each $f_i$ is a rotation through the multiplicity of the same angle. The greatest $n$, which satisfies the above equation, is called a **multiplicity** of R and M specified for this $n$ is called a **minimal motif** of R.

An example of a 6-fold rotational symmetry rosette with its minimal motif is presented in Figure 1.

*Figure 1.* A 6-fold rotational symmetry rosette and its minimal motif

The second type of periodic patterns is a border. Periodic border designs are considered to be enclosed between two parallel lines. Thus, they are located in a strip of finite width and infinite length with the centre line L that is equidistant from the edges of the strip. For each border a translation vector and a translation unit can be defined. The shortest translation vector, which is invariant in respect to the border, is called a ***translation vector*** and is denoted by T. A minimal parallelogram with a base equal to the translation vector and height equal to the width of the border which translated repeatedly by T and –T generates the border is called a ***translation unit***. A motif included in the translation unit is called a ***minimal motif*** of the border. A maximal subpattern which repeats itself in a minimal motif and which acted upon repeatedly by isometries will produce the whole border design is called a ***basic motif*** of a border.

**Definition 2.6**

A periodic pattern B generated by acting repeatedly upon a minimal motif by the translation T and –T is called a **border**.

An example of a border composed of the translated and rotated copies of the same motif, which is used to create the above rosette, is presented in Figure 2. This is a basic motif of the presented border. A translation unit, which is denoted by ABCD, contains a minimal motif composed of two copies of the basic one.

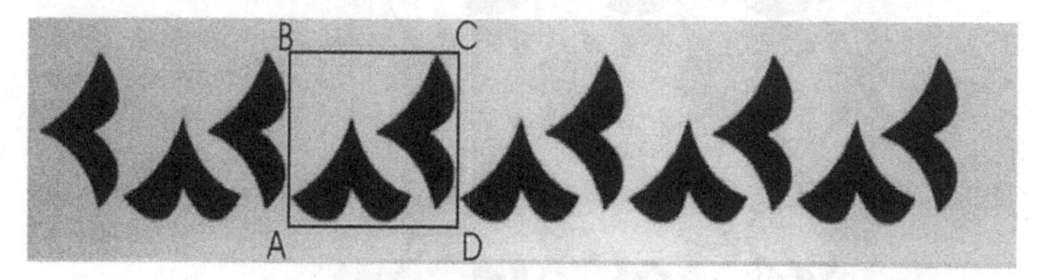


*Figure 2.* An example of a border

The border designs can be classified according to seven distinct one-dimensional symmetry groups. For each group different sets of a minimal number of isometries needed to generate the whole design are specified. They contain different collections of translations with vectors parallel to L, reflections with mirror line L, glide reflections with glide vector on L and half-turns about centres which lie on L.

The last type of the considered patterns is a periodic planar design. Each periodic planar design has in its symmetry group two independent translations $T_1$ and $T_2$ with shortest possible translation vectors. A parallelogram having as its sides the vectors $T_1$ and $T_2$ is called a ***lattice unit***. A ***translation unit*** for the design is a minimal area of the plane which, when acted upon repeatedly by translations $T_1$ and $T_2$ fills out the whole design. The shape of a translation unit can vary but its area is always equal to the area of a lattice unit. A motif included in the translation unit is called a ***minimal motif*** of the planar design. A maximal subpattern which repeats itself in a minimal motif and which acted upon repeatedly by isometries will fill out the whole design is called a ***basic motif*** of the design.

**Definition 2.7**

Let ABCD be a lattice unit. Let the side BC specify a horizontal translation vector $T_1$ and the side AB specify a translation vector $T_2$. Let a translation unit containing a minimal motif be specified in respect to $T_1$ and $T_2$.

A periodic pattern T generated by repeatedly acting upon a given minimal motif by translations $T_1$, $-T_1$, $T_2$ and $-T_2$ is called a **periodic planar design**.

As there are seventeen distinct two-dimensional symmetry groups, seventeen types of periodic planar designs can be distinguished depending on isometries, which are needed to create such designs.

An example of a periodic planar design with a distinguished translation unit is presented in Figure 3. A basic motif of the design is presented on the right-hand side of the figure.

*Figure 3.* An example of a periodic planar design and its basic motif

## 3. An Animated Generative System

A concept of a ***generative system*** as a tool of problem solving can be traced back to Aristotle who described a system for potential animals. Since then generative systems have played important roles not only in philosophy, literature and music but also in engineering and architecture design. Systematic use of these systems in design has been started by Leonardo da Vinci who had used such a system for generating possible forms of central-plan churches (Mitchell 1977).

A generative system produces a variety of potential solutions to a given design task. It consists of a scheme for representation of potential solutions together with a set of elements, which enable us to get different solutions, and a set of rules concerning selection and fitting of these elements into the scheme. In this paper, a generative system consists of structures for periodic designs and a set of motifs, which can be fitted into these structures using geometrical transformations. In this way we can obtain a set of possible periodic designs. Three types of possible structures of periodic designs are considered: a structure of a border, a structure of a planar design and structures of 6, 8, 10 and 12-fold rotational symmetry rosettes. There are eight available motifs, which can be fitted into these structures. Choosing one motif and arranging its copies in a selected structure using isometries create an admissible periodic design.

Structures of a border, 10-fold rotational symmetry rosette and planar design together with eight available basic motifs are shown in Figure 4. The points on the structures indicate places in which motifs will be fitted.

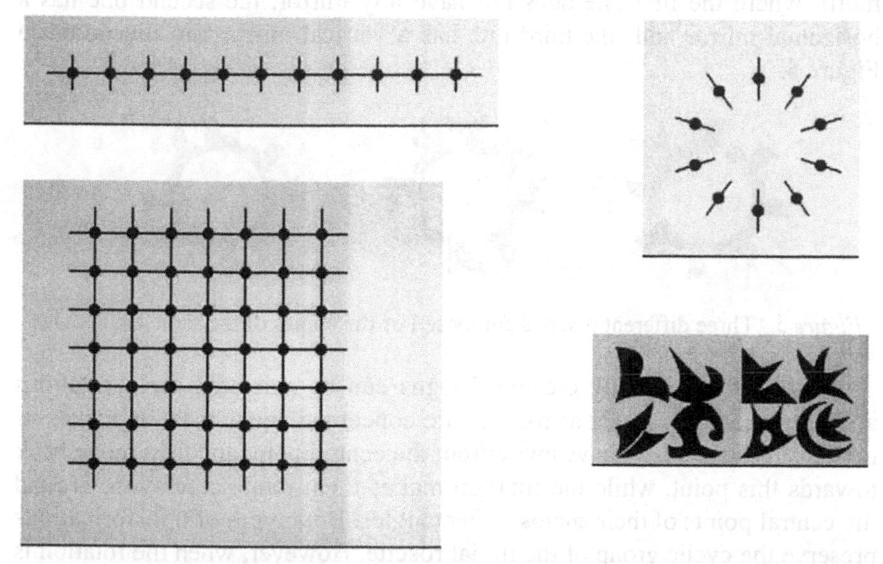

*Figure 4.* **Structures of a border, 10-fold rotational symmetry rosette, planar design and eight available basic motifs**

In this paper we propose to equip this generative system with animation. Due to the animation technique used, each initially obtained periodic pattern is a starting point for many other designs. Rotating and/or translating motifs of the initial solution can capture many interesting effects caused by moving motifs.

Generally, people are used to look at and evaluate static designs. Animation of patterns generated by a generative system enables us to obtain a kaleidoscopic effect. Visual evaluation of changing patterns is different, as not only single key-frames of animation are evaluated but also transition from one pattern to the other makes the designer perceive some new shapes. Thus, the animation gives us a more meaningful description of successive stages of design generation.

In our program all initial designs can be created without any mirror, and then all their motifs face the same direction, or with a horizontal or vertical mirror and then every second motif of a design is a reflection of the basic motif in a horizontal or vertical mirror, respectively. In the last two cases a minimal motif of a rosette is composed of two basic motifs. Thus, a rosette with a mirror and composed of *n* copies of the basic motif has only *n/2*-fold rotational symmetry. It should be noted that a rosette with a vertical mirror has a kaleidoscopic symmetry.

Examples of three rosettes composed of the copies of the same basic motif, where the first one does not have any mirror, the second one has a horizontal mirror and the third one has a vertical mirror, are illustrated in Figure 5.

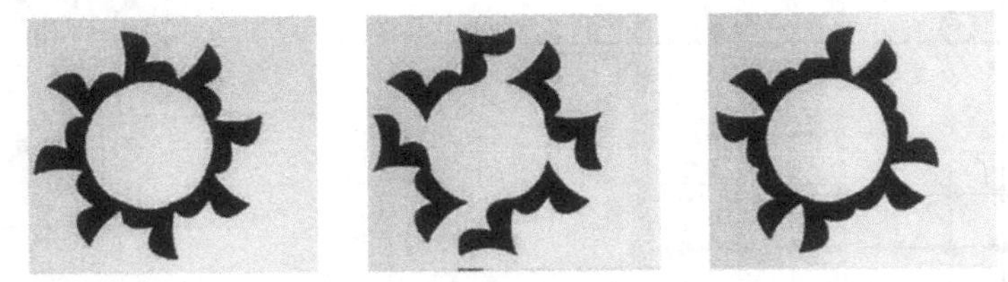

*Figure 5.* Three different rosettes composed of the copies of the same basic motif

Motifs of the initially created designs can be animated using rotations and/or translations. As far as rosettes are concerned, the translation makes its motifs simultaneously move away from the central point and then move back towards this point, while the rotation makes them rotate clock-wise around the central points of their enclosing rectangles. Both types of transformations preserve the cyclic group of the initial rosette. However, when the rotation is used the kaleidoscopic symmetry of an initial pattern disappears. It should be also noted that when motifs which are close together are rotated, or when they approach the central point by translation they can overlap each other and the distinguishable minimal motif does not any longer resemble the basic one.

The first three patterns in Figure 6 present the rosettes from Figure 5 with motifs moved towards their central points. The last one is obtained from the third rosette shown in Figure 5 by rotating its motifs through $30^{o}$.

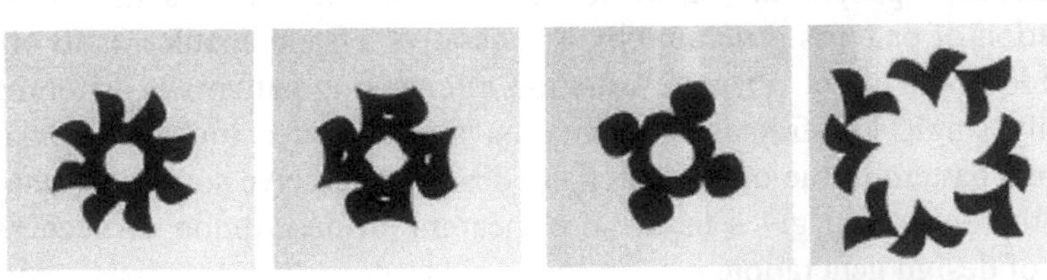

*Figure 6.* Rosettes obtained from the ones shown in Figure 5 by moving their motifs

An initial border design without mirror belongs to the symmetry group ***11*** generated by only one translation, while a border with a horizontal mirror belongs to the group ***1g*** generated by one glide reflection and a border with a vertical mirror belongs to the group ***m1*** generated by one translation and one reflection perpendicular to the border edge.

Examples of three borders, where the first one does not have any mirror, while the second and third one have horizontal and vertical mirrors, respectively, are presented in Figure 7.

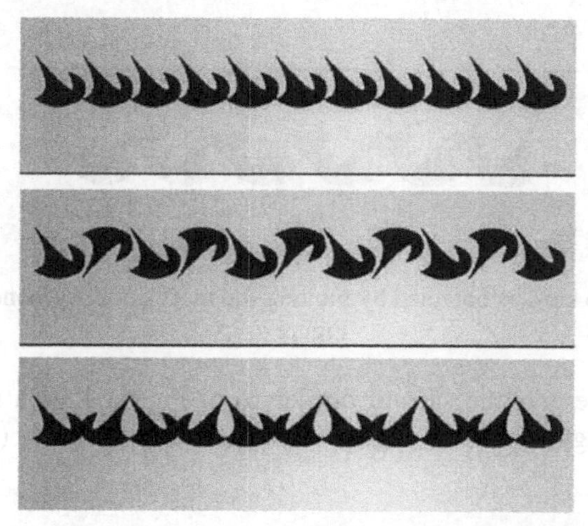

*Figure 7.* Examples of three different borders composed of the same motifs

To each border three types of transformations can be used: rotation, vertical translation and horizontal translation. The horizontal translation leaves the central motif in its position and moves other motifs horizontally away from and then towards the central one parallel to the central line in such a way that all motifs are always equally spaced. This translation applied to the initial border does not change its symmetry group as it only makes the translation unit longer or shorter. However, if the motifs are close together, in case when the horizontal translation or rotation is used, they start to overlap and the translation unit does not contain the basic motif any longer.

The borders obtained by moving the motifs of the second border from Figure 7 away from the central motif and towards this motif are shown in Figure 8.

The rotation preserves the symmetry group of borders of the group ***11***. When the rotation is used to the motifs of borders of the group ***m1*** they become borders of ***1g*** and vice versa, if the angle of rotation equals 90°. In all other cases borders become the patterns of the symmetry group ***11***, where the translation unit includes two consecutive copies of the basic motif. Application of the vertical translation, which moves the alternate motifs up and down in the direction perpendicular to the central line, changes the width of the translation unit. It preserves the symmetry group of borders belonging to the group ***1g***, but in case of patterns from groups ***11*** and ***1m*** it makes their translation units include two consecutive copies of the basic motif, and therefore the animated borders become patterns of the group ***11***.

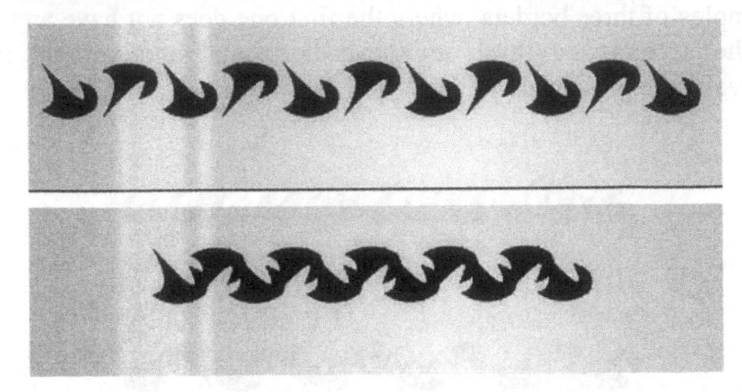

*Figure 8.* Two borders obtained by moving the motifs of the second border from Figure 7

Figure 9 presents the second border from Figure 7 with motifs rotated through 45° (Figure 9a), 90° (Figure 9b) and moved vertically (Figure 9c).

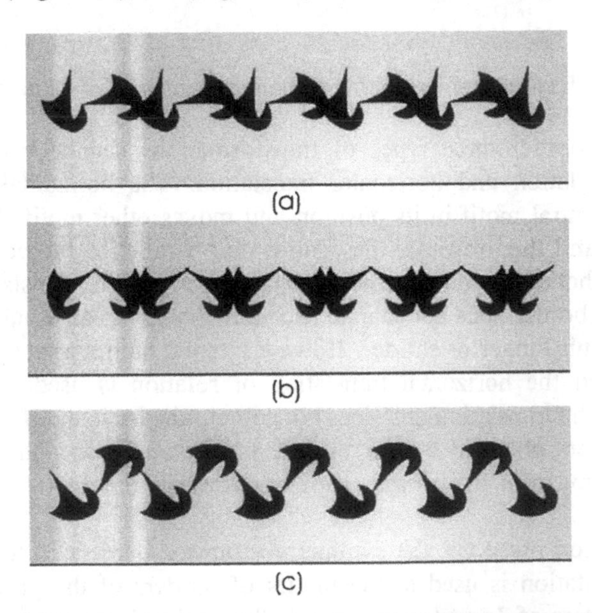


*Figure 9.* Borders obtained from the second border shown in Figure 7 by a) rotating its motifs through 45°, b) rotating its motifs through 90°, c) moving its motifs vertically

Initial periodic planar designs generated by our program have the symmetry group ***p1*** generated by two translations, if they are created without any mirror (Figure 10a), or the symmetry group ***pg*** generated by two glide reflections, if they have horizontal or vertical mirrors (Figures 10b and c).

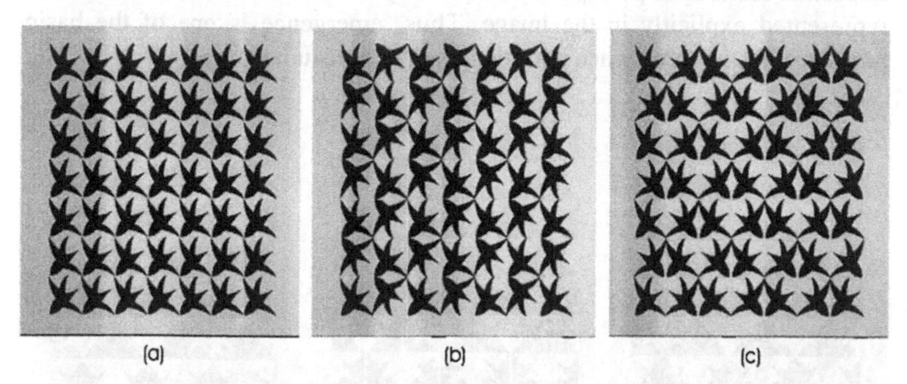


*Figure 10.* Examples of periodic planar designs created a) without any mirror, b) with a horizontal mirror, c) with a vertical mirror

Animation which uses the horizontal translation, which changes the distance between columns of motifs, or the vertical translation, which changes the distance between rows of motifs, does not change the symmetry groups of designs as it only changes the length or width of their translation units. The rotation preserves the symmetry group only for the designs created without any mirror. The other type of translation, which moves alternate columns of motifs up and down, also preserves the symmetry group for designs without mirrors but makes their translation units include two copies of the basic motif. The rotation and the translation, which moves the alternate motifs' columns, applied to designs with mirrors change their symmetry groups to ***p1*** as they make the translation units include four copies of the basic motif. The shape of a translation unit of a design to which the shift of motifs' columns has been applied is shown in Figure 3. Moreover, vertical and horizontal translations as well as rotations of motifs, which are located close to each other, could make the motifs overlap in such a way that a translation unit would not contain any basic motif any longer.

Figure 11 illustrates all types of translations that can be activated to animate periodic designs. An initial periodic planar design with the vertical mirror is shown in Figure 11a. In Figure 11b the columns of motifs are drawn aside, while in Figure 11c the rows of motifs are drawn aside. Figure 11d presents the design with columns and rows of motifs moved towards its central point. In Figure 11e every second column of motifs is moved down, while in Figure 11f all motifs of the initial designed are rotated.

## 4. Perceptual Actions and Emergent Shapes in Periodic Patterns

Ambiguity of human perception enables us to group or segment images in different ways and to create different interpretations of patterns. One of the important features of perception is the ability to discover features that are not represented explicitly in the image. Thus, emergence is one of the basic problems encountered when periodic patterns are studied.

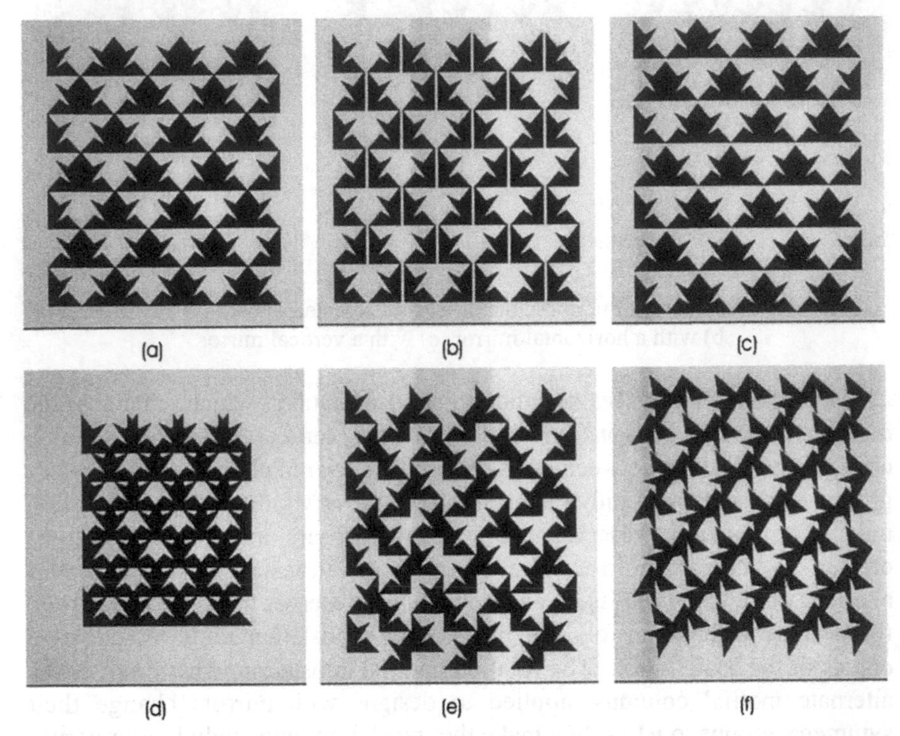


*Figure 11.* a) An initial periodic planar design, b) the same design with the columns of motifs drawn aside, c) with the rows of motifs drawn aside, d) with the columns and rows of motifs moved towards the central point e) with every second column of motifs moved down, f) with the rotated motifs

Perceptual actions of the design refer to attending to features of depicted elements. In these actions the designer not only attends to such visual features of elements as shapes or sizes but also pays attention to spatial relations among elements such as for instance proximity, alignment, intersection. He/she also perceives implicit spaces that exist in-between depicted elements and discovers organization between them like order or similarity.

It is easy to notice that perceptual actions strongly depend on physical ones (drawing, coping, erasing, redrawing). For example drawing motifs of a border in such a way that they touch one another makes the designer pay attention to the connectedness of the motifs. The motifs of a border in Figure 12a are located in such a way that the alignment and connectedness among them are easy to perceive. In Figure 12b intersection of initial motifs, which makes them form a new motif, can be seen. Implicit spaces in-between motifs in Figure 12c make us perceive shapes of arrows.

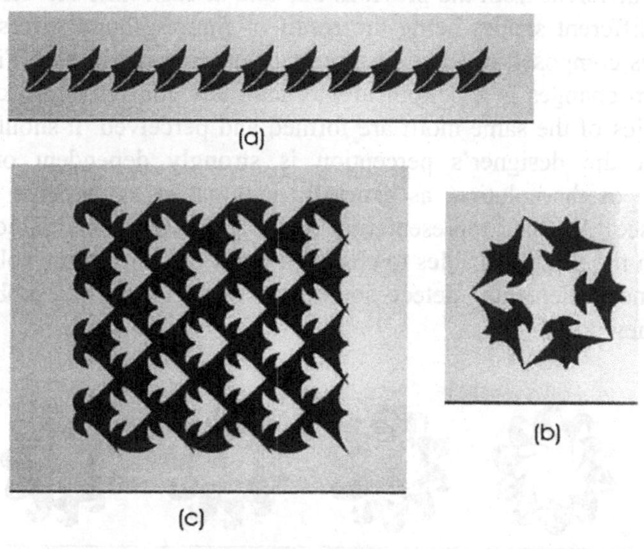


*Figure 12.* a) Alignment and connectedness of the motifs of a border, b) intersection of the motifs of a rosette, c) implicit spaces in-between the motifs of a periodic design

As it was mentioned above, physical actions influence perceptual ones. Therefore using animation to visualize a sequence of physical actions performed in design of a periodic pattern illustrates changes occurring in the pattern and in this way supports perceptual actions.

The changes in the appearance of a single rosette when only rotation of its motifs is activated are illustrated in Figure 13. The numbers under the patterns indicate the numbers of animation frames.

As a result of perceptual actions unintentional shapes can be often discovered. Emergent features in designs are not represented explicitly but can be perceived during analysis and aesthetic evaluation of the solution. While inspecting visualized solutions the designer is able to find new elements and generate new concepts or take decisions about required modifications. Thus, unexpected discoveries can be defined as a class of perceptual actions (Suwa et al. 2000). Emergence that is based on detecting implicit spaces that exist in-between elements is called perception of figure-ground reversal and is one of the characteristic features of human perception.

Finding emergent shapes in designs develops the solution space and stimulates the creativity of designers.

The emergent shapes occurring in periodic designs are easy to discover when the animation of motifs is used. Unexpectedly discovered shapes depend on the current configuration of depicted elements. When motifs are simultaneously translated and/or rotated, at each moment this configuration is slightly different from the previous one and at each time the designer can discover different shapes being the result of figure-ground reversal and/or new shapes composed of some fragments of moving motifs (see Figure 13). The pattern changes in a periodical way and new configurations composed of the copies of the same motif are formed and perceived. It should also be noted that the designer's perception is strongly dependent on his/her conception of the solution, as generally it is easier to perceive what one wants to see. A visual representation of design ideas, which is continually moving on the screen, enables to change a point of view on the solution and perceive new elements, detect some new aspects of the problem and conceive new ideas.

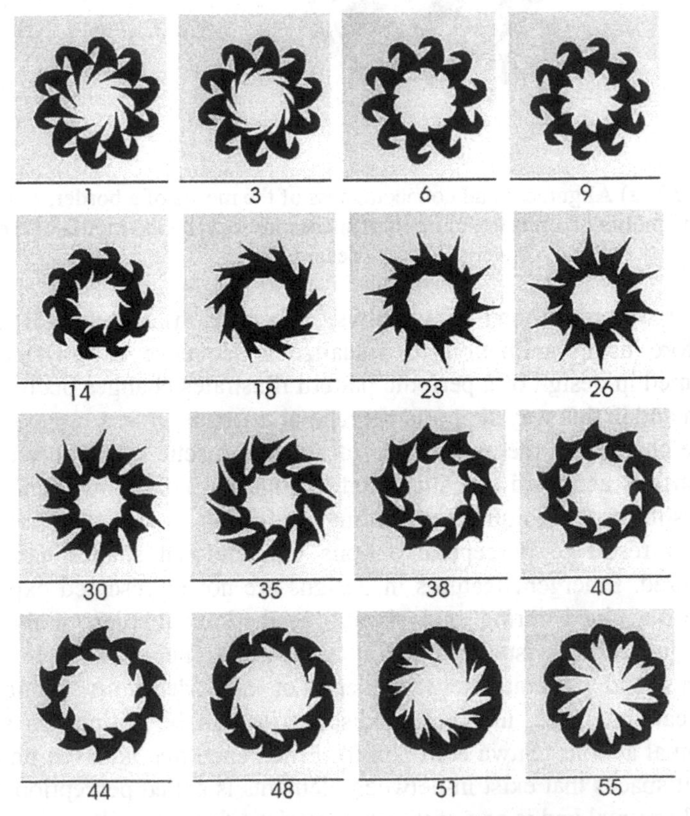


*Figure 13.* Some animation frames of a rosette

Examples of 8-fold symmetry rosettes created of different motifs without mirrors, inside which 'stars' or 'flowers' can be discovered as the result of figure-ground reversal are presented in Figures 14 and 15, respectively.

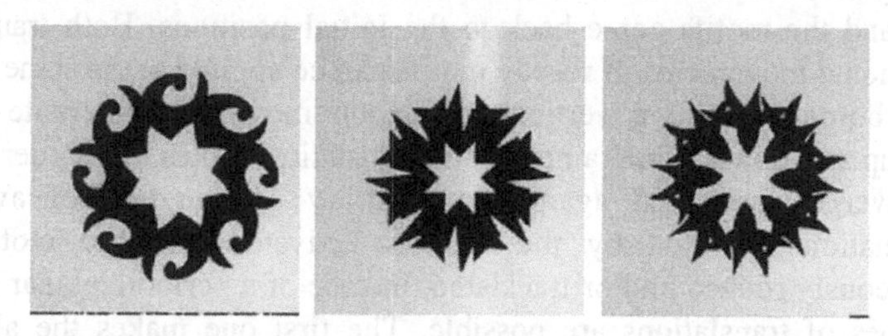

*Figure 14.* Three rosettes with emergent 'stars' inside

## 5. Program Description

A program DARTAN (Decorative ART ANimation) that enables the user to generate periodic patterns with one motif was implemented in Flash at the Institute of Computer Science of the Jagiellonian University (Szłapak 2003) in 2003.

*Figure 15.* Three rosettes with emergent 'flowers' inside

In order to obtain an animated periodic pattern the user should:

- choose a type of the pattern. It can be a rosette design, periodic border design or a periodic planar design. In case of a rosette design also the multiplicity of rotational symmetry should be chosen. Rosettes with 6, 8, 10 or 12-fold symmetry can be generated.
- select a mirror type of the design (vertical mirror, horizontal mirror or without mirror).
- choose a motif of the pattern. The eight available motifs are shown in Figure 4. Any other motif created by the user can be imported and used. The selected motif is automatically arranged into the structure of the required type.
- select types of transformations that are to be activated.

As far as animation is concerned, the motifs of all pattern types can be simultaneously rotated about the centres of their enclosing rectangles. In the case of a rosette design, if the translation is chosen, the motifs are moved away from the centre point of the rosette in such a way that the radius of its enclosing circle increases to some extent. Then an inverse movement is applied and the motifs come back to the initial positions. Both translation and rotational movements of rosette motifs can be applied at the same time.

In a border design, a vertical translation moves the alternate motifs slightly up and down, while a horizontal translation makes the border stretch and converge again and again. An arbitrary combination of available transformations selected by the user is activated and the motifs are simultaneously rotated and/or translated. In case of a periodic planar design, three types of translations are possible. The first one makes the alternate columns of motifs move up and down, while the vertical and horizontal ones move motifs aside and then push together again in vertical and horizontal directions, respectively. For each pattern all types of the available motifs' transformations can be activated at the same time or only some of them can be chosen.

A window for generation and animation of border designs is presented in Figure 16. Some of the available motifs are shown at the top right-hand side. The lowest button activates all the available transformations at once.

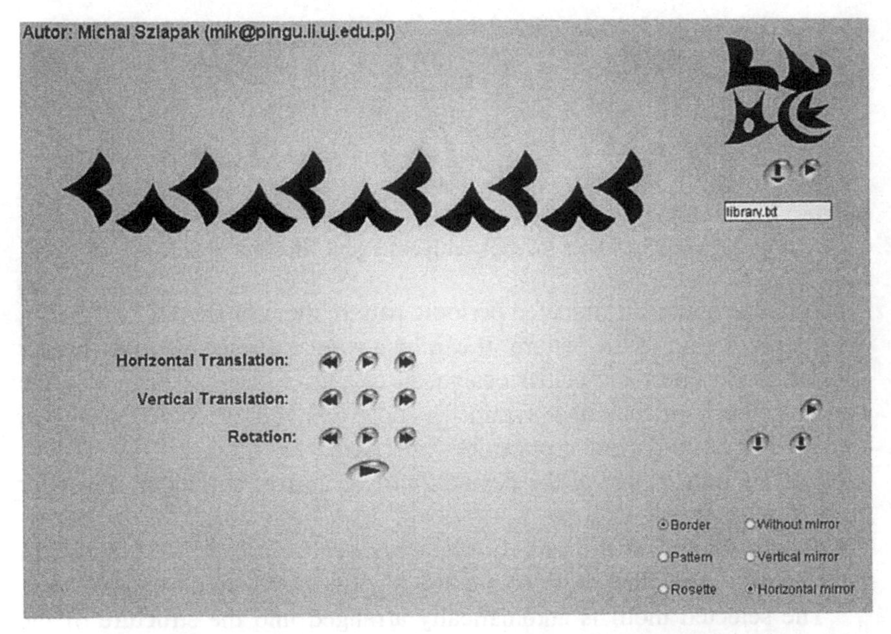


*Figure 16.* A window for generation and animation of border designs

In our program a key-framed animation is used. The location of each basic motif in the global coordinate system is specified only for a few key frames. All other frames are interpolated from the key ones. The rotation of motifs is animated in such a way that the consecutive frames show the motifs rotated through 6°. Thus 60 frames are needed for the whole turn of each motif. The translation is animated in such a way that the distance between motifs shown by successive frames changes by 10 pixels.

The animated pattern continually changes in a periodical way. The animation can be stopped at any time and the current image can be saved. Also each transformation can be performed frame by frame in the forward or backward direction.

Moreover, the system enables us to save parts of the created designs as new motifs. They can be imported to the motifs' library and used to create other periodic patterns. The whole created rosettes, patterns composed of the first two motifs from left of the created borders and patterns composed of four motifs from the top left-hand side corner of the periodic planar designs can be imported as new motifs. Figure 17 presents some motifs formed of previously created rosettes, borders and periodic designs.

*Figure 17.* Some motifs formed of previously created periodic patterns

Figures 18, 19 and 20 present some more interesting rosettes, borders and periodic planar designs obtained in our program.

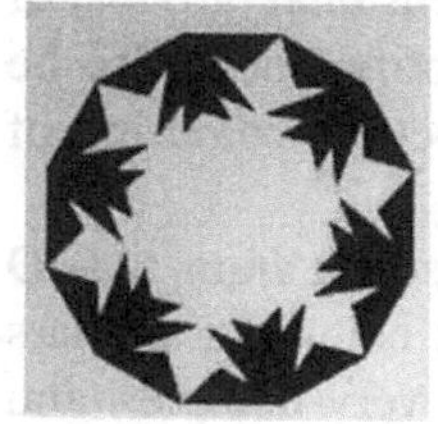

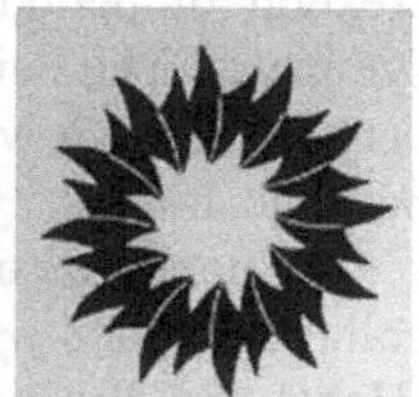

*Figure 18.* Some rosettes generated by DARTAN

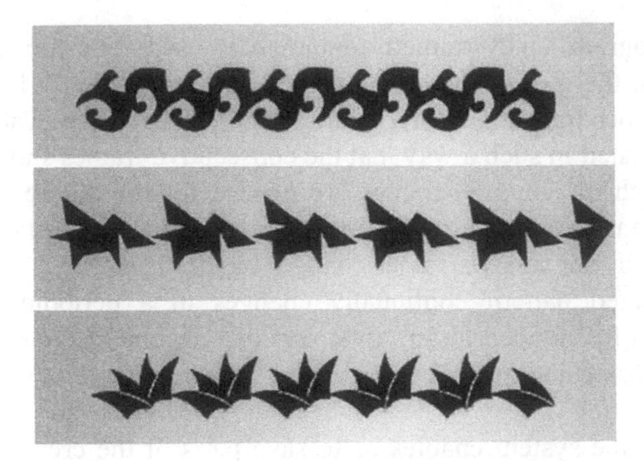

*Figure 19.* Some borders generated by DARTAN

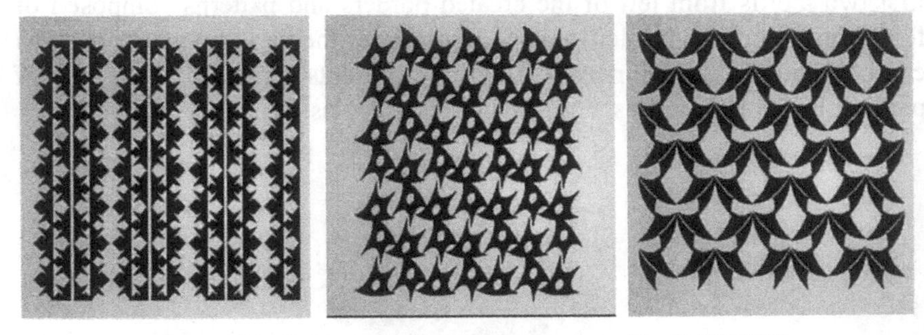

*Figure 20.* Some periodic designs generated by DARTAN

## 6. Conclusions

The proposed method of automated synthesis of decorative patterns, which takes advantage of animation, facilitates perception of new features as well as emergent shapes in periodic designs. Animation of pattern motifs helps the designer to explore their possible arrangements in order to find designs with the required artistic merit. Thus, substituting a computer for a note pad allows for many experiments that would not be possible in design by hand.

In future work we would like to extend our program in order to enable creation of periodic designs of all one and two-dimensional groups of symmetry. The build-in editor of shapes of the basic motifs would also be very useful. We would like to implement the rules of perfect colouring of periodic designs.

The next class of designs under consideration will constitute animated 3D collages created by means of generation systems called collage grammars (Habel and Kreowski 1991). They allow one to design very complex and irregular shapes from simple ones. Our first attempts in this field are illustrated in Figure 21. We also hope to take advantage of animation in such domains of design as engineering and architecture.

The question arises in what extent the presented method supports creative design and facilitates achieving better means of expressing artistic ideas. It can only be answered by conducting empirical studies with graphic designers using the DARTAN system.

*Figure 21.* An animated generation of a 3D object

## Acknowledgements

The authors wish to thank one of the anonymous referees of the paper for his insightful comments and suggestions.

## References

Grabska, E: 2003, Design and reasoning with diagrams, *Machine Graphics and Vision* **12**(1): 5-16.

Habel, A and Kreowski, HJ: 1991, Collage grammars, *Lect. Not. Comp. Sci.* **532**: 411-429.

Mitchell, W: 1977, *Computer-Aided Architectural Design*, Charter Publishers, New York.

Schattschneider, D: 1986, In black and white: How to create perfectly colored symmetric patterns, *Comp. & Maths. with Appls.* **12B**(3/4): 673-695.

Suwa, M, Gero, J and Purcell, T: 2000, Unexpected discoveries and s-invention of design requirements: Important vehicles for a design process, *Design Studies* **21**(6): 539-567.

Szłapak, M: 2003, Zastosowanie animacji w projektowaniu wspomaganym komputerowo, *MSc Thesis*, UJ, Kraków.

Woodbury, RF, Carlson, CN and Heissermann, JA: 1990, Geometric design spaces, *in* H Yoshikawa and T Holden (eds), *Intelligent CAD II*, North-Holland, pp. 337-354.

JS Gero (ed), *Design Computing and Cognition'04*, 337-353


# FORMULATION OF AESTHETIC EVALUATION & SELECTION

*In an Interactive Facial Shape Evolution System*

ZHEN YU GU, MING XI TANG, JOHN HAMILTON FRAZER
*The Hong Kong Polytechnic University, Hong Kong*

**Abstract.** The application of evolutionary algorithms in an aesthetic domain involves human subjective judgment and artificial selection. Evolutionary systems that facilitate this selection strategy are referred to as Interactive Evolutionary Systems (IES). One of the crucial problems for IES is that artificial selection is a time consuming process during which human users are faced with limited scale of the population and high dimensionality of solution space. This paper addresses this problem through an integration of General Regression Neural Network (GRNN) and an IES, using facial character creation as an example domain. This approach formulates designers' aesthetic fitness evaluation in an IES through a learning mechanism provided by the GRNN. Our aim is to build an intelligent and evolutionary system that possesses some empirical knowledge as well as a convergent thinking ability to support human users. In order to study the feasibility of our approach, we have implemented a prototype system for evolutionary facial character creation. The initial results generated by this system are reported in this paper.

## 1. Introduction

Computer modeling in a virtual environment based on aesthetic judgment today is an essential skill for designers and modern artists. Computational support for this paradigm must therefore match this development by moving beyond today's passive tools, towards active and conscious environments.

### 1.1. PROBLEM STATEMENT

Evolutionary computation has been adopted in art and aesthetic design for more than two decades. An evolutionary system typically employs a generate-and-test strategy. Every single run of an evolutionary algorithm involves thousands of evaluations and selections. Most of today's evolutionary systems in the art and design domains employ a strategy of interactive user interface, in which users view, judge, and select the results.

However, the narrow view of a computer screen constrains the population size of each generation that can be displayed at one time. If the search space of the design problem is big and diverse, at the start, one may find that nothing is useful, and nothing can even be partially associated with the expected target solutions in one's mind. Something might even mislead the searching to a zigzag round path. The more serious problem associated with the narrow spectrum of the starting generation is that the algorithms may result in premature results. In this case, the user has rare chance to reach the satisfied point in the search space.

This explains the reason why so many evolutionary art systems did not use the crossover operator. In such systems, only the mutation operator is adopted. Evolution is used more as a continuous novelty generator, but not as an optimizer (Bentley 1999). Inevitably, such systems have to count on increasing much more generate-and-test work in order to get relatively optimized results. If the search activity by a designer or an artist is totally open-minded, then the generators might do well enough. But in design, the problem cannot be formulated easily as an open-ended problem. Even if it can, this kind of search does not necessarily mean a totally mindless search without a direction. Rather it is always directed towards certain conceptual targets. In this case, non-convergence evolutionary systems will not perform well.

In order to solve this dilemma, ideally a divergent or so-called creative evolutionary system should have somewhat a convergent mechanism in the occasion when subtle optimization is appreciated. A convergent process relying on the continuing exhaustive artificial selection is not so welcome. However it is difficult to define explicit fitness functions when the problem of design is not well identified and stated. A computational system supporting the evaluation of the aesthetic quality of a design work remains to be an issue for research. It is one of the major difficult tasks in the development of generative and intelligent design support systems.

### 1.2. RELATED WORK

Interactive evolutionary art or design is normally based on aesthetic appeal. The fitness function defines the shape of the fitness landscape and the phenotype defines the position of an individual on it. It is reasonable and natural to use the neural network approximation as a way for formulating the aesthetic fitness landscape in a generative and evolutionary design system.

Sims (1991) suggested that: "Large amounts of information of all the human selection choices of many evolutions could be saved and analyzed. A difficult challenge would be to create a system that could generalize and 'understand' what makes an image visually successful, and even generate other images that meet these learned criteria."

Fitness approximation has been reported in evolutionary art and design (Biles 1994; Baluja et al 1994; Johanson and Poli 1998). Biles's Genjam is a genetic algorithm for generating jazz solo with a three-layer MLP neural network called Elualuator (Biles 1994). Similarly, Baluja et al. (1994) used back-propagation to train a multi layer feed-forward Artificial Neural Network (ANN) with human aesthetic judgments on the images produced by an IES. This system classifies images into several ranked categories. Using approximate models that embody the opinions of human evaluators were really imaginative attempts. However, the drawback of all of those tries is that the input space is so big (even with a short phrase of melody or small patch of image), whereas the granularity and the number of subspaces with different ranks are so indistinct, their distributions must be very complicated and overlapped with unavoidable errors and contradictions of human judgment. Therefore it is hard to decide the capacity (hidden layer scale) and the minimum size of training set of normal MLP. The ANN would be either over-trained or over-generalized.

Ohsaki and Takagi (1998) proposed using feed forward Neural Network and Euclidean distance measure to predict human evaluation of newborns and display them in order to reduce the burden of a human operator. The second method uses an average fitness value of the past generation weighted by the reciprocals of the distances as the predicted fitness of the present individual. And their experiment showed that the result of this method was quite precise. This was a quite inspirable simple idea, but theoretically it was only an approximating calculation because a linear method was used to solve a nonlinear problem. It was not clear why the neural network approach did not work well and what kind of the structure was adopted for the neural network.

Saunders and Gero (2001) used SOM (self-organizing map) neural networks (Kohonen 1993), which categorized each artwork pieces and produced a map of a typical artwork for the region of the genetic art space. Comparing newborn artwork against this map, potentially interesting artwork can be detected. The novelty in this approach is the distance of the closest category prototype (kernel neuron) to the input pattern on the map. In Saunders' thesis he also suggested to use an ART network as an alternative to SOM (Saunders 2002). Using ART as a novelty sensor is a good idea, but apparently the novelty alone just reflects on one aspect of searching activity, i.e., divergence. It does not help for convergence. For convergence, a system must be able to infer on new points based on the former materials it holds.

Fitness approximation in evolutionary computation has increasingly received attention in Engineering areas, where explicit fitness functions are either hard to formulate or computational expensive, such as structure optimization, aerodynamic design optimization, and protein structure prediction. Yaochu (2003) reported that several feed-forward neural network models have been used for approximating real models in optimization, such as multi-layer perceptrons, radial basis function (RBF) networks and support vector machines.

A general strategy used by this kind of systems is to use an approximated fitness function to accelerate the convergence by reducing the rate of using real fitness evaluation. And the approximated fitness function is dynamically updated from time to time.

## 2. Formulation of Aesthetic Evaluation using Neural Networks

Our approach is, just as we mentioned in the section above, to add an appropriate convergent mechanism to an interactive evolutionary design system by formulating a fitness landscape. The approximation model we employ is the General Regression Neural Networks (Specht 1991), which is a kind of probabilistic neural networks (Specht 1990).

### 2.1. CONVERGENT AND DIVERGENT THINKING

The goal of divergent thinking is to generate many different ideas about the same topic in a short period of time. Divergent thinking typically occurs in a spontaneous, free-flowing manner, such that the ideas are generated in a random, disorganized fashion. Following divergent thinking, the ideas and information must be organized using convergent thinking, i.e., putting various ideas together in an organized and structured way. A creative act involves both divergent and convergent thinking. Divergent thinking is for generating new ideas, whilst convergent thinking is for narrowing all options to one solution, which brings the materials from a variety of sources to bear on a target problem, so as to produce the best answer to the original problem.

As for an IES, the randomness and the tendency for bigger step sized mutation mean divergence. The recombination (crossover) followed by mutation within a stably decreasing space is convergence. Obviously, the divergence and convergence form a couple of forces in reverse directions. The art of evolutionary search is to subtly control the strength of the transferring between these two tendencies.

Usually, at the beginning of solving a design problem divergent thinking is overwhelming. At this stage, there is no clear a target projection in one's mind. During the exploration, there may be many attractive results showing up. But these may not necessarily match one's conceptual intention, because of flaws. It is impossible that one gets exactly what one wants just at the beginning by luck. The convergent thinking means that the system should be able to synthetically reason out the best result based on all this procedural divergent materials in a global view.

In order to capture the tendency in human aesthetic selections in an interactive evolutionary process, in our approach, the information about human choices are saved, analyzed, and then formulated as a fitness landscape for identifying the high performance regions and the significant areas for an evolutionary algorithm. The formulation of aesthetic evaluation is expected to accelerate the tiresome and lengthy convergent search for the potentially optimum solutions. A generative and evolutionary design process with a learning capability is more delightful and efficient.

To achieve this objective, we need an approximation tool to shape the underneath structure of the data, and a way to abstract heuristic information from designers as well as the knowledge for future reuse. We adopt the GRNN to serve this purpose.

## 2.2. GENERAL REGRESSION NEURAL NETWORK (GRNN)

The probability distribution of specific class of objects in an unsearched space is presumed to be even. But, with human selection, some points in the space may be crossed out and some areas may emerge as having more favorite points. When the equilibrium of a search space is disturbed, the probability density in the space becomes uneven. Theoretically specific knowledge is exhibited in such a process. It is usually the probability density and distribution that give a heuristic clue for the convergence and further exploration.

Estimating the Probability Density Function (PDF) from available data is a normal approach in Statistics (Parzen 1962), and in this context it fits the area of Bayesian Statistics. The approach gets closer to the true underlying class density function as the number of the training samples increases. A cluster of cases closely packed together usually indicates an area of high probability density. In kernel-based approximation, simple functions are located at each available case, and added up for estimating the overall PDF. Typically, these kernel functions are Gaussians. If sufficient training points are available, this will indeed yield an arbitrarily good approximation to the true PDF.

We can also estimate a regression function from available data following the ideas of kernel regression. In kernel regression method we seek to estimate the probability density function $p(x, d)$ of the input-desired pairs $(xi, di)$ using the Parzen window method (Parzen 1962). If we know the joint PDF of $x$ and $d$, the conditional PDF and the expected value can be computed:

$$y(x) = \frac{\sum_i d_i \exp(-\|x - x_i\|^2 / 2\sigma^2)}{\sum_i \exp(-\|x - x_i\|^2 / 2\sigma^2)} \tag{1}$$

Basically, the method places a Gaussian at each sample multiplied by the desired response *di* and normalized by the response in the input space. GRNN is based the above probabilistic formula.

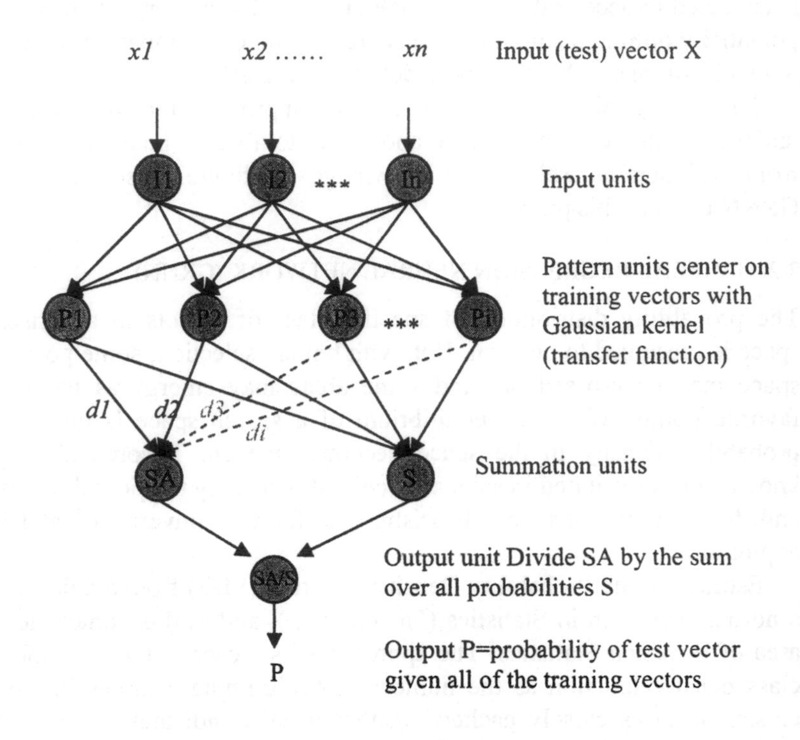


*Figure 1.* A GRNN structure and workflow

The Pattern units are copied directly from the sample data, one per case. Each can model a Gaussian function centered at the sample point. *di* is desired response of *Pi* unit. It represents the belongingness (reciprocal of fitness) to a class A. Suppose that *P1, P2* definitely belong to class A, the desired out put *d1* and *d2* should be set to 1, while the rest (*d3~di*) should be set to zero. The output is the probability of the events that the currently applied test vector belongs to class A. The following is a simplified form of function (1).

$$P(X) = \frac{\sum_i d_i \exp(-Z_i / \sigma)}{\sum_i \exp(-Z_i / \sigma)} \tag{2}$$

*Zi* is the Euclidian distance between test vector *X* and the training vectors *Xi* corresponding to pattern node *Pi*. There are some alternatives to reducing the computation cost of Euclidian distances. For example, the dot product:

$$Z_i = X^T X_i \tag{3}$$

or so called city block distance (Specht 1990):

$$Z_i = \sum_{j=1}^{n} | x_j - x_j^i | \tag{4}$$

The most important parameter is $\sigma$, which determines the smoothness of final regression function. (i.e., the radial deviation of the Gaussian functions). We should notice that if it is too small, the functions lose their generalization ability. The function is not very sensitive to the precise choice of smoothing factor.

Such kind of probabilistic models can be observed in many trivial activities performed by designers. For example, when a designer draws a curve on a paper, the first freehand curve is normally undesirable. Usually the designer may draw a cluster of lines rather than erase the every last drawing. It is easy to imagine an unseen curve by judging the density of the line cluster.

## 3. System Architecture

As a matter of fact, the generate-and-test strategy can be seen as a diverge-and-converge strategy. Briefly speaking, we want to build an interactive evolutionary system, which supports not only the part of 'generating' but also the part of 'testing'. However, if only counting on human subjective evaluation without assistant tools, the convergent process will almost certainly be misled to a zigzag path, especially when searching a broad space through a small view port. To overcome this difficulty, a basic idea is to use a servo module, which observes human selection activities to derive user's intention. The data sets generated are then used to strengthen the convergent power in the evolutionary process.

Our experimental system is called the Genetic Face Creator (GFC), which uses genetic algorithms to produce 3D facial masks defined by certain parameters. This experiment can be extended to other applications in product and multimedia design domains, because such kind of multi variable optimization problem is extensively involved in modeling and shading processes. In order to get an impressive effect on screen, designers have to adjust dozens of parameter for many times. It is unintuitive and much more exhausting than the case of drawing a curve on a paper.

The GFC is a typical IES. It randomly seeds the first generation, and continually reproduces its offspring based on artificial selection. The reproduction strategy used in GFC is simple. The user just simply picks up one individual that seems relatively good or promising from twelve individuals displayed on the screen. Then, the system will mutate it to get twelve newborns, which replace the pervious generation on the screen. This cycle continues until a desirable result is obtained. This route is similar to the evolutionary strategy. But the mutation step size and the direction are not controlled. So it is a totally divergent searching process.

However, we also provide a servo module based on a GRNN, in order to give such a simple IES several novel aspects. The GRNN shapes the PDF of the target object based on all the selected individuals as well as those unselected. The more selections are observed, the more accurate regression it gets. The peak of PDF means the most interesting area.

When we switch on this module, the GFC can perform via another route, i.e., a convergent searching process. In this route, the GFC works in exactly the same way as the most Genetic Algorithm based systems do. In this route, there is a parent population other than just a single parent. Individuals of parent population have the chances to be selected for crossover and mutation based on their PDF values. All the newborns will be evaluated by the GRNN module and then inserted into the parent population. The weakest one in a parent population will be weeded out. The cycle continues until the system reaches a stable point. If the result is not so satisfactory, the process can be restarted, switching between the two alternative routes, depending on the judgment and decisions of the users. Figure 2 illustrates the overall approach of our system architecture.

It's not so hard to imagine that there are some other extended usages of the module. For example, it can sort out twelve best individuals displayed on the screen in certain orders from the individuals memorized in its sample pools. Then the user can select one to start a new search. It also can arrange the newborns on the screen according to their predicted fitness and novelty, omitting those, which are considered not promising.

## 4. Modeling and Representation of Facial Characters

Facial character design is chosen as the application of our system for testing, verifying and modifying the theoretical hypothesis and computational frameworks as proposed in this research.

Perception of faces is a particularly developed ability of humans. Facial character has an extensive application in anthropomorphic agent, animation, game, and toy design. The use of computers for evolving amusing or attractive animated characters is like a star hunter finding a suitable actor/actress, which is totally based on visual impression of strong character traits. The character develops towards a design that will help to tell the story. As a result of these preferences, both Mickey Mouse and Teddy Bear have undergone pronounced evolutionary changes since their first introduction (Baxter 1995). In the 1930's Mickey Mouse had relatively rodent-like facial features: a long snout, sloping forehead and small eyes. In a series of progressive changes Mickey was transformed into his modern version. The size of his eye nearly doubled (relative to his head size), his head grew by 15 percent relative to his body and his fore head bulged by nearly 20 percent.

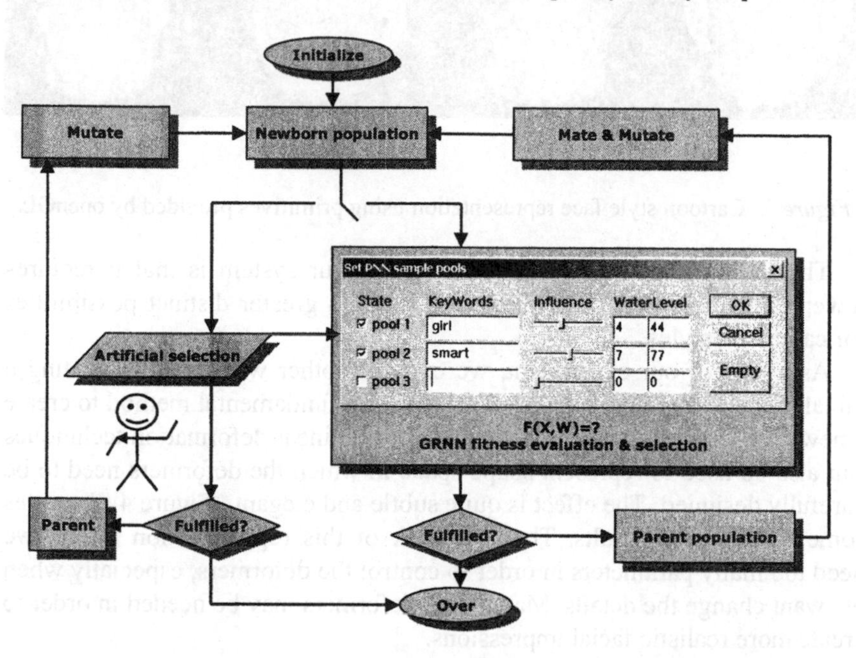


*Figure 2.* The GFC system has two different Genetic Algorithm routes. One is based on artificial selection. Another is based on GRNN fitness evaluation module.

## 4.1. REPRESENTATION

The geometric representation of a human face defines the potential searching space. In our implementation, we define the phenotype of GFC using the parameters of constructive solid geometry (CSG). The facial characters are represented as hierarchies of separate primitive shapes, which can be varied to include basic geometric shapes such as spheres, cylinders, and cones to represent nose, ears, eyes and mouths. A complicated free form NURBS is constructed to represent the basic shape of the face. The solution space for evolutionary exploration is represented by a set of fifty parameters, which controls the NURBS surface and primitives' affine transformations. Figure 3 illustrates the visualization of some face models used in our experiment.

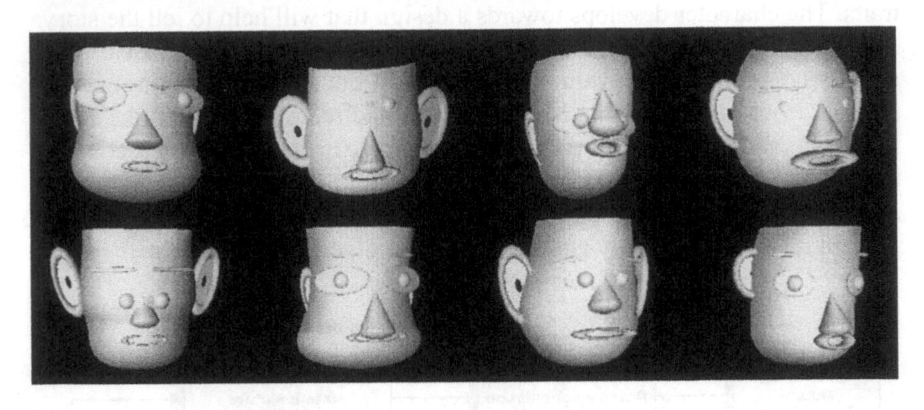

*Figure 3.* Cartoon style face representation using primitives provided by openGL.

The reason why we choose CSG to test our system is that it requires fewer parameters than other models, but offers greater distinct possibilities for caricature styles.

As a part of our experiment, we explored other ways of representing a facial space. Deformer is the most popular and fundamental method to create a new face from an existing one. Linear or nonlinear deformation techniques can also be used to represent shape space in which the deformers need to be carefully designed. The effect is quite subtle and elegant. Figure 4 illustrates some deformation results. The drawback of this representation is that we need too many parameters in order to control the deformers, especially when we want change the details. Many local deformers may be needed in order to create more realistic facial impressions.

## 4.2. THE TRAINING SET FOR GRNN

Integrating GRNN and IES is straightforward, and there is no training process for the GRNN module. The learning of GRNN is just to add the training cases in the form of input-output pairs into a pool. So the only problem for training is to decide what kind of information is taken as input and output. For such kind of GA-GRNN integrated system, the simplest way is to use normalized phenotype of faces directly as the inputs to neural networks.

For corresponding desired output, one option is to let the user input a scale value of the fitness using a slide bar for each individual. That may be a quite informative way and it reduces the number of the training examples required. However we abandoned this approach for two reasons: usability and reliability. First it is hard to give many of the results a scale rank with slide bars or any other ways. Second, the values may not be reliable because the evaluation is based only on the present twelve individuals displayed on the screen, other than the whole candidates in the global view. So in our experiment system, we put the all of new candidates into two categories: selected and unselected. The human evaluation of fitness only requires one click (selected), click again (deselected) or not click (unselected). From the usability point of view, this is satisfactory.

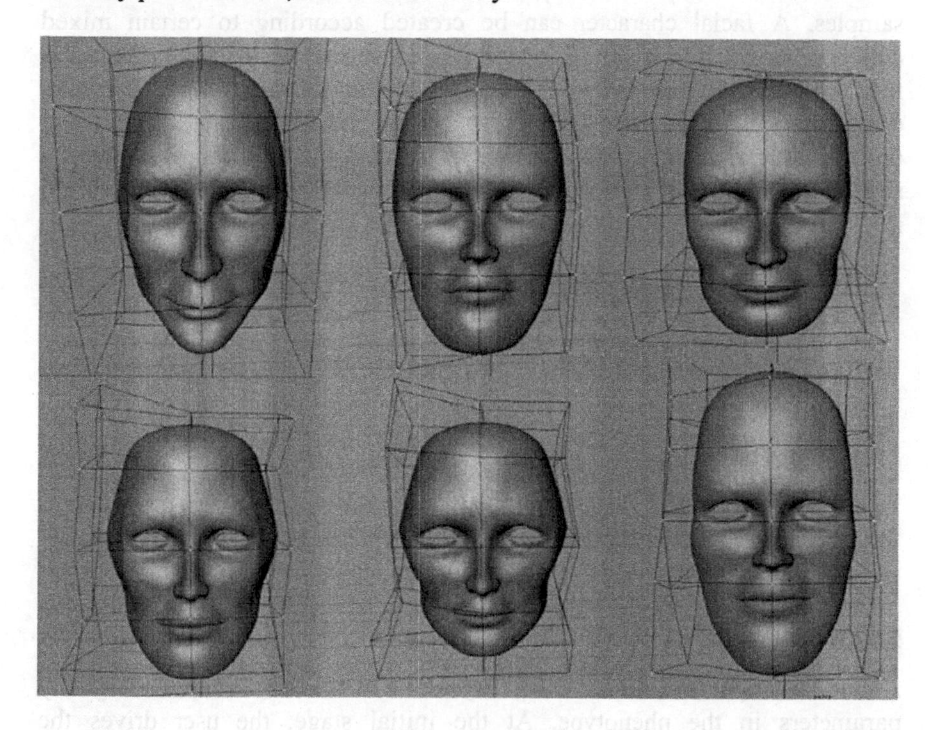

*Figure 4*. A random face deformer written by Maya MEL, which can produce thousands alternatives in a few minutes. The detail variation is not so significant because only one deformer is added.

## 4.3. GRNN LEARNING PROCESS

GRNN module has several pools to record the selections. When it is set to the learning mode, for each step of the hands-on evolution, the module puts the normalized phenotype of individuals into a pool with the mark of true (selected) or false (unselected). We suppose that the concept of aesthetic impression can be represented by such kind of memory based GRNN structure. Then, such kind of empirical knowledge can be retained and reused next time. Furthermore, we can give the pool several linguistic descriptions. In this way, GRNN performs concept formation and learns to distinguish shapes with some concepts, such as soft, lovely, nervous or relax, and as for face shape, such as manlike, womanlike, teenage or adult and so on.

A computational framework, which understands simple semantic concepts, should be capable of evaluating and controlling the evolution process with some specifications from users.

Mixed and complex concepts can then be established based on the samples. A facial character can be created according to certain mixed concepts specified by a user. For example, a user may wish to create an adult face which is manlike but with a foolish looking. This kind of generative design process can indeed be seen as a constrained multi objective optimization problem. As for the GRNN system, the multi objective means several weighted PDFs are added up to provide the highest point. Figure 5 illustrates a dialogue interface for the user to define face types.

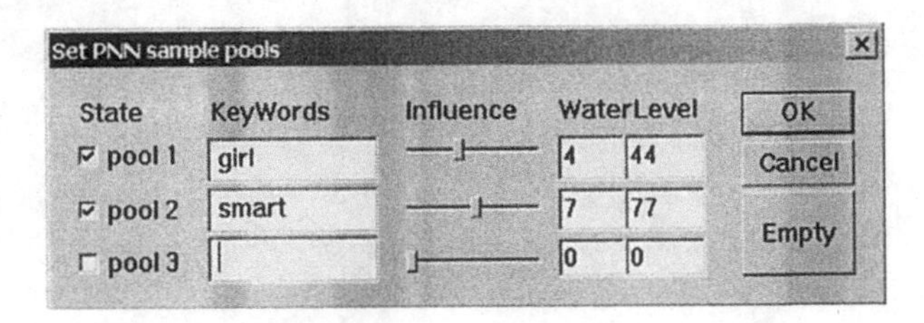


*Figure 5.* The experimental system has several pools to record the training cases. The influence slide bars decide the weights of the objectives.

## 5. Does It Work?

As introduced in Section 3, the GFC system evolves facial shapes using fifty parameters in the phenotype. At the initial stage, the user drives the evolution generation by generation. The present generation is mutated from one parent, which is selected from the last generation. Every gene has one chance to randomly flip one bit in the mutation. We make sure that enough diversity is derived from each new generation. This kind of divergent searching worked well in the experiments conducted so far. However, the key questions to ask are:

- Does the GRNN make the system converge?
- Where does the system converge upon?

It is quite easy to answer the first question. When we disabled the GRNN module and entered an auto run route (this means that the newborns enter the parent population with equal fitness values), we found that the faces at generation 200,000 were still changing at almost the same speed as at initial time. Although after just a few generations, the crossover operator made the individuals in the same population became similar to each other, the mutation power pushed the process to a never-ending path.

On the contrary, when we enabled the GRNN module, which already has some learning cases in its pools, the system rapidly converged to the almost the same shape in just few hundreds of generations, no mater how many times we reinitialized and restarted it again.

If the fitness landscape is multi modal and has several peaks, the convergence point may be different. Theoretically, it depends on the positions of the initial population.

The second question is a little bit harder to answer than the first one, but we have observed interesting results in the initial experiments. In one experiment, A facial character of a young lady was our searching target. This experiment had three steps:

At the first step, we initialized the newborns, and let the GRNN learn until the 8th generation with 7 individuals selected and 89 individuals unselected. The selections were recorded in a pool. Then, at the 9th generation we entered automatic route involving the GRNN module. The searching process became stable between 200~500 generations in one minute or so of running time. The result is shown at the bottom row of Figure 6.

At the second step, we closed the first pool at first, and reinitialized the newborns and had 4 generations with 4 individuals selected and 44 individuals unselected and recorded in the second pool. Then, at the 5th generation we entered automatic route, in which the GRNN exerted its convergent power based on the data in the second pool. 200~500 generations later, the results became stable, as shown at the bottom row of Figure 7.

At the third step, both pools were activated. We entered automatic route from the first generation (right after initialization). The system converged fast. The result is shown in Figure 8.

We found that the third one was a mixture of results observed separately at the first step and second step. The eyes and eyebrows were like in the first one, whereas the mouth and nose were like in the second one. The face silhouette looked more elegant than in the first and second case because some appearance flaws were moderated.

The reason is that, if we simplify our input in just one dimension (this means that only one parameter is taken), statistically, there must be a highest peak of PDF at the area, where is relatively the densest place of the selected points under the consideration of distribution of the unselected. Finally the GA will find the peak, i.e., the most favorable point.

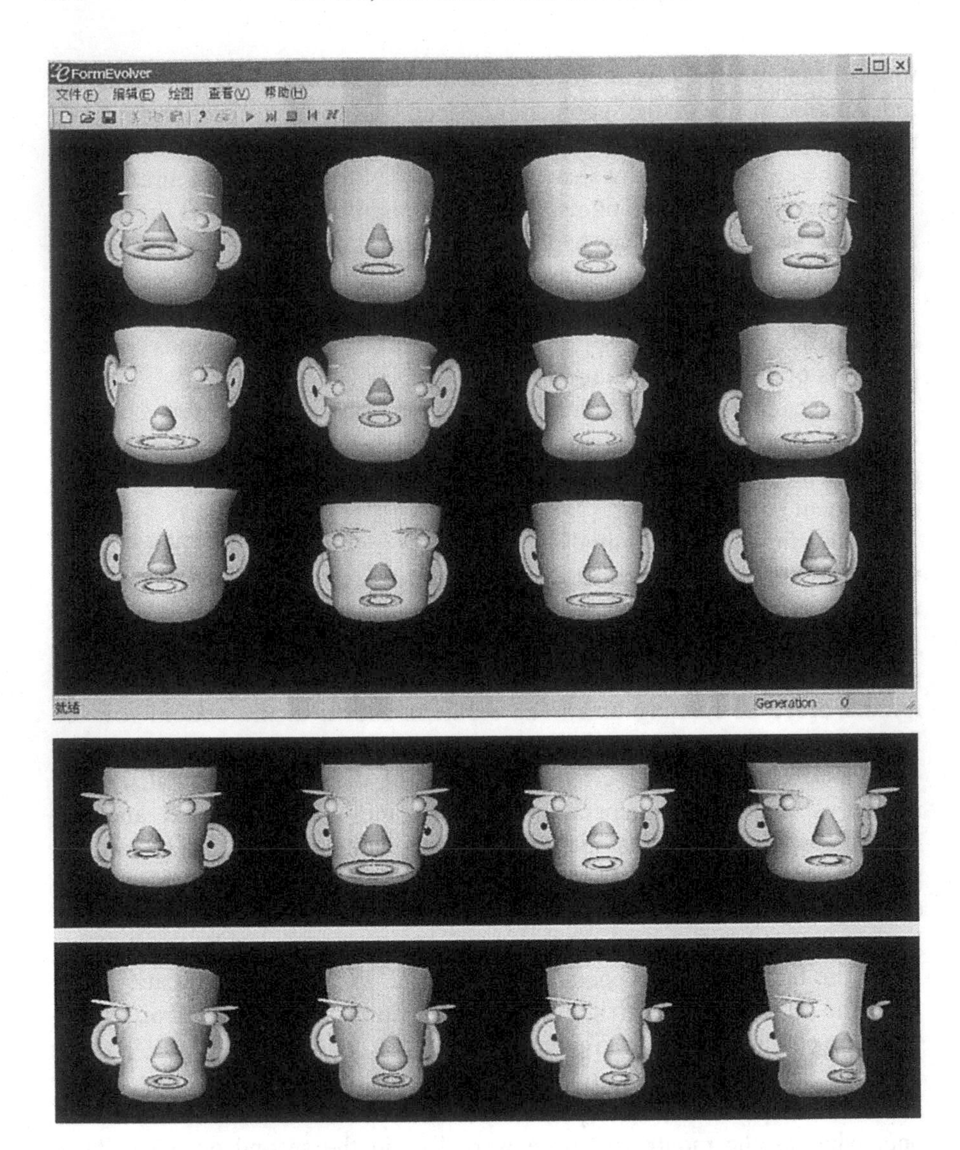


*Figure 6.* First step of the experiment: The upper is the screen shot of first generation. The middle is cut from fifth generation. The bottom is convergent state based on first pool, which records first hands-on evolution.

## 6. Conclusions and Further Work

The formulation of aesthetic evaluation and selection ability in a computational shape evolution system provides both theoretical and technical challenges. It is useful for accelerating the convergence of an IES using neural networks as well as for knowledge reuse. This paper focuses on the feasibility of using a General Regression Neural Network to formulate human aesthetic fitness evaluation in an IES through its case based learning mechanism. A system called GFC has been implemented and some initial experiments have been conducted for testing how the integration of GRNN and IES in an aesthetic domain can support evolutionary facial character creation. The results obtained so far showed that the GRNN module worked well in this domain. Based on an initial analysis of the results obtained, we suggest that GRNN can support human users when it possesses empirical knowledge abstracted from users. The system as a whole has a convergent exploration capability.

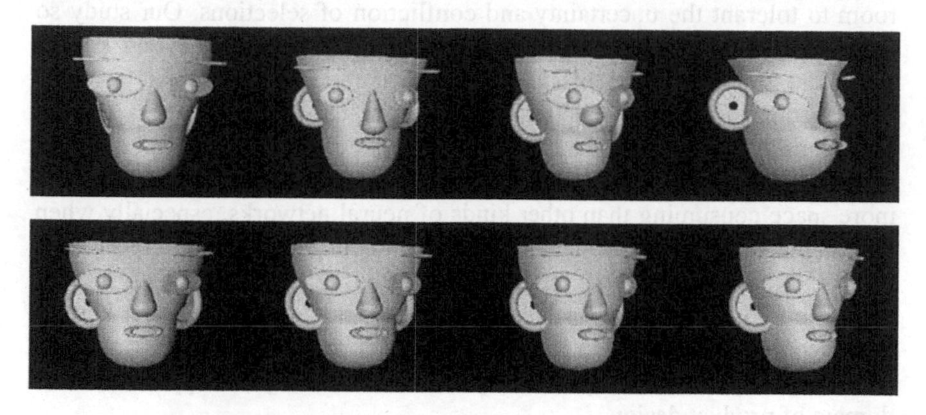

*Figure 7.* Second step of the experiment: The upper is the fourth generation of second hands-on evolution. The bottom is the convergent state based on second pool which records second hands-on evolution.

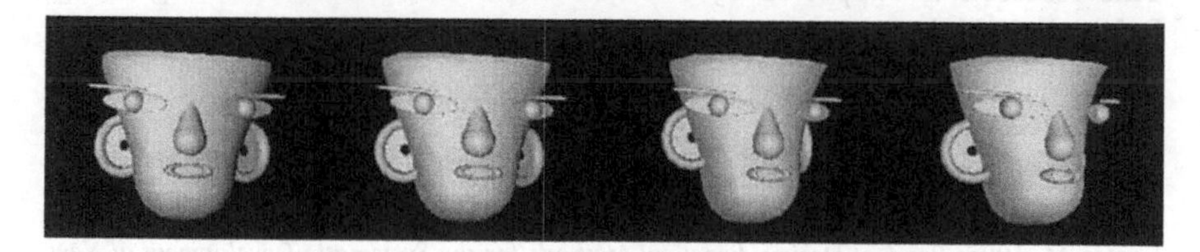

*Figure 8.* Convergent result based on both pools

In forthcoming studies, we will continue to test and adjust our prototype system. We will then try to extend the use of GRNN to different kind of evolutionary systems, such as Evolutionary Strategy (ES). In an ES approach, the algorithms operate directly on the solution parameters, and the search space is more continual and coherent. The mutation step size and direction are controllable. A solution space can be related to its mutation operator, step size and so on. So some parameters and processes in IES must influence the neural networks' learning results. This relation needs to be explored further.

In order to reduce the dimensionality of face space, a new face can be represented by a linear sum of several eigenfaces, which are obtained from principal components analysis (PCA). Another experiment is to mimic the human perception by extracting universal information of the feature points and proportions as the input of networks, so that the empirical knowledge is reusable in different modeling methods.

GRNN has a quite simple structure but fast learning speed suitable for online dynamic learning systems. It is robust to noisy data. So, it has enough room to tolerant the uncertainty and confliction of selections. Our study so far also implies that GRNN is not inclined to extrapolate beyond known data. The response drops off towards zero if the data points far from the training data used. Extrapolation too far from the training data is usually dangerous and unjustified. And it usually results in convergence failure.

The only drawback of GRNN is that it tends to be slower to execute and more space consuming than other kinds of neural networks, especially when the sample set becomes too large. So we will identify a method to reorganize the GRNN, or give GRNN a temporary substituting component. This may be some other kind of Radial Basis Function networks, which use fewer kernels but place them at optimal places. This issue is currently under investigation.

And finally, we will explore the possibility of using this approach in the domain of product design.

## Acknowledgements

This research is supported by a Ph.D studentship from the Hong Kong Polytechnic University.

## References

Baluja, S, Pomerleau, DA and Jochem, T: 1994, Towards automated artificial evolution for computer generated images, *Connection Science* **6**(2&3): 325-354.

Baxter, M: 1995, *Product Design, Practical Methods for the Systematic Development of New Products*, Chapman & Hall, pp. 45-46.

Bentley, PJ: 1999, An introduction to of evolutionary design by computers, *in* PJ Bentley (ed), *Evolutionary Design by Computer,* Morgan Kaufmann, San Francisco, California, pp. 1-73.

Biles, JA, 1994: Genjam: A genetic algorithm for generating jazz solos, *Proceeding of International Computer Music Conference*, pp. 131-137.

Johanson, B and Poli, R: 1998, GP-music: An interactive genetic programming system for music generation with automated fitness raters, *in* JR Koza, W Banzhaf, K Chellapilla, K Deb, M Dorigo, DB Fogel, MH Garzon, DE Goldberg, H Iba, and R Riolo (eds), *Proceedings of the Third Annual Conference on Genetic Programming*, pp. 181-186.

Kohonen, T: 1993, Things you haven't heard about self-organizing map, *Proceedings of the International Conference on Neural Networks*, III 1147-1156.

Ohsaki, M and Takagi, H: 1998, Improvement of presenting interface by predicting the evaluation order to reduce the burden of human interactive EC operators, *IEEE International Conference on System, Man, and Cybernetics (SMC'98)*, pp. 1284-1289.

Parzen, E: 1962, On estimation of a probability density function and mode, *Annals of Math. Stat.* **33**: 1065-1076,

Saunders, R: 2002, *Curious Design Agents and Artificial Creativity*, PhD Thesis, University of Sydney.

Saunders, R and Gero, JS: 2001, The digital clockwork muse: A computational model of aesthetic evolution, *in* G Wiggins (ed), *Proceedings of the AISB'01Symposium on AI and Creativity in Arts and Science*, SSAISB, York, UK, pp. 12-21.

Sims, K: 1991, Artificial Evolution for Computer Graphics, *Computer Graphics* **25**(4): 319-328.

Specht, DF: 1990, Probabilistic neural networks, *Neural Networks* **3**: 109-118.

Specht, DF: 1991, A general regression neural network, *IEEE Transactions on Neural Networks* **2**: 568-576.

Yaochu, J: 2003, A comprehensive survey of fitness approximation in evolutionary computation. *Soft Computing Journal* (in press).

# DESIGNING WITH SHAPES AND FEATURES

JS Gero (ed), *Design Computing and Cognition'04*, 357-376


# EVALUATION OF A 3D SHAPE GRAMMAR IMPLEMENTATION

HAU HING CHAU, XIAOJUAN CHEN, ALISON McKAY, ALAN de PENNINGTON
*University of Leeds, UK*

**Abstract.** The geometric design of the exterior appearance of consumer products is a principal consideration to retain brand identity. Architectural and engineering shape grammars had demonstrated shape computation as a formal and viable approach for supporting style conformance. However, most existing shape grammar implementations operated in limited experimental domains, and lacked support for complex three dimensional geometry, shape emergence and parametric shape rules. The aim of the reported research is to address these issues. Shape algebras are reduced in terms of shape operations with their basic elements which is generally applicable to shape computation. Specifically for algebra $U_{13}$, exhaustive cases were enumerated for shape sum and shape difference operations. A $U_{13}$ shape grammar implementation was developed to support both rectilinear and curvilinear basic elements in three dimensional space. Mathematical representations of basic elements were based on non-uniform rational b-splines and their reduced form. This allowed a simple yet exact notation which simplified support for maximal representation and its computation. Two cases studies, a Coca-Cola bottle grammar and a Head & Shoulder bottle grammar, were used to test the implementation.

## 1. Introduction

Architectural and engineering shape grammars have demonstrated shape computation as a formal and viable approach for supporting style conformance. There are at least twenty shape grammars in the literature, Figure 1. It has been more than thirty years since Stiny and Gips (1972) introduced the first shape grammar. Some early shape grammars were used to analyse paintings and decorative arts. Shape grammars were gradually used to study architectural plans and historical houses and they became the mainstream. The last ten years has seen an increasing interest in engineering shape grammars. Realising the enormous commercial potential for enforcing brand identity for consumer products, some pioneer work on brands like

Dove (Chau 2002), Harley-Davidson (Pugliese and Cagan 2002) and General Motor Buick (McCormack et al. 2004) were conducted.

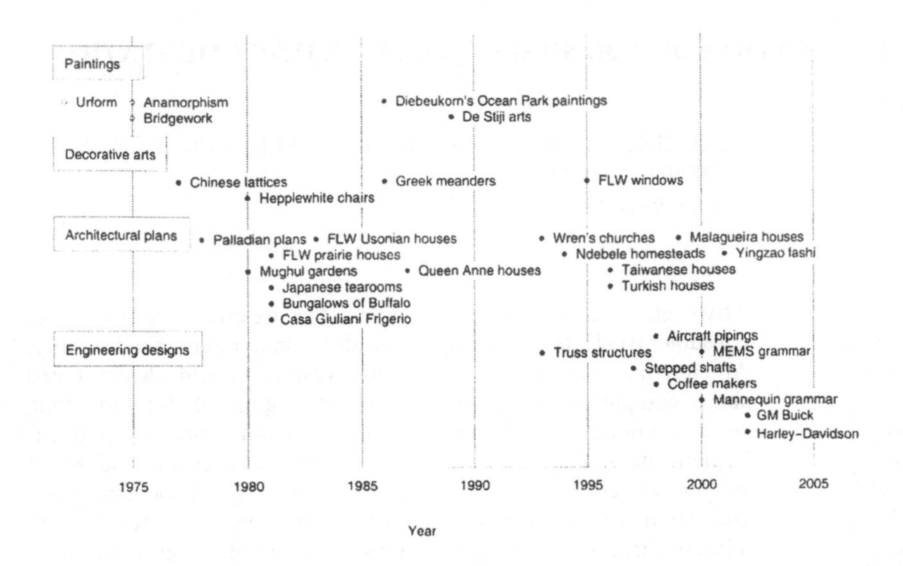


*Figure 1.* Applications of shape grammars

Many of these shape grammars did exceptionally well in capturing existing corpora of designs, albeit many were using pencil and paper. Tapia (1999) rightly pointed out that computer implementations of shape grammars should make machines to handle bookkeeping tasks and allow designers to focus on creative activities. However, most existing shape grammar implementations operated in limited experimental domains and fell short in support for real designs listed in Figure 1. They used only rectilinear basic elements and mostly confined to two dimensional space. The aim of this research was to reconcile this mismatch between the promise that shape grammars offer and the limitation of their current implementations with an emphasis on supporting brand identify conformance for the design of consumer products.

## 2. Review of Existing Shape Grammar Implementations

A shape grammar interpreter by Krishnamurti (1982) was the first shape grammar implementation in a form prescribed by Stiny and Gips (1972). This implementation featured maximal representation of straight lines and used homogeneous coordinates. It introduced the notion of rational shape as an attempt to overcome the problem of non exact arithmetic which had an influence on other shape grammar implementations. It subsequently had been rewritten (Krishnamurti and Giraud 1986) but the key ideas remain unchanged.

A set grammar implementation of Queen Anne houses by Flemming (1987) had introduced three dimensional geometry into shape computation research and implementation. He also pioneered the use of parametric shape rules for a substantial shape grammar. However, advantage was not taken of shape emergence which is a key feature in shape grammars.

Heisserman produced two Genesis boundary solid interpreters (Heisserman 1991; Heisserman 1994). The later one which he produced at Boeing was applied to aircraft piping design. Although the details of the implementation is not publicly available, it marked an important milestone in commercial interest in the shape grammars research.

Mark Tapia's GEdit (Tapia 1999; Tapia 1996) is arguably the best shape grammar implementation available. GEdit implemented maximal representation of basic elements. Furthermore, it supported subshape detection and enables shape emergence. Given a shape and a shape rule, it worked out automatically all possible shape transformations which the shape rule could apply to that shape. Then, a user could select from a list of possible derived shapes. Great care was also taken in the considering the user interface design. Nevertheless, basic elements were limited straight lines on a two dimensional plane only. It severely restricted on the range of designs can be expressed in GEdit. Rotational transformations are further limited to multiples of 90°. Since the values of sines and cosines for arbitrary angles are likely to be irrational numbers, exact representation for rational numbers is not possible. Hence, the use of approximation or other representations is inevitable. In the computional geometry community, the use of floating point representation to approximate real numbers is widely adopted.

A shape grammar interpreter by Piazzalunga and Fitzhorn (1998), although it was indeed a set grammar implementation using oblongs in three dimensions, was the first to use a solid kernel. It opened up the interaction between shape computation and computational geometry researches, and substantially reduced the coding effort.

Under Cagan's leadership at Carnegie Mellon University, Michalek produced an implementation of Agarwal's coffee maker grammar (Agarwal and Cagan 1998), and Agarwal produced a simplified implementation of their original MEMS grammar (Agarwal et al. 2000).

There are a number of computer implementations listed in Table 1 that did not use maximal representation (Krishnamurti 1992). Hence they were not able to perform subshape detection nor to enable shape emergence (Stiny

TABLE 1. Comparison of shape grammars implementations

| | Name | Reference | Tool(s) used | Shape emer-gence | 2D/3D |
|---|---|---|---|---|---|
| 1 | Simple interpreter | Gips1975 | SAIL[1] | No | 2D |
| 2 | Shepard-Metzler analysis | Gips 1974 | SAIL[1] | No | 2D/3D |
| 3 | Shape grammar interpreter | Krishnamurti 1982 | conventional language | Yes | 2D |
| 4 | Shape generation system | Krishnamurti and Giraud 1986 | PROLOG[2] | Yes | 2D |
| 5 | Queen Anne houses | Flemming 1987 | PROLOG | No | 2D |
| 6 | Shape grammar system | Chase 1989 | PROLOG | Yes | 2D |
| 7 | Genesis (CMU) | Heisserman 1991 | C/CLP(R)[3] | No | 3D |
| 8 | GRAIL | Krishnamurti 1992 | | Yes | 2D |
| 9 | Grammatica | Carlson 1993 | | No | |
| 10 | | Stouffs 1994 | | Yes | 2D/3D |
| 11 | Genesis (Boeing) | Heisserman 1994 | C++/CLP(R)[3] | No | 2D/3D |
| 12 | GEdit[5] | Tapia 1996 | LISP[4] | Yes | 2D |
| 13 | Shape grammar editor | Shelden 1996 | AutoLISP | Yes | 2D |
| 14 | Implementation of basic grammar | Simondetti 1997 | AutoLISP | No | 3D |
| 15 | Shape grammar interpreter | Piazzalunga and Fitzhorn 1998 | ACIS Scheme | No | 3D |
| 16 | SG-Clips | Chien et al 1998 | CLIPS | No | 2D/3D |
| 17 | 3D Shaper | Wang 1998 | Java/Open Inventor | No | 3D |
| 18 | Coffee maker grammar[6] | Michalek 1998 | Java | No | 2D/3D |
| 19 | MEMS grammar | Agarwal et al 2000 | LISP | | 2D |
| 20 | Shaper 2D[7] | McGill 2001 | Java | No | 2D |
| 21 | $U_{13}$ shape grammar implementation | Chau 2002 | Perl | Yes | 3D |

[1] Stanford Artificial Intelligence Language
[2] SeeLog developed at EdCAAD
[3] IBM CLP(R) compiler
[4] Macintosh Common LISP
[5] http://www.shapegrammar.org
[6] http://www.andrew.cmu.edu/org/CDL/
[7] http://www.architecture.mit.edu/~miri/shape2d/

1994). They were best described as set grammars in line with the definition coined by Stiny (1982). Nevertheless, they demonstrated the generative power of Post production systems (Gips and Stiny 1980) on which shape grammars were based.

During the NSF/MIT Workshop on Shape Computation in 1999, Manish Agarwal and Mark Tapia performed live demonstrations of two shape grammar implementations, namely the Coffee Maker Grammar and GEdit respectively. There followed a discussion on how best to move forward research on computer implementation of shape grammars, which would benefit not only shape grammarians but more importantly practising designers. Three key issues raised were support for shape emergence, adoption of parametric shape rules and an intuitive user interface. Although that was four years ago, this assessment still holds true today. Chase (2002) envisioned different models of shape grammar implementations ranging from full user control to fully autonomous.

The following list is an attempt to characterise an idealised general shape grammar implementation in the context of supporting the geometric design of consumer products:

- Using maximal representation, thus enabling subshape recognition and shape emergence.
- Enabling automatic shape recognition under Euclidean transformations.
- Allowing parametric shape rules.
- Enabling automatic shape recognition for parametric shape grammars.
- Allowing three dimensional shapes.
- Allowing curvilinear basic elements.
- Incorporating an intuitive user interface.
- Providing aesthetic measures for ranking designs for automated selection.
- Supporting surfaces and solids.
- Providing unambiguous interpretation of resulting designs to their physical realisation.

This is very challenging. Thanks to Krishnamurti and Earl (1992), automatic shape recognition for straight line type basic elements in algebra $U_{13}$ is now a solved problem. Tapia (1996) gave much insight into user interface design. This paper will focus on enabling shape computation, including shape emergence, for complex three dimensional rectilinear and curvilinear basic elements. It also outlines out a rigorous mathematical foundation which simplifies computer coding.

### 3. Shape Emergence and Parametric Shape Recognition

Krishnamurti and Earl (1992) enumerated an exhaustive account of shape recognition of straight line type basic elements in algebra $U_{13}$. It can be readily adapted to include circular arc type basic elements. Extension to freeform curves in general requires approximation. They were the first to introduce the use of homogeneous coordinates in shape computation. Krishnamurti also introduced the concept of rational shapes and used exact arithmetic. While it eased the effort of computer coding, it severely hindered the range of possible shape transformations. For example, rotation of shapes other than multiples of 90° was not possible. The implementation reported in this paper adapted homogeneous coordinates, floating point arithmetic and tolerance zones which are standard in computational geometry and commercial CAD/CAM systems.

Parametric shape grammars are more versatile than standard shape grammars. Each shape rule $A \rightarrow B$, can be used to derive a family of shape rules $g(A) \rightarrow g(B)$ where g is a parameterisation function. Hence the standard shape computation $C' = [C - t(A)] + t(B)$ becomes $C' = [C - t(g(A))] + t(g(B))$ Methods of automatic enumeration of transformation t were available, but automatic generation of parameterisation g remains a difficult research question. McCormack and Cagan (2002) provided some insights on parametric shape recognition.

There are two modes of operation for obtaining a transformation t of parameterisation g of shapes A and B such that subshape relation $t(g(A)) \leq t(g(B))$ can be satisfied. In manual mode, the user of a shape grammar implementation provides parameters required by a parameterisation function g. Users are required to input fairly detailed information. At each step of the shape computation, human intervention is required. In automatic mode, possible parameterisations are generated by a computer in a fashion similar to generation of shape transformations.

### 4. Algebra $U_3$

Every *shape* is a finite but possibly empty set of basic elements that are maximal with respect to one another. The empty shape, which consists of no basic elements, is denoted by the symbol $s_{\varnothing}$. *Basic elements* are points, curves, surfaces or solids that are defined in three dimensional space. A basic element is said to be *maximal* with respect to another basic element if and only if it cannot be combined to form a single basic element. Maximal representation of shape is a central theme in shape grammar formalism. It ensures a unique representation of physically identical shapes, which is essential to shape recognition during shape computation and to shape emergence.

A shape A consisting of an unordered set of basic elements, $a_1, a_2, \ldots, a_i, \ldots, a_m$, which are maximal with respect to one another, is denoted by using a pair of braces.

$$A = \{a_1, a_2, \ldots, a_i, \ldots, a_m\} \quad (1)$$

The original definitions of *subshape* relation, *shape sum*, *shape difference* and *shape product* by Stiny (1980) were adopted which are denoted by the symbols $\leq$, $+$, $-$ and $\cdot$ respectively.

### 4.1. ALEGBRA $V_{03}$

Labelled points in three dimensions were also used in the implementation. Since algebra $V_{03}$ did not interact with algebra $U_{13}$ directly, labelled points could be manipulated separately.

## 5. Expressing Algebra $U_{13}$ using Basic Elements

In this section algebra $U_{13}$ operations between two shapes are reduced in terms of their basic elements. This forms the basis of the $U_{13}$ shape grammar implementation reported in this paper. Shape algebra operations are subshape relation, shape product, shape difference and shape sum.

In the last section, it was stated that a unique representation of physically identical shapes is desirable. These shapes could be curves, surfaces or solids. Krishnamurti and Stouffs (1997) outlined solutions for subshape detection of planar surfaces in $U_{23}$ which was based on their earlier work (Krishnamurti and Stouffs 1993). However, in general, subshape detection for surfaces and solids in three dimensions is an unsolved problem and remains a very challenging research question in its own right. Nevertheless, it is common practice that industrial designers produce control drawings to accompany their artistic renderings. These control drawings allow precise definition of geometric forms and are based on three dimensional wireframe models (Chau et al. 2000). In turn these wireframes inform the definition of surfaces and solids.

A shape rule $A \rightarrow B$ applies to a working shape C where there is a shape transformation t such that $t(A) \leq C$. The resultant shape C' is given by $C' = [C - t(A)] + t(B)$.

### 5.1. SUBSHAPE RELATION $\leq$

The subshape relation between two shapes can be expressed in terms of their basic elements. For shapes t(A) and C, where $t(A)=\{a_1,a_2,...,a_i,...,a_m\}$ and $C=\{c_1,c_2,...,c_j,...,c_n\}$, subshape relation $t(A)\leq C$ is equivalent to,

$$\begin{aligned} t(A)\leq C &\Leftrightarrow a_1\leq C\wedge\cdots\wedge a_i\leq C\wedge\cdots\wedge a_m\leq C \\ &\Leftrightarrow \begin{bmatrix} (a_1\leq c_1 & \vee\cdots\vee & a_1\leq c_j & \vee\cdots\vee & a_1\leq c_n)\ \wedge \\ \vdots & & \vdots & & \vdots \\ (a_i\leq c_1 & \vee\cdots\vee & a_i\leq c_j & \vee\cdots\vee & a_i\leq c_n)\ \wedge \\ \vdots & & \vdots & & \vdots \\ (a_m\leq c_1 & \vee\cdots\vee & a_m\leq c_j & \vee\cdots\vee & a_m\leq c_n) \end{bmatrix} \\ &\Leftrightarrow \bigwedge_i\bigvee_j a_i\leq c_j \end{aligned} \tag{2}$$

The subshape relation between two basic elements can be further reduced with the aid of counting number of endpoints coinciding with the other basic elements. This in turn can be used to compute whether two basic elements are relatively maximal to one another — columns 3, 4 and 5 in Table 2.

## 5.2. SHAPE PRODUCT ·

For the shape product between two shapes, it is necessary to introduce the concepts of singleton shape and singleton sum. A singleton shape is a shape consisting of exactly one basic element. Singleton sum $+^{\text{singleton}}$ is an operation between two shapes, singleton shape or otherwise, that are maximal to one another. For more than two shapes, it can be written in shorthand as $\sum^{\text{singleton}}$.

The shape product of two shapes can be computed in terms of the shape products of their basic elements. For shapes A and C, where $A=\{a_1,a_2,...,a_i,...,a_m\}$ and $C=\{c_1,c_2,...,c_j,...,c_n\}$, the shape product $A\cdot C$ is given by,

$$\begin{aligned} A\cdot C &= \{a_1,a_2,\cdots,a_i,\cdots,a_m\}\cdot C \\ &= \sum_i{}^{\text{singleton}} a_i\cdot\{c_1,c_2,\cdots,c_j,\cdots,c_n\} \\ &= \sum_{\substack{i\\j}}{}^{\text{singleton}} a_i\cdot c_j \end{aligned} \tag{3}$$

By considering straight lines and circular arcs (less than 180° span) as basic elements for algebra $U_{13}$, all possible results of a shape product of two basic elements or two singleton shapes, are listed in column 6 of Table 2.

A straight line type basic element is represented by two points in a four dimensional homogeneous space (Faux and Pratt 1979) as its endpoints which can be mapped then into a three dimensional physical space. A circular arc type basic element is represented by three points in homogeneous coordinates. Two points denote endpoints and the third is a control point. Two straight lines, formed by joining the control point with the endpoints, are tangent at their respective endpoints.

Both types of basic elements described here are a reduced form of non-uniform rational b-spline (NURBS). A straight line type basic element is a parametric (not in parametric shape grammars terms) straight line, with a parameter varying between 0 to 1 from one endpoint to the other. Circular arc type basic elements behave in a similar fashion. There is work underway to generalise circular arcs of any span including circle and free from curve in the form of NURBS.

### 5.3. SHAPE DIFFERENCE –

The shape difference between two shapes can be computed in terms of the shape differences of their basic elements. For shapes t(A) and C, where $t(A)=\{a_1,a_2,\ldots,a_i,\ldots,a_m\}$ and $C=\{c_1,c_2,\ldots,c_j,\ldots,c_n\}$, let shape $C^*$ be the shape difference C – t(A) given by,

$$\begin{aligned} C^* &= C - t(A) \\ &= (((((\underbrace{\underbrace{\underbrace{\underbrace{\underbrace{\underbrace{C}_{C_0} - a_1)}_{C_1} - a_2)}_{C_2} - \cdots) - a_i)}_{C_i} - \cdots) - a_m)}_{C_m = C^*} \end{aligned} \tag{4}$$

which can be written in a recursive form,

$$\begin{cases} C_0 = C \\ C_i = C_{i-1} - a_i = \sum^{\text{singleton}}{}_j c_j - a_i \\ C - t(A) = C^* = C_m \end{cases} \tag{5}$$

All possible results of shape difference between a basic element from another basic element, or two singleton shapes, are listed in column 7 of Table 2.

TABLE 2. Shape difference and shape sum of two straight line basic elements or two circular basic elements of same radius

<table>
<tr><th>number of endpoints coincided</th><th>case[1]</th><th>$a_i \leqslant c_j$</th><th>$c_j \leqslant a_i$</th><th>$a_i$ and $c_j$ are maximal to one another</th><th>$c_j \cdot a_i$</th><th>$c_j - a_i$</th><th>$c_j + b_i$</th></tr>
<tr><td>0 or 1</td><td>1a 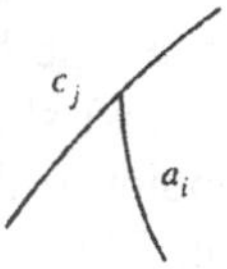</td><td rowspan="5">false</td><td rowspan="4">false</td><td rowspan="2">true</td><td rowspan="3">$s_\varnothing$</td><td rowspan="3">$c_j$</td><td rowspan="2">$\{c_j, b_i\}$</td></tr>
<tr><td rowspan="2">2</td><td>1b 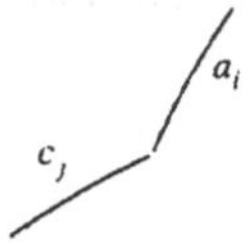</td></tr>
<tr><td>1c 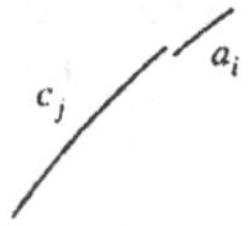</td><td rowspan="4">false</td><td>$\{c_j + b_i\}$</td></tr>
<tr><td rowspan="3">2, 3 or 4</td><td>2 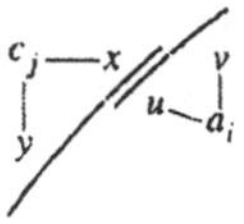</td><td>$x$</td><td>$y$</td><td>$\{\overbrace{y + \underbrace{x + v}_{b_i}}^{c_j}\}$</td></tr>
<tr><td>3 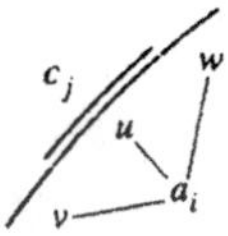</td><td>true</td><td>$c_j$</td><td>$s_\varnothing$</td><td>$b_i$</td></tr>
<tr><td>4 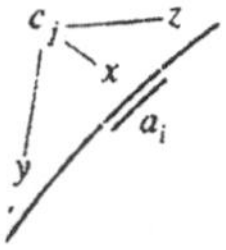</td><td>true</td><td>false</td><td>$a_i$</td><td>$\{y, z\}$</td><td>$c_j$</td></tr>
</table>

[1] $a_i$ should read as $b_i$ for column $c_j + b_i$.

### 5.4. SHAPE SUM +

The shape sum of two shapes can be computed in terms of shape sums of their basic elements. For shapes $C^*$ and t(B), where $C^* = \{c_1, c_2, \ldots, c_j, \ldots, c_n\}$ and $t(B) = \{b_1, b_2, \ldots, b_i, \ldots, b_m\}$, let C' be the shape sum $C^* + t(B)$ and is given by,

$$\begin{aligned} C' &= C^* + t(B) \\ &= (((((\underbrace{\underbrace{\underbrace{\underbrace{\underbrace{\underbrace{C^*}_{C_0^*} + b_1)}_{C_1^*} + b_2)}_{C_2^*} + \cdots) + b_i)}_{C_i^*} + \cdots) + b_m)}_{C_m^* = C'} \end{aligned} \tag{6}$$

which can be written in a recursive form,

$$\begin{cases} C_0^* = C^* \\ C_i^* = C_{i-1}^* + b_i = \sum_{j=k+1}^{\text{singleton } n+1} c_j \\ C' = C_m^* \end{cases} \tag{7}$$

where a singleton shape $c_{n+1} = \{c_1, c_2, \ldots, c_j, \ldots, c_k\} + b_i$.

For each $C_{i-1}^*$ there are n basic elements, these basic elements are reordered such that $c_1, c_2, \ldots, c_k$ can be combined with $b_i$ to form a single basic element, whereas $c_{k+1}, c_{k+2}, \ldots, c_n$ are relatively maximal to $b_i$. All possible results of the shape sums of two basic elements, or two singleton shapes, are listed in column 8 of Table 2.

## 6. Note on the Programming Tools used

The $U_{13}$ shape grammar implementation, Figure 2, developed in this research was implemented in the Perl programming language. There are a large number of standard modules available, which are roughly equivalent to function libraries in languages such as C or C++. The Perl Data Language (PDL) and Perl/Tk were two key modules used in this prototype implementation. They were used for matrix manipulation and the graphical user interface respectively.

## 7. Applications

### 7.1. A TWO DIMENSIONAL COCA-COLA BOTTLE GRAMMAR

Coca-Cola (Beyer and McDermott 2002; Schaeffer and Bateman 1995) has always understood the value of its shapely bottle, which was an icon of their brand image. Although the curved bottle design has been altered and refined over the years from the original contour glass bottles to the plastic bottles of today, the Coca-Cola brand identity has been maintained as its designs evolved. Figure 3 shows six typical bottles that have been produced over the years.

By studying these bottles, it can be seen that they preserved certain characteristics. A Coca-Cola bottle grammar was designed to capture these common features, which are described using shape rules. Shape computation was then carried out using the computer implementation described in last Section to reproduce the existing designs as well as to generate new designs.

The design of a shape grammar started from dividing the shape of a bottle into five parts: cap, upper part, label region, lower part, and bottom, Figure 4. Not all the bottles are comprised of the five parts, and the parameters for same rule may vary. These characteristics were reflected in the design of the shape rules as shown in Figure 5.

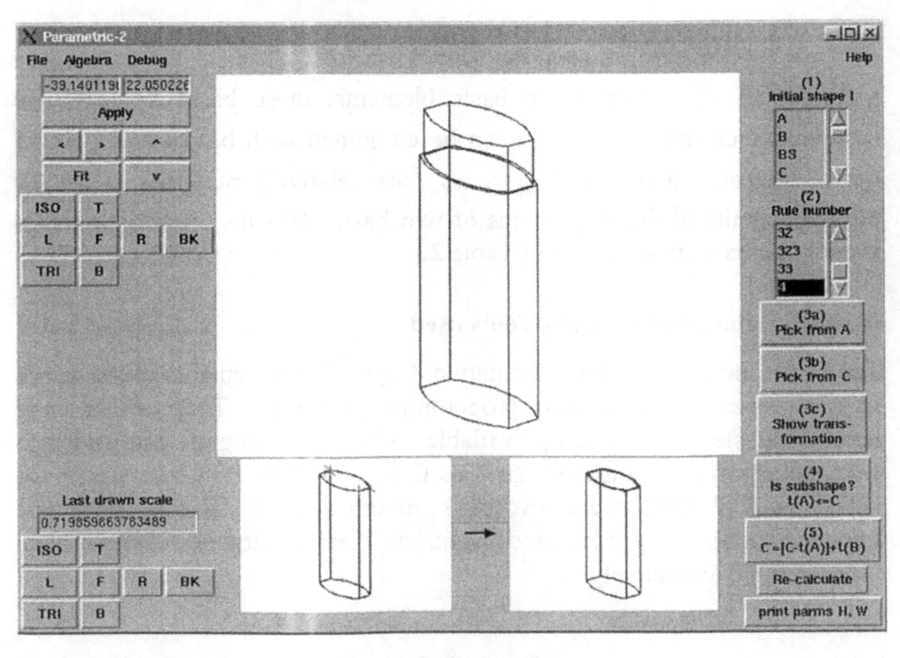


*Figure 2.* Screen shot

*Figure 3.* Evolution of Coca-Cola bottles

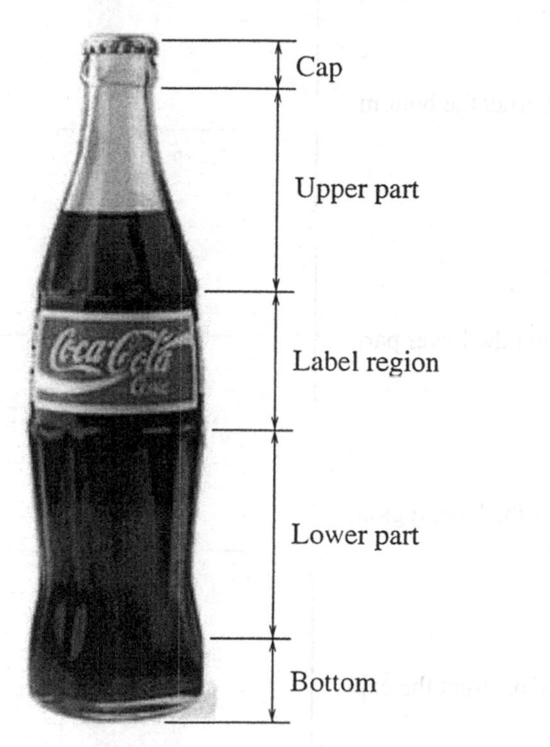


*Figure 4.* Division of a Coca-Cola bottle

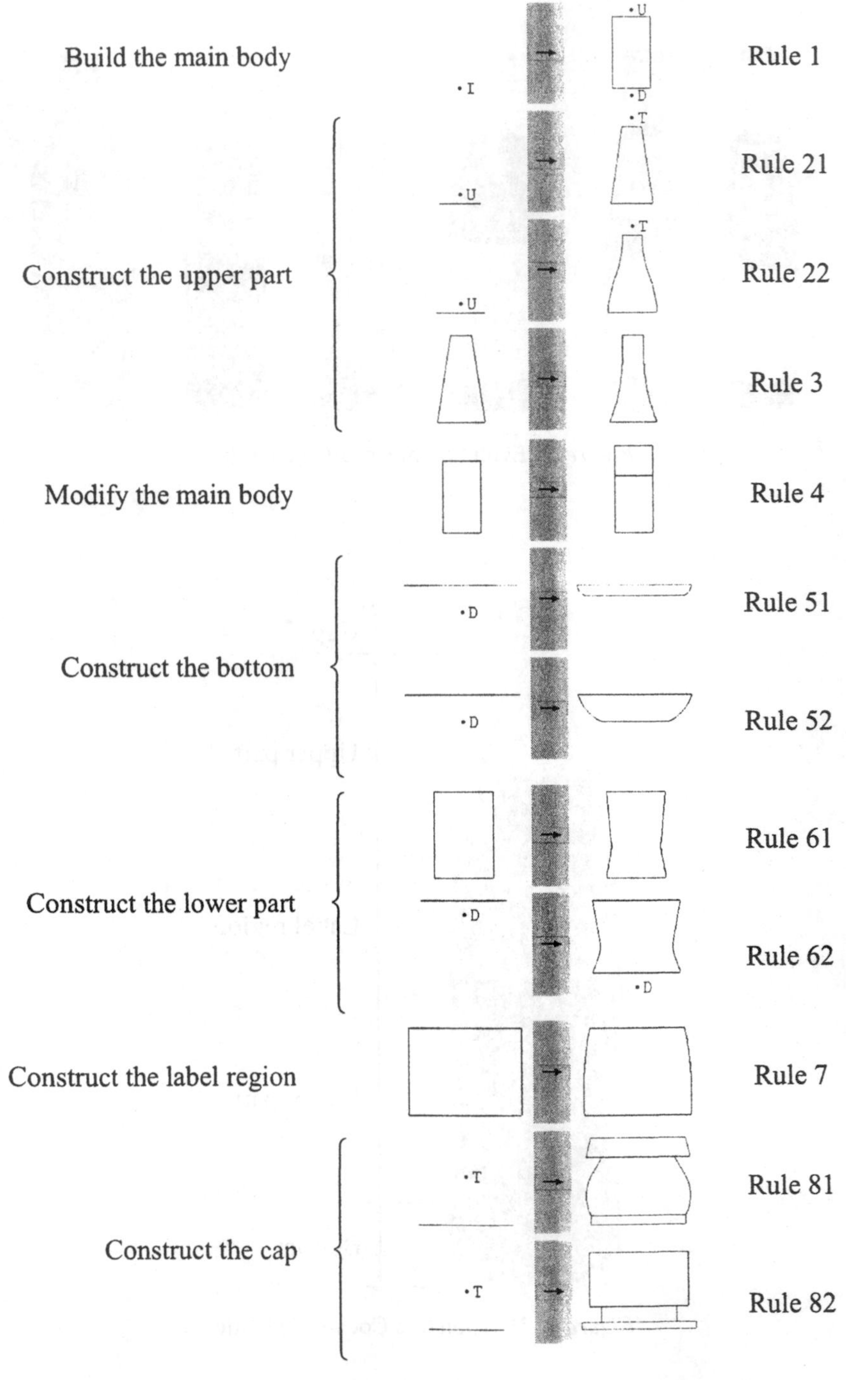


*Figure 5.* Shape rules for the Coca-Cola bottle grammar

The Coca-Cola bottle grammar consisted of 12 shape rules that were categorised according to the part of the bottles they generate. The shapes in the grammar consisted of two dimensional rectilinear and curvilinear basic elements, some of them with predefined parameters. The computation of the grammar was carried out using a $U_{13}$ shape grammar implementation.

Before conducting a computation, the geometric data of shapes and rules were put into a data file. Straight lines and circular arcs were constructed in three-dimensional coordinate with maximal representation. Initial shape and current shapes were displayed in the middle of the screen. Each shape rule was represented using the left hand side (lhs) shape and the right hand side (rhs) shape separated by an arrow. If there were any predefined parameters, values are required to be input to represent the defined shape. Each step of the shape computation with a rule consisted of five stages.

The first stage was the selection of an initial shape or carrying on from any existing working shape. The second stage was the selection of a shape rule which would be displayed on the bottom of the screen. The third stage was to choose three coordinates each from the lhs shape and working shape, thus defining the shape transformation. The fourth stage was to check whether the transformation of the lhs of the shape rule is a subshape of the current shape. If the subshape relation was satisfied, then the fifth stage was to carry out a step of shape computation using shape difference and shape sum operations.

Figure 6 shows a computation to generate one of the existing bottles. In this computation, after rule 81 had been applied, a straight sided bottle was reproduced. Carrying on applying rule 4, the straight sided body was divided into two rectangles. At this point, in order to carry on the computation, it required that the programme recognising both two rectangles, and then rule 7 and rule 61 could be applied to produce the label region and the curved lower part.

The Coca-Cola bottle grammar was used to reproduce six existing designs, and Figure 7 shows two new designs generated by the grammar.

## 7.2. A THREE DIMENSIONAL HEAD & SHOULDER BOTTLE GRAMMAR

Head & Shoulders shampoo has a family of 200ml, 400ml and 500ml bottles. The geometric form of the 200ml and 400ml bottles are identical except different parameters for different capacities. The 500ml bottle re-used the cap of the 400ml bottle and the body was expanded to increase its capacity.

A shape grammar for Head & Shoulders bottles was designed based on the above observations. Shape computation of this grammar required working on three-dimensional shapes with predefined parameters. Figure 8 shows a computation of the 500ml bottle using the implementation.

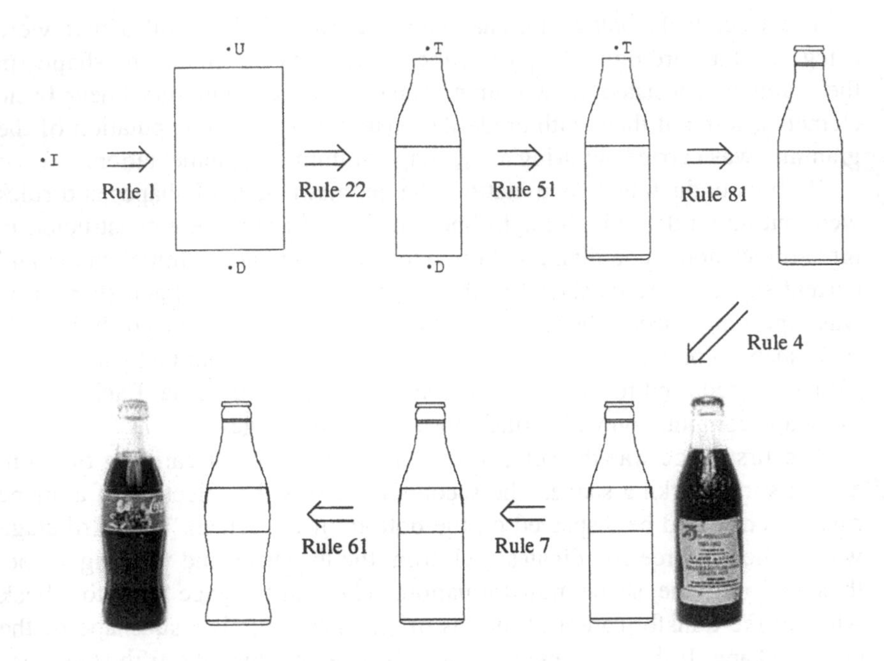


*Figure 6.* Shape computation for two existing Coca-Cola bottles

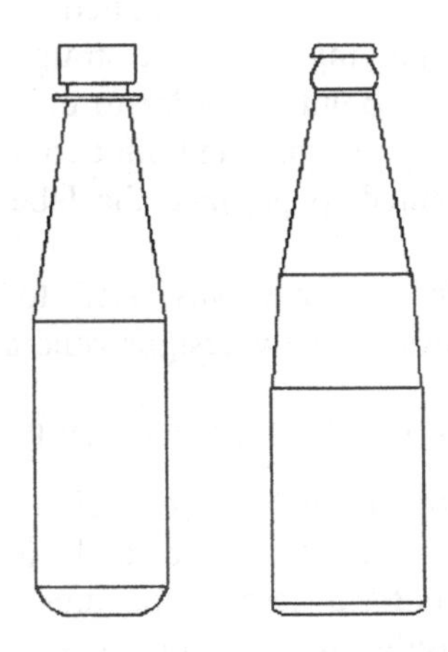

*Figure 7.* Grammar generated new Coca-Cola bottles

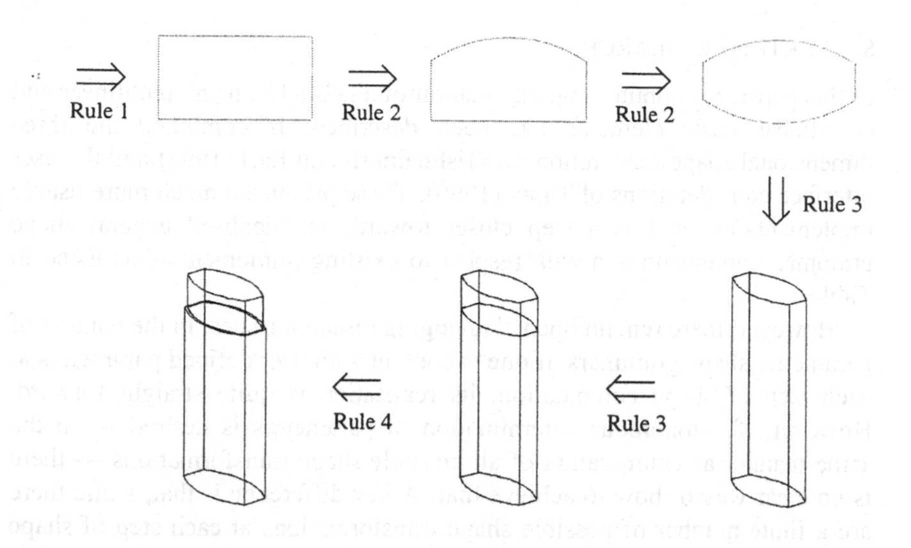


*Figure 8.* Shape computation of 500ml Head & Shoulders bottle

## 8. Discussions and Conclusions

With the aim of working towards a shape grammar implementation for practising designers, this paper examined recent advances and identified challenges in shape emergence, parametric shape recognition, and the use of complex three dimensional geometries as basic elements. Two different modes of parametric shape recognition were discussed, namely, user defined parameters and autonomously determined parameters.

Using straight lines and circular arcs in algebra $U_{13}$, and labelled points in algebra $V_{03}$ as basic elements, shape operations between two shapes were reduced in terms of shape operations on their basic elements. In algebra $U_{13}$, exhaustive cases were enumerated for shape operations between two basic elements which were readily used in producing of a computer implementation. The resultant $U_{13}$ shape grammar implementation was tested against two shape grammars designed to imitate two well known styles of mass produced consumer products.

The Coca-Cola grammar was computed using the implementation, which was able to incorporate curvilinear basic elements in their maximal representation and to support shape emergence. It was used to generate designs in the original corpus and also new designs using the same shape rules, arguably in the same style too.

The Head & Shoulders grammar was also computed using the implementation. The ability to manipulate shapes in a true 3D fashion was demonstrated in contrast to 2½D.

8.1. FURTHER RESEARCH

In this paper, a computer implementation of algebra $U_{13}$ using rectilinear and curvilinear basic elements has been described. If combined the three dimensional shape recognition of Krishnamurti and Earl (1992) and the user interface considerations of Tapia (1996). These promise a much more usable implementation and is a step closer towards an idealised general shape grammar implementation with respect to existing implementations listed in Table 1.

However, there remain open challenging research issues. In the context of parametric shape grammars, if one is content with user defined parameters at each step of shape computation, its realisation is quite straight forward. However, if autonomous determination of parameters is desired — in the same manner as enumeration of all possible shape transformations — there is no clear way of how to achieve that. A key difference is that, while there are a finite number of possible shape transformations at each step of shape computation, there are potentially an infinite number of possible parameterisations, thus exhaustive enumeration is not a practical strategy.

Brand owners are attracted to the notion of shape grammars and the prospect of computer implementations. A frequently asked question is how one tells if a grammar generated design conforms to a given style. The approach used by Chau et al. (2000) was to analyse the pattern of shape rule applications. A much neglected approach originated by Stiny and Gips (1978a), soon after their invention of shape grammars, was the aesthetic measure. An evaluation of Palladian plans (Stiny and Gips 1978b) and shape annealing (Shea 1997) were effectively manifestations of aesthetic measure, which allowed automatic ranking of designs. However, there did not seem to be a generally applicable approach.

Chase (2002) has commented on different models of shape grammar implementation, given a corpus of designs, translating them into a set of shape rules remains a task requiring great skills. There is suggested that no agreed way how best to proceed. There are also debates on whether shape emergence is essential in the design of shape rules and hence shape computation. On one level, shape emergence is indispensable and is a unique feature which differentiates shape grammars from traditional set theoretical computation. However, there are some shape grammarians who contest that non-trivial shape grammars can be made without the use of shape emergence.

Shape grammars and their computer implementations have been used for different purposes such as analysis of existing designs and pedagogic tools. In the context of supporting brand identity through the geometric design of the exterior appearance of consumer products, shape computation remains a formal and viable approach for supporting style conformance. Although challenging research issues remain, we are now closer than any time in the past thirty years towards a shape grammar implementation that can be used to assist practising designers.

## Acknowledgements

The authors would like to thank Mr Jim Rait, Mr Richard Parker and Mr David Raffo for their insightful comments on the reported research. Thanks are to Dr Terry Knight for being the first author's mentor during the period he was a visiting scholar at MIT, and also to Prof George Stiny for the second author. The first two authors would like to express their appreciation to Universities UK for Overseas Research Students Awards and the University of Leeds for Tetley & Lupton Scholarships. The first author also wishes to thank the School of Mechanical Engineering for its financial support and the Keyworth Institute for a Keyworth Scholarship. The authors would like to thank the anonymous reviewers for their constructive comments.

## References

Agarwal, M and Cagan, J: 1998, A blend of different tastes: The language of coffeemakers, *Environment and Planning B: Planning and Design* **25**(2): 205-226.

Agarwal, M, Cagan, J and Stiny, G: 2000, A micro language: Generating MEMS resonators by using a coupled form-function shape grammar, *Environment and Planning B: Planning and Design* **27**(4): 615-626.

Beyer, H and McDermott, C: 2002, *Classics of Designs,* The Brown Reference Group/Bookmart, London.

Chase, SC: 2002, A model for user interaction in grammar-based design systems, *Automation in Construction* **11**(2): 161-172.

Chau, HH: 2002, *Preserving Brand Identity in Engineering Design using a Grammatical Approach,* PhD thesis, School of Mechanical Engineering, University of Leeds.

Chau HH, de Pennington, A and McKay, A: 2000, Shape grammars: Capturing the quintessential aspects of industrial design for consumer products, *Greenwich 2000: Digital Creativity Symposium,* The University of Greenwich, London, UK pp. 27-31.

Faux, ID and Pratt, MJ: 1979, *Computational Geometry for Design and Manufacture*, John Wiley & Sons, New York.

Flemming, U: 1987, More than the sum of parts: The grammar of queen Anne houses, *Environment and Planning B: Planning and Design* **14**(3): 323-350.

Gips, J and Stiny, G: 1980, Production systems and grammars: A uniform characterization, *Environment and Planning B* **7**(4): 399-408.

Heisserman, J: 1994, Generative geometric design, *IEEE Computer Graphics and Applications* **14**: 37-45.

Heisserman, JA: 1991, *Generative Geometric Design and Boundary Solid Grammars,* PhD dissertation, Department of Architecture, Carnegie-Mellon University, Pittsburgh.

Krishnamurti, R: 1982, SGI: A shape grammar interpreter, *Technical report,* Design Discipline, The Open University, Walton Hall, Milton Keynes MK7 6AA.

Krishnamurti, R: 1992, The maximal representation of a shape, *Environment and Planning B: Planning and Design* **19**(3): 267-288.

Krishnamurti, R and Earl, CF: 1992, Shape recognition in three dimensions, *Environment and Planning B: Planning and Design* **19**(5): 585-603.

Krishnamurti, R and Giraud, C: 1986, Towards a shape editor: The implementation of a shape generation system, *Environment and Planning B: Planning and Design* **13**(4): 391-404.

Krishnamurti, R and Stouffs, R: 1993, Spatial grammars: Motivation, comparison and new results, *in* U Flemming and S Van Wyk (eds), *CAAD Futures '93: Proceedings of the Fifth International Conference on Computer-Aided Architectural Design Futures,* Amsterdam: North-Holland, Pittsburgh, PA. pp. 57-74.

Krishnamurti, R and Stouffs, R: 1997, Spatial change: Continuity, reversibility , and emergent shapes, *Environment and Planning B: Planning and Design* **24**(3): 359-384.

McCormack, JP and Cagan, J: 2002, Supporting designers' hierarchies through parametric shape recognition, *Environment and Planning B: Planning and Design* **29**(6): 913-931.

McCormack, JP, Cagan, J and Vogel, CM: 2004, Speaking the Buick language: Capturing, understanding, and exploring brand identity with shape grammars, *Design Studies* **25**(1): 1-29.

Piazzalunga, U and Fitzhorn, P: 1998, Note on a three dimensional shape grammar interpreter, *Environment and Planning B: Planning and Design* **25**(1): 11-30.

Pugliese, MJ and Cagan, J: 2002, Capturing a rebel: Modelling the Harley-Davidson brand through a motorcycle shape grammar, *Research in Engineering Design* **13**(3): 139-156.

Schaeffer, R and Bateman, B: 1995, *Coca-Cola: A Collectors Guide to New and Vintage Coca-Cola Memorabilia,* The Apple Press, London.

Shea, K: 1997, *Essays of Discrete Structures: Purposeful Design of Grammatical Structures by Directed Stochastic Search,* PhD dissertation, Department of Mechanical Engineering, Carnegie-Mellon University, Pittsburgh.

Stiny, G: 1980, Introduction to shape and shape grammars, *Environment and Planning B* **7**(3): 343-351.

Stiny, G: 1982, Letter to the editor: Spatial relations and grammars, *Environment and Planning B* **9**(1): 113-114.

Stiny, G: 1994, Shape rules: Closure, continuity and emergence, *Environment and Planning B: Planning and Design* **21:** s49-s78.

Stiny, G and Gips, J: 1972, Shape grammars and the generative specification of painting and sculpture, *in* CV Freiman (ed), *Information Processing 71: Proceedings of IFIP Congress,* North-Holland, Amsterdam, pp. 1460-1465.

Stiny, G and Gips, J: 1978a, *Algorithmic Aesthetics: Computer Models for Criticism and Design in the Arts,* University of California Press, Berkeley.

Stiny, G and Gips, J: 1978b, An evaluation of Palladian plans, *Environment and Planning B* **5**(2): 199-206.

Tapia, M: 1999, A visual implementation of a shape grammar system, *Environment and Planning B: Planning and Design* **26**(1): 59-73.

Tapia, MA: 1996, *From Shape to Style, Shape Grammars: Issues in Representation and Computation, Presentation and Selection,* PhD dissertation, Department of Computer Science, University of Toronto, Toronto.

JS Gero (ed), *Design Computing and Cognition'04*, 377-396


# A GRAMMAR BASED APPROACH FOR FEATURE MODELLING IN CAD

EGON OSTROSI, MICHEL FERNEY
*Université de Technologie de Belfort-Montbéliard, France*

**Abstract.** This paper presents a feature recognition method based on the use of a Feature Grammar. Given the complexity of feature recognition in interactions, the basic idea of the method is to find the latent and logical structure of features in interaction. The method includes five main phases. The first phase, called Regioning, identifies the potential zones for the birth of features. The second phase, called Virtual Extension, builds links and virtual faces. The third phase, called Structuring, transforms the region into a structure compatible with the structure of the features represented by the Feature Grammar. The fourth phase, called Identification, identifies the features in these zones. The fifth phase, called Modeling, represents the model by features. The FMS software (Feature Modeling System) is developed based on this method.

## 1. Introduction

The introduction of solid modelers made it possible to unambiguously define the geometric models for parts. This type of modeler provides a complete representation of the shape of the part, offering sufficient information about its geometry and topology. Thus this type of representation is increasingly popular in various applications used in the product development process.

Despite these advantages, these modelers take into account the product shape, but do not include the various types of knowledge required to build and utilize it. The introduction of the concept of features made it possible to associate shape and knowledge in understanding a CAD model. Features are generic or specific shapes with which engineers associate certain attributes and application knowledge used in various development phases. In order to use features as a means for integration, we must find ways to transform their representation when moving between applications. This problem involves on the one hand transforming a geometric model for the part into a feature-based model adapted to the desired view, and on the other hand enabling the transformation of features between views (Bronsvoort and Jansen 1993; De Martino and Gianini 1994). The transformation, which is developed based on a geometric model to other models representing different application views, is automatic feature recognition. Automatic feature recognition is a process for transforming one representation into another. More specifically, we can say that the process defines the transformation of all the lower-level CAD model entities such as primitives, faces, edges, and nodes into features. Automatic feature recognition must resolve the following problems: Feature Representation and Feature Recognition.

The first problem involves choosing or developing a method suitable for representing features so that their representation is unique. The second problem involves developing inference procedures able to perform the most complete recognition possible. The main automatic feature recognition methods are based on graph theory (Marefat and Kashyap 1990; Joshi and Chang 1988); expert systems (Bond and Chang 1988; Henderson and Anderson 1984); volume based decomposition (Menon and Kim 1994; Woo 1982; Kim and Wilde 1992; Kim 1992); syntactic method (Falcidieno and Gianini 1989; Li 1988; Srinavasan et al 1985) and neural networks (Henderson 1994). Automatic feature recognition is a complex process. Despite major developments in this area, several problems remain. Thus feature recognition in interaction remains an area for research.

In the case of design, a finite set of canonical features can produce an unlimited number of configurations of features in physical interaction. The physical interaction between canonical features can deform their initial representation. Indeed, the representation of a feature, resulting from the physical interaction between canonical features, is dissimilar from the representation of canonical features. Furthermore, the representation of a canonical feature as a component of a resulting feature may be different from its initial representation. Listing all the features in interaction appears to be a utopian task of very little interest.

This paper will discuss the problem of recognizing canonical features and features in interactions. In the second section, using the hypothesis that a product has a final structure that is the result of an ideal evolution from a set of significant structures, we will propose a Feature Grammar for their representation. Consideration of the semantic and uncertain aspects generalized the Feature Grammar, by producing the Conditional and Fuzzy Feature Grammar. The third section presents a recognition method based on the Conditional and Fuzzy Feature Grammar. Examples and the application illustrate the steps involved and the advantages of this method.

## 2. Feature Representation

### 2.1. TOPOLOGIC AND GEOMETRIC ENTITY GRAPH

A *feature* is a geometric entity defined by its shape and technological characteristics, typically represented by a set of topologically associated faces. Given two finite, non-empty sets $D^{tpl} = \{ D_1^{tpl}, D_2^{tpl}, \cdots D_m^{tpl} \}$ and $D^{geo} = \{ D_1^{geo}, D_2^{geo}, \cdots D_n^{geo} \}$ called the set of topologic domains and the set of geometric domains, respectively, two sets of attributes $A^{tpl} = \{ a_1^{tpl}, a_2^{tpl}, \cdots a_m^{tpl} \}$ and $A^{geo} = \{ a_1^{geo}, a_2^{geo}, \cdots a_n^{geo} \}$, called the set of geometric attributes and the set of topologic attributes, respectively, where each attribute is associated with each domain, and $X = \{ X_1, X_2 \cdots X_i \cdots X_m \}$ a set of features. Then any shape feature $X_i$ can be characterized by a set of faces $F = \{ f_1, f_2 \cdots f_i \cdots f_m \}$ that satisfy a set of topologic and geometric relations. These relations are defined for domains corresponding to the set of topologic and geometric attributes, respectively. Table 1 shows typical cases of those attributes and their respective domains. These relations may be represented by the Topologic and Geometric Entity Graph, defined as follows:

***Definition**: For two given sets:*

*- $F^* = \{ ( f_i, e_i^* ) \mid f_i \in F \}$ where: $F = \{ f_1, f_2 \cdots f_i \cdots f_m \}$ is a set of faces; $e_i^* = ( a_2^{tpl}, a_3^{geo} )$ is an entity associated with each face $f_i$;*

*- $E = \{ ( f_i, f_j, e_{ij} ) \mid f_i, f_j \in F \}$ where: $e_{ij} = ( a_1^{tpl}, a_1^{geo}, a_2^{geo} )$*

*we will call $G = (F^*, E)$, the **Topologic and Geometric Entity Graph.***

TABLE 1. Domains and associated attributes

| | Codes | Attributes | | | | |
|---|---|---|---|---|---|---|
| | | Topology | | Geometry | | |
| | | *Relative positions* | *Type of face* | *Angle* | *Type of adjacency* | *Type of face* |
| | | $a_1^{tpl}$ | $a_2^{tpl}$ | $a_1^{geo}$ | $a_2^{geo}$ | $a_3^{geo}$ |
| Domains | 0 | *adjacent,* | *base* | *convex* | *line* | *plane* |
| | 1 | *non-adjacent* | *side* | *concave* | *non-straight line* | *non-plane* |
| | 2 | *parallel* | *frontal* | *flat* | *other* | |
| | 3 | *virtual adjacent* | | *other* | | |
| | 4 | *same prolongable support* | | | | |
| | 5 | *same non-prolongable support* | | | | |

In the graphical representation of the Topologic and Geometric Entity Graph, the nodes associated with the label $e_i^*=\left(a_2^{tpl},a_3^{geo}\right)$ represent the faces and their Topologic and Geometric relation, and the edges associated with the label $e_{ij}=\left(a_1^{tpl},a_1^{geo},a_2^{geo}\right)$ represent the Topologic and Geometric relation between a pair of faces $\left(f_i,f_j\right)$. Figure 1 shows the representation of the <Slot> feature by the Topologic and Geometric Entity Graph.

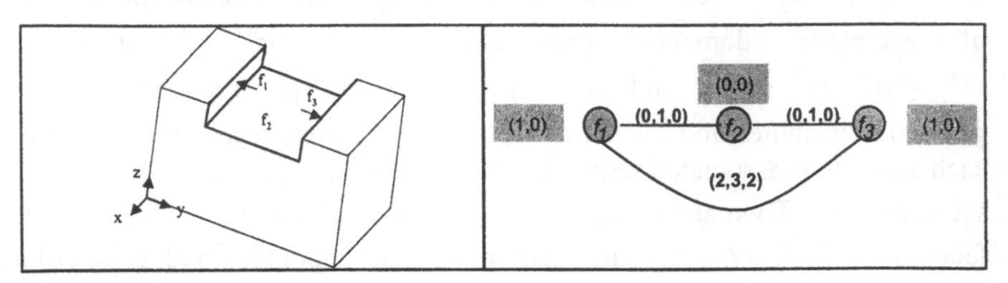


*Figure 1.* <Slot> and its topologic and geometric entity graph

## 2.2. FEATURE GRAMMAR

A *feature language* describes the generation of feature structures, joint elements and attaching elements. A *grammar* provides the finite generic description of this language. Thus we will focus on finding a *feature grammar*, which provides the generic and productive description of the *feature language*. In these conditions, a Feature Grammar can be defined as an 8-plet:

$$G_{Feature}=\left\{V_{structure}^T,V_{joint-tie}^T,V_{structure}^N,V_{joint-tie}^N,S,\nabla,\Lambda,P\right\} \quad (1)$$

where:

- $V_{structure}^T=\{a,b,c\cdots\}$ is the terminal vocabulary of structures, a non-empty finite set;
- $V_{joint-tie}^T=\{0,1,2,\cdots j\cdots m\}\quad m\in N$ is the terminal vocabulary of joint-tie elements, a non-empty finite set;
- $V_{structure}^N=\{A,B,\cdots S\cdots\}$ is the non-terminal vocabulary of structures, a non-empty finite set;
- $V_{joint-tie}^N=\{O,I,II,III\cdots\nabla,\Lambda\}$ is the non-terminal vocabulary of joint-tie elements, a non-empty finite set;
- $S,\nabla,\Lambda$ are respectively the structure, joint, and connection axioms.
- $P:\left\{\begin{bmatrix}\alpha\\ \Gamma_\alpha\\ \Delta_\alpha\end{bmatrix}\rightarrow\begin{bmatrix}\beta\\ \Gamma_\beta\\ \Delta_\beta\end{bmatrix}\right\}$ is a set of production rules

- $V^T_{structure} \cap V^N_{structure} = \varnothing$ ; - $V^T_{jonction-connexion} \cap V^N_{jonction-connexion} = \varnothing$ ;

- $(V^T_{structure} \cup V^N_{structure}) \cap (V^T_{jonction-connexion} \cup V^N_{jonction-connexion}) = \varnothing$

The production rules of the Feature Grammar have the following format:

$$\begin{bmatrix} \alpha \\ \Gamma_\alpha \\ \Delta_\alpha \end{bmatrix} \rightarrow \begin{bmatrix} \beta \\ \Gamma_\beta \\ \Delta_\beta \end{bmatrix} \qquad \begin{matrix} Level\ 1 \\ Level\ 2 \\ Level\ 3 \end{matrix} \tag{2}$$

where:

- $\alpha$ is called the **left-side component matrix**, $\alpha = [\alpha_{ij}]$, $i=1; j=1$;

- $\beta$ is called the **right-side component matrix**, $\beta = [\beta_{ij}]$, $i=1; j=1,2\cdots m$ ; $m$ is the number of components;

- $\Gamma_\alpha$ is called the **left-side joint matrix**, $\Gamma_\alpha = [\Gamma_{\alpha ij}]$, $i=1,2,\cdots n; j=1$; $n$ is the number of attaching elements;

- $\Gamma_\beta$ is called the **right-side joint matrix**, $\Gamma_\beta = [\Gamma_{\beta_{ij}}]$, $i=1,2,\cdots n;\ j=1,2\cdots m$

- $\Delta_\alpha$ is called the **left-side tie-point matrix**, $\Delta_\alpha = [\Delta_{\alpha ij}]$, $i=1,2,\cdots s; j=1$; $s$ is the number of attaching elements;

- $\Delta_\beta$ is called the **right-side tie-point matrix**, $\Delta_\beta = [\Delta_{\beta_{ij}}]$, $i=1,2,\cdots s;$ $j=1,2\cdots n$;

There are three levels of production rules for the Feature Grammar. The first is the component level. These rules have the following format: $[\alpha_{ij}] \rightarrow [\beta_{ij}]$. Or the following format:

$$[A] \rightarrow [\,v_1 \quad v_2 \quad \cdots \quad v_j \quad \cdots \quad v_m\,] \tag{3}$$

where: $\alpha_{ij} = A \in V^N_{structures}$; $\beta = v_1, v_2, \cdots v_j \cdots v_m$; this matrix defines an order relation for these components; $v_j \in V^T_{structures} \cup V^N_{structures}$, is a terminal or non-terminal structure called the *component structure*; $m$ is the number of structures.

The second level is the joint level. These rules have the following format: $[\Gamma_{\alpha ij}] \rightarrow [\Gamma_{\beta_{ij}}]$. Or the following format (4):

where:

- $y_i$ is a joint element

- $t_{ij}$ is an attaching element of the component $j$, defined according to the order in the right-side component matrix that participates in forming the joint element $y_i$.

The third level is the connection level. These rules have the following format: $[\Delta_{\alpha ij}] \rightarrow [\Delta_{\beta_{ij}}]$. Or the following format (5),

where:

- $z_i$ is an attaching element
- $t_{ij}$ is an attaching element of the component j, defined according to the order in the right-side component matrix, that participates in forming the attaching element $z_i$.

$$\begin{bmatrix} y_1 \\ y_2 \\ \vdots \\ y_i \\ \vdots \\ y_n \end{bmatrix} \rightarrow \begin{bmatrix} t_{11} & t_{12} & \cdots & t_{1j} & \cdots & t_{1m} \\ t_{21} & t_{22} & \cdots & t_{2j} & \cdots & t_{2m} \\ \vdots & \vdots & \ddots & \vdots & \ddots & \vdots \\ t_{i1} & t_{i2} & \cdots & t_{ij} & \cdots & t_{im} \\ \vdots & \vdots & \ddots & \vdots & \ddots & \vdots \\ t_{n1} & t_{n2} & \cdots & t_{nj} & \cdots & t_{nm} \end{bmatrix} \quad (4) \quad \begin{bmatrix} z_1 \\ z_2 \\ \vdots \\ z_i \\ \vdots \\ z_s \end{bmatrix} \rightarrow \begin{bmatrix} t_{11} & t_{12} & \cdots & t_{1j} & \cdots & t_{1m} \\ t_{21} & t_{22} & \cdots & t_{2j} & \cdots & t_{2m} \\ \vdots & \vdots & \ddots & \vdots & \ddots & \vdots \\ t_{i1} & t_{i2} & \cdots & t_{ij} & \cdots & t_{im} \\ \vdots & \vdots & \ddots & \vdots & \ddots & \vdots \\ t_{n1} & t_{n2} & \cdots & t_{nj} & \cdots & t_{nm} \end{bmatrix} \quad (5)$$

### *2.2.1. Conditional Feature Grammar*

The Feature Grammar represents the purely *syntactic* side. It does not always allow to express the full complexity of structural relations between the primitive elements composing a feature. If a syntax rule must meet mandatory conditions before being applied, then a Conditional Feature Grammar is defined as follows:

$$G^C_{Features} = \{G_{Features}, A^{geo-tpl}, D^{geo-tpl}, C\} \quad (6)$$

where: - $G_{Features}$ is the Feature Grammar; $C = \begin{bmatrix} C_{\alpha \rightarrow \beta} \\ C_{\Gamma_\alpha \rightarrow \Gamma_\beta} \\ C_{\Delta_\alpha \rightarrow \Delta_\beta} \end{bmatrix}$ are the three levels of semantic conditions; $A_{geo-tpl}$ is the set of geometric and topologic attributes; $D_{geo-tpl}$ is the set of geometric and topologic domains.

### *2.2.2. Fuzzy Feature Grammar*

The elements of $V^T_{structure} = \{a, b, c \cdots\}$ that have a certain property, such as the *non-existence of virtual edges in a terminal a*, make up a subset of $V^T_{structure}$. If some elements of $V^T_{structure}$ do not have this property in an absolute manner, we may choose to indicate the extent to which each element has the property. Thus we define a fuzzy subset of $V^T_{structure}$. The fuzzy subset of $V^T_{structure}$ is defined by a membership function that associated with each element $a$ of $V^T_{structure}$, the extent (between 0 and 1) to which $a$ is a member of this subset: $\mu_V : V^T_{structure} \rightarrow [0,1]$. In the presence of a rule in format $[\alpha_{ij}] \rightarrow [\beta_{ij}]$, if $[\beta_{ij}]$ is characterized by the membership function $\mu_\beta = min(\mu_{v_1}, \mu_{v_2}, \cdots \mu_{v_j} \cdots \mu_{v_n})$, and if the rule $[\alpha_{ij}] \rightarrow [\beta_{ij}]$ is characterized by the membership function $\mu_{\alpha \to \beta}$, then $[\alpha_{ij}]$ is defined by the membership function $\mu_\alpha = min(\mu_\beta, \mu_{\alpha \to \beta})$. In this way we can determine the membership function of each non-terminal, and therefore of axiom *S*. The Feature Grammar that benefits from these characteristics is called the Fuzzy Feature Grammar.

*2.2.3. Application*

Given a set of features, Figure 2:

$X = \{Step, Slot, Blind\ Slot, Hole, Pocket, Blind\ Step, Simple\ Blind\ Slot, Partial\ Hole, Hole\}$

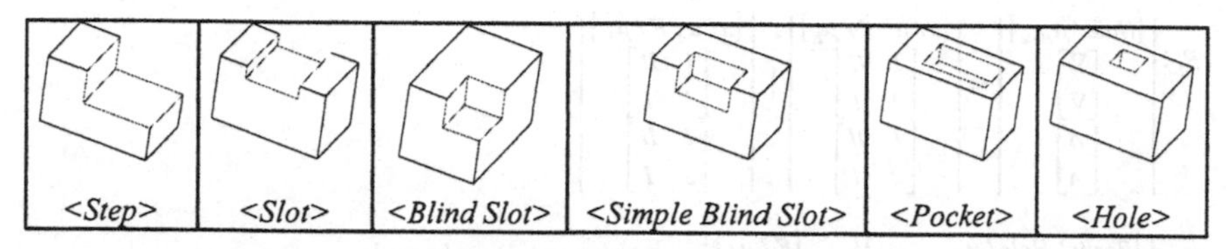


*Figure 2.* Features of the set X

A feature can be represented by the Selected Topologic and Geometric Entity Graph, defined as follows:

***Definition:*** *For two given sets:*

$F^* = \{(f_i, e_i^*) \mid f_i \in F\}$ *where:* $F = \{f_1, f_2 \cdots f_i \cdots f_m\}$ *is a set of faces;*

$e_i^* = (a_2^{tpl}, a_3^{geo})$ *is an entity associated with each face* $f_i$ *(see Table 1);*

$E = \{(f_i, f_j, e_{ij}) \mid f_i, f_j \in F\}$ *where:* $e_{ij} = (adjacentes, concave, a_2^{geo})$ *(see Table 1)*

*we call* $G = (F^*, E)$ *the **Selected Topologic and Geometric Entity Graph**.*

For example, the representation of *<Slot>* and *<Blind Slot>* features by the Selected Topologic and Geometric Entity Graph is illustrated in Figure 3.

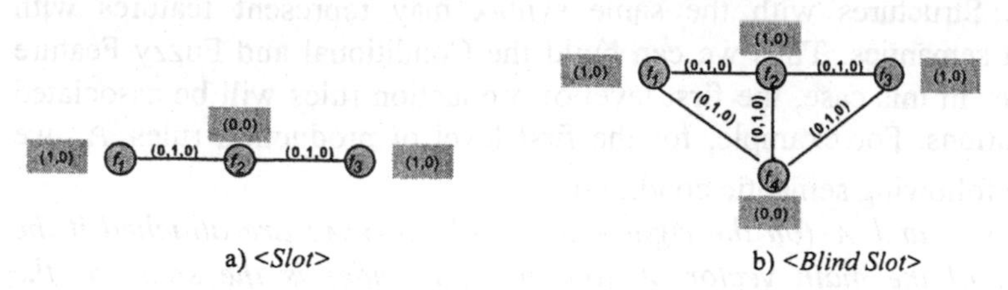


*Figure 3.* Selected Topologic and Geometric Entity Graph

Two Fuzzy Feature Grammars $G^C_{Features}$ are inferred for the feature classes: $C_1^X = \{Step,\ Slot,\ Partial\ Hole,\ Hole\}$ and $C_2^X = \{Blind\ Step,\ Simple\ Blind\ Slot, Blind\ Slot,\ Pocket\}$ respectively. For the feature class $C_1^X$, we have: $G_{Features} = \{V^T_{structure}, V^T_{jonction-connexion}, V^N_{structure}, V^N_{jonction-connexion}, S, \nabla, \Lambda, P,\}$ where:

- $V^{T}_{structure} = \{a/\mu_a\}$ a/1 = (1,0) 1 —(0,1,0)— 2 (1,0)

- $V^{T}_{jonction-connexion} = \{0,1,2\}$ $m \in N$ ;

- $V^{N}_{structure} = \left\{ \begin{matrix} D/\mu_D, E/\mu_E, F/\mu_F, Step/\mu_{Step}, Slot/\mu_{Slot}, Partial\ Hole/\mu_{Partial\ Hole}, \\ Hole/\mu_{Hole}, Feature/\mu_{Feature} \end{matrix} \right\}$

- $V^{N}_{jonction-connexion} = \{O, I, II, \nabla, \Lambda\}$; - $S = Feature/\mu_{Feature}, \nabla, \Lambda$ ; - $P$ :

$$P_0: \begin{vmatrix} [Feature/\mu_{Feature}] \\ [\varnothing] \\ [\varnothing] \end{vmatrix} \rightarrow \begin{vmatrix} [Step/\mu_{Step}] \\ [\varnothing] \\ [\varnothing] \end{vmatrix} \mid \begin{vmatrix} [Slot/\mu_{Slot}] \\ [\varnothing] \\ [\varnothing] \end{vmatrix} \mid \begin{vmatrix} [Partial\ Hole/\mu_{Partial\ Hole}] \\ [\varnothing] \\ [\varnothing] \end{vmatrix} \mid \begin{vmatrix} [Hole/\mu_{Hole}] \\ [\varnothing] \\ [\varnothing] \end{vmatrix}$$

$$P_1: \begin{vmatrix} [Hole/\mu_{Hole}] \\ \begin{bmatrix} \nabla \\ \nabla \end{bmatrix} \\ \begin{bmatrix} \Lambda \\ \Lambda \end{bmatrix} \end{vmatrix} \rightarrow \begin{vmatrix} [a/\mu_a\ E/\mu_E] \\ \begin{bmatrix} I & II \\ 2 & I \end{bmatrix} \\ \begin{bmatrix} I & II \\ 2 & I \end{bmatrix} \end{vmatrix} \mid \begin{vmatrix} [a/\mu_a\ F/\mu_F] \\ \begin{bmatrix} I & II \\ 2 & I \end{bmatrix} \\ \begin{bmatrix} I & II \\ 2 & I \end{bmatrix} \end{vmatrix}$$

$$P_2: \begin{vmatrix} [Partial\ Hole/\mu_{Partial\ Hole}] \\ [\varnothing] \\ \begin{bmatrix} \Lambda \\ \Lambda \end{bmatrix} \end{vmatrix} \rightarrow \begin{vmatrix} [F/\mu_F] \\ [\varnothing] \\ \begin{bmatrix} I \\ II \end{bmatrix} \end{vmatrix} \quad P_3: \begin{vmatrix} [Slot/\mu_{Slot}] \\ [\varnothing] \\ \begin{bmatrix} \Lambda \\ \Lambda \end{bmatrix} \end{vmatrix} \rightarrow \begin{vmatrix} [E/\mu_E] \\ [\varnothing] \\ \begin{bmatrix} I \\ II \end{bmatrix} \end{vmatrix}$$

$$P_4: \begin{vmatrix} [Step/\mu_{Step}] \\ [\varnothing] \\ \begin{bmatrix} \Lambda \\ \Lambda \end{bmatrix} \end{vmatrix} \rightarrow \begin{vmatrix} [D/\mu_D] \\ [\varnothing] \\ \begin{bmatrix} I \\ II \end{bmatrix} \end{vmatrix} \quad P_5: \begin{vmatrix} [F/\mu_F] \\ [\nabla] \\ \begin{bmatrix} I \\ II \end{bmatrix} \end{vmatrix} \rightarrow \begin{vmatrix} [a/\mu_a\ E/\mu_E] \\ [2\ \ I] \\ \begin{bmatrix} I & 0 \\ 0 & II \end{bmatrix} \end{vmatrix} \mid \begin{vmatrix} [a/\mu_a\ F/\mu_F] \\ [2\ \ I] \\ \begin{bmatrix} I & 0 \\ 0 & II \end{bmatrix} \end{vmatrix}$$

$$P_6: \begin{vmatrix} [E/\mu_E] \\ [\nabla] \\ \begin{bmatrix} I \\ II \end{bmatrix} \end{vmatrix} \rightarrow \begin{vmatrix} [a/\mu_a\ D/\mu_D] \\ [2\ \ I] \\ \begin{bmatrix} I & 0 \\ 0 & II \end{bmatrix} \end{vmatrix} \quad P_7: \begin{vmatrix} [D/\mu_D] \\ [\varnothing] \\ \begin{bmatrix} I \\ II \end{bmatrix} \end{vmatrix} \rightarrow \begin{vmatrix} [a/\mu_a] \\ [\varnothing] \\ \begin{bmatrix} 1 \\ 2 \end{bmatrix} \end{vmatrix}$$

The Fuzzy Feature Grammar represents the syntactic and fuzzy aspect of features. Structures with the same syntax may represent features with different semantics. Thus we can build the Conditional and Fuzzy Feature Grammar. In this case, the first level of production rules will be associated by conditions. For example, for the first level of production rules $P_6$, we have the following semantic condition:

*structures $b$ and $A$ (on the right side of rule $B \rightarrow bA$) are attached if the direction of the main vector $A$ (on the right side) is the same as the direction of the vector $\vec{n}_1 \wedge \vec{n}_2$ of $b$, where $1$ and $2$ represent the attaching elements of $b$.*

The previous condition is used in a similar fashion for rules $P_1, P_3, P_5$. We will have the following condition for the first level of production rules:

*the direction of the main vector of $A$ (left side of the rule $A \rightarrow b$) is initialized from $\vec{n}_1 \wedge \vec{n}_2$ of $b$, where $1$ and $2$ represent the attaching elements of $b$.*

There are no semantic conditions to be satisfied for the other rules. For the second class, we have: $G^{C}_{Features} = \{V^{T}_{structure}, V^{T}_{jonction-connexion}, V^{N}_{structure}, V^{N}_{jonction-connexion}, S, \nabla, \Lambda, P\}$ where:

- $V^{T}_{structure} = \{b/\mu_b\}$; - $V^{T}_{jonction-connexion} = \{0,1,2,3\}$ $m \in N$

- $V^{N}_{structure} = \left\{ \begin{matrix} A/\mu_A, B/\mu_B, \textit{Blind Step}/\mu_{Blind\ Step}, \\ \textit{Blind Slot}/\mu_{Blind\ Slot}, \textit{Pocket}/\mu_{Pocket}, \textit{Feature}/\mu_{Feature} \end{matrix} \right\}$



- $V^{N}_{jonction-connexion} = \{O, I, II, III, \nabla, \Lambda\}$; - $S = \textit{Feature}/\mu_{Feature}, \nabla, \Lambda$; - $P$ :

$P_0$ :

$$\left|\begin{matrix} [\textit{Feature}/\mu_F] \\ [\varnothing] \\ [\varnothing] \end{matrix}\right| \rightarrow \left|\begin{matrix} [\textit{Blind Step}/\mu_{B.Step}] \\ [\varnothing] \\ [\varnothing] \end{matrix}\right| \,|\, \left|\begin{matrix} [\textit{Simple Blind Slot}/\mu_{SBS}] \\ [\varnothing] \\ [\varnothing] \end{matrix}\right| \,|\, \left|\begin{matrix} [\textit{Blind Slot}/\mu_{B.Slot}] \\ [\varnothing] \\ [\varnothing] \end{matrix}\right| \,|\, \left|\begin{matrix} [\textit{Pocket}/\mu_P] \\ [\varnothing] \\ [\varnothing] \end{matrix}\right|$$

$$P_1 : \left|\begin{matrix} [\textit{Pocket}/\mu_{Pocket}] \\ \begin{bmatrix} \nabla \\ \nabla \\ \nabla \end{bmatrix} \\ \begin{bmatrix} \Lambda \\ \Lambda \\ \Lambda \end{bmatrix} \end{matrix}\right| \rightarrow \left|\begin{matrix} [b/\mu_b B/\mu_B] \\ \begin{bmatrix} 1 & I \\ 2 & III \\ 3 & II \end{bmatrix} \\ \begin{bmatrix} 1 & I \\ 2 & III \\ 3 & II \end{bmatrix} \end{matrix}\right| \,|\, \left|\begin{matrix} [b/\mu_b C/\mu_C] \\ \begin{bmatrix} 1 & I \\ 2 & III \\ 3 & II \end{bmatrix} \\ \begin{bmatrix} 1 & I \\ 2 & III \\ 3 & II \end{bmatrix} \end{matrix}\right|$$

$$P_2 : \left|\begin{matrix} [\textit{Blind Slot}/\mu_{Blind\ Slot}] \\ [\varnothing] \\ \begin{bmatrix} \Lambda \\ \Lambda \\ \Lambda \end{bmatrix} \end{matrix}\right| \rightarrow \left|\begin{matrix} [C/\mu_C] \\ [\varnothing] \\ \begin{bmatrix} I \\ II \\ III \end{bmatrix} \end{matrix}\right| \quad P_3 : \left|\begin{matrix} [\textit{Simple Blind Slot}/\mu_{Simple\ Blind\ Slot}] \\ [\varnothing] \\ \begin{bmatrix} \Lambda \\ \Lambda \\ \Lambda \end{bmatrix} \end{matrix}\right| \rightarrow \left|\begin{matrix} [B/\mu_B] \\ [\varnothing] \\ \begin{bmatrix} I \\ II \\ III \end{bmatrix} \end{matrix}\right|$$

$$P_4 : \left|\begin{matrix} [\textit{Blind Step}/\mu_{Blind\ Step}] \\ [\varnothing] \\ \begin{bmatrix} \Lambda \\ \Lambda \\ \Lambda \end{bmatrix} \end{matrix}\right| \rightarrow \left|\begin{matrix} [A/\mu_A] \\ [\varnothing] \\ \begin{bmatrix} I \\ II \\ III \end{bmatrix} \end{matrix}\right| \quad P_5 : \left|\begin{matrix} [C/\mu_B] \\ \begin{bmatrix} I \\ \nabla \end{bmatrix} \\ \begin{bmatrix} I \\ II \\ III \end{bmatrix} \end{matrix}\right| \rightarrow \left|\begin{matrix} [b/\mu_b B/\mu_A] \\ \begin{bmatrix} 1 & I \\ 3 & II \end{bmatrix} \\ \begin{bmatrix} 1 & I \\ 2 & 0 \\ 0 & III \end{bmatrix} \end{matrix}\right| \,|\, \left|\begin{matrix} [b/\mu_b C/\mu_C] \\ \begin{bmatrix} 1 & I \\ 3 & II \end{bmatrix} \\ \begin{bmatrix} 1 & I \\ 2 & 0 \\ 0 & III \end{bmatrix} \end{matrix}\right|$$

$$P_6 : \left|\begin{matrix} [B/\mu_B] \\ \begin{bmatrix} I \\ \nabla \end{bmatrix} \\ \begin{bmatrix} I \\ II \\ III \end{bmatrix} \end{matrix}\right| \rightarrow \left|\begin{matrix} [b/\mu_b A/\mu_A] \\ \begin{bmatrix} 1 & I \\ 3 & II \end{bmatrix} \\ \begin{bmatrix} 1 & I \\ 2 & 0 \\ 0 & III \end{bmatrix} \end{matrix}\right| \quad P_7 : \left|\begin{matrix} [A/\mu_A] \\ [\varnothing] \\ \begin{bmatrix} I \\ II \\ III \end{bmatrix} \end{matrix}\right| \rightarrow \left|\begin{matrix} [b/\mu_b] \\ [\varnothing] \\ \begin{bmatrix} 1 \\ 2 \\ 3 \end{bmatrix} \end{matrix}\right|$$

The preceding production rules are associated by conditions similar to the previous class. In the case of this feature class, the mixed product $n_1 \cdot (n_2 \wedge n_3)$ is considered. Thus the structures (in the right part) are attached if the sign of the mixed product $n_1 \cdot (n_2 \wedge n_3)$ does not change.

## 3. Feature Recognition Method

Using the principles discussed above, we have developed a new feature recognition method. The flowchart in Figure 4 shows the main phases of this method.

The first phase, called ***Regioning***, consists in identifying the potential zones for the birth of features.

The second phase, called ***Virtual Extension***, consists in building links and virtual faces.

The third phase, called ***Structuring***, consists in transforming the region into a structure compatible with the structure of the features represented by the Feature Grammar.

The fourth phase, called ***Identification***, consists in identifying the features in these zones.

The fifth phase, called, ***Modeling***, consists in representing the model either by regions or by features.

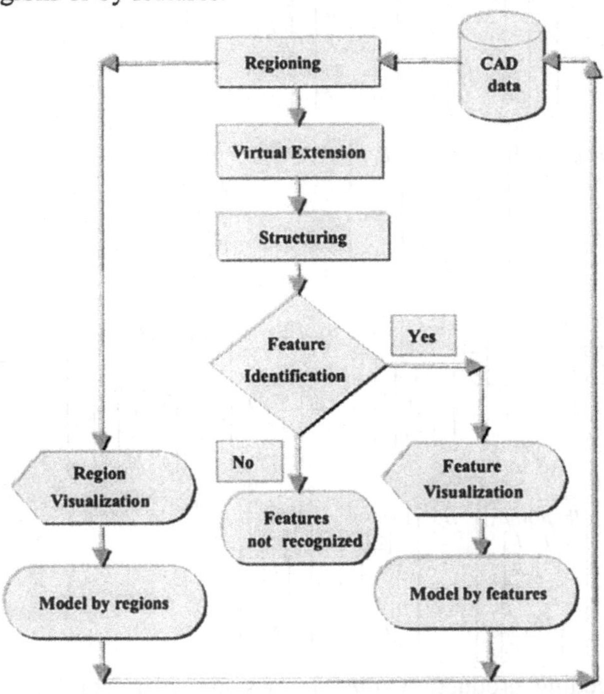


*Figure 4.* Flowchart of the Feature Recognition method

### 3.1. REGIONING

A *region* defines a potential area of a part where either canonical features or features in interaction may be recognized. During the interaction, features may lost their concavity. As a result, some faces of features in interaction are not identified during the recognition phase. In this case, the potential region for feature recognition is expanded with concave border faces ***(local expansion principle)***. Thus we can define a region of the part as a set of connected faces characterized by their concavity or their convexity that may be transformed into concavity. Furthermore, the interaction between features may produce neighboring regions that may be either adjacent or recoverable. If $R_1,\cdots R_i,R_{i+1}\cdots R_n$ is a set of regions, then a macro-region $R$ may be defined by grouping a set of neighboring regions ***(global expansion principle)***. For example, the part shown in Figure 5a contains two regions. The first region comprises the concave faces 1, 2, 3, and the convex face 7, which may be transformed to concave by the virtual extension towards face 1. The second region comprises the concave faces 4, 5, 6, and the convex face 7. In this case, the convex face 7 can be transformed to concave by virtual extension towards face 6. These two regions share face 7. As a result, a macro-region is defined by $\langle \textit{macro}-\textit{région}_1 \rangle \rightarrow \langle \textit{région}_{11} \rangle \langle \textit{région}_{12} \rangle$ where $\langle \textit{région}_{11} \rangle$ and $\langle \textit{région}_{12} \rangle$ are the first and second regions, respectively, of the first macro-region $\langle \textit{macro}-\textit{région}_1 \rangle$. The part, Figure 5b, comprises a region that includes concave faces 1, 2, 3, 4, 5, 6.

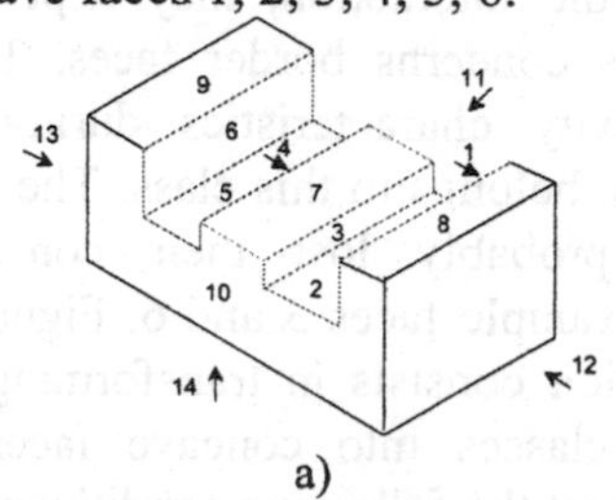

a)

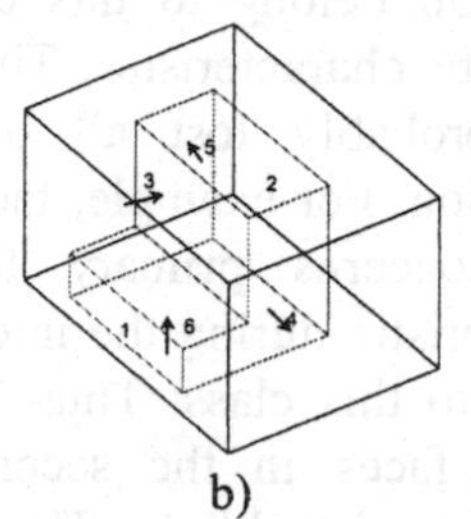

b)

*Figure 5.* Example of regions

Thus the Regioning procedure involves three subphases:

*1. Search for regions.* The faces characterized by concavity are called the primary faces of the region. This phase consists in extracting the set of primary faces from the part and grouping them into regions.

*2. Local expansion.* Based on the local expansion principle, the border faces, called the secondary faces, characterized by their convexity susceptible to be transformed into concavity, are assigned to the region in question.

*3. Creating macro-regions.* Based on the global expansion principle, regions are grouped together into macro-regions.

***Example 1.*** Given two parts, Figures 5a and 5b. For the part in Figure 5a, we can write:

$$\langle \textit{pièce} \rangle \rightarrow \langle \textit{macro}-\textit{région}_1 \rangle \tag{7}$$

$$\langle macro-région_1\rangle \rightarrow \langle région_{11}\rangle\langle région_{12}\rangle \tag{8}$$

$$\begin{aligned}&\langle région_{11}\rangle \rightarrow \langle faces-primaires_{11}\rangle\langle faces-secondaires_{11}\rangle\\&\langle région_{12}\rangle \rightarrow \langle faces-primaires_{12}\rangle\langle faces-secondaires_{12}\rangle\end{aligned} \tag{9}$$

$$\begin{aligned}&\langle faces-primaires_{11}\rangle \rightarrow \langle f_1\rangle\langle f_2\rangle\langle f_3\rangle\\&\langle faces-primaires_{12}\rangle \rightarrow \langle f_4\rangle\langle f_5\rangle\langle f_6\rangle\end{aligned} \tag{10}$$

$$\begin{aligned}&\langle faces-secondaires_{11}\rangle \rightarrow \langle f_7\rangle\\&\langle faces-secondaires_{12}\rangle \rightarrow \langle f_7\rangle\end{aligned} \tag{11}$$

### 3.2. VIRTUAL EXTENSION

The faces in a region can be divided into three classes:

***Class 1:*** *primary faces characterized by concavity only;*

***Class 2:*** *secondary faces characterized by convexity only;*

***Class 3:*** *primary faces characterized by both concavity and convexity.*

The first class concerns faces that resist to interaction. For example, faces 1, 2, 3 and 4, 5, 6 in the part in Figure 5a and faces 1, 2, 4, 5 of the part in Figure 5b belong to this class. Despite the interaction, they kept their concavity characteristic. The second class concerns border faces. These faces probably lost all of their concavity characteristics during the interaction. For example, face 7, Figure 5a, belongs to this class. The third class concerns primary faces, which probably lost their concavity characteristic during the interaction. For example faces 3 and 6, Figure 5b belong to this class. Thus Virtual Extension consists in transforming the convex faces in the second and third classes into concave faces by generating virtual links. These links must meet the following conditions:

***Condition 1:*** *A pair of virtually extended faces belongs to a group of extended faces, if an only if their virtual adjacency is concave, Figure 6a.*

***Condition 2:*** *Given two convex edges $e_i$ and $e_j$. Faces X and Y are adjacent in $e_i$, and faces Z and V are adjacent in $e_j$. If the virtual extension of faces X and Y and the virtual extension of faces Z and W create two edges, called virtual-virtual edges, then those edges are considered simultaneously, Figure 6b.*

***Condition 3:*** *If condition 2 is false, then between two adjacent convex faces, one and only one face can be virtually extended by forming an edge, called a virtual-real edge. This edge will jointly belong to the extended virtual face and the real virtually intersected face, Figure 6c.*

***Condition 4:*** *In a set, if each of the virtually extended faces forms virtual-real edges, and if the virtual extensions intersect, then the selection of one of those faces penalizes the others, Figure 6d.*

***Condition 5:*** *The virtually extended face does not intersect the interior parts of the part faces, Figure 6e.*

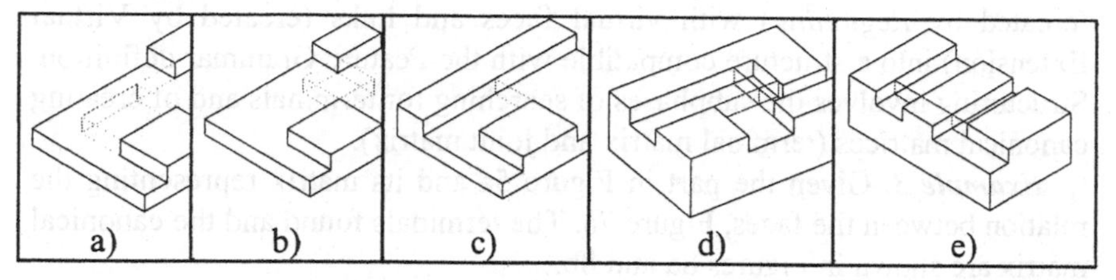


*Figure 6.* Illustrations of conditions

The recomposition of the virtual face is a special case for the generation of virtual links. The interaction between the canonical features may break a face down into a group of small faces. Then those small faces may be unified into a virtual face. The minimal conditions for a group of faces to be unified are as follows (Marefat and Kashyap 1990):

***Condition 1:*** *The faces must have the same support surface;*

***Condition 2:*** *The normals of the faces must have the same directions;*

***Condition 3:*** *The unified face must not intersect the interior parts of the part faces.*

The first and second conditions show that the faces have the same geometry, while the third shows that the unified face cannot be destroyed by the other faces of the part. We define an order relation between the generation of virtual faces and virtual links: first, we try to generate the virtual faces by unifying the groups of faces that meet the minimal conditions. If a virtual face is generated from the unification of a group of faces, then any face in that group must be virtually extended to form virtual links.

***Example 2.*** Given the parts in Figure 5a and Figure 5b. For the part in Figure 5a, face 7 may be transformed into a concave face by generating virtual links with faces 6 and 1 matrix, Figure 7a , while for the part in Figure 5b, face 3 and face 6 may be transformed into concave faces by generating virtual links with faces 1 and 2, respectively, Figure 7b .

a)

| Faces | 1 | 2 | 3 | 4 | 5 | 6 | 7 |
|---|---|---|---|---|---|---|---|
| 1 | | 0.1.0 | | | | | |
| 2 | 0.1.0 | | 0.1.0 | | | | |
| 3 | | 0.1.0 | | | | | |
| 4 | | | | | 0.1.0 | | |
| 5 | | | | 0.1.0 | | 0.1.0 | |
| 6 | | | | | 0.1.0 | | |
| 7 | 3.1.0 | | | | | 3.1.0 | |

b)

| Faces | 1 | 2 | 3 | 4 | 5 | 6 |
|---|---|---|---|---|---|---|
| 1 | | 0.1.0 | | 0.1.0 | 0.1.0 | |
| 2 | 0.1.0 | | | 0.1.0 | 0.1.0 | |
| 3 | 3.1.0 | | | 0.1.0 | 0.1.0 | |
| 4 | 0.1.0 | 0.1.0 | 0.1.0 | | | 0.1.0 |
| 5 | 0.1.0 | 0.1.0 | 0.1.0 | 0.1.0 | | 0.1.0 |
| 6 | | 3.1.0 | | 0.1.0 | 0.1.0 | |

*Figure 7.* Matrix representing the real or virtual concavity between faces

### 3.3. STRUCTURING

Structuring consists in converting the representation of macro-regions (created by *Regioning*) with virtual faces and links (created by Virtual Extension) into a structure compatible with the Feature Grammar definition. Structuring involves the subphases of searching for terminals and of creating canonical matrices (terminal matrix and joint matrix).

***Example 3.*** Given the part in Figure 5a and its matrix representing the relation between the faces, Figure 7a. The terminals found and the canonical matrix are shown in Figures 8a and 8b.

| No. | Faces | terminals | Coloring | μ(terminal) |
|---|---|---|---|---|
| 1 | 4,5 | a | 1,2 | 1 |
| 2 | 5,6 | a | 1,2 | 1 |
| 3 | 6,7 | a | 1,2 | 0.8 |
| 4 | 1,7 | a | 2,1 | 0.8 |
| 5 | 1,2 | a | 2,1 | 1 |
| 6 | 2,3 | a | 2,1 | 1 |

a)

| | | *a* | *a* | *a* | *a* | *a* | *a* |
|---|---|---|---|---|---|---|---|
| $\mu$ | | *1.0* | *1.0* | *0.8* | *0.8* | *1.0* | *1.0* |
| | *1* | 2 | *1* | | | | |
| | *2* | | 2 | *1* | | | |
| | *3* | | | 2 | *1* | | |
| *jonction* | *4* | | | | 2 | *1* | |
| | *5* | | | | | 2 | *1* |
| | *6* | *1* | | | | | |
| | *7* | | | | | | 2 |

b)

*Figure 8.* Terminals and the canonical matrix of the recognized macro-region

These terminals are of type *a*. The terminal with faces 4 and 5 is colored first. Face 4 is colored with color 1 and face 5 with color 2. The terminal with faces 5 and 6 is colored second. As face 5 is colored with color 2 in a previous terminal, in this terminal it is colored with color 1. As a result, face 6 is colored with color 2. The membership function $\mu_V(a)$ associated with each terminal shows the extent to which that terminal is similar to terminal *a*. The terminal with a virtual concave edge is associated with $\mu_V(a)=0.8$.

***Example 4.*** Given the part in Figure 5b and its matrix representing the relation between the faces, Figure 7b. The terminals found and the canonical matrix are shown in Figure 8.

The transformation of the maximum subrelations into colored terminals is shown in Figure 9b. These are type *b* terminals. For any terminal, the face considered as the base is colored with color 1. Thus face 1 is colored with color 1. The other faces are colored using the same procedure. For example, the terminal with faces 1, 2 and 4 is colored first (face 1 is already colored). Face 2 is colored with color 2 and face 4 with color 3. The terminal with faces 1, 2, and 5 is colored second. As face 2 was colored with color 2 in a previous terminal, in this terminal it is colored with color 3. As a result, face 5 is colored with color 3. Thus we continue coloring for all terminals. Terminals 5 and 6, with faces 6, 5, 2 and 6, 2, 4 respectively, are not colored because faces 5, 2, 4 are already colored with colors 2 or 3. Furthermore, we consider that there is only one face with the characteristic base in a macro-region or a region. The membership function $\mu_V(a)$ associated with each terminal shows the extent to which the terminal is similar to terminal $b$. In the case of non-colored faces, this function takes the value of $\mu_V(b)=0.3$. The symbol (*) shows that terminals 5 and 6 participate in forming joints 2,4,5,6, but those terminals are not colored.

| No. | Faces | terminals | Coloring | μ(terminal) |
|---|---|---|---|---|
| 1 | 2,1,4 | b | 2,1,3 | 1 |
| 2 | 2,1,5 | b | 3,1,2 | 1 |
| 3 | 5,1,3 | b | 3,1,2 | 0.8 |
| 4 | 3,1,4 | b | 3,1,2 | 0.8 |
| 5 | 5,6,2 | b | * | 0.3 |
| 6 | 2,6,4 | b | * | 0.3 |

a)

| | | *b* | *b* | *b* | *b* | *b* | *b* |
|---|---|---|---|---|---|---|---|
| $\mu$ | | *1.0* | *1.0* | *0.8* | *0.8* | *0.3* | *0.3* |
| | *1* | *1* | *1* | *1* | *1* | | |
| | *2* | *3* | *2* | | | * | * |
| *joint* | *3* | | *3* | *2* | | | |
| | *4* | | | *3* | *2* | * | |
| | *5* | *2* | | | *3* | * | * |
| | *6* | | | | | * | * |

b)

*Figure 9.* Terminals and the canonical matrix of the recognized macro-region

### 3.4. IDENTIFICATION

The representation and the resolution of the recognition problem, for either canonical features or features in interaction, is shown in a state graph. Consider Feature Grammar G and a feature structure to be analyzed, represented by the canonical matrix (terminals matrix and joint matrix). A state is any possible rewriting of canonical matrices (terminal matrix and joint matrix) by one or more production rules of Feature Grammar G. States will constitute the nodes of the state graph. An arc will connect two states if we can pass from one state to another by a single application of a production rule of Feature Grammar G or if the structure is updated after the recognition of a canonical feature in a region of features in interaction. The structure update consists in:

- *virtually removing the faces and edges that are exclusively part of the recognized feature and considering the Virtual Extension conditions;*
- *reusing the faces shared by several features, knowing that if n virtual links are created to a real face, then that face will be shared by n+1 features (sharing principle);*
- *creating new terminals knowing that if a face and/or edges are virtually erased in an order 3 terminal, then the terminal may be transformed into an order 2 terminal(s) (embedding principle).*

Each arc is labeled with a letter representing either the production rule applied, or the notation for updating the structure. A state can be in one of the following conditions:

Condition 0: It is initial.
Condition 1: It has not been built yet.
Condition 2: It was just built by applying a production rule or by updating the structure.
Condition 3: It is a dead-end. Some nodes are dead-end structures, for which no production rules are applicable.
Condition 4: It is terminal. A state is terminal if it represents either the recognized canonical feature, or features in interaction.
By applying the *Virtual Extension conditions, the sharing principle, the embedding principle, and the union principle*, we have developed the following heuristic for identifying canonical features and/or features in interaction:

Step 1: The initial structure represented by the canonical matrix defines the initial state (Condition 0). The initial state is added to an ordered list of states according to the order relation of their structures.
Step 2: If the list is empty, then failure; the procedure halts. Otherwise, we choose a state from the list using the following rule: *last entered in the list, first out*. It represents the current state $w_i$. We remove it from the list and add it to the group of processed states W.
Step 3: If the current state $w_i$ is a terminal state (Condition 4), then success. The production rules can be found using the pointers created in Step 5.
Step 4: If production rules, set in an order relation, can be applied, then we develop the current state $w_i$ by creating the set of successor states *Y*, otherwise if a feature is recognized, then the structure can be updated *(according to the **Virtual Extension conditions**, the **sharing principle**, the **embedding principle**, the **union principle**)* by creating only a successor state in *Y* (Condition 2), otherwise to 2 (Condition 3).
Step 5: For all states Y $w_j$, we put $w_j$ in list if it does not already belong there and if it does not belong to the group of processed states. If we put $w_j$ in the list, then we create a pointer $w_j$ to $w_i$ along with the production rule used.
Step 6: Go to 2.

In the list, *the last* represents a *maximal state*. A state is maximal if it is hierarchically greater than the other states. Between two states at the same hierarchical level, the state from which a new state is formed and that is characterized by a value greater of the membership function $\mu$, is considered as a maximal state.

***Example 5.*** Given the part in Figure 5a, its canonical matrix, Figure 7a and the inferred Conditional and Fuzzy Feature Grammar $G^{C}_{Features}$ for the feature class $C_2^X = \{Step, Slot, Partial\ Hole, Hole\}$. We find three "Slot" features, each of which is comprised by the following terminals, respectively: 5 and 6 for the first, 1 and 2 for the second, and 3 and 4 for the third. Each of these features is associated with the respective value of the membership function: $\mu = 1$ for the first and second features and $\mu = 0.8$ for the third.

***Example 6.*** Given the part in Figure 5b, its canonical matrix, Figure 8b, and the inferred Conditional and Fuzzy Feature Grammars $G^{C}_{Features}$ for the feature classes: $C_2^X = \{Step, Slot, Partial\ Hole, Hole\}$-$C_2^X = \{Blind\ Step, Simple\ Blind\ Slot, Blind\ Slot, Pocket\}$. The canonical matrix (Figure 8b) represents the initial state. By successively applying rules $P_7, P_6, P_5, P_1, P_0$ (developed on the right) in iterations 1, 2, 3, 4, 5, we obtain the "Pocket" recognition feature with $\mu_{Pocket} = 0.8$. This feature is comprised of faces 1, 2, 3, 4, 5. The structure is updated in the ***sixth iteration***. According to Condition 3 of the **Virtual Extension Conditions,** the virtual extension of face 3 penalizes the virtual extension of face 6. Thus virtual edge 6, 2 will no longer exist in the fifth and sixth terminals. According to the **embedding principle,** the fifth terminal of type *b* comprising faces 2, 5, 6 will be transformed into two type *a* terminals comprising faces 2, 5 and 5, 6, respectively. Similarly, the sixth terminal of type *b* comprising faces 2, 4, 6 will be transformed into two type *a* terminals comprising faces 2, 4 and 6, 4, respectively. As both terminals comprising faces 2, 5 and faces 2, 4, respectively, are already used in the two terminals comprising faces (1, 2, 5) and faces (1, 2, 4), respectively of the "Pocket" feature, then they will not be considered for the recognition of a new feature. Thus the two terminals comprising faces 5, 6 and 6, 4, respectively, remain to be considered. According to the **sharing principle**, face 1 will be shared with other features because a virtual link is built to it. Faces 2 and 3 will be erased because these faces do not exist in the remaining terminals. As a result, according to the embedding principle, we will have two terminals comprising faces 1, 5 and 4, 1, respectively. Finally, we have four type *a* terminals, comprising faces (1, 5), (5, 6), (6, 4), (4, 1), respectively. The matrix in Figure 10 shows the updated structure.

| | terminals | | | | |
|---|---|---|---|---|---|
| | a | a | a | a | |
| μ | 1 | 1 | 1 | 1 | |
| | 2 | 1 | | | 1 |
| | | 2 | 1 | | 2 |
| joint | | | 2 | 1 | 3 |
| | 1 | | | 2 | 4 |

*Figure 10.* Updated structure

By applying in succession rules $P_7, P_6, P_5, P_1, P_0$ of the Conditional and Fuzzy Feature Grammar $G^C_{Features}$, inferred for class $C^X_2 = \{Step, Slot, Partial\ Hole, Hole\}$, we finally recognize the "Hole" feature with $\mu_{Hole} = 1$.

3.5. MODELING

The Regioning phase transforms the representation B.Rep of the part into a representation by Regions. The feature recognition in the Identification phase transforms either the representation by macro-regions into a representation by features – macro-regions, or the representation by regions into a representation by features – regions. For example, for the part, Figure 5a, we have:

$$\langle macro-région_1\rangle \rightarrow \langle Slot\rangle \langle Slot\rangle \langle Slot\rangle \tag{12}$$

$$\begin{aligned} \langle Slot\rangle &\rightarrow \langle f_1\rangle \langle f_2\rangle \langle f_3\rangle \\ \langle Slot\rangle &\rightarrow \langle f_4\rangle \langle f_5\rangle \langle f_6\rangle \\ \langle Slot\rangle &\rightarrow \langle f_1\rangle \langle f_7\rangle \langle f_6\rangle \end{aligned} \tag{13}$$

Equation 12 shows that the macro-region comprises three <Slots>; equation 13 shows the faces that make up each <Slot>.

## 4. Application

To give a comparative example, consider the part in (Marefat and Kashyap 1990), Figure 11. Table 2 gives the results of the feature recognition.

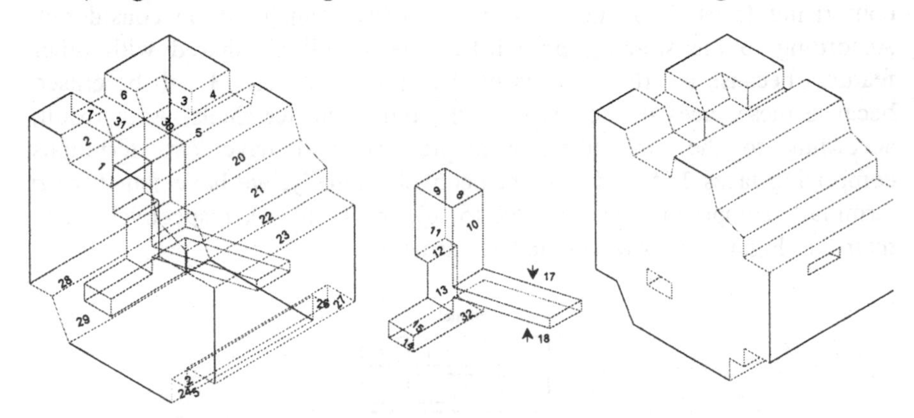


*Figure 11.* Part with features

If we compare our results with those of (Marefat and Kashyap 1990), we can see several differences. For example, in the second region of the first macro-region, first we recognized the two pockets and then the two holes:

| *Pocket* | *8,13,14,10-32,15* | *0.8* | *Hole* | *10-32,11,13,8* | *0.9* |
|---|---|---|---|---|---|
| *Pocket* | *12,10-32,11,9,8* | *0.8* | *Hole* | *8,17,11,18* | *1* |

TABLE 2. Recognition results

| *Macro - Régions* | *Régions* | *Features reconnus* | *Faces* | $\mu$ |
|---|---|---|---|---|
| *1* | *1* | *Blind Step* | *1-30-31,2-33,3,5* | *0.8* |
| | | *Slot* | *7,1-30-31,6* | *0.9* |
| | | *Slot* | *4,1-30-31,5* | *0.9* |
| | *2* | *Pocket* | *8,13,14,10-32,15* | *0.8* |
| | | *Pocket* | *12,10-32,11,9,8* | *0.8* |
| | | *Hole* | *10-32,11,13,8* | *0.9* |
| | | *Hole* | *8,17,11,18* | *1* |
| *2* | *1* | *Step* | *20,21* | *1* |
| | *2* | *Step* | *22,23* | *1* |
| *3* | *1* | *Step* | *24,25* | *1* |
| | *2* | *Step* | *26,27* | *1* |
| *4* | *1* | *Step* | *28,29* | *1* |

Thus our method implicitly follows the idea of extracting the volume of a recognized feature. In the method of (Marefat and Kashyap 1990), the recognition of the preceding two pockets had no influence on subsequent feature recognitions. Thus after recognizing those two pockets, in (Marefat and Kashyap 1990), we recognize another pocket that is defined by faces 14, 13, 11, 10-32, 8. The FMS (**F**eature **M**odeling **S**ystem) software is developed in a CAD environment (CATIA from DASSAULT SYSTEM on an IBM RS 6000 workstation).

## 5. Conclusions

This paper has proposed a method for recognizing either canonical features or features in interaction. The proposed recognition method includes various phases. The first phase, called *Regioning*, consists in identifying the potential zones for the birth of features The creation of macro-regions, as a subphase in Regioning, can be used later to define relations between features. The second phase, called *Virtual Extension*, consists in building links and virtual faces. Here we used a possibilistic method, given the problem for recognizing features in interaction. The third phase, called *Structuring*, consists in transforming the region into a structure compatible with the structure of the features represented by the Feature Grammar. The fourth phase, called *Identification*, consists in identifying the features in these zones. The heuristic attempts to generate a set of possible solutions. We must stress that the number of solutions can be reduced if we consider a predefined knowledge context. The fifth phase, called *Modeling*, consists in representing the model by features. The FMS (Feature Modeling System) software using this method is implemented in the Catia CAD-CAM environment. The proposed method provides a global framework for the feature recognition problem. Feature recognition is highly dependent on the representation by the Feature Grammar. In cases where the features are not formalized by a grammar and as a result are not recognized, the method proposes potential regions by differentiating macro-regions and the constituent regions. Recognized features are used for various applications.

## References


Bond, AH and Chang, KJ: 1988, Feature based process planning for machined parts, *Proc. ASME Inter. Computers in Engineering Conf.,* San Diego, CA, Vol. 1, pp. 571-576.

Bronsvoort, W and Jansen, FW: 1993, Feature modelling and conversion - key concepts to concurrent engineering, *Computers in Industry* **21**: 61-68.

Choi, BK Barabash, MM and Anderson, DC: 1988, Automatic recognition of machined surfaces from a 3-D solid model, *Computer Aided Design* **20** (2): 58-64.

De Martino, T and Giannini, F: 1994, The role of feature recognition in the future CAD systems, *Proc. of IFIP Inter. Conf. "Feature Modelling and Recognition in Advanced CAD/CAM Systems",* Valenciennes, France, pp. 343-356.

Falcidieno, B and Gianini, F: 1989, Automatic recognition and representation of shape-based features in a geometric modelling system, *Computers Vision, Graphics and Image Proceedings* **48**: 92-123.

Henderson, MR: 1994, Manufacturing feature identification, *Artificial Neural Networks for Intelligent Manufacturing,* London: Chapman &Hall, pp. 229-264.

Henderson, MR and Anderson, DC: 1984, Computer recognition and extraction of form features: A CAD/CAM link, *Computers in Industry* **5**: 329-339.

Joshi, S and Chang, TS: 1988, Graph based heuristics for recognition of machined features from 3-D solid model, *Computer Aided Design* **20**: 58-66.

Kim, YS and Wilde, DJ: 1992, A convex decomposition using convex hulls and local cause of its non-convergence, *ASME Journal of Mechanical Design* **114**: 459-467.

Kim, YS: 1992, Recognition of form features using convex decomposition, *Computer Aided Design* **24**(9): 461-476.

Laako, T and Mäntylä, M: 1993, Feature modelling by incremental feature recognition, *Computer Aided Design* **25**(8): 479-492.

Li, RK: 1988, A part-feature recognition system for rotational parts, *Int. J. Prod.Res.* **26** (9): 1451-1475.

Marefat M and Kashyap RL: 1990, Geometric reasoning for recognition of three dimensional object features, *Trans. of 8th Army Conf. on Applied Mathematics and Computing,* pp. 705-731.

Menon, S and Kim, YS: 1994, Cylindrical features in form feature recognition using convex decomposition, *Proc. of IFIP Inter. Conf. "Feature Modelling and Recognition in Advanced CAD/CAM Systems",* Valenciennes-France, pp. 295-314.

Srinavasan, R Liu, CR Fu, SK: 1985, Extraction of manufacturing details from geometric models, *Computers and Industrial Engineering* **9**: 125-134.

Woo, T: 1982, Feature extraction by volume decomposition, *Proc. CAD/CAM Technology in Mechanical Enginering Conference,* MIT, Cambridge, pp. 76-94.

JS Gero (ed), *Design Computing and Cognition'04*, 397-416


# COMPUTING WITH ANALYZED SHAPES

DJORDJE KRSTIC
*Alcatel, USA*

**Abstract.** Shapes play an important role in many human activities, but are rarely seen in their natural form as raw and unanalyzed. Shapes are usually analyzed or structured in terms of their certain parts. Analyzed shapes or shape decompositions are central to this paper. Different shape decompositions are developed together with their algebras. The most interesting decompositions are the ones that could successfully be used as shape approximations. Two kinds of such decompositions: *discrete* and *bounded* are examined in greater detail.

## 1. Introduction

Shapes are part of our everyday experiences and play important roles in many human activities. Shapes come without apparent structure therefore rendering any division into parts possible. However, any attempt to (verbally) describe a shape inevitably leads to structuring the shape in terms of its certain parts or to a shape *decomposition*.

"These are defined in a variety of ways, but most simply as sets of shapes that show how their sums are divided into parts of certain kinds." (Stiny 1991)

For example, the 3D shape in Figure 1 (a) may be described as a *table* having *four legs* and a *board*, which leads to the decomposition (b).

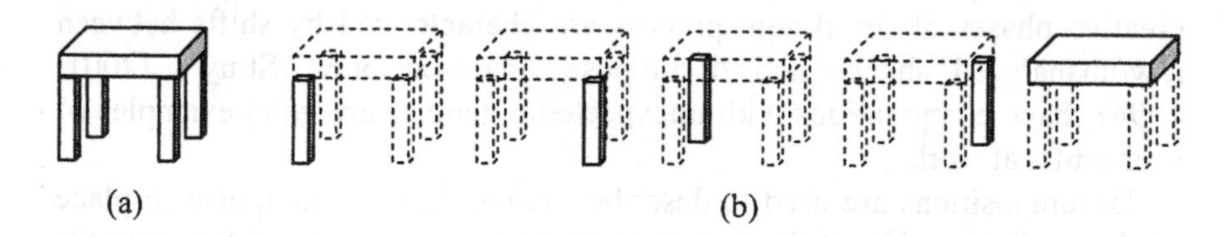

(a) (b)

*Figure 1.* Decomposition of a table: the original shape (a), and its decomposition (b)

The fact that we can name things around us structures our surroundings in a shape decomposition; while a simple naming of a shape structures it by recognizing the shape as being a part of itself.

For example, the original shape (a) in Figure 1 is named a *table* in the description above. It should also be included in the decomposition so that all of the shapes in Figure 1 become the elements of the latter.

Shapes are rarely perceived in their natural form as raw/unanalyzed, but rather approximated by their decompositions. An unknown shape may at first appear as raw/unanalyzed, but gradually it acquires a structure as one tries to understand it. As our understanding grows we move from raw/unanalyzed shapes to structured/decomposed shapes.

Although important in many human activities shapes are prime to design and fine arts. Designers use shapes to specify things for making, to provide information on how things function or how they should be built. To do so designers differentiate some parts of their designs, they name and relate the parts. Shapes in designs are utilitarian in nature. They come structured in decompositions with ambiguities removed in order to avoid misunderstanding.

Ambiguities though dismissed in designs may well be at home in fine arts. Shapes of art may serve different purpose than those of design. They need not be rationally understood. An art object may appear pure, unanalyzed and open to any kind of interpretation by both the public and critics. This kind of ambiguity allows for multiple interpretations and adds to the richness of the object.

Although the outcomes of designer's and artist's effort may be different in nature--analyzed vs. raw shape--the design process may well be the same.

In the course of design an unanalyzed shapes may be seen in a certain way and accordingly decomposed into a certain set of parts. These may then be used to compose new shapes. The latter may again be seen as unanalyzed only to be decomposed in some new way leading to new sets of parts. The process may be repeated until a design or an art object is created. Moving from analyzed to raw shapes creates an opportunity to see things in a different way, which then moves shapes back from raw to analyzed. The creative phases of the design process are characterized by shifts between raw/unanalyzed shapes and decomposed/analyzed ones. Stiny's (2001; 1994) shape computations with unexpected outcomes are good examples of such shifts at work.

Decompositions are used to describe shapes. They often appear in place of shapes functioning as their approximations. The purpose of this paper is to investigate how good such approximations are.

The investigation is conducted in the context of formal design theory that involves shape grammars and related shape algebras. Central to the theory are shape computations that attempt to formalize what designers do when they design. In order to use decompositions as shape approximations tools for computing with decompositions are developed. In particular algebras for decompositions similar to algebras for shapes are introduced. The properties of such algebras are investigated and compared to the properties of shape algebras. The measure of their agreement determines how computations with decomposition may compare to the similar computations with shapes. This ultimately shows how well shapes are approximated by their decompositions.

### 1.1. A NOTE ON NOTATION

Symbols like $\varnothing$, $\cap$, $\cup$, $\subseteq$, $\subset$, $\in$, $\notin$ and $\wp$ will be used in their standard set-theoretic meaning, that is, as the empty set, intersection, union, subset, proper subset, an element of, not an element of, and power-set, respectively. Some standard set-theoretic notions will be assumed, such as the ordered set with related maximal and minimal elements, bounds and closures, as well as the direct product and related functions and relations.

Computations will be carried out in different *algebras*, where an algebra is seen as a set of objects that is closed under a set of operations. The elements of an algebra may satisfy certain equations, like distributive or associative laws, which are *axioms* of that algebra. Axioms may serve to distinguish classes of algebras such as rings, lattices, groups, Boolean algebras, etc....

New algebras, in particular, subalgebras and quotient algebras will be constructed from the old ones. The former are subsets of an algebra that are algebras themselves. The latter are algebras whose elements are equivalence classes of some algebra, which has been partitioned with respect to an equivalence relation of a certain kind: *congruence*. Congruence preserves the operations of the algebra so that the result of computations carried on with the equivalence classes of some elements of the algebra is the equivalence class of the result of the same computation carried on with the elements themselves.

We shall also make use of shape algebras (Stiny 2001; 1992; 1991). These compute with shapes in a Boolean fashion and are also closed under geometric transformations that act on shapes. Shape algebras shall be seen as two-sorted--operating on objects of two different sorts--conveniently keeping both shapes and transformations in a single algebraic structure (Krstic 2001; 1999; 1996). Shape algebra $\boldsymbol{U_{ij}}$ computes with $i$-dimensional shapes occupying $j$-dimensional space forming the set $U_{ij}$ ($i, j$ = 0, 1, 2, 3 and $i \leq j$), and $j$-dimensional geometric transformations forming the set $T_j$.

The set of shapes $U_{ij}$ has a lower bound, which is the empty shape denoted by 0, but has no upper bound. Shapes from $U_{ij}$ together with Boolean operations of sum +, product $\cdot$, and difference - form a *generalized*

*Boolean algebra*--a relatively complemented distributive lattice with a smallest element (Birkhoff 1993). This establishes the *Boolean part* of $\boldsymbol{U}_{ij}$.

Transformations of $T_j$ form a group establishing the *group part* of $\boldsymbol{U}_{ij}$. Transformations are combined with the aid of group operations and act on shapes of $U_{ij}$ by the binary operation of group action which takes a shape $\boldsymbol{a}$ and a transformation $\boldsymbol{t}$ to produce a transformed shape $\boldsymbol{t}(\boldsymbol{a})$.

1.2. DEFINITION

Decompositions are used as shape descriptions that emphasize certain properties of shapes. There are different ways to define a decomposition of a shape depending mainly on the kind of information that is to be highlighted.

For example, decompositions defined as directed graphs containing shapes and spatial relations among them expose the details of how shapes are generated with a shape grammar (Knight 1988).

Decompositions defined simply as sets of shapes will be used exclusively in this work. Although very elementary, the definition is general. Such decompositions do only the minimum decompositions should do: they structure unanalyzed shapes in terms of their certain parts and nothing else. That is, certain properties (parts) of a shape are chosen to represent the shape while others are neglected. This renders decompositions, shape approximations that suite certain purposes.

Shape parts are central to decompositions and any shape, with the exception of shapes consisting of points, has infinitely many parts. A picture is literally worth *infinitely many* words. The set of all parts of a shape is the upper bound for all decompositions of that shape. This set has some interesting properties that reveal shape structure.

Let $\mathrm{B}(\boldsymbol{a})$ denote the set that includes all of the parts of a shape $\boldsymbol{a}$. Any Boolean combination of parts of $\boldsymbol{a}$ carried out in the algebra in which $\boldsymbol{a}$ is defined yield another part of $\boldsymbol{a}$. Set $\mathrm{B}(\boldsymbol{a})$ is a subalgebra of this algebra closed under its Boolean operations. If $\boldsymbol{a}$ is a shape from $\boldsymbol{U}_{ij}$, then $\mathrm{B}(\boldsymbol{a})$ is a Boolean algebra (Earl 1997; Stiny 1980). Because $\boldsymbol{a}$ is not changed by the transformations of its symmetry group $\mathrm{S}(\boldsymbol{a})$, the set $\mathrm{B}(\boldsymbol{a})$ is closed under $\mathrm{S}(\boldsymbol{a})$, which motivates the following definition.

The *maximal structure of a shape* $\boldsymbol{a}$ denoted by $\mathbf{B}(\boldsymbol{a})$ is a two-sorted shape algebra with $\mathrm{B}(\boldsymbol{a})$ as the Boolean part and $\mathrm{S}(\boldsymbol{a})$ as the group part.

The maximal structure of $\boldsymbol{a}$ is clearly a proper subalgebra of $\boldsymbol{U}_{ij}$, which shows that shapes like their algebras have a dual nature. They have parts that are shapes from $U_{ij}$ closed under transformations from $T_j$, where $\mathrm{B}(\boldsymbol{a})$ and $\mathrm{S}(\boldsymbol{a})$ are respective upper bounds. The interplay of structure and symmetry, which is central to design emerges here in the very nature of the building blocks of designs: shapes.

Although it is possible to have decompositions containing infinitely many parts of a shape, finite decompositions only will be examined in this work. The former may be appealing to the mathematicians, but the latter are central to the design practice.

*Definition 1* A finite non-empty set of shapes is a *decomposition*.

The set of all decompositions denoted by $\Delta_{ij}$ is a proper subset of $\wp(U_{ij})$. Set $\Delta_{ij}$ is clearly without an upper bound when ordered by $\subseteq$. It also lacks a lower bound, since the empty set is not a decomposition.

*Definition 2* Let $\boldsymbol{a}$ be a shape, the set $A$ is a *decomposition of* $\boldsymbol{a}$ whenever it is a decomposition and the sum of its elements is $\boldsymbol{a}$, or $\boldsymbol{a} = \sum A$.

Any decomposition is a decomposition of a shape because finite sums of shapes are guaranteed to be shapes.

The set of finite subsets of B($\boldsymbol{a}$) that sum to $\boldsymbol{a}$ is the set of all decompositions of $\boldsymbol{a}$. Like $\Delta_{ij}$, this set does not have an upper or a lower bound and is closed under the symmetry group of $\boldsymbol{a}$.

Decompositions structure shapes, but also structure their parts. In Figure 2 the square (a) is structured as a consequence of the structure of the double square (b) that contains it. Note that the location of the shapes in some coordinate system is indicated by cross marks.

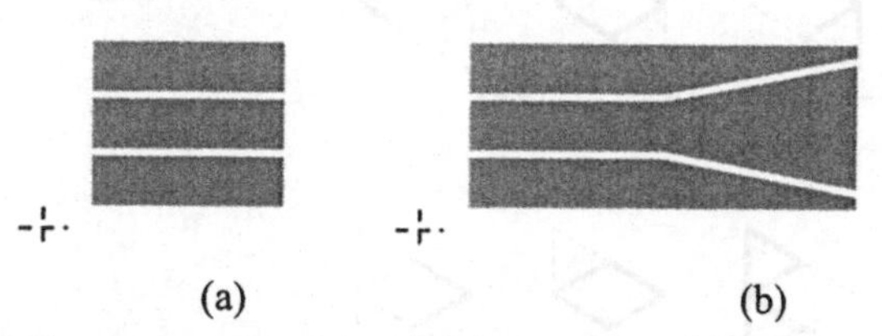

(a) (b)

*Figure 2*. Decompositions (a) and (b): structure of (b) recognized in (a) while its elements are not.

The structure of an analyzed shape is relativized to each of its parts. If $\boldsymbol{a}$ is a shape, $A$ its decomposition, and shape $\boldsymbol{b}$ a part of $\boldsymbol{a}$, than $\boldsymbol{b}$ is implicitly structured in decomposition $B = \{\boldsymbol{b}\cdot\boldsymbol{x} \mid \boldsymbol{x} \in A\}$which is the *relativization* of structure of $\boldsymbol{a}$ to $\boldsymbol{b}$ or the relativization of $A$ to $\boldsymbol{b}$.

## 2. Algebras of Decompositions

Computations with decompositions may be carried on in the similar way as ones with shapes

For example, in Figure 3 the two shapes (a) and (b) when combined in the framework of $U_{12}$ yield shapes (c), (d) and (e), which are their respective sum, product, and difference. It should be possible to duplicate these computations with sets (f) and (g) which are respective decompositions of shapes (a) and (b).

Computations with decompositions should be carried on in the framework of appropriate algebras belonging to the same family with shape algebras $U_{ij}$. Such an algebra should be two-sorted, having decompositions as well as transformations that act on decompositions. The operations should include sum, product, and difference for decompositions, group operations for transformations as well as a group action relating the two.

The group action is extended to decompositions so that: $t(A) = \{t(x) | x \in A\}$, where $A$ is a decomposition and $t$ is a transformation.

The way decompositions are combined and ordered distinguishes one algebra of decompositions from another. We shall examine two of such algebras: the set algebra of decompositions and the complex algebra of decompositions. Based on the two algebras some other algebras will be constructed as well.

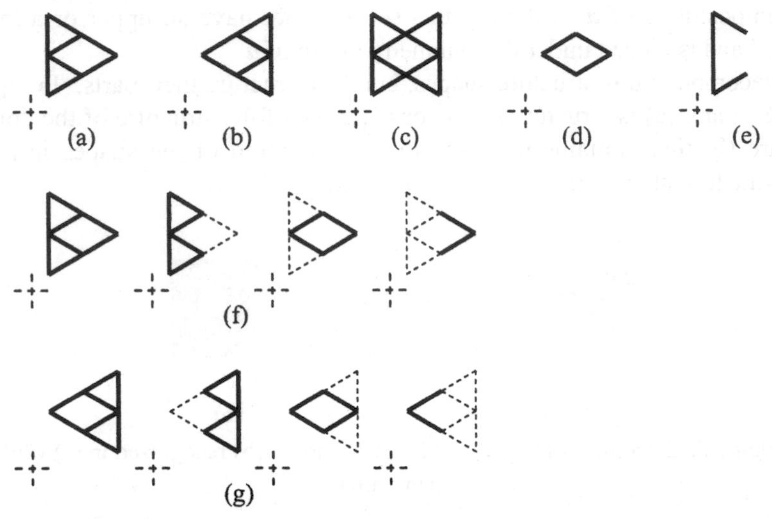


*Figure 3.* Computing in $U_{12}$: shapes (a) and (b) and their sum (c), product (d), and difference (e), as well as their respective decompositions (f) and (g).

## 2.1. SET ALGEBRAS OF DECOMPOSITIONS

If parts recognized in decompositions are paramount in computations combining decompositions as sets is a viable option. Set operations of union, intersection, and difference may be taken as sum, product, and difference of decompositions. The operations preserve the set elements and set inclusion establishes the ordering on $\Delta_{ij}$. However, set $\Delta_{ij}$ of decompositions is not closed under $\cap$ and -. The intersection or difference of two decompositions may be empty, but no decomposition is empty containing at least one shape, which may be 0. The set $\Delta_{ij}$ has to be augmented with the empty set to become an algebra.

A set algebra of decompositions $D_{ij}$ is in the family of shape algebras. The set $\Delta_{ij} \cup \{\varnothing\}$ together with operations $\cup$, $\cap$, and – forms its Boolean part, while its group part is as in $U_{ij}$. Note that the operation of group action of $D_{ij}$ has, been extended to decompositions.

This algebra--introduced by Stiny (1990; 1994)--treats shapes that are elements of decompositions as symbolic objects without further dividing or summing them. The algebra works well if shapes have predetermined parts that are to remain fixed throughout a computation. The result of the computation contains only shapes that are elements of decompositions entering the computation. These are the atoms of the computation.

For example, decompositions in Figures 4 (a), (b), and (c) are respective sum, product, and difference of decompositions in Figure 3 computed in the framework of $D_{ij}$. Note that all of the resulting shapes are elements of the decompositions entering computations. No new shape has been created in the computations.

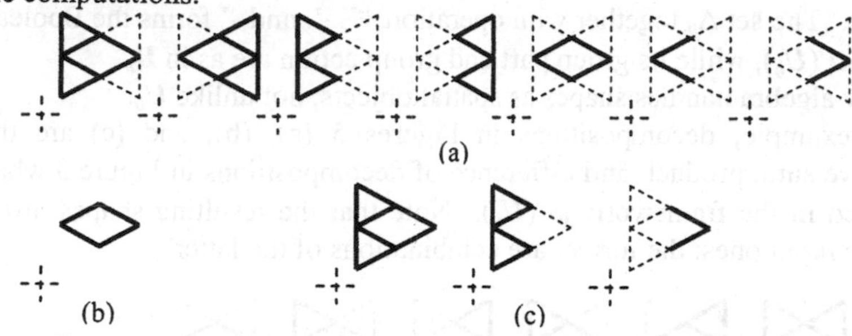


*Figure 4.* Computing in $D_{12}$: sum (a), product (b), and difference (c) of decompositions in Figure 3.

## 2.2. COMPLEX ALGEBRAS OF DECOMPOSITIONS

Set algebras compute with shapes as they would with any other objects. Properties of shapes are irrelevant in $D_{ij}$. On the other hand, shape properties, in particular, spatial properties are of the most importance in design. An algebra of decompositions which aspires to be an algebra of analyzed shapes should take spatial properties of shapes that are elements of decompositions into account. This can be achieved by letting the elements of one decomposition combine with the elements of another one in computations. With no special preference these combinations may be taken exhaustively. This renders operations of such an algebra mere extensions of shape operations to direct products of decompositions.

Let $A$ and $B$ be two decompositions, then their sum +' is defined as $A$ +' $B = +(A \times B)$. "Polish" notation used here is an elegant way of writing $A$ +' $B = \{x + y | x \in A, y \in B\}$. The other two operations product ·' and difference -' follow as $A$·'$B = \cdot(A \times B)$ and $A$ -' $B = -(A \times B)$.

Because decompositions are finite subsets (complexes) of some algebra of shapes $U_{ij}$, they may be seen as elements of a complex algebra $\wp(U_{ij})$ constructed on $U_{ij}$. The algebra operates on the set $\wp(U_{ij})$ - {∅} with the set of all decompositions $\Delta_{ij}$ as a subset. The latter set is precisely the set of all finite complexes and forms a subalgebra in $\wp(U_{ij})$. The general rule for operations of complex algebras (Gratzer 1979) yields operations +', ·' and -' as defined above.

Decompositions are ordered by ≤' defined in accordance with the general definition for relations of complex structures (Gratzer and Withney 1978). That is, $A \leq' B$, if for every $x \in A$ there exists $y \in B$, such that $x \leq y$, and for every $y \in B$ there exists $x \in A$ such that $x \leq y$.
The set $\Delta_{ij}$ has a lower bound under the ordering ≤'. It is the decomposition of the empty shape {0}. However, $\Delta_{ij}$ does not have an upper bound.

Complex algebra of decompositions $\wp'(U_{ij})$ is in the family of shape algebras. The set $\Delta_{ij}$ together with operation +', ·', and -' forms the Boolean part of $\wp'(U_{ij})$, while its group part and group action are as in $D_{ij}$.
The new algebra handles shapes as spatial objects, not unlike $U_{ij}$.

For example, decompositions in Figures 5 (a), (b), and (c) are the respective sum, product, and difference of decompositions in Figure 3 when computed in the framework $\wp'(U_{ij})$. Note that the resulting shapes differ from the input ones: the former are combinations of the latter.

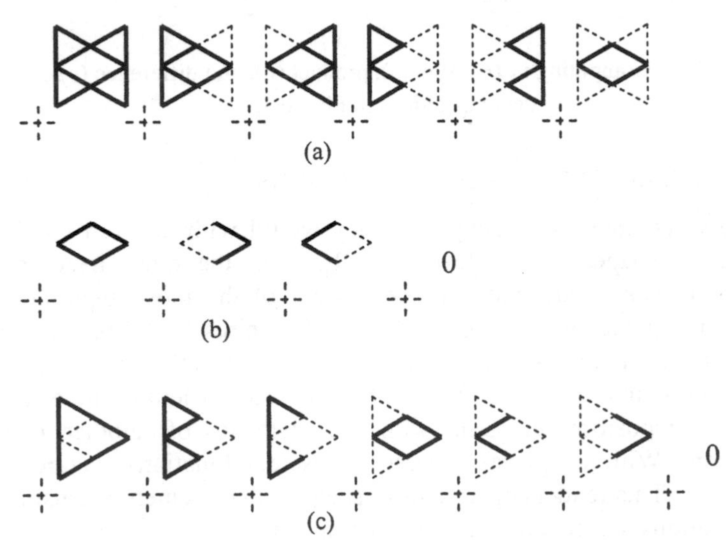


*Figure 5.* Computing in $\wp'(U_{ij})$: sum (a), product (b), and difference (c) of decompositions in Figure 3.

## 3. Decompositions as Shape Approximations

Shape grammars and later shape algebras have a history of over three decades as formal design theory tools. They have been established as formalizations of design practice. The algebras model what designers do when they draw and erase shapes, build and modify models, or just move shapes around to produce new spatial relations. Grammars and algebras are capable of handling the emergence and ambiguity of shapes, which are important in the most creative, exploratory phase of a design process (Knight 2003a; 2003b). On the other hand, algebras of decompositions operate on sets of shapes, but these may or may not be seen as approximations of shapes. If decompositions are seen as shape approximations, the properties of their algebras should, as closely as possible, match those of shape algebras.

At minimum, an algebra for decompositions should be homomorphic to a shape algebra. This guaranties that the result of a computation carried on with decompositions of shapes is a decomposition of the result of the same computation carried on with shapes themselves. Such symmetry between the two algebras is important if shapes in computations are to be approximated by their decompositions. Note that the mapping defining the homomorphism is $\Sigma$ from definition 2.

At maximum, algebras for shapes and decompositions should be isomorphic. At this point symmetry between the algebras is complete and decompositions behave as shapes. Such an algebra of decompositions becomes a true algebra of analyzed shapes.

As an additional requirement, an algebra of decompositions should have enough power to inform the decision on whether two of its elements are analyzing the same shape.

### 3.1. PROPERTIES OF SET ALGEBRAS OF DECOMPOSITIONS

Set algebras for decompositions compare favorably with shape algebras. All of the important properties of the latter such as distributive and idempotent are preserved by the former algebras. However, set algebras are less successful in satisfying minimum and the additional requirements.

It is not possible to decide--by using operations and the relation of $\boldsymbol{D}_{ij}$--whether two of its different elements are decompositions of the same shape. The algebra sees decompositions as sets of shapes (definition 1), rather then decompositions of shapes (definition 2). It manipulates elements of decompositions as symbols disregarding their spatial properties.

The minimum requirement holds only if products and differences are ignored.

This renders $\boldsymbol{D}_{ij}$ an unlikely choice if computations with analyzed shapes are an objective. However, $\boldsymbol{D}_{ij}$ is not without design applications.

Set algebras work well in situations where design is restricted to a finite kit of discrete parts, such that all shapes of interest are simple sums of the parts. Designing in the framework of a building system is a good example. Such systems usually offer kits of prefabricated building parts that may be assembled to produce variety of buildings. Parts may range from structural members, like bearing walls or ceiling slabs, to architectural elements, like partitions or facade articulations.

The set of all subsets of such a kit is a subalgebra of $\boldsymbol{D}_{ij}$, which is isomorphic to a certain subalgebra of $\boldsymbol{U}_{ij}$.

If $K_{ij}$ is a finite set of discrete shapes--no two of its elements share parts--and $s(K_{ij})$ is the set of all of the shapes that could be created with elements of $K_{ij}$, then $\wp(K_{ij})$ is the set of decompositions of all of the shapes from $s(K_{ij})$, or $s(K_{ij}) = \sum\wp(K_{ij})$. Sets $\wp(K_{ij})$ and $s(K_{ij})$ are closed under set operations of $\boldsymbol{D}_{ij}$, and shape operations of $\boldsymbol{U}_{ij}$, respectively. They are Boolean parts of the two new algebras: decomposition algebra $\wp(\boldsymbol{K}_{ij}) \subset \boldsymbol{D}_{ij}$, and shape algebra $s(\boldsymbol{K}_{ij}) \subset \boldsymbol{U}_{ij}$. The two new algebras are isomorphic.

The new *kit of parts* algebra $\wp(\boldsymbol{K}_{ij})$ computes with analyzed shapes in a meaningful way even though it does not see them as spatial objects. In formal design theory $\wp(\boldsymbol{K}_{ij})$ algebras are used in systems like set grammars (Stiny, 1982) or structure grammars (Carlson et al 1991), which work with predefined vocabularies of shapes. These systems have the same computational power as shape grammars and produce languages of designs similar to that generated by shape grammars. However, they cannot handle the emergence and ambiguity of shapes, which are important in creative phases of a design process. Shape grammars, on the other hand, do that well (Knight 2003a; 2003b).

### 3.2. PROPERTIES OF COMPLEX ALGEBRAS OF DECOMPOSITIONS

Unlike set algebras, complex algebras for decompositions do not preserve important properties of $\boldsymbol{U}_{ij}$ algebras. There are two reasons for this.

First, it is known in Universal Algebra that the identities $r = s$ valid in an algebra are also valid in its complex counterpart if and only if individual variables occur only once in both $r$ and $s$ (Shafaat 1974; Bleicher at al 1973; Gautam 1957). However, most of the important identities of $\boldsymbol{U}_{ij}$ are not of this kind. In particular, these defining idempotent, $\boldsymbol{a} + \boldsymbol{a} = \boldsymbol{a}$, $\boldsymbol{a}\cdot\boldsymbol{a} = \boldsymbol{a}$, and distributive property, $\boldsymbol{a}\cdot(\boldsymbol{b} + \boldsymbol{c}) = (\boldsymbol{a}\cdot\boldsymbol{b}) + (\boldsymbol{a}\cdot\boldsymbol{c})$, $\boldsymbol{a} + (\boldsymbol{b}\cdot\boldsymbol{c}) = (\boldsymbol{a} + \boldsymbol{b})\cdot(\boldsymbol{a} + \boldsymbol{c})$ are valid in $\boldsymbol{U}_{ij}$, but not in $\wp'(\boldsymbol{U}_{ij})$.

Second, the relation ≤' is not a partial order, but a preorder. It is reflexive and transitive, but not anti-symmetric. Consequently, $\Delta_{ij}$ is not a partially ordered set so that identities that are valid in $\boldsymbol{U}_{ij}$, under some condition(s) expressed in terms of ≤, may not be valid in $\wp'(\boldsymbol{U}_{ij})$. In particular, if $\boldsymbol{a} \leq \boldsymbol{b}$ then $\boldsymbol{a} + \boldsymbol{b} = \boldsymbol{b}$ and $\boldsymbol{a} \cdot \boldsymbol{b} = \boldsymbol{a}$, and if $\boldsymbol{a} \leq \boldsymbol{b}$ and $\boldsymbol{b} \leq \boldsymbol{a}$, then $\boldsymbol{a} = \boldsymbol{b}$ are valid in $\boldsymbol{U}_{ij}$, but not in $\wp'(\boldsymbol{U}_{ij})$.

Algebra $\wp'(\boldsymbol{U}_{ij})$, does better than $\boldsymbol{D}_{ij}$ when it comes to satisfying the minimum and the additional requirements above. The former holds if the differences are ignored while the latter holds with no constrains.

Because it is possible to decide by using operations and the relation of $\wp'(\boldsymbol{U}_{ij})$ whether its two elements are decompositions of the same shape, the algebra sees decompositions as in definition 2, that is as decompositions of shapes.

Computations involving sums and products satisfy the minimum requirement.

For example, decompositions in Figures 5 (a) and (b) analyze shapes in Figure 3 (c) and (d). The latter are respective sum and product of shapes (a) and (b), while the former are respective sum and product of their decompositions (f) and (g).

Unfortunately, the minimum requirement is not satisfied by computations involving difference.

For example, decomposition in Figure 5 (c) analyzes shape in Figure 3 (a). The latter is not the difference of shapes (a) and (b), although the decomposition is the difference of their decompositions (f) and (g).

Although $\wp'(\boldsymbol{U}_{ij})$ treats shapes as spatial objects--the same way $\boldsymbol{U}_{ij}$ does--$\wp'(\boldsymbol{U}_{ij})$ is still a poor choice for computing with analyzed shapes. It does not satisfy the minimum requirement and most of the important identities (properties) of $\boldsymbol{U}_{ij}$.

However, as with $\boldsymbol{D}_{ij}$, special kinds of decompositions form a subalgebra in $\wp'(\boldsymbol{U}_{ij})$, which is capable of handling analyzed shapes in a meaningful way. The decompositions are singletons containing the shape they analyze and nothing else. Approximating a shape with itself does not add much to the shape description, so that an algebra of such decompositions should behave as a shape algebra. This is a due to the fact that any identity $r = s$ valid in $\boldsymbol{U}_{ij}$ is valid in $\wp'(\boldsymbol{U}_{ij})$ if to each individual variable occurring more than once in $r$ or $s$ a singleton from $\wp'(\boldsymbol{U}_{ij})$ is assigned. Note that this holds for any algebra and its complex counterpart.

The algebra for singleton decompositions can now be constructed as follows.

The set $A_{ij} \subset \Delta_{ij}$ of singleton decompositions together with operations +', ·', and -' forms the Boolean part of a new algebra $\boldsymbol{A}_{ij}$. The group part and the operation of group action of $\boldsymbol{A}_{ij}$ are the same as in $\wp'(\boldsymbol{U}_{ij})$. The new algebra is a subalgrebra of $\wp'(\boldsymbol{U}_{ij})$ isomorphic to $\boldsymbol{U}_{ij}$.

### 3.3. ALGEBRAS OF ANALYZED SHAPES

Even though $\wp'(U_{ij})$ treats shapes as spatial entities the only analyzed shapes it can meaningfully handle are (trivial) singleton decompositions. This is rather disappointing because analyzed shapes play an important role not only in design, but also in our everyday lives. However, there is still some room for improvement of $\wp'(U_{ij})$ left.

One can *normalize* operations of $\wp'(U_{ij})$ in order to make it homomorphic to a shape algebra--the minimum requirement. Normalization relays on the relativization of the structure of a shape to its subshape. In $\wp'(U_{ij})$ the relativization can be expressed as $A = \{\boldsymbol{a}\}\cdot' D$, where $\boldsymbol{a}$ is a shape and $D$ is a decomposition such that $\boldsymbol{a}$ is a part of $\sum D$. Set $A$ is a decomposition of $\boldsymbol{a}$ which is the relativization of $D$ to $\boldsymbol{a}$.

If $D$ is the result of a computation with decompositions and $\boldsymbol{a}$ is the result of the same computations carried on with the corresponding shapes, then $A$ is the result of the normalized operation. If $D$ is the result of a sum or a product, then $A = D$, which means that normalization does not affect these operations. The normalized sum $+^*$ and product $\cdot^*$ are the same as $+'$ and $\cdot'$. Normalization affects only the difference which becomes $A -^* B = \{\sum A - \sum B\}\cdot'(A -' B) = A -' \{\sum B\}$, where $A$ and $B$ are decompositions. Set $A -^* B$ is clearly a decomposition of the difference of shapes analyzed by decompositions $A$ and $B$. It has the structure of $A$, but not of $B$.

It is now possible to define a new *normalized* complex algebra of decompositions $\wp^*(U_{ij})$ which is the same as $\wp'(U_{ij})$ except for the Boolean operations which are now normalized.

For example, decompositions in Figure 6 (a), (b), and (c) are the respective sum, product, and difference of decompositions in Figure 3 when computed in the framework $\wp^*(U_{ij})$. Note that sum (a) and product (b) are the same as the ones computed in the framework of $\wp'(U_{ij})$, Figure 5 (a) and (b).

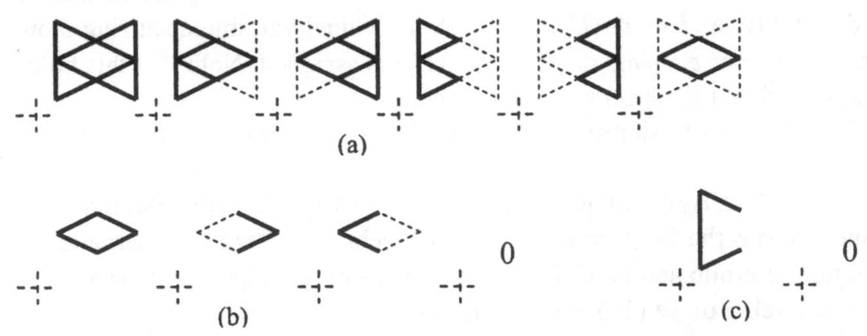


*Figure 6.* Computing in $\wp^*(U_{ij})$: sum (a), product (b), and difference (c) of decompositions in Figure 3

The new algebra satisfies the minimum requirement: it is homomorphic to $U_{ij}$. However, the two algebras still have different properties.

These differences could be largely reduced if the ordering on decompositions is changed to reassemble that on shapes. Relation $\leq'$ is a preorder on $\Delta_{ij}$ while $\leq$ is partial a order on $U_{ij}$. Universal Algebra comes in handy with a procedure for turning a preordered set into a partially ordered one. The partially ordered set can then be extended to an algebra to handle analyzed shapes.

Let an equivalence relation $\equiv$ be defined on $\Delta_{ij}$ by $A \equiv B$ if and only if $A \leq' B$ and $B \leq' A$, where $A, B \in \Delta_{ij}$. The quotient set $\Delta_{ij}/\equiv$ of equivalence classes $\Delta_{ij}$ modulo $\equiv$ is a partially ordered one (Vickers 1989). Each equivalence class of $\Delta_{ij}/\equiv$ contains decompositions that have the following common properties.

Two decompositions are equivalent modulo $\equiv$ if and only if they share the sets of minimal and maximal elements and (consequently) analyze the same shape.

Equivalence $\equiv$ also satisfies an additional condition required if $\Delta_{ij}/\equiv$ is to be extended to a quotient algebra: the equivalence is a congruence on $\wp^*(U_{ij})$.

It is now possible to construct a quotient algebra $\wp^*(U_{ij})/\equiv$, with elements that are classes of decompositions that share the shape they analyze and sets of minimal and maximal elements. The set of such classes is partially ordered with a smallest element, which is a one-element equivalence class of $\{0\}$, but with no greatest element.

Computations in $\wp^*(U_{ij})/\equiv$ are carried on as in $\wp^*(U_{ij})$. For example, computations in Figure 6 may be seen as computations in $\wp^*(U_{ij},)/\equiv$ except that the decompositions are now seen as representatives of their equivalence classes. Consequently, any other representative of the same class may be used in the place of any of the decompositions. For example, decomposition in Figure 6 (a) may be replaced with decomposition in Figure 7 (a) because they belong to the same equivalence class. That is, they share sets of maximal and minimal elements, represented in Figures 7 (b) and (c), respectively.

Although all decompositions of an equivalence class are guarantied to analyze the same shape, there are other classes in $\wp^*(U_{ij})/\equiv$ with decompositions that analyze that very shape.

For example, in Figure 7 decompositions (a) and (d) both analyze shape (b), but do not belong to the same equivalence class. Decomposition (a) has a singleton containing (b) as the set of maximal elements and (c) as the set of minimal elements. On the other hand, (d) has (c) as the set of maximal elements and a singleton containing an empty shape as the set of minimal elements.

There is no one to one correspondence between shapes of $\boldsymbol{U}_{ij}$ and classes of $\wp^*(\boldsymbol{U}_{ij})/\equiv$ so that the two algebras are not isomorphic. However, any subalgebra of $\wp^*(\boldsymbol{U}_{ij})/\equiv$ whose elements correspond to shapes in a one to one fashion is isomorphic to $\boldsymbol{U}_{ij}$.

For example, the algebra $\boldsymbol{A}_{ij}$ of singleton decompositions can be constructed as a subalgebra of $\wp^*(\boldsymbol{U}_{ij})/\equiv$. For each singleton $\{\boldsymbol{a}\}$ sets of maximal and minimal elements are equal and both are equal to $\{\boldsymbol{a}\}$. This places $\{\boldsymbol{a}\}$ into an equivalence class of $\wp^*(\boldsymbol{U}_{ij})/\equiv$ with $\{\boldsymbol{a}\}$ the only element.

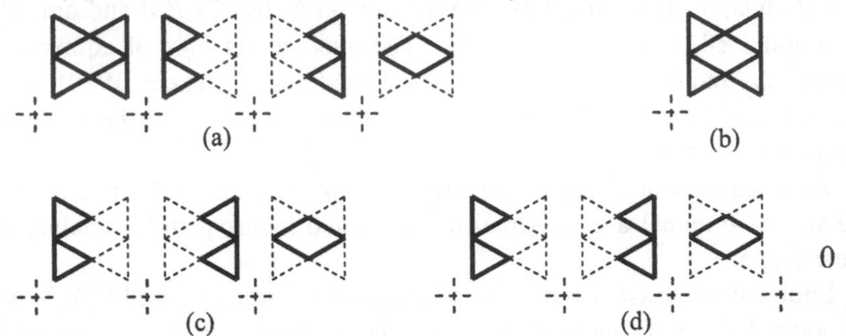


*Figure 7.* Decompositions (a) and Figure 6 (a) belong to the same equivalence class. Decompositions (a) and (d) both analyze shape (b), but do not belong to the same equivalence class. Their sets of maximal and minimal elements are different: (b) and (c) for (a), and (c) and {0} for (d), respectively.

Another example of a subalgebra of $\wp^*(\boldsymbol{U}_{ij})/\equiv$ isomorphic to $\boldsymbol{U}_{ij}$ is based on equivalence classes of decompositions that recognize the shape they analyze as well as the fact that the empty shape is a part of any shape. Such a decomposition may contain other parts too, but has, at least, to recognize the two. For each shape $\boldsymbol{a} \in U_{ij}$ there is exactly one equivalence class $[\{\boldsymbol{a}, 0\}]_\equiv$ of such decompositions that corresponds to it. This is due to the fact that for any such decomposition of $\boldsymbol{a}$ the set of its maximal elements is $\{\boldsymbol{a}\}$ and the set of its minimal elements is $\{0\}$, so that the decompositions belongs to $[\{\boldsymbol{a}, 0\}]_\equiv$. Set $C_{ij} \subset \Delta_{ij}/\equiv$ of such equivalence classes is closed under the operations of $\wp^*(\boldsymbol{U}_{ij})/\equiv$.

For example, for sum: $[\{\boldsymbol{a}, 0\}]_\equiv +_\equiv [\{\boldsymbol{b}, 0\}]_\equiv = [\{\boldsymbol{a}, 0\} +^* \{\boldsymbol{b}, 0\}]_\equiv = [\{\boldsymbol{a}+\boldsymbol{b}, \boldsymbol{a}, \boldsymbol{b}, 0\}]_\equiv = [\{\boldsymbol{a}+\boldsymbol{b}, 0\}]_\equiv \in C_{ij}$, where $\boldsymbol{a}, \boldsymbol{b} \in U_{ij}$.

The new algebra for decompositions $\boldsymbol{C}_{ij}$ with set $C_{ij}$ and operation $+^*$, $\cdot^*$, and $-^*$ as the Boolean part is isomorphic to $\boldsymbol{U}_{ij}$.

For example, each of the decompositions in Figure 3 may be augmented with an empty shape and used to compute in the framework of $C_{ij}$ as illustrated in Figure 8.

Algebra $C_{ij}$ is the most interesting of the algebras constructed here. It computes with decompositions that are nontrivial shape approximations and behaves as a shape algebra. The shape approximations of $C_{ij}$ are rather general ones. They may contain any set of parts of a shape provided that the shape itself and the empty shape are included. This renders $C_{ij}$ an excellent choice as a framework for computations with decompositions seen as shape approximations. Computations carried on with shapes in the framework of $U_{ij}$ algebra can now be repeated with analyzed shapes in the framework of $C_{ij}$ algebra.

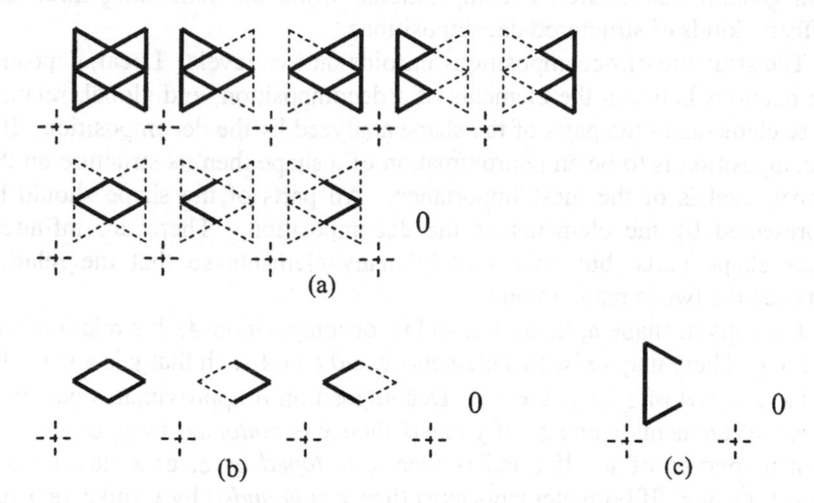


*Figure 8.* Computing in $C_{ij}$: sum (a), product (b), and difference (c) of decompositions in Figure 3, each augmented with an empty shape

For example, the result $[\{a + b, a, b, 0\}]_{\equiv}$ of the sum above seems like a reasonable explanation of the sum of shapes $a$ and $b$. One should expect both $a$ and $b$ to be preserved by the sum. If we add two new parts $c \leq a$ and $d \leq b$ to the decompositions above the same computation becomes:

$$[\{a, c, 0\}]_{\equiv} +_{\equiv} [\{b, d, 0\}]_{\equiv} = [\{a + b, a, b, c + d, c, d, 0\}]_{\equiv}$$

The new parts are preserved, while their sum $c + d$ --also included-- extends the computation to the new parts. The two principles, preservation and extension, inform computations in $C_{ij}$. A computation preserves as much as possible of the original parts, but also extends (to) these parts.

Algebra $C_{ij}$ is a true algebra of analyzed shapes, but is it *the* true algebra of analyzed shapes? Unfortunately, the answer to this question is affirmative. Algebras $A_{ij}$ and $C_{ij}$ are the only two subalgebras of $\wp^*(U_{ij})/\equiv$ isomorphic to $U_{ij}$. This result singles out both: algebra $C_{ij}$, as the only nontrivial algebra for analyzed shapes, and also decompositions that recognize the shape they analyze as well as the empty shape as the only meaningful shape approximations.

3.4. PROPERTIES OF DECOMPOSITIONS

So far different algebras of decompositions were constructed and analyzed. Now the attention will be shifted to different decompositions of shapes emerging with these algebras. Seven algebras $\boldsymbol{D}_{ij}$, $\wp'(\boldsymbol{U}_{ij})$, $\wp^*(\boldsymbol{U}_{ij})$, $\wp^*(\boldsymbol{U}_{ij})/\equiv$, $\boldsymbol{A}_{ij}$, $\wp(\boldsymbol{K}_{ij})$, and $\boldsymbol{C}_{ij}$, have been defined. The first four compute with general/unstructured decompositions, while the remaining three use different kinds of structured decompositions.

The structure of decompositions unfolds on two levels. Local, exposing the relations between the elements of a decomposition, and global, relating these elements to the parts of the shape analyzed by the decomposition. If a decomposition is to be an approximation of a shape then its structure on the global level is of the most importance. All parts of the shape should be represented by the elements of the decomposition. There are infinitely many shape parts, but only finitely many elements so that the relation between the two is many to one.

For a given shape $\boldsymbol{a}$, its part $\boldsymbol{x}$, and its decomposition $A$, this relation is as follows. There may exist two elements $\boldsymbol{y}$ and $z$ in $A$ such that $\boldsymbol{y}$ is a part of $\boldsymbol{x}$ and $\boldsymbol{x}$ is a part of $z$, or $\boldsymbol{y} \leq \boldsymbol{x} \leq z$. Decomposition $A$ approximates part $\boldsymbol{x}$ by means of elements $\boldsymbol{y}$ and $z$. If $\boldsymbol{y}$ exists then $\boldsymbol{x}$ is *bottomed* by $\boldsymbol{y}$, or $\boldsymbol{x}$ has at least properties of $\boldsymbol{y}$. If $z$ exists then $\boldsymbol{x}$ is *toped* by $z$, or $\boldsymbol{x}$ has at most properties of $z$. If both elements exist then $\boldsymbol{x}$ is *bounded* by $\boldsymbol{y}$ and $z$, or $\boldsymbol{x}$ has at least properties of $\boldsymbol{y}$ and at most properties of $z$. If neither $\boldsymbol{y}$ nor $z$ exist then $\boldsymbol{x}$ is not recognized by decomposition $A$. All elements of $A$ when seen as parts of $\boldsymbol{a}$ are bounded by themselves.

Algebra $\boldsymbol{A}_{ij}$ works with *singleton* decompositions that are singleton sets of $\Delta_{ij}$. These contain only the shape they decompose and nothing else. Each part of the shape is trivially represented, (toped), by the shape itself (i.e. it has at most properties of the shape).

In set algebras--where shape parts are of the most importance--singletons inform that the shape is a part of itself. In complex algebras singletons are less informative. The former depend on shape structures, but singletons by themselves do not structure shapes. Singletons and their algebras simply demonstrate that complex algebras behave well in a boundary case. They do so by validating all of the identities of $\boldsymbol{U}_{ij}$ .

Algebra $\wp(\boldsymbol{K}_{ij})$ uses *discrete* decompositions, or subsets of a finite set $K_{ij}$ of pairwise discrete shapes. No two elements of such a decomposition have common parts. Approximating shapes with discrete decompositions is thrifty with no redundancy or ambiguity. The only shape parts that are bounded are the elements of a decomposition. All other parts are either toped by only one element, or bottomed by one and possibly more elements, or not recognized at all. Any part of the shape that consists of proper parts of two or more elements of the decomposition is not recognized.

The set in Figure 9 (a) is an example of a discrete decomposition of the shape in Figure 3 (a), while shapes (b), (c), and (d) are its proper parts. Shape (b) is toped by the first element of decomposition (a), shape (c) is bottomed by both the second and the third element of (a), while shape (d) is not recognized by (a).

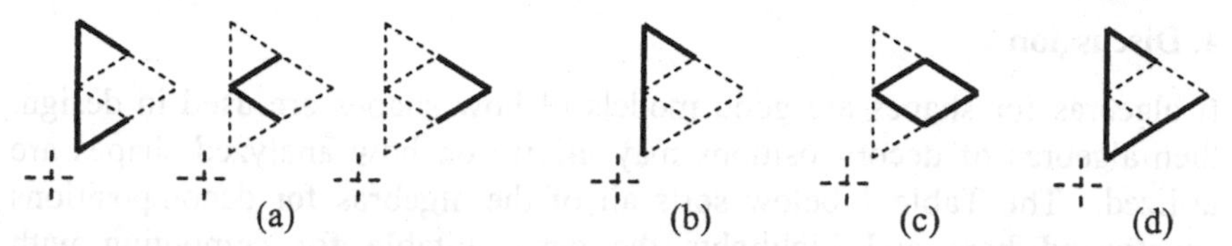


*Figure 9.* A discrete decomposition (a) of the shape in Figure 3 (a), with shape parts that are toped (b), bottomed (c), or not recognized (d) by the decomposition (a).

Discrete decompositions are good representations of finished designs ready to be assembled from the parts that are elements of the decompositions.

By including the sums of the original elements, a discrete decomposition could extend to a hierarchical structure. Hierarchies are often used in design and elsewhere, because of their straightforward structure that shows not only the parts of a shape, but also the way the parts are put together to assemble the shape.

The algebra of analyzed shapes $C_{ij}$ computes with *bounded* decompositions, which are as shown earlier the only meaningful shape approximations. These recognize the shape they analyze and the empty shape and can include other shape parts besides the two. At minimum no other parts are included. Every shape part is guaranteed to be bounded (represented) by such a decomposition. However, at minimum this representation is a rather trivial one. A part has at most properties of the shape itself and at least no properties at all. The other elements a bounded decomposition may have amount to a finer shape part representation. Shapes in Figure 5 (b) and (c) are examples of bounded decompositions.

The two shapes recognized in a minimum bounded decomposition are important from the algebraic point of view. They work in computations to preserve structures and shape parts recognized by decompositions.

For example, if two shapes $\boldsymbol{a}$ and $\boldsymbol{b}$ are analyzed by their respective bounded decompositions $A$ and $B$ and if the product $A{\cdot}{*}B$ is taken then shape $\boldsymbol{a}$ from $A$ combines with elements of $B$ to preserve the structure recognized by $B$, while shape $\boldsymbol{b}$ from $B$ combines with the elements of $A$ to preserve the structure recognized by $A$. If $\boldsymbol{b} \leq \boldsymbol{a}$, then $\boldsymbol{a}$ preserves elements of

$B$ as well. Similarly, $\boldsymbol{b}$ preserves elements of $A$ if $\boldsymbol{b} \leq \boldsymbol{a}$. The empty shape has the same function in sums: the one in $A$ preserves elements of $B$ and vice versa. Both $\boldsymbol{a}$ and $\boldsymbol{b}$ assure that the product, sum, or difference of $A$ and $B$ contains the shape it analyzes: $\boldsymbol{a} + \boldsymbol{b}$, $\boldsymbol{a}\cdot\boldsymbol{b}$, or $\boldsymbol{a} - \boldsymbol{b}$, respectively.

Bounded decomposition may serve as the basis for creating more structured decompositions of shapes. They may satisfy additional conditions to become lattices, topologies, groups, and Boolean algebras.

## 4. Discussion

If algebras for shapes are good models of how shapes are used in design, then algebras of decompositions may inform on how analyzed shapes are utilized. The Table 1 below sorts all of the algebras for decompositions constructed here and highlights the ones suitable for computing with analyzed shapes. It also exposes relations among the algebras and shows how they relate to shape algebras.

TABLE 1. Algebras for Decompositions

| Algebra | | Decompositions | | Properties | | | | | |
|---|---|---|---|---|---|---|---|---|---|
| Name | Symbol | Type | Structure | Subalgebra of | Homomorphic to | Isomorphic to | Additional condition | Identities of $U_{ij}$ preserved | Shape approximations |
| Set Algebra | $D_{ij}$ | general | none | N/A | N/A | N/A | N | Y | N |
| Complex Algebra | $\wp'(U_{ij})$ | general | none | N/A | N/A | N/A | Y | N | N |
| Normalized Complex Algebra | $\wp^*(U_{ij})$ | general | none | N/A | $U_{ij}$ | N/A | Y | N | N |
| Quotient Algebra | $\wp^*(U_{ij})/\equiv$ | equivalence class | none | N/A | $U_{ij}$ | N/A | Y | Y | N |
| Kit of Parts Algebra | $\wp(K_{ij})$ | discrete | discrete elements | $D_{ij}$ | $U_{ij}$ | $s(K_{ij})$ | N | Y | Y |
| Singletons Algebra | $A_{ij}$ | singleton | $\{a\}$ $[\{a\}]_\equiv$ | $\wp'(U_{ij})$ $\wp^*(U_{ij})$ $\wp^*(U_{ij})/\equiv$ | $U_{ij}$ | $U_{ij}$ | Y | Y | Y trivial |
| Analyzed Shapes Alg. | $C_{ij}$ | bounded | $\{a,\ldots 0\}$ $[\{a, 0\}]_\equiv$ | $\wp^*(U_{ij})/\equiv$ | $U_{ij}$ | $U_{ij}$ | Y | Y | Y |

Unstructured decompositions are not suitable as approximations of shapes. Neither set nor complex algebras of general decompositions could be used to duplicate computations with shapes. This could only be achieved with decompositions structured in some ways. The two of such structures were distinguished: discrete decompositions and bounded decompositions.

The structure of a discrete decomposition is minimum in the sense that there is no overlap between the elements. The same amount of material would be used for making all of the elements of a discrete decomposition as for making the shape itself.

Bounded decompositions, however, are far from being minimum as they always contain the shape they analyze and may contain some other parts as well. Although their structure comes from purely algebraic considerations, it is not without intuitive appeal. The presence of both, the shape analyzed by a decomposition and the empty shape provides certain context for the other shape parts recognized by the decomposition.

The former shape assures that the parts are always seen in the context of the whole. This local context is given explicitly by the fact that the whole is a member of the decomposition. If a description is associated with a bounded decomposition then it contains the name of the shape analyzed by the decomposition. Bounded decompositions are *named* in this sense.

The empty shape in a bounded decomposition implies the global context in which the analyzed shape is placed. It shifts attention from shape parts to the shape surroundings, which are implied by the absence of the parts. In an associated description the empty shape may map to a description of the shape surroundings.

Algebra $C_{ij}$ suggests that both local and global contexts are necessary for decompositions to be successfully used as shape approximations. How intuitive is this? Do we analyze (see) shapes with their parts always related to both the whole and its surroundings? These questions are interesting in their own right, but need to be decided by cognitive psychologists.

For our part, research should concentrate on moving from general decompositions to ones tailored to specific applications. This leads to decompositions with more elaborate structures.

As mentioned earlier, both discrete and bounded decompositions could be extended to more structured entities. The elements of such decompositions may form different algebras. The latter could then be combined in the framework of algebras capable of preserving the algebraic structure of decompositions. Certain closure operations facilitate the preservation of the structure in computations.

For example, Stiny (2001; 1994) computes with decompositions that are topologies, and Boolean algebras, while Krstic (1996) does it with decompositions that are lattices, hierarchies, topologies, Boolean algebras, and Boolean algebras with operators. Such decompositions are able to handle a variety of problems from continuity of computations in shape grammars to recasting spatial computations with shapes into nonspatial ones with symbols.

## References

Birkhoff, G: 1993, *Lattice Theory*, American Mathematical Society, Providence, Rhode Island.

Bleicher, MN, Schneider, H and Wilson, RL: 1973, Permanence of identities on algebras, *Algebra Universalis* **3**: 72–93.

Earl, C: 1997, Shape boundaries, *Environment and Planning B: Planning and Design* **24**: 668–687.

Guatam, N: 1957, The validity of equations of complex algebras, *Arch. Math. Logik Grundlagenforch* **3**: 117–124.

Gratzer, G: 1979, *Universal Algebra*, American Mathematical Society, Providence, Rhode Island.

Gratzer, G and Whitney, S: 1978, Infinitary varieties of structures closed under the formation of complex structures (Abstract), *Notices American Mathematical Society* **25**: A–224.

Knight, T: 2003 a, Computing with emergence, *Environment and Planning B: Planning and Design* **30**: 125–155.

Knight, T: 2003 b, Computing with ambiguity, *Environment and Planning B: Planning and Design* **30**: 165–180.

Knight, T: 1988, Comparing designs, *Environment and Planning B: Planning and Design* **7**: 73–110.

Krstic, D: 2001, Algebras and grammars for shape and their boundaries, *Environment and Planning B: Planning and Design* **28**: 151–162.

Krstic, D: 1999, Constructing algebras of design, *Environment and Planning B: Planning and Design* **26**: 45–57.

Krstic, D: 1996, *Decompositions of Shapes* PhD thesis, University of California Los Angeles.

Shafaat, A: 1974, On Varieties closed under construction of power algebras, *Bulletin Australian Mathematical Society* **11**: 213–218.

Stiny, G: 2001, How to calculate with shapes, *in* EK Antonson and J Cagan (eds), *Formal Engineering Design Synthesis*, Cambridge University Press, Cambridge, pp. 24–60.

Stiny, G: 1994, Shape rules: closure, continuity, and emergence, *Environment and Planning B: Planning and Design* **21**: S49–S78.

Stiny, G: 1992, Weight, *Environment and Planning B: Planning and Design* **19**: 413–430.

Stiny, G: 1991, The algebras of design, *Research in Engineering Design* **2**: 171–181.

Stiny, G: 1990, What is a design, *Environment and Planning B: Planning and Design* **17**: 97–103.

Stiny, G: 1980, Introduction to shape and shape grammars, *Environment and Planning B* **7**: 243–251.

Vickers, S: 1989, *Topology Via Logic*, Cambridge University Press, London, New York.

JS Gero (ed), *Design Computing and Cognition'04*, 417-436


# EXTENDING SHAPE GRAMMARS WITH DESCRIPTORS

HALDANE LIEW
*Massachusetts Institute of Technology, USA*

**Abstract.** One of the problems in using shape grammars is being able to describe succinctly and precisely the conditions necessary to perform the appropriate transformation in a drawing. This paper proposes new descriptors in shape grammars that can control rule selection and the matching conditions between a schema and a drawing. The framework for the descriptors is based on the decision making process when applying a rule. This process is divided into six phases: rule selection, drawing state, parameter requirements, transformation requirements, contextual requirements, and application method. The characteristic of each phase is described and some examples are presented.

## 1. Introduction

Using the descriptors introduced in this paper, the author of a shape grammar can (1) explicitly determine the sequence in which a set of rules is applied, (2) restrict rule application with a filtering process, and (3) use context to guide the rule matching process. The framework for developing the new descriptors is based on the rule application process. There are four basic actions to apply a rule: rule selection, drawing state, matching conditions, and application method. These four actions can be expanded into six phases specific to shape grammars as shown in Figure 1. The descriptors in each phase are characteristic of their respective phases.

The first step in applying a rule is to determine which rule to apply. The rule selection phase controls the derivation of the grammar by determining the availability and sequencing of rules. This phase has two descriptors: directive and rule-set. The directive controls what rule should be applied next based upon the success or failure of the given rule to apply in the drawing. The rule-set descriptor uses a parallel description to determine what set of rules is available to the user at any point during the derivation of the grammar.

Once a rule is selected, the user has to determine where to apply the rule.

The drawing state phase determines what portions of the drawing can be used for rule application. The two descriptors in this phase are label-filter and focus. Label-filter controls which labeled shapes are applicable based on the labeled shapes in the left-hand schema of a rule. The labeled shapes in the drawing that are not part of the left-hand schema are temporarily removed. The focus descriptor restricts the application of a rule to specific areas of the drawing demarcated by a special focus labeled polygon. A rule can only apply to the areas inside the polygons.

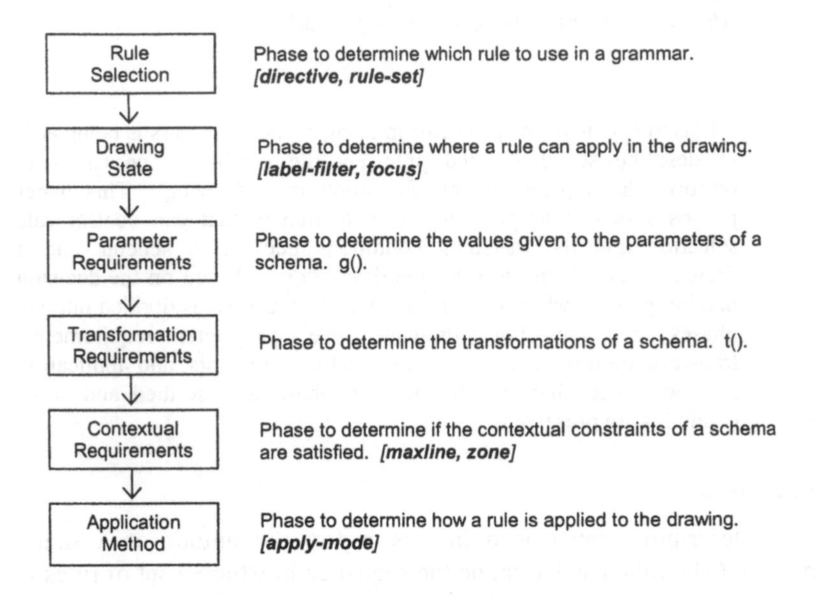


*Figure 1.* The six phases of the rule application process. The corresponding descriptors for each phase are shown in italics.

The next three phases, parameter requirements, transformation requirements, and contextual requirements, combine together to determine the matching conditions for finding a subshape in the drawing. In order for there to be a match, the requirements of all three phases must be fulfilled. The parameter requirements phase determines what values are assigned to the parameters of a schema in order to match a subshape in the drawing. The transformation requirements phase determines what combinations of transformations are necessary to match a subshape in the drawing. There are no new descriptors in these two phases.

The contextual requirements phase adds an additional constraint that determines if the context of the subshape satisfies a predicate. This phase has two descriptors: maxline and zone. The maxline descriptor specifies that a line in the subshape has to be a maximal line in the drawing. The zone descriptor associates an area of the schema with a predicate function which can be used to test portions of a drawing relative to the subshape. A commonly used predicate function is the void function which tests if the area relative to the subshape is void of all shapes.

The application method phase is the last phase in the rule application process. This phase determines how a set of subshapes can be applied to the drawing. Typically, the user applies the rule to one selected subshape, but there are other possibilities such as a parallel application where the rule applies to all the subshapes in a drawing. The only descriptor in the application method phase is apply-mode.

There are descriptors for every phase except the parameter requirements and transformation requirements phases. The other phases have one or more descriptors for a total of seven new descriptors: directive, rule-set, label-filter, focus, maxline, zone, and apply-mode. The rest of this paper will provide details of each phase and any corresponding new descriptors. The explanations will start in reverse order from the application method phase to the rule selection phase.

## 2. Application Method Phase

The application method phase determines how a set of subshapes is applied to a drawing. There can be many different subshape matches between the left-hand schema of a rule and the drawing. Typically only one subshape is used but there are other possibilities such as a parallel application which would apply all subshapes at the same time. The descriptor, apply-mode, allows the author of a grammar to specify what type of application a rule should have.

### 2.1. APPLY-MODE

The apply-mode descriptor has three options named single, parallel, and random. The single option allows the user to select one of the subshapes for application. Figure 2B shows the selection of a single 2x2 square in the upper right hand corner using the rule in Figure 2A.

The second option is parallel. This option will apply the rule to all possible subshapes in the drawing. In a 3x3 square grid, there are 14 different square subshapes. A derivation showing the effects of a parallel application using the rule in Figure 2A is shown in Figure 2C.

The third option is named random. With this option, any subshape can be applied to the drawing. The subshape can be randomly selected by the user or if used in conjunction with a shape grammar interpreter, a computer can randomly select the subshape (Chase 1989; 2002). This option can also be used when all subshapes are known to produce an equivalent shape in spite of differences in parameters and transformations. An example of this is shown in Figure 2D where the application of the rule in Figure 2A will produce the same result regardless of which transformation is selected.

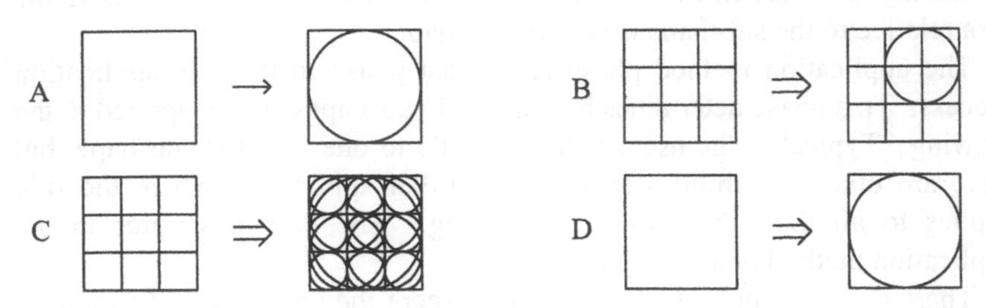


*Figure 2.* (A) Rule that finds a square and places a circle in the middle of it (B) The derivation of the rule when the apply-mode is single. Here the user has selected the upper right-hand square. (C) The derivation of the rule when the apply-mode is parallel. A circle is placed in all 14 squares. (D) The derivation of the rule when the apply-mode is random. All possible matches produce the same result.

## 3. Contextual Requirements Phase

The matching conditions between a schema and the drawing are traditionally determined by the parameter function g() and the transformation function t() which correspond to the parameter requirements and transformation requirements phases. The third phase, contextual requirements, provides additional matching constraints for the determination of a subshape based on the context of the subshape in the drawing. These three phases, in combination, determine the matching conditions of a schema.

One of the difficulties in using shape grammars is that the matching conditions between the schema and the drawing works only on the shapes found. It is not easy to define a schema that uses the surround conditions of the subshape as part of the matching constraint. For instance, suppose the author of a grammar wanted to define a schema that found rectangles that were clear of any shapes on the inside.

This is difficult to define in the shape grammar language because there is no direct convention that states that the inside of a subshape found in the drawing must be empty of shapes. Instead such a condition could be defined using a series of compound rules that combine shapes along with a parallel description grammar (Knight 2003). The function in the parallel description grammar is the algorithm that would determine if the inside of the rectangle is void of any shapes. Other techniques include, adding additional geometries or labels, constraining the transformations, or varying the parametric values in order to isolate the desired condition.

Notice that these techniques complicate the situation and deviate from the simplicity of the original condition which is succinctly stated as finding a rectangle that is clear of any shapes on the inside. Instead of relying on compound rules with parallel description grammars or using transformation and parameter restrictions, the descriptors in the contextual requirements phase rely on the visual properties of the subshapes. Determining if the inside of a rectangle subshape is empty is a visual property of the subshape relative to the drawing.

There are two descriptors in the contextual requirements phase: maxline and zone. Maxline constrains a line in the subshape to be a maximal line. The zone descriptor evaluates a predicate function against an area of the drawing defined relative to the subshape. This method allows subshapes in a drawing to assess the surrounding conditions. The zone descriptor can directly define a schema that finds rectangles with no shapes on the inside.

### 3.1. MAXLINE

The maxline descriptor adds an additional constraint that the matching subshape line must be a maximal line. By definition, a maximal line is a line that can not be embedded in another line. The use of the maxline descriptor allows the author of a grammar to specify that a line in the subshape is a line that is not a smaller portion of a larger line in the drawing.

For example, Figure 3A shows a rule where the left-hand schema is a square composed of maxline lines. By using this rule only one subshape, the outer square, can be found because of the restriction that all the lines in the square must be maximal lines, Figure 3B. A parallel application of the rule without the maxline descriptor would result with a circle in all 14 possible squares as shown in Figure 2C. The use of the maxline descriptor therefore defines a schema that differentiates the larger outer square from the smaller inner squares in the drawing.

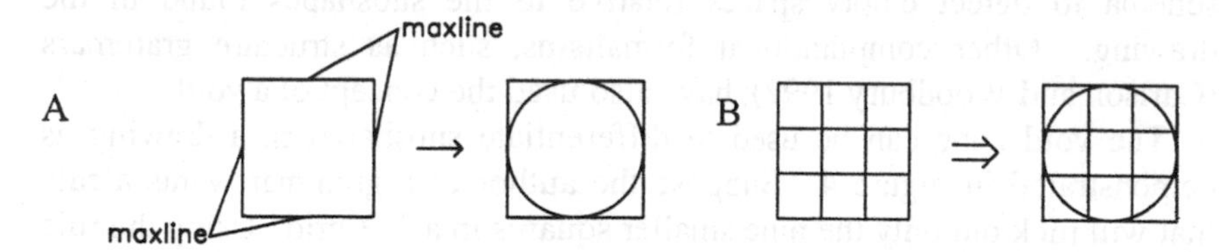


*Figure 3.* (A) Rule where the left-hand shape is composed of lines with the maxline descriptor (B) The parallel application of the rule on a 3x3 grid. Only the larger outer square is possible for selection.

Another key use of the maxline descriptor is to define schemata that are determinate. A determinate schema is one that has a finite number of subshapes in a drawing. An indeterminate schema has an infinite number of subshapes. A schema composed of only one line is indeterminate because there are an infinite number of lengths that can be embedded in a line. Often times, a determinate shape is desired to fix the number of possibilities. The maxline descriptor can be used to make a line a determinate shape.

The descriptors are based on the shape grammar language. Therefore, the same effect can be achieved without the use of the new descriptors. To describe a schema that finds a maximal line without the use of the maxline descriptor requires thinking about how to specify the same conditions using the transformation and parameter components. The maxline descriptor is the equivalent to restricting the scaling factor of the line to be at 100%. In other words, the line must match the entire line.

The descriptors in this phase can be considered shortcuts for the equivalent definitions in the shape grammar language. But more importantly the descriptors describe a rule by emphasizing certain visual conditions as opposed to emphasizing the descriptive methods of the shape grammar language. For the maxline descriptor it is the difference between seeing a line as a whole versus seeing the line as a scaling factor of 100%.

### 3.2. ZONE

The zone descriptor adds an additional constraint on the matching conditions of a schema in the form of a predicate function. With the use of the zone descriptor, not only must the geometry of the schema be embedded in the drawing, the predicate must also be true. The predicate function evaluates a marked area of the schema to determine if the predicate is true or false. By using the zone descriptor, the schema can use the context of the drawing in which the subshape is found as part of the constraints.

A commonly used function is the void function which states that the demarcated area must be void of any shape. The void zone enables the schema to detect empty spaces relative to the subshapes found in the drawing. Other computational formalisms, such as structure grammars (Carlson and Woodbury 1992), have also used the concept of a void.

The void zone can be used to differentiate subshapes in a drawing as demonstrated in Figure 4. Suppose the author of a grammar wants a rule that will pick out only the nine smaller squares in a 3x3 grid. Using the rule in Figure 2A will find all 14 different types of squares. In order to find only the nine smaller squares, the rule in Figure 4A is used which has a void zone to demarcate that the area inside the square must be empty. A parallel application of the rule is shown in Figure 4B.

The use of the void zone therefore differentiates the smaller squares from the larger squares. This occurs because whenever a larger square is selected, the void zone constraint rejects the subshape since there are lines in the interior of the square. In other words, the void predicate function returns false. The void zone can also be used to distinguish other types of squares in the grid. To define a schema that selects only the corner squares, one could use the rule in Figure 4C.

The difference between Figure 4A and 4C is that the void zone in Figure 4C has been enlarged to cover one corner of the square. A corner square has the characteristic that one corner of the square does not have any protruding lines. By expanding the void zone to cover one corner, the rule is guaranteed to select only the corner squares. A parallel application is shown in Figure 4D. A similar approach is used to define a rule that finds only the edge squares as shown in Figures 4E and 4F.

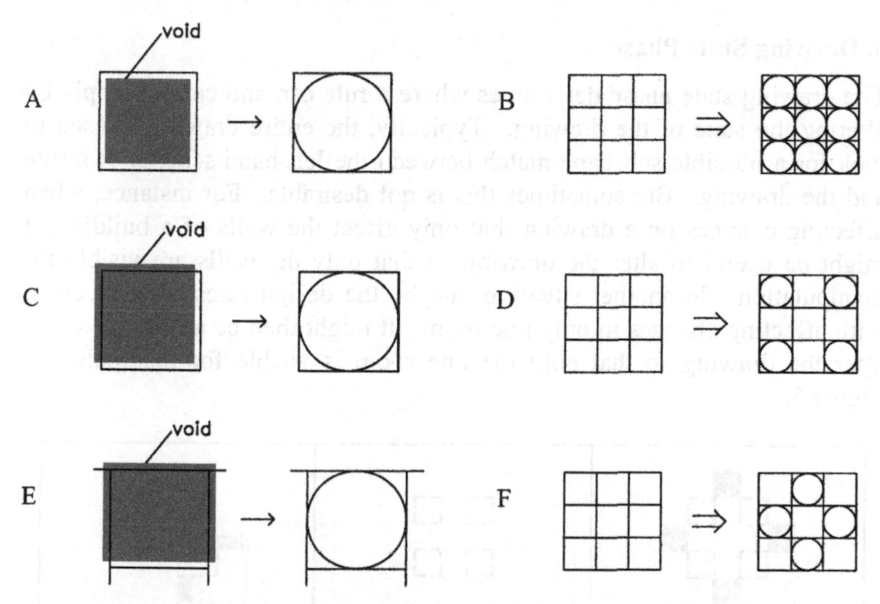


*Figure 4*. (A) A rule that finds a square such that the inside of the square is void of all shapes (B) The parallel application of Figure 4A (C) A rule that finds a corner square (D) The parallel application of Figure 4C (E) A rule that finds an edge square (F) The parallel application of Figure 4E

## 4. Transformation Requirements Phase

The transformation requirements phase determines what transformations are necessary in order to have a subshape match in the drawing. This phase corresponds with the t() function. There are no new descriptors in this phase. The transformations include translation, rotation, reflection and scaling. Restrictions can be stipulated by specifying a value such as setting the scaling factor to be 100%, which is the equivalent of the maxline descriptor, or by specifying the type of transformations to use such as all transformations except for reflection.

## 5. Parameter Requirements Phase

The parameter requirements phase corresponds with the g() function and determines what values are given to the parameters of a schema in order to have a subshape match in the drawing. There are no new descriptors in this phase. A common use of the g() function is to determine the parameters for the size of a shape. Restrictions can be placed on the parameters to match certain types of subshapes in the drawing.

## 6. Drawing State Phase

The drawing state phase determines where a rule can and can not apply by altering the state of the drawing. Typically, the entire drawing is used to look for a possible subshape match between the left-hand schema of a rule and the drawing. But sometimes this is not desirable. For instance, when affecting changes on a drawing that only affect the walls of a building, it might be useful to alter the drawing so that only the walls are visible for manipulation. In another situation, maybe the design rules are concerned with affecting changes in only one room. It might then be advantageous to alter the drawing so that only the one room is visible for manipulation, Figure 5.

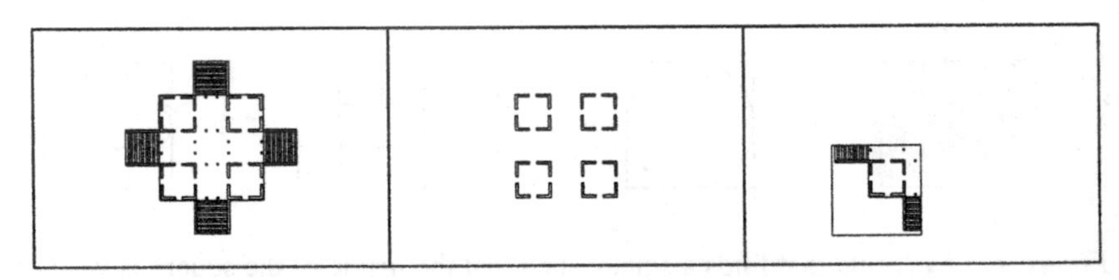

*Figure 5.* A plan drawing of a building by Durand (left) with alternative views where only the walls are visible (middle) and where only one room is visible (right)

The examples above all alter or "see" the drawing in a different context. If we are concerned with only walls, we see only walls. If we are concerned with seeing a particular room, we see only that particular room. This process of seeing a drawing as something else is part of the design process. Schön and Wiggins (1992) has characterized this seeing as a part of the cyclical see-move-see process in design. The designer sees the drawing as something and evaluates the design to make a move. This in turn generates a new drawing which can be re-interpreted and re-evaluated to make new moves.

In a similar fashion, the modifications permitted by the drawing context phase gives a rule the ability to see the drawing as something else. What we see varies greatly and depends, in part, on the context in which the rule is applied. The ultimate goal of this phase is to view the drawing in such a way as to isolate the parts of the drawing that are important. A drawing can be quite complex and the drawing state phase is a means to manage that complexity by removing parts of the drawing that are not of interest at the time.

The descriptors in the drawing state phase provide two generic methods for altering the drawing. The first method is information filtering which filters out unnecessary shapes from the drawing. What shapes to filter out is dependant on the labeled shapes in the left-hand schema. This method has the same effect as looking at only a specific labeled shape, such as walls, in a drawing, Figure 5 middle. The name for this descriptor is label-filter.

The second method is based on visual attention. When a person pays attention to a particular object in a drawing, they produce what is commonly called the searchlight of attention (Posner 1980). The location and scope of the searchlight determines what the person is focused on. This effect is abstractly mimicked using the focus descriptor. An area of focus is demarcated by placing an enclosed polygonal shape in the drawing. All areas outside of the polygonal shape are ignored. This method has the same effect as looking at a specific room in a drawing, Figure 5 right.

### 6.1. LABEL-FILTER

The label-filter descriptor filters out any labeled shapes in the drawing that does not have the same labels as those used in the left-hand schema of a rule. In an architectural setting, the label-filter option could be used to filter out background information in a drawing. For instance, a drawing can consist of walls where the square grid pattern is used as the centerline, Figure 6B. The black lines represent wall labeled lines and the gray lines represent the grid labeled lines. To select an empty room, the schema in Figure 6A, which looks for a rectangular room that is void of shapes on the interior, could be used.

Unfortunately this schema will find no rooms in Figure 6B because the void zone will detect the grid labeled lines whenever it finds a room. To fix this problem, the label-filter option in the schema is turned on, Figure 6C. This has the effect of removing any labeled shapes in the drawing that are not part of the labeled shapes used in the schema, Figure 6D. Since the schema does not have any grid labeled lines in it, the grid labeled lines in the drawing are temporarily removed before searching for an empty room. With the grid lines removed, the three empty rooms can now be selected.

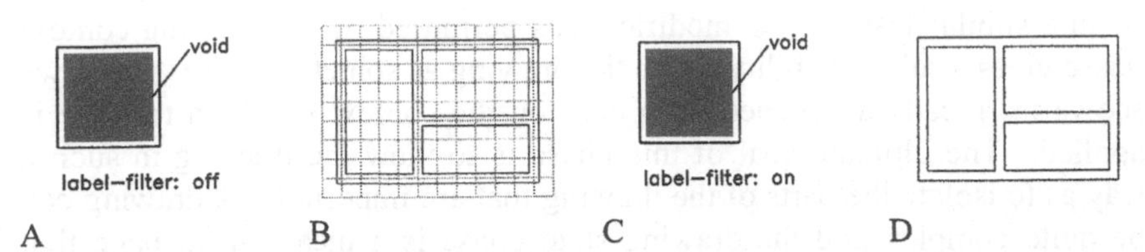


*Figure 6.* (A) A schema that looks for an empty rectangular room with the label-filter option turned off (B) A drawing where the grid lines are used as the centerline for the wall (C) A drawing showing the effects of the label-filter option. Only the wall labeled lines, shown in black, remain. (D) A schema that looks for an empty rectangular room with the label-filter turned on

## 6.2. FOCUS

The focus descriptor controls what area or areas of the drawing can be used to find a subshape match. These areas are demarcated by enclosed polygons composed of a special labeled line named "focus". The focus lines can be altered and erased, just like any other labeled line and the enclosed polygons can be of any shape, concave or convex, as long as the lines do not intersect themselves or another focus polygon.

When a drawing contains an enclosed polygon composed of focus lines, all lines outside of the enclosed area are temporarily removed from the drawing. This temporary state is persistent until the focus lines are erased. Therefore, once a focus line is placed, all subsequent rules can only apply inside the demarcated areas. To apply a rule outside of the focus area, the focus lines have to be erased.

In an architectural setting, the focus lines can be used to define where changes can occur. For instance, suppose the task is to replace a wall with a series of columns at grid intersections, Figure 7. The straightforward method is to write rules that transform walls of various sizes into a series of columns. The problem with this method is that you will need an infinite number of rules to accommodate the infinite number of possible wall and grid sizes.

Another method is to develop a procedural series of rules that can vary according to the size of the wall and grid. A derivation of this process is shown in Figure 9 using the rules in Figure 8. The first step is to define the area where the columns will be placed. In this particular case, it is in a wall. We can use rule A1 to place a focus rectangle around the desired wall. The next step is to place columns at the grid intersections. This can be achieved by applying rule A2 in parallel. If the focus rectangle did not exist, then the rule would have applied to all the grid intersections in the drawing as opposed to only the grid intersections inside the focus rectangle. The last step is to replace the walls and remove the focus polygon so that subsequent rules can apply to the entire drawing using rule A3.

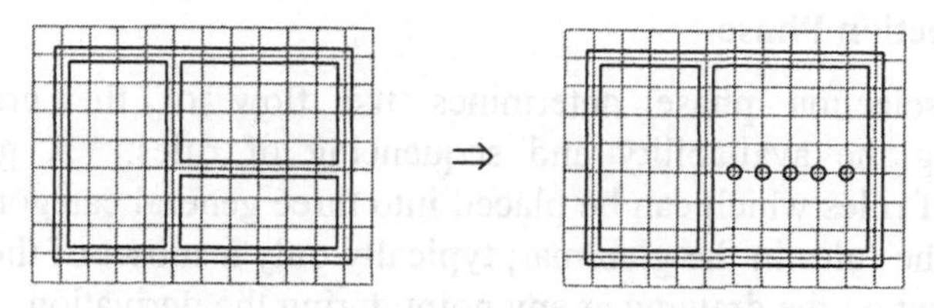

*Figure 7.* The derivation shows one wall of the floor plan transformed into a series of columns where the columns are placed at the grid intersections

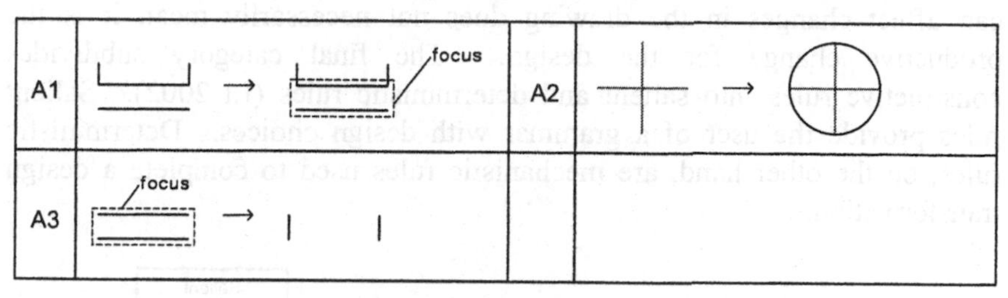


*Figure 8.* Rule A1 places a focus rectangle around a wall. Rule A2 places a column at a grid intersection. Rule A3 removes the focus lines and caps off the walls.

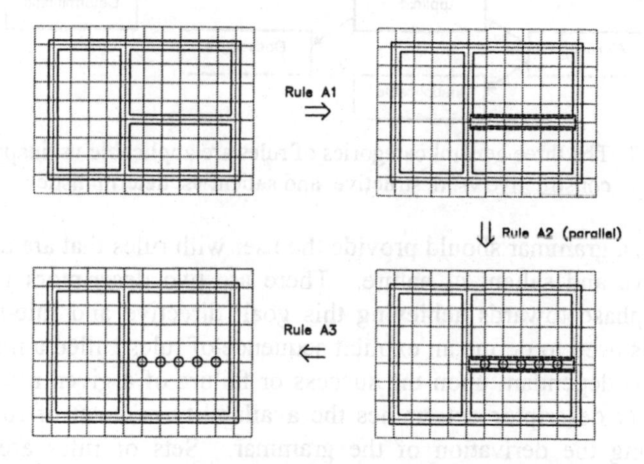


*Figure 9.* Derivation to replace a wall with a series of columns using the rules in Figure 8. Rule A1 adds a rectangular focus to the selected wall. Rule A2 is applied in parallel to place a column at every grid intersection inside the focus area. The last rule applied is rule A3 which removes the focus lines and patches up the walls.

The three step process I have just described requires that the rules are applied sequentially and in the correct order. Typically this is achieved by using state labels. The next section will show some alternative methods for controlling the execution of a grammar.

## 7. Rule Selection Phase

The rule selection phase determines the flow of the grammar by manipulating the availability and sequencing of rules. A grammar is composed of rules which can be placed into three general categories, Figure 10. Of all the rules in the grammar, typically only a subset of the rules will have an effect on the drawing at any point during the derivation. This is the first category: applicable vs. inapplicable. The applicable rules can be further divided into constructive or destructive rules. Just because a rule can affect changes in the drawing does not necessarily mean it is the productive change for the design. The final category subdivides constructive rules into salient and deterministic rules (Li 2002). Salient rules provide the user of a grammar with design choices. Deterministic rules, on the other hand, are mechanistic rules used to complete a design transformation.

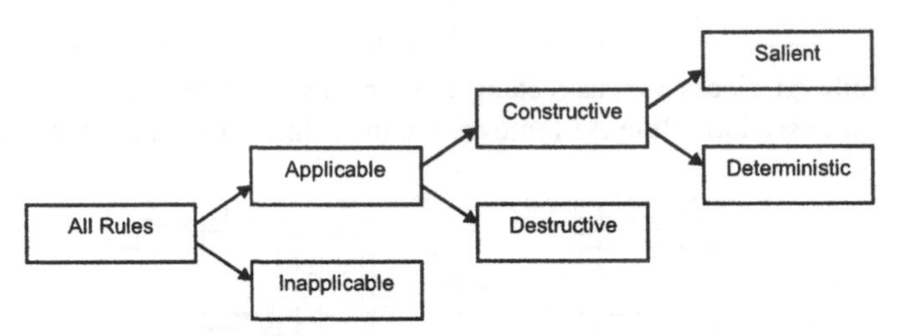


*Figure 10.* The three general categories of rules are applicable vs. inapplicable, constructive vs. destructive, and salient vs. deterministic

Ideally, a grammar should provide the user with rules that are applicable, constructive and salient in nature. There are two descriptors in the rule selection phase towards achieving this goal: directive and rule-sets. The directive is used to define an explicit sequence of rules called a macro. The sequence is dependant upon the success or failure of a given rule to apply. The rule-set descriptor determines the availability of a set of rules at any point during the derivation of the grammar. Sets of rules are typically associated with stages of a grammar.

### 7.1. DIRECTIVE

The directive descriptor adds an additional component to the rule that dictates which rule to apply next depending upon the success or failure of the current rule to apply. There are two options to the directive descriptor: success rule and failure rule, Figure 11. The success rule determines which rule to apply next if the rule was successfully applied. The failure rule determines which rule to apply next if the rule fails to apply. This occurs when the left-hand schema of a rule does not exist in the drawing. The rule specified in the success rule or failure rule can be any rule in the grammar including itself.

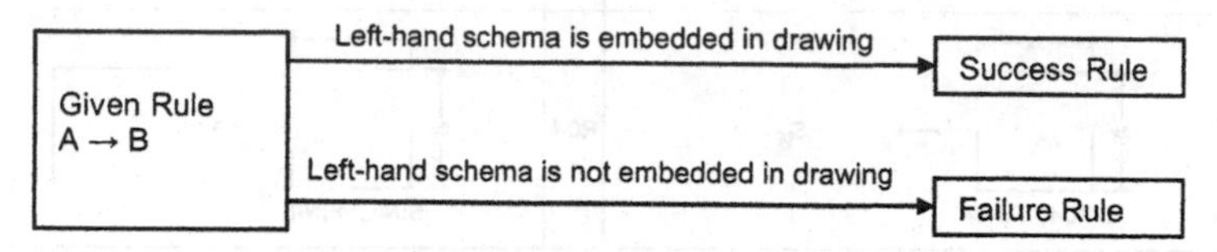


*Figure 11.* Diagram showing the components of the directive descriptor

By using the directive descriptor, the author of a grammar is able to link a series of rules together to create a macro. A macro is composed of a primary rule and one or more secondary rules. The primary rule is the first rule in the macro. The secondary rules are the rules that succeed the primary rule.

An example of how the directive can be used in a grammar is shown in Figures 12 to 14. Suppose the design task is to generate a series of walls using an underlying pattern as the centerline, Figure 12. One method to achieve this effect is to offset the centerlines half the thickness of the walls and then trim off lines at the intersections where the walls join. This process can be achieved by applying the four rules in Figure 13 in sequential order.

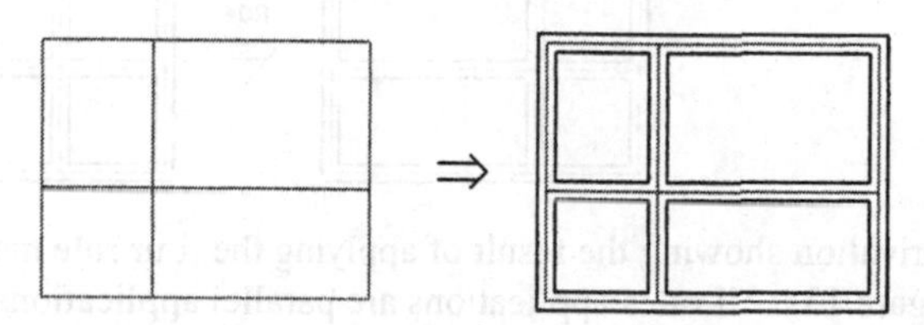

*Figure 12.* The derivation shows orthogonal centerlines (shown in gray) used as a basis to generate the walls (shown in black)

Each rule is linked to the next rule through the use of the directive. Rule R01 is applied in parallel first then rule R02, R03, and R04. R01 is the primary rule and R02, R03, and R04 are the secondary rules. Since rule R03 and R04 are subshapes of R02, it is possible for rule R03 and R04 to apply in situations where rule R02 should have been applied instead. For this reason, if the rules are applied in a different order, the results could potentially be incorrect and produce a dead-end state (Liew 2002). The derivation of the four rule macro is shown in Figure 14.

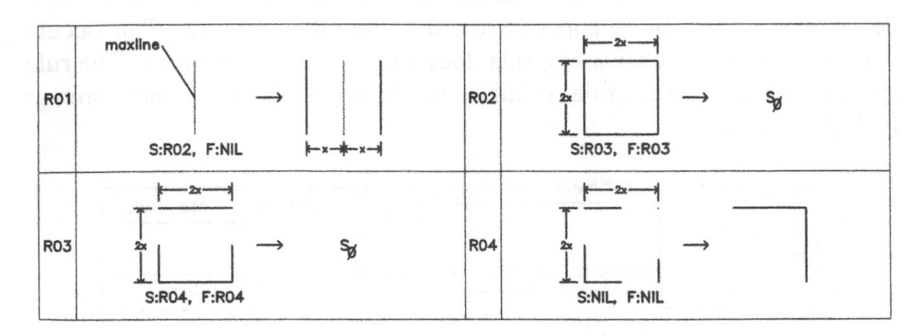


*Figure 13.* A macro composed of four rules linked together using the directive descriptor. Rule R01 offsets a pair of lines by distance x. R02 trims off the excess lines that occur in a cross intersection. R03 does the same for T intersections. And R04 trims and connects lines in an L intersection.

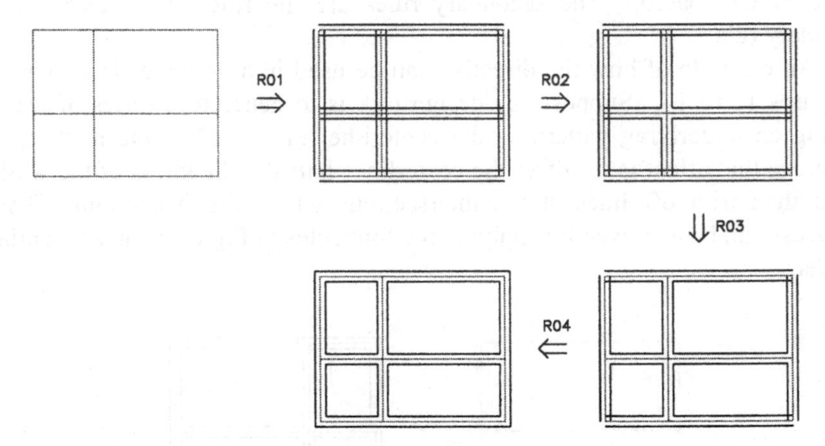


*Figure 14.* Derivation showing the result of applying the four rule macro shown in Figure 13. All rules applications are parallel applications.

To achieve the same effect without the use of the directive descriptor requires the use of labeled points which act as state labels. Each rule would need to have a unique state label so that there is no confusion as to which rule to apply. For a grammar with few rules, this is a manageable task. But if there are hundreds of rules with hundreds of unique state labels, the grammar becomes daunting and unmanageable. The directive manages the complexity of having numerous state labels by placing the information of which rule to apply within the rule.

The main difference between using state labels and the directive descriptor are: (1) The mechanism for determining the next rule, when using the directive, is not in the drawing but in the rule; (2) The directive creates a macro which can not be subverted by adding a rule with the same state label; (3) The directive has a failure component which allows an alternative rule to apply if a rule fails. The failure component can be used as a terminal case for recursive rules.

A macro is designed to start from the primary rule. If the user picks a secondary rule to apply, the macro will not necessarily work as intended. Often times, the author of a grammar wants to prevent the user from selecting the secondary rules of a macro because it can have negative effects on the design. One means of controlling this restriction can be accomplished by using the rule-set descriptor described in the next section.

### 7.2. RULE-SET

The rule-set descriptor provides the user of a grammar with a set of rules at any point during the derivation of a grammar. This is achieved with a parallel description that contains a group of rules called a rule-set. A rule-set usually corresponds to one stage of a grammar. A change in the rule-set is the same as going from one stage of the grammar to another. A rule is able to modify the rule-set with three control options: set-rule, which defines the set of rules that are available, add-rule, which inserts additional rules into the rule-set and sub-rule, which removes rules that exist in the rule-set. An illustrative derivation is shown in Figure 15 where the rule-set of each step is complete different from the previous step.

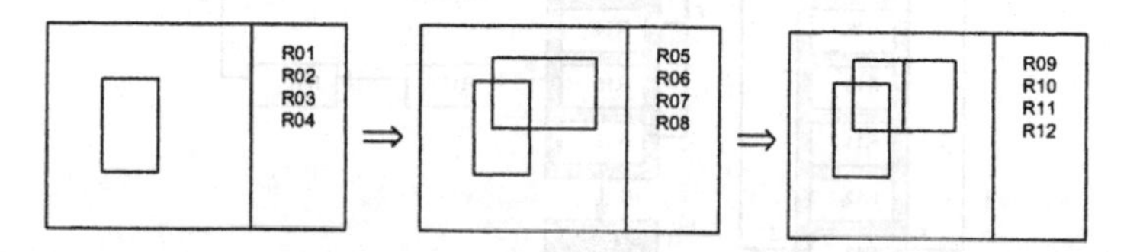


*Figure 15.* An illustrative derivation of a grammar showing the changes in the rule-sets. Each step is composed of a drawing on the left and the parallel rule-set description on the right. In the first step, the rule-set is composed of R01, R02, R03, and R04. In the second step, the rule-set changes to R05, R06, R07 and R08.

The appropriateness of the rules in the rule-set is determined by the author. Ideally, the rule-set should include only rules that are applicable, constructive and salient. By presenting the user with all viable options, the rule-set mechanism alleviates the need for the user to determine which rules can and can not apply at any stage of the grammar.

The rule-set descriptor helps the author to organize a complex set of rules when designing a grammar. The author has the flexibility to use the same rule in different stages of the grammar. He can also prevent the user from using a rule, even if it is applicable, by not including the rule in the rule-set. This method is commonly used in conjunction with macros. By including only the primary rule of a macro in the rule-set, the author of a grammar can prevent the user from selecting the secondary rules.

Figure 16 is a diagram of a sample grammar that illustrates how rule-sets in combination with directives can be used to define and manipulate stages of a grammar. The diagram shows the first three stages of a sample grammar. In the first stage, the user has four options: R01, R03, R06, and R07. This stage also has two macros: R01+R02 and R03+R04+R05. Although the actual number of rules in the first stage is seven, the user can only select four of the rules. The rule-set descriptor acts as a filter to differentiate the salient rules from the deterministic rules of a macro.

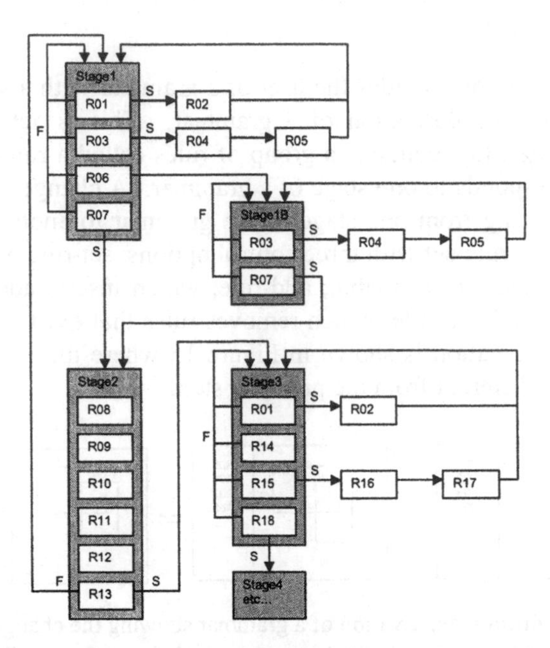


*Figure 16.* Diagram showing how stages of grammar can be defined using the rule-set and directive descriptors. Each stage created by the rule-set is represented by a gray box and the white rectangles inside are rules. The user can select any of the rules in the gray boxes. The lines connecting the rules together represent a directive link. A success rule is denoted by the letter "S" and a failure rule is denoted by the letter "F". If there is no letter then both the success and failure rules are the same.

Two of the rules in the first stage change the rule-set with the rule-set controls. Rule R06 uses the sub-rule control to remove rule R01 and R06 from stage1. This might occur when a rule affects some changes to the drawing which makes other rules no longer useful. Rule R07 changes the rule-set to go to stage2 which is composed rules R08 through R13. This can be achieved by using set-rule or combining both the sub-rule and add-rule controls to define the new rule-set. Rule R13 changes the rule-set in stage2 to go to stage3 if it is applied successfully to the drawing. If rule R13 can not be applied to the drawing then the rule-set is changed to go back to stage1. Such a move might occur if certain conditions do not exist in the drawing and requires rules in another stage to implement the necessary changes.

The rule-set can also reuse rules from other stages of the grammar as shown in the stage3 of the diagram. Stage3 has seven rules but the rule-set only allows the user to select four of the rules. Two of the rules in the rule-set are the primary rules of a macro. One of the two macros, R01+R02, is a macro from stage1. The rule-set description allows a rule to be used in any rule-set without having to modify the rule.

The rule-set descriptor simplifies the process of modifying state labels in rules by providing a separate mechanism that explicitly groups rules together. The key difference between using the rule-set descriptor and using state labels is the parallel description in rule-sets that succinctly displays for the user what rules are available for application. Both methods are control flow mechanisms and may be combined together to obtain even more detailed control over the execution of a grammar.

## 8. Implications of the Six Phases

I have described characteristics of the six phases in the rule application process and their corresponding descriptors. Three of the phases, parameter requirements, transformation requirements, and application method are based on the original formulas (equations 1-2) which are concerned with the mechanics of applying a rule (Stiny 1980; 1990; 1991).

$$t(g(A)) \leq C \quad (1)$$

$$C' = C - t(g(A)) + t(g(B)) \quad (2)$$

The new phases, rule selection, drawing state, and contextual requirements complement the original formula by addressing the decision making process when applying a rule. These phases can be incorporated together to generate a new set of formulas.

Traditionally, the matching conditions between a schema and the drawing are determined by the parameters, the transformations, and the parts relation. The contextual requirements phase introduces an additional matching constraint based on a predicate function. In order for a schema to have a subshape match in a drawing, all three components must be true. The parameters and transformations must produce a shape that is embedded in the drawing and the predicate function must be true. This changes the original formula as follows:

$$p(t(g(A))) \leq C \quad (3)$$

Here p() is the additional predicate function of the contextual requirements phase. There are two descriptors in this phase: maxline and zone. The maxline descriptor tests to make sure the subshape lines used for embedding are maximal lines. The zone descriptor tests a predicate function against a demarcated area of the subshape in the drawing. One commonly used predicate is void which checks if an area of the drawing is void of all shapes.

The contextual requirements phase modifies the matching condition between the schema and the drawing by altering the schema. The drawing state phase, on the other hand, modifies the matching condition by altering the drawing. The goal of the drawing state phase is to isolate portions of the drawing for rule application. This is achieved by hiding elements or areas of the drawing. This phase is composed of functions that "see" the drawing in a different context. The "see" function can be incorporated into the original formula in the follow manner:

$$t(g(A)) \leq s(C) \tag{4}$$

Here s() is the function that changes what the rule sees in the drawing. The drawing state phase has two descriptors: label-filter and focus. The label-filter descriptor determines what elements of a drawing can be used for a subshape match. When the label-filter option is used, all labeled shapes in the drawing that are not in the left-hand schema of a rule will be removed. The focus descriptor, on the other hand, determines what areas of the drawing can be used for a subshape match. The applicable areas are defined by enclosed polygons marked by special focus labeled lines. Any shape outside of the enclosed polygons can not be used for a subshape match.

Putting the two modifications together we get the following formulas:

$$p(t(g(A))) \leq s(C) \tag{5}$$

$$C' = s(C) - p(t(g(A))) + t(g(B)) \tag{6}$$

Here p() is associated with the contextual requirements phase and s() is associated with the drawing state phase. And finally, to have a set of all possible subshapes:

$$\text{For all } t \text{ and } g \text{ such that } p(t(g(A))) \leq s(C) \tag{7}$$

And to apply the entire set of subshapes to drawing C:

$$C' = \sum(s(C) - p(t(g(A))) + t(g(B))) \tag{8}$$

The formulas mentioned so far deal with the mechanics of the rule application process. This includes the drawing state, parameter requirements, transformation requirements, contextual requirements and application method phases. The rule selection phase does not apply to the above formulas because it deals with the overall control of the grammar.

The two descriptors in the rule selection phase are rule-set and directive. Rule-sets are an explicit mechanism for grouping rules in a grammar. The descriptor uses a parallel description to show the user which rules are available to choose from at different stages in the grammar. The directive is another control mechanism that dictates which rule to apply next depending upon the success or failure of a given rule to apply. Both descriptors are alternatives to the use of state labels which is the traditional method used in shape grammars to control the execution of the grammar.

The six phases of the rule application process is the framework for developing the new descriptors, Figure 17. The characteristic of each phase deals with the decisions necessary to apply a rule. By developing descriptors that can manipulate those decisions, greater control is obtained over how a rule is selected and what the matching conditions are between the left-hand schema of a rule and a drawing. The descriptors provide mechanisms for rule selection, a method to incorporate context as part of the schema definition, and a means to filter information in a drawing. The set of new descriptors creates a type of meta-language for the shape grammar language (Liew 2003).

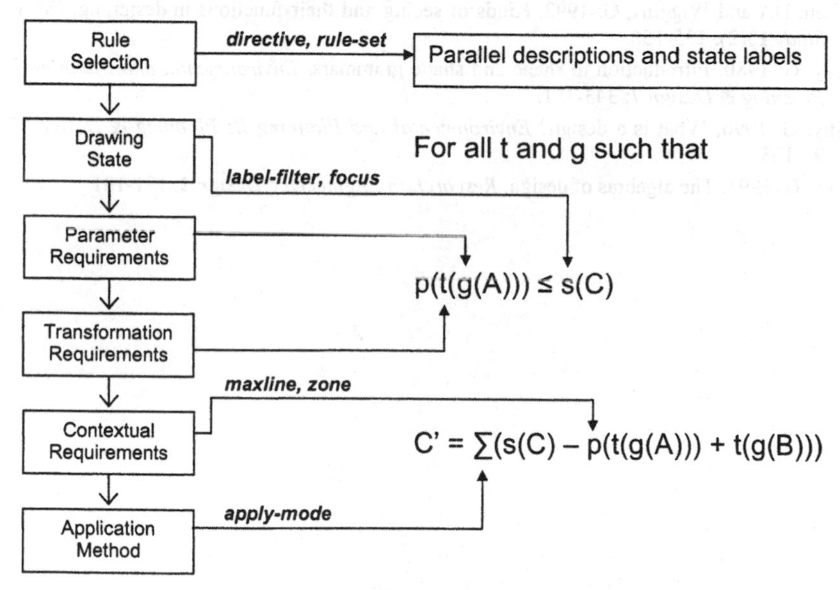


*Figure 17.* The diagram shows how the seven descriptors relate to the six phases of the rule application process and the new formulas in the shape grammar language

## References

Carlson, C and Woodbury, R: 1992, Structure grammars and their application to design, *in* DC Brown, M Waldron, and H Yoshikawa (eds), *Intelligent Computer Aided Design*, Elsevier Science Publishers, Amsterdam, pp. 107-132.

Chase, S: 1989, Shapes and shape grammars: From mathematical model to computer implementation, *Environmental and Planning B: Planning & Design* **16**: 215-242.

Chase, S: 2002, A model for user interaction in grammar-based design systems, *Automation in Construction* **11**: 161-172.

Knight, T: 2003, Computing with emergence, *Environmental and Planning B: Planning & Design* **30**: 125-155.

Li, A: 2002, A prototype interactive simulated shape grammar, *in* K Koszewski and S Wrona (eds), *Connecting the Real and the Virtual - Design Education, Proceedings of the 20th Conference on Education in Computer Aided Architectural Design in Europe*, eCAADe, Warsaw, pp. 314-317.

Liew, H: 2002, Descriptive conventions for shape grammars, *in* G Proctor (ed), *Thresholds - Design, Research, Education and Pràctice, in the Space Between the Physical and the Virtual, Proceedings of the 2002 Annual Conference of the Association for Computer Aided Design In Architecture*, ACADIA, Pomona, pp. 365-378.

Liew, H: 2003, SGML: A shape grammar meta-language, *in* W Dokonal and U Hirschberg (eds), *Digital Design, Proceedings of the 21st Conference on Education in Computer Aided Architectural Design in Europe*, eCAADe, Graz, pp. 639-647.

Posner, M: 1980, Orienting of attention, *Quarterly Journal of Experimental Psychology* **32**: 3-25.

Schön, DA and Wiggins, G: 1992, Kinds of seeing and their functions in designing, *Design Study* **13**(2): 135-156.

Stiny, G: 1980, Introduction to shape and shape grammars, *Environmental and Planning B: Planning & Design* **7**: 343-351.

Stiny, G: 1990, What is a design? *Environmental and Planning B: Planning & Design* **17**: 97-103.

Stiny, G: 1991, The algebras of design, *Research in Engineering Design* **2**: 171-181.

## AGENTS IN DESIGN

JS Gero (ed), *Design Computing and Cognition'04*, 439-457


# DESCRIBING SITUATED DESIGN AGENTS

GREGORY J SMITH, JOHN S GERO
*University of Sydney, Australia*

**Abstract.** Situated design agents are agents built using concepts from situated cognition. As situated design agents are constructive and interactive, we desire a formalism that starts with interaction and works backwards to what representations of structure, behaviour and function such interaction requires. This paper begins the process of providing a formal underpinning to agency that better corresponds to existing informal descriptions of designing as interactive, situated reflection.

## 1. Introduction

Dissatisfaction in the design science community with designing being cast by the AI community as problem solving has led to some researchers recasting designing in terms of situated reflection. A situated approach to agency generally holds that agents are social, embodied, concrete, located, engaged and specific (Wilson and Keil 1999). They are social in the sense of being located in a society of agents. Embodied means that actions by the agent are part of a dynamic with the world and results in immediate sensory feedback (Brooks 1991). Concrete, located and specific mean that actions by the agent constrain its behaviour and provide a context within which it reasons and acts. Autonomy is taken to mean that each agent decides by itself what actions to take. A crucial difference between an agent and an object is that an object encapsulates state and behaviour realisation, but not behaviour activation or action choice (Jennings 2000). Engaged means that the agent has an ongoing interaction with the environment; that planning and acting are not separated in time. So when Schön describes designing as reflection-in-action (Schön and Wiggins 1992) he is describing a dialectic view of designing by a situated agent.

For the most part, the recasting of designing into situated reflection and interaction has been informal. Attempts to describe agents less informally tend to start by representing the "mental attitudes" of agents. The FIPA agent communication language (FIPA 2002), for example, has a semantics based on an underlying realism and BDI modalities of belief and desire. Multi-agent phenomena are then defined in terms of agent beliefs.

Contrary to this, we believe that design agents should be situated, constructive and interactive. We therefore desire a formalism that starts with interaction and works backwards to what representations of structure, behaviour and function such interaction requires. The motivation behind this paper, then, is to start with interactions between situated design agents and work backwards to a formal model of distributed design agents that is suited to design. The research described in this paper is founded on two concepts. First, the situated version of Gero's FBS paradigm (Gero and Kannengiesser 2002), here called sFBS. The sFBS paradigm approaches situatedness by introducing three different kinds of worlds that interact with one another: the external world, the interpreted world, and the expected world. The external world is the world that is outside of the agent. The interpreted world is constructed inside the design agent in terms of sensory experiences, percepts and concepts. The expected world is that which the agent imagines its actions will produce. Second, a model of the computational processes that has been the basis of much of our recent work (Maher, Smith and Gero 2003, Smith, Maher and Gero 2004). The long term aim of this work is to formalise notions such as reflection and common ground. This paper is a start along that path.

Figure 1 shows a world as viewed by an agent called Agent1. Reasoning consists of five processes: sensation, interpretation, hypothesiser, action activation, and effection. Interpretation uses sense-data and expectations to interpret what the agent believes it's world to be. Throughout this paper we use the word "world" to mean that part of the system that the agent is aware of, and "environment" to mean the whole system. The hypothesiser monitors the interpretations of the world, and asserts goals associated with the agent's view of itself in the world. The action activator reasons about the steps to achieve a goal and triggers the effectors to make changes to the environment.

Three levels of action are possible: reflexive, reactive and reflective. Reflexive action is where sense-data triggers action activation directly. Reflexive actions do not involve beliefs. Reactive actions are where interpretations trigger action activation. These do not involve explicit reasoning with goals and expectations. Reflective action involves the hypothesiser explicitly reasoning about expectations and alternative goals.

To avoid confusion between mathematical function and FBS function, in this paper the word "function" on its own means "mathematical function" and the phrase "FBS function" refers to the ascribed function of an artifact or agent. The theory described here is interactive rather than algorithmic. Agents achieve their goals through situated interaction rather than algorithmic planning. The attitude is one of "everything is allowed that is not forbidden" rather than the algorithmic attitude of "everything is forbidden that is not allowed" (Wegner and Goldin 1997).

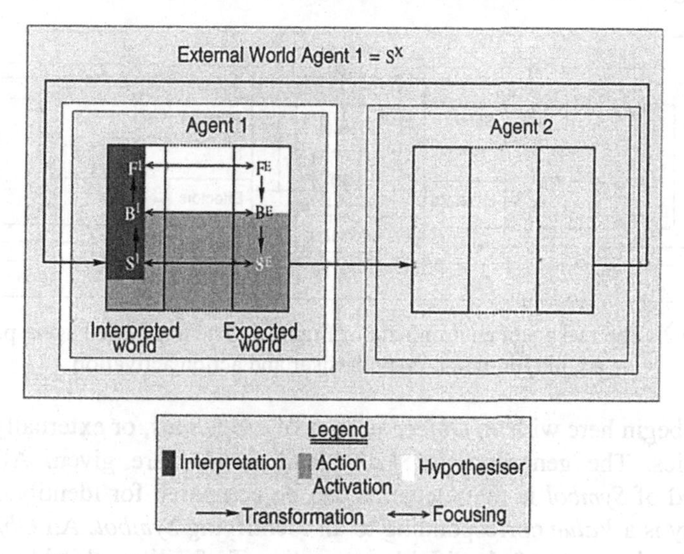


*Figure 1.* Example world as viewed by an agent Agent1

In this paper we present a formal model of the agents introduced above. In Section 2.1 we describe a system that to an external omnipotent observer appears as a structured set of objects, where the objects are or belong to agents. Section 2.2 then describes the environment and some possible views of it. Section 2.3 reviews situated FBS in this context, with Section 2.4 describing situated action. We then describe an example system in Section 3. We begin by developing symbolic representations of objects and agents. We then represent the environment in similar terms. This representation is used to represent situated FBS before concluding with a representation for situated action.

## 2. A World of Agents

### 2.1. OBJECTS AND AGENTS

In this paper we consider a system that to an external omnipotent observer appears as a structured set of objects, where the objects belong to agents. Figure 2 is an example. We call the entire system the *environment*, we call the environment as is viewed by an agent the *external world* of that agent. External world may be shortened to *world* in such cases as will not cause confusion with *interpreted world* or *expected world*. Agents sense the environment and construct an interpretation of that environment as their interpreted world. The way that they wish the environment to be is their expected world. Agents take actions through effectors by which they attempt to change their external world to make it like their expected world.

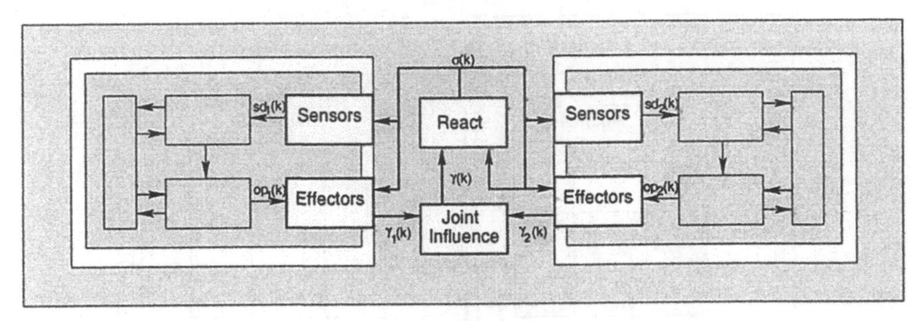


*Figure 2.* The two agent environment of Figure 1. The unlabelled agent processes are interpretation, hypothesiser and action activation.

We begin here with an *Object* as a set of *exogenous*, or externally visible, properties. The generic types *Value* and *Symbol* are given. All that is assumed of *Symbol* is that elements can be compared for identity. A single property is a *Value* corresponding to an identifying *Symbol*. An *Object* is an entity that has a set of identifiable properties. To facilitate this identification we define *Object* as a partial function from *Symbol*s to *Value*s.

$Object: Symbol \nrightarrow Value$

An *Object* may or may not have concealed, autogenous properties. The observable structure of a table, for example, is determined by its atomic structure but that atomic structure is not observable: the observable structure is exogenous, the atomic structure is autogenous. The same distinction applies to humans and to artificial agents. This definition does not require that the set of properties be fixed or even finite. Together with the generic types *Value* and *Symbol*, this allows for *Object* of arbitrary complexity. It can denote an instance of an object oriented language, or it can denote an object sketched by a human designer: the former has a finite number of properties, the latter does not.

The environments we consider here are structured as a distributed set of communicating agents. Each object is constructed by an agent and contains exogenous properties that can be read by any agent that is aware of it's identity. For uniformity we regard agents as accessing the properties of *Object*s with message passing.

An agent is an object that autonomously senses the environment and acts so as to achieve its goals. The environment[1] contains a set of agents *Agents* $= \{ a_i{:}Agent | i=1..N\}$ where *Agent* is the following tuple:

[1] Some of the following is based on the formalisations in (Wooldridge and Lomuscio 2001, Fagin et al. 1997, Ferber 1999).

*Agent*: *Exog* × *Autog* × *Sensor* × *Reason* × *Effector*
*Exog*: *Object*
*Autog*: $\wp$ *Object*
*Sd*: seq *Object*
$Sensor_i: \sum_0 \times \sum_i \nrightarrow Sd$
$Reason_i: Sd_i \times \sum_i \longrightarrow Op_i$
$Effector_i: Op_i \times \sum_i \longrightarrow \Gamma_i$

In this paper we use $\wp X$ to denote the powerset of $X$ and seq $X$ to denote a sequence of $X$. Each agent $a_i$ contains one object that corresponds exogenous properties constructed from *Exog*. Autogenous properties of the agent are internally constructed and cannot be accessed by other agents. For convenience we let $ex(a_i)$ be the exogenous object of agent $a_i$ and let $au(a_i)$ be the set of autogenous objects of agent $a_i$.

*ex*: *Agent* ⟶ *Exog*
*au*: *Agent* ⟶ *Autog*

$\forall a_i : Agent \bullet$
$ex(a_i) = first(a_i)$
$au(a_i) = second(a_i)$

Function *first* finds the first element of an n-tuple, *second* finds the second element of an n-tuple, and $x \mapsto y$ denotes a 2-tuple. Functions $Sensor_i$, $Reason_i$ and $Effector_i$ sense the environment, handle inference and memory, and effect the environment respectively. $Sd_i$ are sense-data of agent $i$ and is a sequence of objects on which functions *push* and *pop* are defined. Each effector can push one operator to the environment at a time and so is not a sequence.

$Op_i$ are operations that can be effected by agent $i$. $\Gamma$ is the set of influences on the environment by the agents. $\sum$ is the set of global system states and $\sum_i$ is the set of local states of agent $i$. As $Sensor_i$, $Reason_i$ and $Effector_i$ are the program of the agent, current local state of an agent $i$ is a point in a $k$ dimensional *Value* space:

$$\sum_i = Value^k$$

where $k$ is the cardinality of the sets of exogenous and autogenous properties, or $k = \#ex(a_i) + \#(\#au(a_i))$, where $\#X$ is the cardinality of a set $X$.

This formalism is simplified if all objects are considered to be contained within an agent. For this reason, all objects not a part of $\{a_i\}$ are assumed to be a part of an imaginary, predefined agent $a_0$ that represents the causality of the environment. Causality is discussed in Section 2.3. An example is a message sent but not yet received. Agent $a_0$ is a mathematical construct only, hence the use of a subscript from outside the *1..N* range. No claim that the environment is an agent is being made. If a message is sent from $a_i$ to $a_j$, then a message object originates in $\sum_i$, becomes a part of $\sum_0$ in-transit, and finishes in $\sum_j$. The global system state $\sum$ therefore depends on the states of each of the agents.

$$\sum = \sum_0 \times \sum_i \times \ldots \times \sum_N$$

$Sensor_i$ constructs sense-data as a function of the environment and the current local state of the agent. This definition allows for perception that may be incomplete, in error, noisy, or that biases the sensors such as by changing the focus of attention. An external observer may regard the environment as being only partly visible to an agent because some $\sum$ states are treated as being equivalent and so are not viewed as distinct by that agent. The agent's internal view, though, is that perception abstracts away from sense-data to patterns of invariance over interactive experiences.

Agents change properties of external world objects by executing a *push* operation from $Op_i$.

$$Op_i \supseteq \{a_j : Agent \mid a_i \neq a_j \bullet push(a_i, a_j, o)\}$$

where $o$ is a message object. The messages can be of two kinds: $a_i$ *informs* $a_j$ that a property has a value, and $a_i$ *requests* of $a_j$ that a property has some value. Agent $a_i$ can push a message to $a_j$ an *inform* of content $\varphi$ if (FIPA 2002):

1. $a_i$ believes that $\varphi \subset \wp\, Object$ is true:

$$inform(a_i, a_j, a\_j, \varphi) \Rightarrow (\forall s{:}Symbol;\ v{:}Value \mid (s \mapsto v) \in \varphi \bullet ((s \mapsto v) \in ex(a_i) \vee (\exists x{:}Object \mid x \in au(a_i) \wedge (s \mapsto v) \in x)))$$

2. $a_i$ has a goal that $a_j$ believe $\varphi$
3. $a_i$ does not believe that $a_j$ believes $\varphi$

A *request* of content $\varphi$ pushed from $a_i$ to $a_j$ is defined similarly. $Effector_i$ is discussed in the next section.

### 2.2. ENVIRONMENT

An environment is a tuple $\mathfrak{E} : Objects \times \Gamma \times Op \times React$ where

$$Objects{:}\ \wp\, Object$$

$map: \mathscr{E} \times Object \nrightarrow Agent \cup \{null\}$
$ob: \mathscr{E} \longrightarrow Objects$

The initial state of the environment is $\sigma(0) \in \sum$, with $\sum$ the global state as defined above. Function *map* is a partial function on an environment that identifies which agent contains an object, or null otherwise. Function *ob* finds the set of *Objects* in an environment. An agent $a_i$'s effector pushes a message object to the environment as a *push* operator. At time *k* an agent $a_i$'s influence on the environment is $\Gamma(k) \in \Gamma_i$. Now, just as $Sensor_i$ senses the environment differently in different local $\sum_i$ states, so the same operator effected by $Effector_i$ may influence the environment differently in different environmental $\sum_0$ states. The influences of each agent acting simultaneously are combined, and the environment reacts to the joint influence.

$$React: \Gamma \times \textstyle\sum_0 \longrightarrow \textstyle\sum_0$$

The new environment state, therefore, will be (Ferber 1999)

$$\sigma(k+1) = React\left(\bigcup_{i=1}^{N} Effector_i(Reason_i(sd_i(k), \sigma_i(k)))\right)$$

This equation describes the causality in the environment, and is "computed" by "agent" $a_0$. Each agent senses and acts on a subset of the environment. We call a subset of the environment that is visible or constructed for some purpose a view. The obvious view is the omnipotent one. In the following let $e \in \mathscr{E}$ be a particular environment.

$$omnipotent(e) = \{\, o{:}Object \mid o \in ob(e) \,\} \cup \{\, a_i{:}Agent \mid (\exists\, o{:}Object \bullet map(e,o) = a_i) \,\}$$

The multi-agent system (MAS) view is of the environment viewed solely as a multi-agent system. It therefore consists of agents and exogenous objects.

$$MAS(e) = \{\, a_i{:}Agent \mid ((\exists\, o{:}Object \bullet map(e,o) = a_i \,\} \cup \{\, o{:}Object \mid \exists\, a{:}Agent \bullet map(e,o) \wedge ex(a) = o \,\}$$

The *world* is everything between the effectors and sensors of an agent. It is the external world of that agent, or $S_i^X$ from Figure 1.

$externalworld(e,a_i) \subset$

$$\{o{:}Object \mid (\exists a_j{:}Agent \mid a_j \in MAS(e) \wedge o\ ex(a_j))\ \}$$

The view of the environment by a particular agent is of its world plus its autogenous properties.

$$agentview(e,a_i) = world(e,a_i) \cup au(a_i)$$

The view of a person such as a designer interacting with the environment is of a set of objects.

$$personview(e) \subset \{\ o{:}Object \mid (\exists a_i{:}Agent \mid a_j \in MAS(e) \wedge o = ex(a_j)\ \}$$

2.3. SITUATED FBS

Agents represent their environment using the situated FBS (sFBS) formalism (Gero and Kannengiesser 2002; Maher, Smith and Gero 2003a) as beliefs of the structure, behaviour and function of objects. Some object properties are obviously structural, such as location. Other properties must be interpreted by an agent. Structure, then, is an interpretation by an agent of what it believes a sensed object is. It is an interpretation of sense-data; sense-data themselves are uninterpreted inputs to the process of interpretation. In Figure 1, $S_i^X$ is the actual external world structure that is visible by $a_i$ (the subscripts *i* are not shown in Figure 1). As the only access to $S_i^X$ by $a_i$ is via sensors, it includes all exogenous objects other than the agent's own. An agent need not sense its own exogenous properties. $S_i^I$ is the interpretation by $a_i$ of what it believes the set of all structures of objects in the environment to be, and $S_i^E$ is the set of expectations of structure.

$$\forall a_i{:}Agent \bullet$$

$$S_i^X = world(e,a_i) - ex(a_i)$$

$$S_i^I \subset (au(a_i) \cup ex(a_i))$$

$$S_i^E \subset au(a_i)$$

Structure properties can contain values such as location, or can contain relations to other objects. $S_i^I$ can be viewed as a graph of what $a_i$ believes that the world currently is and $S_i^E$ can be viewed as a graph of what $a_i$ believes that the world will be or should be.

In general, behaviour is determined from structure according to some causation. Causation is the relation between two things where the first is thought of as somehow bringing about the second (Lacey 1996). In the natural world we regard nature as doing the bringing about in the form of the laws of physics. With an artificial world the bringing about is from computations by the agents, so behaviour is determined by whatever the "virtual physics" are. Regardless of whether the causation is natural or artificial, an agent's representations of interpreted behaviour are computed from its expectations of behaviour and from interpreted structure. These interpretations are computed from either encoded interpretation rules or are learned from experience.

Behaviour is an interpretation by an agent of what it believes an object does, and FBS function is what the agent believes that an object is for. $B_i^I$ is the set of interpretations of behaviours of objects by $a_i$, and $F_i^I$ are interpretations by $a_i$ of FBS functions that may have been ascribed to other objects. $B_i^E$ is the set of expectations of behaviours by $a_i$ and $F_i^E$ are expectations of function.

$\forall$ *$a_i$:Agent* •

$$interpretedworld(a_i) = S_i^I \cup B_i^I \cup F_i^I$$

$$expectedworld(a_i) = S_i^E \cup B_i^E \cup F_i^E$$

$$au(a_i) = (interpretedworld(a_i) \cup expectedworld(a_i)) - ex(a_i)$$

The process *interpretation* re-computes interpreted structure and behaviour whenever either new sense-data arrive or expectations change. The triggering mechanism can be event-driven, polled, or use a combination of both. For polled interpretation a time sense triggers the pulling of sense-data from the environment. Event-driven interpretation occurs when the environment pushes sense-data into sensors. Regardless of the mechanism, interpretation consists of three partial functions. The first computes interpreted structure from sense-data but biased expectations of behaviour. The second computes interpreted behaviour from interpreted structure and expectations of behaviour. The third computes interpreted FBS function from *requests* from other agents and from chat with persons such as designers.

$$strInterp: Sd_i \times B_i^E \twoheadrightarrow\!\!\!\!\!\!\!+\;\; S_i^I$$

$$behInterp: S_i^I \times B_i^I \twoheadrightarrow\!\!\!\!\!\!\!+\;\; B_i^I$$

$$funInterp: Sd_i \times F_i^E \twoheadrightarrow\!\!\!\!\!\!\!+\;\; F_i^I$$

Computing interpreted structure and behaviour are sequential; computing interpreted FBS function can be in parallel.

$$interpretation = (strInterp \circ behInterp) \,||\, funInterp$$

Figure 3 shows reflexive, reactive and reflective reasoning as a Petri net of an agent $a_i$, including *interpretation*. Function *Reason* for $a_i$ is everything between *Sensor* and *Effector*.

There are four types of FBS function that can be ascribed by an agent (Qian and Gero 1996): FBS functions that map to static behaviours, those that map to dynamic behaviours, those that map to a set of concurrent behaviours, and those that map to sequential behaviours. Reflective agents explicitly reason over FBS functions; reactive agents have FBS functionality implicit in action rules. Reflective agents use the *hypothesiser* process to compute expectations of behaviour from expected FBS function, and then compare those expectations against the interpreted world. Newly detected differences between expected and interpreted behaviour are asserted as goals for action activation to satisfy. The hypothesiser consists of two sequential processes. The first formulates expectations of behaviour. The second compares expected and interpreted behaviour.

$$Goals \subset Autog$$
$$formulation\colon\ F_i^E \twoheadrightarrow B_i^E$$
$$evaluation\colon\ B_i^E \times B_i^I \twoheadrightarrow Goals$$
$$hypothesiser = formulation \circ evaluation$$

Action activation synthesises changes in expected structure for goals asserted from differences between expected and interpreted behaviour.

$$action\colon Goals \times S_i^I \longrightarrow S_i^E$$
$$effection\colon S_i^E \longrightarrow Op \cup Exog$$

## 2.4. SITUATED ACTION

For a situated agent, deciding when and how to act is of primary importance. Differences between situations should cause the application of the same knowledge to result in differing behaviour, and the result of such behaviour should allow the agent to learn.

A reflective agent explicitly represents the *interpretedworld($a_i$)*, the *expectedworld($a_i$)*, and encoded knowledge of all of the known effects on *interpretedworld($a_i$)* of each change in structure that the agent can make.

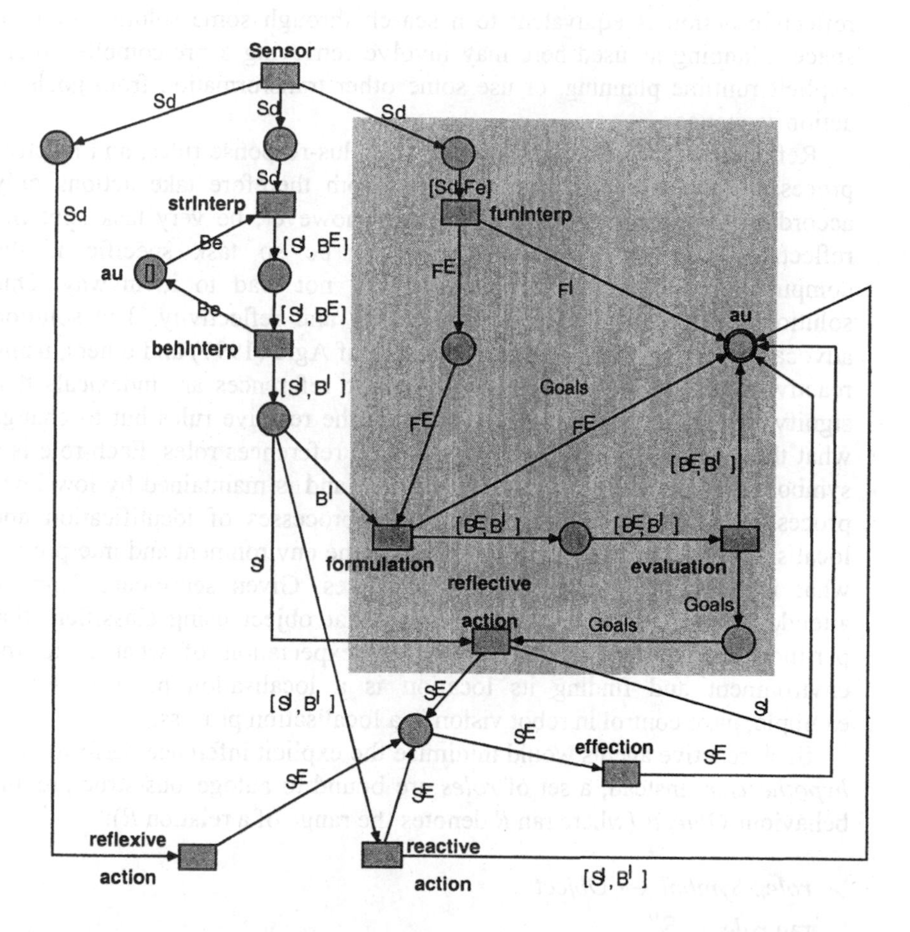


*Figure 3.* Petri net of reflexive, reactive and reflective reasoning of an agent $a_i$. It is drawn for a single agent and so all data and function symbols should be read as being subscripted with i. Each transition corresponds to the execution of a function. The double circles are virtual places: the second *au* virtual place is actually the same place as the single *au* place, being drawn that way only to simplify the diagram. The shaded background shows functions only executed during reflective reasoning and not during reflexive or reactive reasoning.

$B_i^E$ is determined from $F_i^E$ by *formulation* using techniques such as constraint satisfaction and abduction, and goals are then determined by comparing $B_i^E$ against $B_i^I$. These goals, current interpreted structure $S_i^I$ and the encoded knowledge enable reflective *action* to determine a partial order of changes to structure that $a_i$ believes will remove the differences between $B_i^E$ and $B_i^I$. This is a partial order on $B_i^E$, and so deciding on reflective action is equivalent to a search through some solution or plan space. Planning as used here may involve retrieving a pre-compiled plan, explicit runtime planning, or use some other transformation from goals to action sequences.

Reflexive processes use hardwired stimulus-response rules, and reactive processes have no long term memory. Both therefore take actions only according to the current state. They can, however, be very task specific; reflective processes would not need to be so task specific if the computational limitations of planning did not lead to it anyway. One solution is to adopt a hybrid of reactivity and reflectivity. The solution advocated by Horswill (1998) extends that of Agre (1997) and others, using reactive rules on deictic references. Deictic references are indexicals that signify FBS function. The idea is to keep the reactive rules but to change what they signify. Horswill calls his deictic references roles. Each role is a symbol that is bound to a set of properties and is maintained by low level processes. This separates the perceptual processes of identification and localisation. Looking at a specific place in the environment and interpreting what is there is an identification processes. Given sense-data from an attended object, identification identifies that object using classifiers that partition the sensory space. Having an expectation of what is in the environment and finding its location is a localisation process. As an example, gaze control in robot vision is a localisation process.

Such reactive agents would minimise the explicit inference performed by *hypothesiser*. Instead, a set of *roles* are bound to autogenous structure and behaviour *Objects* (where ran *R* denotes the range of a relation *R*):

$$role_i\colon Symbol \leftrightarrow Object$$
$$\text{ran } role_i \subseteq S_i^I$$

Example roles may be a door agent with a role `the-person-that-needs-clearance` that is bound to an avatar object in a virtual world, or a wall object in a CAD system that binds a role `the-object-that-needs-moving` to picture objects whenever the wall needs to move. Structure interpretation *strInterp* simply observes whatever object the role is bound to and maintains those properties. Because roles have functional meaning, explicit *formulation* and *evaluation* are not needed. Instead, knowledge of FBS function is implicit in reactive rules. Similarly, reactive action rules need not plan a partial order of structure changes from knowledge of the object being acted on. Instead, it just manipulates the object bound to the role.

For such reactive actions to work, though, requires either that the designer of the agent encode the reactive rules such that FBS functions of the agent are maintained with changing structure, or we allow the agent to learn from reflections such that in the future it can react in similar situations. That is, we either implement agents such that they react appropriately in different situations, or we implement learning such that it can recognise how to react appropriately in different situations. Analytical learners such as explanation based learning allows for action sequences constructed from reflection to be used to learn new reactive actions as macro actions. But learning new reactive actions from successful reflective ones may not be enough: to communicate requires common ground (Gero and Kannengiesser 2003). What if an agent is added dynamically to a system at runtime, and that new agent communicates content that is not understood. For simple communicated properties the receiver could respond with a `not understood` message and the initiator could describe the space of that variable. In general, though, to learn common ground in this way requires complex language learning that is beyond the scope of this work.

## 3. An Example

Figure 4 shows a view of a world on the Active Worlds platform[2]. In this section we describe an example using agents running on this platform. The use here of virtual worlds as a test environment is for convenience; we do not require or intend the theory only apply to virtual worlds. Interested readers should, therefore, refer to Maher, Smith and Gero (2003) and Smith, Maher and Gero (2004) for details of the AWAgent package and its use with the Active Worlds platform.

Often when citizens enter a virtual world for a meeting they all arrive at the same specified location and then stand “on top of each other”. Chair agents self-organise so as to relocate the avatars appropriately. When slides are being shown on a wall or the whiteboard being used they will reorganise, teleporting their assigned avatars with them, around the slide display or whiteboard. Afterwards they may reorganise around a central table in the meeting room.

The environment contains the agents listed in Table 1. Chair agents are constructed dynamically as required. Consider the chair agent $a_1$, labelled as "chair0" in the table (the others are identical). The FBS function of $a_1$ is to maintain the equilibrium location of the chair and of a citizen's avatar that is

[2] The virtual world platform Active Worlds, http://www.activeworlds.com, is one that we use for agent testing.

allocated to that chair. Chairs are implemented reactively here, though, and so do not explicitly represent or reason about FBS function.

*Figure 4.* Meeting room in Active Worlds

TABLE 1. Agents in the example environment.

| Agent No. | Symbol | Type of agent | Categories |
|---|---|---|---|
| 1 | "chair0" | Chair | *obstacle* |
| 2 | "chair1" | Chair | *obstacle* |
| 3 | "chair2" | Chair | *obstacle* |
| 4 | "chair3" | Chair | *obstacle* |
| 5 | "wall0" | Wall | *wall, goal* |
| 6 | "wall1" | Wall | *wall, goal* |
| 7 | "wall2" | Wall | *wall, goal* |
| 8 | "wall3" | Wall | *wall, goal* |
| 9 | "table" | Table | *obstacle, goal* |

Expected structure is a set of roles and a function from sensed object properties to an object category. For this implementation the categories are determined from the Active Worlds *Models*.

*Category: { obstacle, wall, goal, null }*

$$S_1^E = \{ classify: Model \leftrightarrow Category \mid$$
$$(\forall x: \text{dom } role \bullet \exists_1 y \bullet classify(x,y)) \}$$

Chair reactivity is implemented as subsumption, encoded as expectations of behaviour. They are a partial order on the priority of each subsumption behaviour contribution to resulting action vector, and so each expected behaviour applies to a set of sensed objects. Each expected behaviour is a function of the following type:

*Subsumption: Vector × Gain × Gain ⟶ Vector*

where the first *Gain* is a linear gain and the second is a radial gain. The linear and radial *Gain* are tunable parametrically, and so can be adapted should equilibrium not be found.

$B_i^E$ = < *{calcs ↦ obstacle, linear ↦ 4, radial ↦ 2,*
*inhibitedBy ↦ { } },*
*{calcs ↦ goal, linear ↦ 3, radial ↦ 100,*
*inhibitedBy ↦ {obstacle} },*
...>

Behaviour *obstacle* computes an exponentially decreasing (with distance) vector of repulsion from a sensed 3D object such as another chair. Behaviour *wall* and *goal* are similar, except that wall avoidance is like a ball bouncing off a hard surface. Goal attraction computes an exponentially increasing (with distance) vector of attraction a 3D object (the goal, such as a table, wall or whiteboard) towards the chair. Behaviour *random* computes a vector that is a random vector step. Behaviour *anger* maintains the anger of the chair. Every time that the chair is forced to move (the action vector, described below, is above a threshold), the chair gets a little more angry; every time it does not move it gets a little less angry. Obstacles compete and the strongest repulsion vector wins. Some vectors are then subsumed by others if their magnitude is large enough and superposition is used to arrive at a single movement vector for an agent.

*Sensor₁* is a pseudo-sonar sensor: it senses properties that are interpreted by the agent as an object category and vector from the agent to the closest point on each sensed object. Interpreted structure and behaviour compute properties of each sensed object. For example,

Ran $role_l$ = $S_1^I$

$S_1^I$ = <..., *{vector ↦ (50,0), category ↦ goa}*, ...>

$B_1^I$ ={ *{anger ↦ 0}*,
<..., *{vector ↦ (0,0), threshold ↦ 8, inhibits ↦ false }*,
...> }

$S_1^I$ is a sequence of structure objects, with each object being represented as a set of object properties. The particular sequence element shown is for the goal object. $B_1^I$ contains a behaviour object for this agent, holding the *anger* property, plus a sequence of behaviour objects.

Structure interpretation computes $Sd_l \times B_1^E \twoheadrightarrow S_1^I$ such that the following holds.

$$
\begin{aligned}
&pop(Sd_l) = o \wedge c{=}classify(o(model)) \wedge c \neq null \wedge \\
&(|ex(a_l)(location) - o(location)| < |role(c)(vector)|) \\
&\quad \Rightarrow O(o{=}role(c) \wedge c{=}o(category) \wedge \\
&\qquad\qquad\qquad\qquad o(vector){=}ex(a_l)(location) - o(location))
\end{aligned}
$$

where $O$ is the temporal "next" operator and $|\ |$ is the Euclidean distance metric. Structure interpretation therefore maintains vector and category properties for the nearest sensed object of each category. Behaviour interpretation similarly maintains $B_1^I$ as a behaviour vector for each $S_1^I$ object corresponding to *calcs* from $B_1^E$, and maintains *inhibits* from the behaviour vector and threshold. So interpreted structure is inferred from sense-data, and interpreted behaviour is interpreted from interpreted structure and expected behaviour. Reactive action then performs the subsumption inference, ensuring that the following holds.

$$
\begin{aligned}
&\forall\, n,m{:}1..\#\,B_1^I\,;\ \forall\, c_n,c_m{:}Category\ | \\
&\quad n \neq m \wedge c_n{=}S_1^I\,(n)(category) \wedge c_m{=}S_1^I\,(m)(category) \bullet \\
&\quad c_m \in B_1^E\,(n)(inhibitedBy) \wedge B_1^I\,(n)(inhibits) \\
&\qquad\qquad\qquad \Rightarrow B_1^I\,(m)(inbibits) = true
\end{aligned}
$$

The action vector is set to the sum of $B_1^I\,(n)(vector)$ over all $B_1^I$ for which $B_1^I\,(n)(inhibits){=}false$. Effection changes $ex(a_l)$ to relocate chair according to action vector.

Figure 5 shows one trial run of a set of chair agents around a table object that is the goal. This output is from a simulation written to capture and so better understand and illustrate the behaviour of the agents. In Figure 5 boundaries correspond to walls, the small rectangles correspond to the chairs, and the large rectangle corresponds to the table. In this trial the table is the goal and the chairs are assigned random initial locations. As can be seen, the chairs move towards the goal until an equilibrium is reached between attraction to the goal, repulsion from the object that is the goal, and repulsion from other chairs.

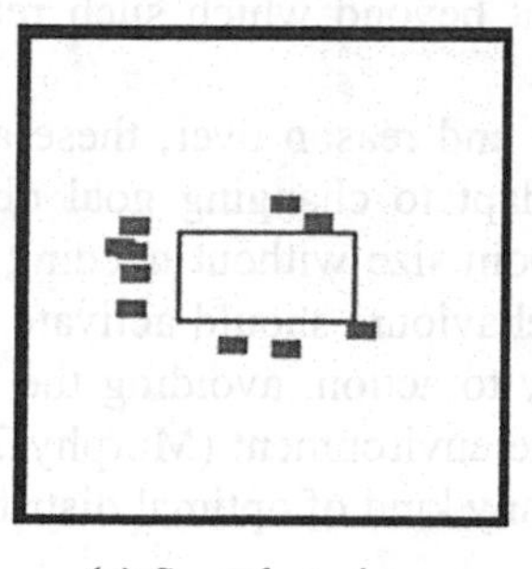

(a) Snapshot view.

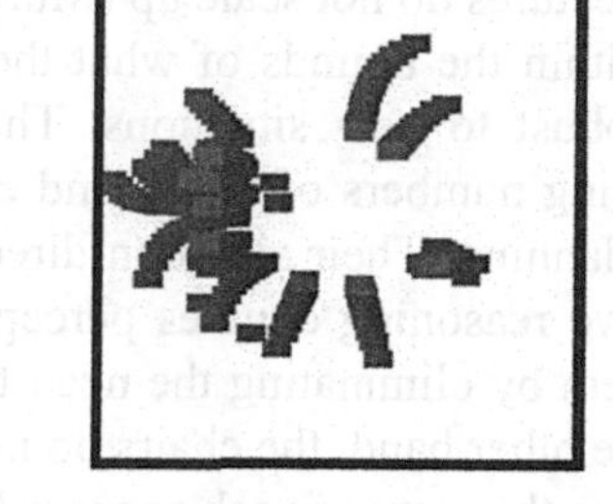

(b) Timelapse view.

*Figure 5*. Trial 1 simulation of the subsumption model of chair self-organisation: view after 40 iterations. (a) is a snapshot view, (b) is a timelapse view. The black rectangle around the boundary are the four walls listed in Table 1, the large polygon in the centre shows where the goal is *(table)*, and the small polygons show the positions of the chairs. The darker small polygons on the timelag view shows a timelag view of the chairs, indicating their movement towards the goal.

Notice however that they do not line up evenly around the goal. One price paid for a flexible, situated system of computationally efficient, independent agents is that their behaviour is not globally constrained. That is, some agents reach what they believe to be a minima when in fact it is only locally so. Whenever *anger* reaches a threshold the agent gets "upset" and moves off a random amount in a random direction. So some chairs quickly reach an equilibrium position around the goal, at which time their anger reduces. Some, however, will be prevented from reaching equilibrium by others. The slower ones get repelled by the faster ones, get angry, and take a random step. There are a number of ways of handling this, such as by decaying the behaviour parameters, or by having walls adapt to anger so as to change room geometry.

The subsumption implementation is reactive and situated but uses no concepts. A chair does not reason about what it is attracted to or repelled from and does not reason about the nature of the space it occupies. We could consider a reflective version as a planning problem using constraint satisfaction, with the constraint properties being the spatial locations of chairs and the constraints being both the geometry and on the set of citizens to be seated. One difference would be that it would now be a non-distributed task rather than a distributed reactive one. Here it must first pre-allocate spaces, whereas the reactive version treats the space as continuous. On the other hand we can extend the reflection of the agent by adding constraints. For example, that adjacent chairs should not chair citizens that do not like each other. A similar effect could be achieved by the reactive implementation by adding more subsumption modules, but it is well known from robotics work that there comes a point beyond which such reactive architectures do not scale up (Murphy 2000).

Within the bounds of what they perceive and reason over, these agents are robust to new situations. The chairs adapt to changing goal objects, changing numbers of chairs and changing room size without needing to do any planning. Their situation directs what behaviours should activate. Such reactive reasoning couples perception tightly to action, avoiding the frame problem by eliminating the need to model the environment (Murphy 2000). On the other hand, the chairs do not achieve any kind of optimal distribution because the view of each agent is local.

The reactive example here can be extended to facilitate designing from within the design. Consider, for instance, a set of room agents together with wall, floor, ceiling, furniture agents and so on. If the designer decides to change the shape of a room then adjoining walls, floors, ceiling, and furniture would automatically shift to new equilibrium locations. Combining it with explicit communication[3]. would allow adjoining rooms to negotiate to decide which wall(s) should move.

## 4. Conclusion

We have defined an environment as a distributed system of agents that communicate via message passing. Agents need not be distributed and need not explicitly use message passing. We defined multiple views of an environment because the world viewed by one agent need not be the same as another. This allows for situated agents to not only sense different subsets of objects but to then construct their own interpretations. We defined reasoning and communication by agents in sFBS terms so as to facilitate the description of such reasoning and interaction as it applies to designing without prescribing what that reasoning should necessarily look like. That is, instead of starting with representations of mental entities and then describing communication in those terms, we start with interaction and describe what needs to be represented so as to facilitate it.

This paper marks the beginning of a larger enterprise: to provide a formal underpinning to agency that better corresponds to the informal casting of designing as interactive, situated reflection. Future work needs to further constrain relations between structure, behaviour and function of agents with respect to situated agents and this framework, to refine the semantics of

---

[3] See (Maher, Smith and Gero 2003, Smith, Maher and Gero 2004) for further discussions of agent communication and Active Worlds.

*inform* and *request* messages in sFBS terms, and to describe what common ground between agents means in these terms.

## Acknowledgements

This work was supported in part by an Australian Postgraduate Award at the University of Sydney, Australia and by a University of Sydney Sesqui R&D grant.

## References

Agre, PE: 1997, *Computation and Human Experience*, Cambridge University Press, Cambridge, UK.

Brooks, RA: 1991, Intelligence without reason, *Proceedings of the 12th International Conference on Artificial Intelligence*, pp.569-595.

Fagin, R, Halpern, JY, Moses, Y and Vardi, MY: 1997, Knowledge-based programs, *Distributed Computing* **10**: 199-225.

Ferber, J: 1999, *Multi-Agent Systems: An Introduction to Distributed Artificial Intelligence*, Addison Wesley, Harlow.

FIPA: 2002, *Agent Communication Language Specifications*. Version H.
http://www.fipa.org/repository/aclspecs.html

Gero, JS and Kannengiesser, U: 2002, The situated function-behaviour-structure framework, *in* JS Gero (ed), *Artificial Intelligence in Design '02*, Kluwer, Dordrecht, pp. 89-102.

Gero, JS and Kannengiesser, U: 2003, Towards a framework for agent-based product modelling, *in* K Gralen and U Sellgren (eds), *International Conference on Engineering Design, ICED 03, Stockholm, Sweden*, The Design Society, August 2003, pp. 1621-1622. Abstract - full paper on CD-ROM, ISSN/ISBN: 1-904670-00-8.

Horswill, ID: 1998, Grounding mundane inference in perception, *Autonomous Robotics* **5**(1): 63-77.

Jennings, NR: 2000, On agent-based software engineering, *Artificial Intelligence* **117**: 277-296.

Lacey, AR: 1996, *A Dictionary of Philosophy*, Routledge, London.

Maher, ML, Smith, GJ and Gero, JS: 2003, Design agents in 3D virtual worlds, *IJCAI Workshop on Cognitive Modeling of Agents and Multi-Agent Interactions*, pp. 92-100.

Maher, ML, Smith, GJ and Gero, JS: 2004, Situated agents in virtual worlds, *Working Paper*, Key Centre of Design Computing and Cognition, University of Sydney (in preparation).

Murphy, RR: 2000, *An Introduction to AI Robotics*, MIT Press, Cambridge, MA and London.

Qian, L and Gero, JS: 1996, Function-behaviour-structure and their roles in analogy-based design, *Artificial Intelligence in Engineering Design, Analysis and Manufacture* **10**: 289-312.

Schön, DA and Wiggins, D: 1992, Kinds of seeing and their functions in designing, *Design Studies* **13**(2): 135-156.

Wegner, P and Goldin, D: 1997, Interaction as a framework for modeling, *in* PPS Chen (ed), *Conceptual Modeling: Current Issues and Future Directions*, Springer, Berlin and New York, pp. 243-257.

Wilson, RA and Keil, FC: 1999, *The MIT Encyclopedia of the Cognitive Sciences*, MIT Press, Cambridge, MA.

Wooldridge, M and Lomuscio, A: 2001, A computationally grounded logic of visibility, perception, and knowledge, *Logic Journal of the IGPL* **9**(2): 273-288.

JS Gero (ed), *Design Computing and Cognition'04*, 459-478


# AN ONTOLOGICAL AND AGENT BASED APPROACH TO KNOWLEDGE MANAGEMENT WITHIN A DISTRIBUTED DESIGN ENVIRONMENT

CAMELIA CHIRA, THOMAS ROCHE
*Galway Mayo Institute of Technology, Ireland*
and
DAVID TORMEY, ATTRACTA BRENNAN
*National University of Ireland, Galway*

**Abstract.** Design activities are becoming an increasingly global, complex, multidisciplinary, information and knowledge intensive activity. This paper proposes to address the problems associated with communication and knowledge management activities within a distributed design environment by adopting an Ontological-oriented Multi-Agent-based System (MAS) approach to knowledge management within a distributed design environment. This approach will utilize a Multi-Agent System architecture that will interactively assist in the communication and coordination of data, information and knowledge among multidisciplinary distributed/collocated design teams.

## 1. Introduction

Representing new models of the enterprise, the extended enterprise involves multiple organisations cooperating together to design, manufacture and market products. In the context of design, multiple users with concurrent access to multiple system resources generally have to collaborate in a distributed design environment (DDE) in order to achieve 'global' optima. As problems become more complex, teamwork is becoming increasingly important (Patel et al. 1997). From a design perspective, "complex design problems generally require more knowledge than any one single person possesses because the knowledge relevant to a problem is usually distributed among stakeholders" (Arias et al. 2000). Furthermore, designers working on one specific area of a project generally have a limited awareness of how the work of other designers working in the same project affects their own work. One immediate benefit of collaborative work is the coming together of participants with diverse skills, who, on sharing their knowledge, expertise and insight, create what is known as distributed cognition (Edmonds and Candy 1994). The collaboration of individuals with different insights, knowledge and expertise generally results in the generation of new insights, new knowledge and new expertise. Basically, someone viewing the problem from an alternative perspective can help in uncovering tacit aspects previously hidden. Therefore, cooperative multidisciplinary design teams dispersed across the enterprise have to be supported and, the management of distributed information and knowledge has to be facilitated. However, current DDE models fall short of providing an effective solution to this distributed cognition.

This paper proposes an agent-based architecture, entitled IDIMS (Intelligent Design Information Management System), to address the knowledge management and communication problems that can arise during the collaboration process between distributed individuals and teams in a virtual environment. Software agents, which are characterized by autonomy, pro-activeness and reactivity, and multi-agent systems, represent a potential solution to support knowledge management and communication activities with a DDE. Ontologies, which formalize concepts related to engineering design in an appropriate context, will be used for knowledge sharing, reuse and integration, and multi-agent systems, for enabling interoperation among distributed resources. The proposed architecture is intended to facilitate the access, retrieval, exchange and presentation of data, information, and knowledge to distributed design teams, in such a manner that their collective conceptual space is expanded, learning and creativity strategies are supported, and design solutions are enhanced. It should be stressed that the authors acknowledge that the engineering design ontology, which was built, is not exhaustive to any degree. It represents a generic engineering design ontology, which was created with the express goal of testing and validating the IDIMS MAS architecture[1].

The organization of this paper is as follows. Section two will review some of the problems that exist with relation to knowledge management activities within a DDE. Section three will describe the overall IDIMS system approach that will facilitate the effective and efficient communication and management of information and knowledge within a DDE. It will also detail the ontological element and the MAS architecture of the IDIMS system. The ontology is populated with information and knowledge relating to the engineering design of a smoke alarm. Finally, Section four will present conclusions and future work.

---

[1]Given the current state of work on the IDIMS project, testing and validation of the IDIMS MAS architecture will not commence for at least another year.

## 2. Knowledge Management Problems within a DDE

Manufacturers are continually coming under pressure to develop increasingly complex products in ever decreasing times, in order for them to remain competitive within dynamic global markets (Griffin 1997; Cooper and Kleinschmidt 1995). As a result, the design of products has become a multidisciplinary and knowledge intensive activity that can be geographically, temporally, functionally and semantically dispersed across the new extended enterprise environment. As a consequence, the practice of distributed design is ever increasing, yet is often inefficient. Recent research has shown (Lindemann et al. 2000; Vadhavkar 2001; Larsson et al. 2002) that some of the problems that are associated with collocated design such as poorly structured information and knowledge management strategies, and social problems, are intensified by the nature of distribution. This whole new problem set, related to the nature of distribution, adds to the significant challenges that already exist within the engineering design domain.

Former research conducted to address some of the socio-technical and knowledge mangement problems that distributed design environments present have included both ontological and non-ontological approaches. The PACT experiment (Cutkosky et al. 1993), looked at applying ontolgies and agent-based systems to support information and knowledge sharing among multidisciplinary design teams engaged in concurrent engineering activities. The SHADE methodology (Toye et al. 1993) also examined the possibility of applying agent technology to facilitate a "shared understanding" of the design process among design teams. Non-ontological approaches to knowledge management activities have looked at utilizing case bases to support design at the conceptual stages (Wood and Agogino1996). Song focuses on modeling design information that is based on tacit knowledge (Song et al. 2002), while Dong presents a technique that facilitates the representation of design that is based on a 'text analysis' of design documents (Dong and Agogino 1997).

All this research supports a common theme i.e. the importance of the effective communication of knowledge between distributed and collocated design teams, in order to achieve successful and effective design solutions. However, it should be noted that dispersed design teams, and the effective communication of knowledge between dispersed design teams also represent one of the largest impediments for design in the future (Pena-Mora et al. 2000; Agah and Tanie 2000). Moreover, the rapid growth of Information Communication Technologies (ICT) and the World Wide Web, have lead to a significant increase in the volume and dispersion of knowledge, thus creating obstacles for the collaboration of distributed teams. These obstacles include firstly, how to synthesis potentially disparate perspectives on a problem, and secondly how to manage the large amounts of information and knowledge that has been created as a result of virtual team collaboration (Fischer and Ostwald 2002). With regard to the latter issue, designers are more often than not bombarded by the plentiful supply of readily available information and knowledge. As a result, the increasingly critical problem for them is to find that information and knowledge which is pertinent to their task (Vianni et al 2000). This is particularly augmented in modern design environments, where products are so complex that externalised information and knowledge are not always readily accessible to the designer, during the critical stages of the design process. Research has shown that design engineers only engage 47% of their time in actually doing design work, while the other 53% of their time was spent at activities that included information retrieval, cost estimating, planning reviews, and social contact (Hales 1987).

Polanyi and the Japanese organization-learning theorist Ikijuro Nonaka indicate that knowledge has two forms i.e. implicit and explicit (Polanyi 1966; Nonaka and Takeuchi 1995; Nonaka and Konno 1998). *Implicit* or *tacit* knowledge represents personal knowledge stored in the individual's mental structures, (e.g. subjective insights, intuitions and hunches). This kind of knowledge is difficult to formalise, hence it is difficult to communicate or share. *Explicit knowledge* is the knowledge codified and systematically expressed in formal structures compatible with human language (e.g. libraries, archives, databases). Nonaka and Tagushi advocate an organisational management approach towards knowledge creation and management. Following this approach, DDE can be viewed as a whole that creates knowledge, disseminates it and incorporates it into products. It can be considered that a dynamic exists between tacit and explicit knowledge and that the process of generating new knowledge consists in the movement and management of knowledge between the two.

Given that design can be described as a problem solving process and considering that engineers tend to solve problems based on '*available knowledge*', it is important to ensure that appropriate knowledge is available at the correct time in the process (Lawson 1990; Cross 1994; Hubka and Eder 1996; Pahl and Beitz 1996). Research has also shown that design engineers depend largely on the information and knowledge to which they have 'ready' access (Kolb 1984; Finger and Dixon 1989; Coyne et al. 1990; Brennan 1996; Hubka and Eder 1996; Roche 1999). Thus, if information and knowledge is not easily accessible to design engineers, they are unlikely to seek or share knowledge and expertise and as a result, at best are likely to generate local rather than global 'optima' (Coyne et al. 1990; Lawson 1990; Roche 1999).

These problems are particularly augmented, when it comes to virtual collaboration within a distributed design environment. This is mainly due to the distribution of design information and knowledge, the distinct non-human interactive nature of distributed design activities, and the inherent dynamic nature of design information and knowledge. In summary, the lack of access to information and knowledge within the DDE can restrict the 'optimality' of the end solution, so from a designer perspective, the support for information and knowledge management is vital. Hence, designers need appropriate support mechanisms and tools to synthesise and manage appropriate information and knowledge from across all spectrums of the enterprise.

## 3. IDIMS System

Traditionally technologies that support virtual collaborative work have dealt with the key elements of infrastructure, human and knowledge on an individual basis. The overall IDIMS system architecture proposes to take a more holistic approach to distributed design by using ontologies that create an explicitly specified shared understanding of a DDE as a whole. Figure 1 illustrates the two main components of the IDIMS architecture, which are the ontology library and the Multi-Agent System. The ontology library forms a machine-readable pool of shared (i.e. design semantics which are the same for all agents that commit to the ontology library) data, information and knowledge representing the environment in which multiple agents act within the IDIMS system. The interoperation of multiple agents within the IDIMS system is achieved through an Agent Communication Language (ACL) and a common shared ontology library. The ontology library enables multiple agents within the IDIMS system to communicate, cooperate, negotiate and support each other.

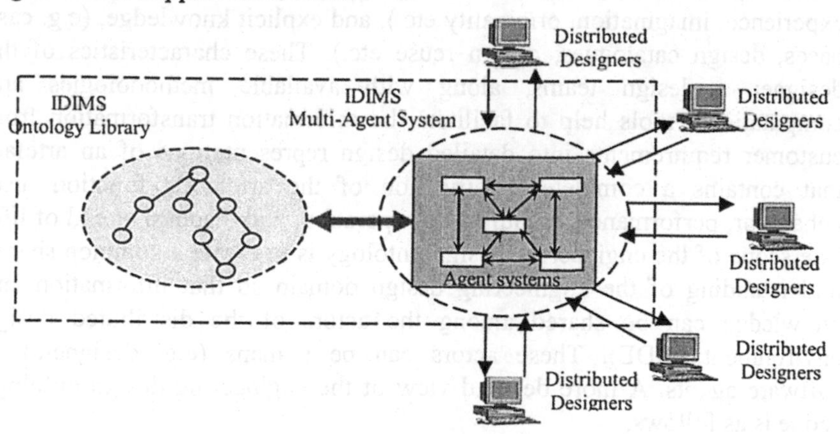


*Figure 1.* Holistic view of the IDIMS system

This cooperation process between software agents supported by the ontology library facilitates the efficient capture, storage, retrieval and dissemination of information and knowledge within a DDE. The IDIMS system approach therefore enables the management of the information and knowledge facets of the DDE as a whole.

### 3.1. IDIMS ONTOLOGY FOR ENGINEERING DESIGN

Ontologies are currently being utilised extensively within areas that require a knowledge-intensive approach to their methodologies and system development, such as knowledge engineering (Gruber 1993; Uschold and Gruninger 1996; Gaines 1997), knowledge representation (Artala et al. 1996), information integration (Bergamaschi et al. 1998; Mena et al 1998) knowledge management and organization, and agent-based design (Nwana 1996; Odell 2000; Chaib-draa and Dignum 2002).

The methodology that was used to create our ontology for the engineering design domain was based on the specification of the Methontology approach (Gomez-Perez 1999), which concentrates on three main steps to creating ontology; identify the *domain* of the ontology; identify the scope; build the ontology. The authors also defined ontological objectives, before building the ontology. A definition of engineering design is required to clearly identify the domain of the ontology, hence the authors have proposed a definition of the engineering design domain, based on an extensive literature review of well established design definitions (Finkelstein and Finkelstein 1983; Coyne et al. 1990; Hubka and Eder 1996; Pugh 1991; Nvbuomwan et al. 1996). Engineering design is a process where designers and design teams make decisions to solve specific problems, based on their own implicit knowledge, (e.g. creativity, knowledge, experience, imagination, originality etc.), and explicit knowledge, (e.g. case bases, design catalogues, design reuse etc.). These characteristics of the designer / design teams, along with available methodologies and computational tools help to facilitate the information transformation from customer requirements into detailed design representations of an artefact that contains a complete specification of the artefact's function, use, behaviour, performance, manufacture, operation, and disposal at end of life. The scope of the engineering design ontology is to create a common shared understanding of the engineering design domain so that information and knowledge can be shared among the actors of the distributed design environment (DDE). These actors can be humans (e.g. designers) or software agents. A more detailed view of the engineering design ontology scope is as follows:

- To support agent interoperations (besides an agent communication language, a common understanding of the concepts used among agents is necessary for a meaningful agent communication) as well as human-to-agent and agent-to-human interactions;

- To support knowledge management activities such as, gathering (capturing), organization (structuring, storing), distribution (sharing, dissemination), using (data information and knowledge use);
- To support design engineer users of the ontology;
- To support the formal and formalizable concepts of the engineering design domain and data structures from the design software tools;
- To describe the processing environment e.g. concepts (environment objects), relationships between concepts, axioms, inference rules.

The overall objective of the resultant ontology is to form a library that defines and describes infrastructural specific, human specific and knowledge specific concepts and relations (of a DDE). In this manner, any aspect of a DDE becomes system knowledge and therefore, the management of the divergent DDE facets is reduced to the management of commonly defined system knowledge.

The next step after identifying the ontology domain, scope and objectives was to build the ontology. The ontology for the engineering design domain was built using a software tool called Protégé-2000, which was developed at Stanford University for the purpose of building domain ontologies (Protégé 2000). The ontology is supported by generic information classes, or sub- ontologies, such as *material, product knowledge, design process knowledge*, (e.g. design tools, methodology, and standards); *manufacture processes*, and *resource information* (e.g. legislation, specification and standards). The ontology tree structure in Figure 2 also presents the hierarchical relationships between the main information classes at the top level and their related subclasses.

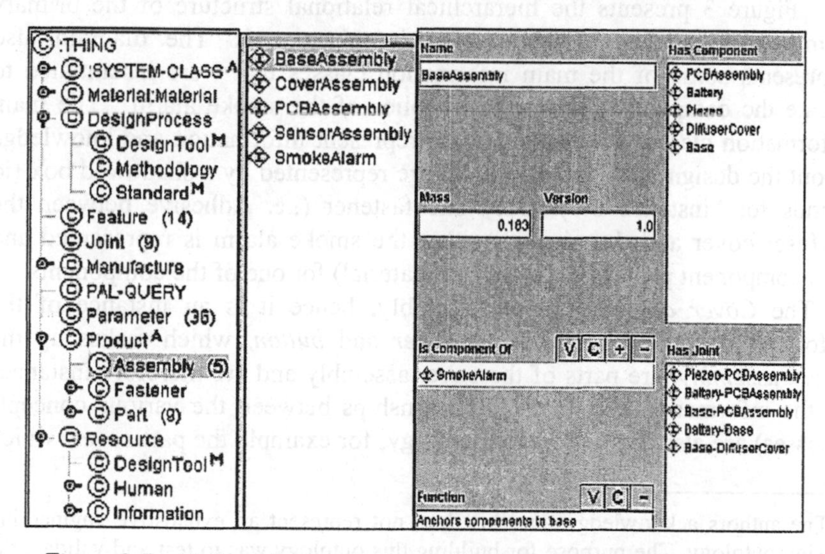


*Figure 2.* Protégé ontology environment displaying class tree structure, selected *Assembly* instances and associated slots

The engineering design ontology[2] categorises a product (the product that was used to populate the ontology with design information and knowledge was a smoke alarm) as an Assembly, Fastener, or Part. These classes are instantiated; for example the ontology has an instance of the *Assembly* class called *BaseAssembly.* Every instance of the class *Assembly* has assigned attributes, or slots, such as *name, mass, version, and function.* The slots *IsComponentOf,* and *HasComponent* are inversed. This means that if an assembly has a specific part, then that part must be a component of that assembly. Therefore when specifying relationships with either of the slots, you automatically generate the other slot relationship. In this way, these slots allow you to assign instances from the *Part* and *Assembly* classes, which are in corresponding relationship with the inherent instance. The *HasJoint* slot allows you to specify fasteners, from the *Fastener* class, that are used to join instances of *Part* and *Assembly* together for a specific instance of *Assembly.* These slots enable the definition of the structural relationships between all instances of the *Assembly* and *Parts* classes for the overall ontology. The Protégé tool also facilitates the application of constraints into the ontology, by the use of the Protégé Axiom Language (PAL). PAL is a superset of first-order logic, which can be used to express constraints about a knowledge base. Numerous constraints, which were needed for the ontology development, were defined and enforced in to the ontology. For example, a constraint had to be written in PAL that would prohibit an instance of the class *Assembly* from having a component of the same instance (i.e. an assembly cannot have a sub-assembly of itself).

Figure 3 presents the hierarchical relational structure of the primary components contained within a smoke alarm unit. The diagram also represents some of the main information classes that were instantiated to create the components and subassemblies of the smoke alarm. The main information classes that were used to represent information and knowledge about the design of the smoke alarm are represented by a shadowed box (io stands for "instance of"). Only one fastener (i.e. Adhesive between the diffuser cover and the piezeo) within the smoke alarm is represented and two component properties (mass and material) for one of the components.

The *Cover assembly* is an assembly; hence it is an instance of the information class *Assembly*. The *cover* and *button,* which makes up the *cover assembly*, are parts of the cover assembly and are therefore instances of the information class *Part*. Relationships between the various concepts (classes) are also defined in the ontology; for example the part *cover,* which

[2] The authors acknowledge that this does not represent an exhaustive engineering design ontology. The purpose for building this ontology was to test and validate the IDIMS MAS architecture.

is made from a certain material, is in a relationship *hasMaterial* with a specific instance of information class *Material*. In this way information and knowledge about the component *cover* can be defined by various material properties (e.g. name, type) represented by the information class *Material*.

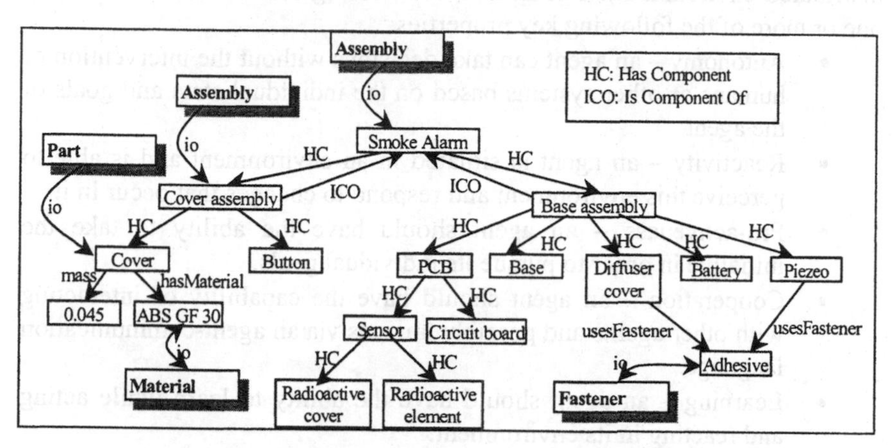


*Figure 3*. Decompositional mapping of smoke alarm design to ontology classes

Engineering design is a knowledge intensive activity, particularly within a DDE. It is for these reasons that we believe that representing the knowledge that resides within DDE in an ontological format can prove very beneficial to the overall activities of a DDE. Our overall ontology for engineering design forms a machine-readable pool of shared data, information and knowledge that aims to maximise the quality of information, knowledge sharing and reuse, the flexibility of the user interfaces, designer's learning and creativity, and the openness to the adoption of new strategies.

## 3.2. IDIMS MULTI AGENT SYSTEM

The functionality of the IDIMS system architecture greatly relies on the multi-agent system component that supports the system. Considered an important new direction in software engineering (Jennings 2000; Wooldridge and Ciancarini 2001), agents and multi-agent systems represent techniques to manage the complexity inherent in software systems and are appropriate for domains in which data, control, expertise and/or resources are inherently distributed (Jennings et al. 1998; Oliveira et al. 1999). A software agent is a computer system situated in an environment (and able to perceive that environment) that autonomously acts on behalf of its user, has a set of objectives and takes actions in order to accomplish these objectives (Nwana 1996; Jennings 2000; Wooldridge and Ciancarini 2001). Under the IDIMS architecture, autonomous software agents able to cooperate and communicate with each other and/or with the user enable interoperation among distributed resources in a virtual environment (and this further translates into more efficient design process operation and management in a distributed environment). The IDIMS software agents are characterised by one or more of the following key properties:

- Autonomy – an agent can take decisions without the intervention of humans or other systems based on the individual state and goals of the agent.
- Reactivity – an agent is situated in an environment and is able to perceive this environment and respond to changes that occur in it.
- Pro-activeness – an agent should have the ability to take the initiative in order to pursue its individual goals.
- Cooperation – an agent should have the capability of interacting with other agents and possibly humans via an agent-communication language.
- Learning – an agent should have the ability to learn while acting and reacting in its environment.
- Mobility – a mobile agent has the ability to move around a network in a self-directed way.

The ontological component of the IDIMS architecture plays a crucial role in supporting the multi-agent system. All agents within the IDIMS multi-agent system submit to the same engineering design ontology to facilitate knowledge sharing and reuse. From a technical perspective, the multi-agent element of the IDIMS architecture is supported by the following implementation strategies, Figure 4:

- The programming language selected for implementation is Java due to its rich library of functions tackling concurrency as well as security (Huget 2002), support for object-oriented programming techniques, code portability, native support for multithreading and introspection of object properties and methods (Bigus et al. 2002).
- Implementation of agent interoperation within the IDIMS system is supported by the Java Agent DEvelopment Framework (JADE), which is a software framework that facilitates the development of multi-agent systems and is compliant with the FIPA[3] specifications and performs all agent communication through message passing (using FIPA ACL to represent messages). Therefore, the communication language for agent interoperability within IDIMS is FIPA ACL and the content language is FIPA SL.

[3] FIPA (Foundation for Intelligent Physical Agents) is a standards organisation in the area of software agents whose goal is to develop specifications that maximize interoperability within and across agent-based systems (Labrou et al. 1999; Poslad et al 2000; http://www.fipa.org).

- The agents commit to a common shared engineering design ontology for which the RDF/RDFS model is available. The IDIMS software agents can manipulate the RDF/RDFS model of the ontology using a Java API developed at HP Labs called the Jena Semantic Web Tool Kit [http://www.hpl.hp.com/semweb/jena] which features statement and resource centric methods for manipulating an RDF model as a set of RDF triples or as a set of resources with properties respectively.

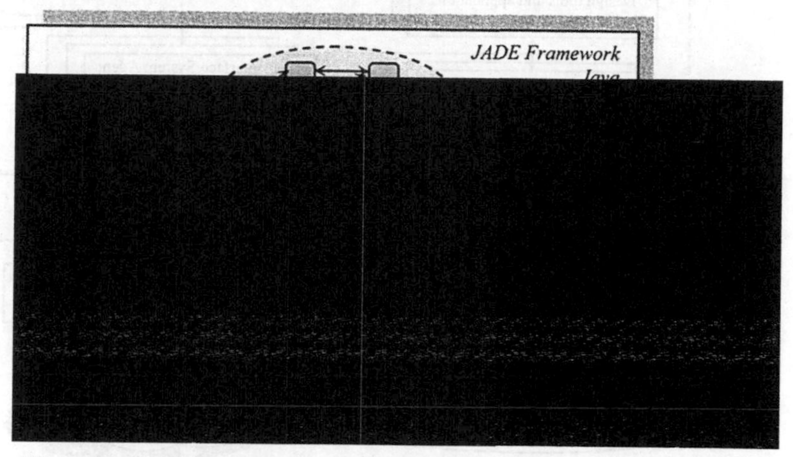


*Figure 4*. JADE Framework support for IDIMS architecture.

The overall MAS architecture of the IDIMS system is presented in Figure 5. Incorporating all strategies presented in Figure 4, the multi-agent component of the IDIMS system facilitates the access, retrieval, exchange and presentation of data, information and knowledge to distributed design teams and individuals.

These functions are achieved through agent systems such as:

- *Application Interface System Agent (AISA)* – captures application specific data and forwards it to be stored in the knowledge base in the format as defined by the ontology library. It deals with the retrieval of data, information and knowledge.
- *Middle Agents (MA)* – handle knowledge management activities e.g. store, structure, update, maintain, and retrieve with regard to Distributed Data, Case and Knowledge Bases.
- *Information Agents (IA)* – semantically explore the World Wide Web to provide the designer with relevant information.
- *User Interface System Agent* (UISA) – handles any user specific aspect within a distributed design environment e.g. collaboration with other distributed design participants, requests for browsing or

searching for design information. It deals with the access, exchange and presentation of data, information and knowledge.

- *Information Management Centre (IMC)* – coordinates the actions of all agent systems within the IDIMS multi-agent system.

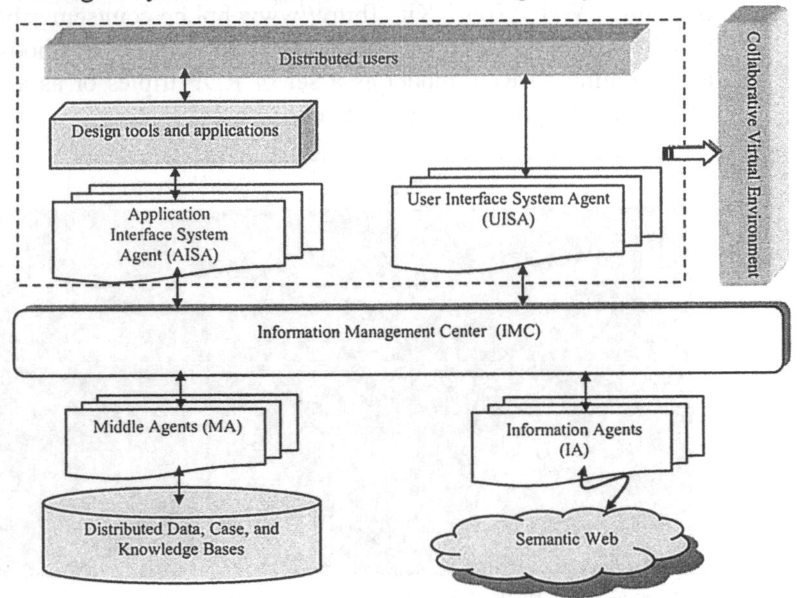


Figure 5. Multi-Agent System architecture for IDIMS system

The *Information Management Centre* supervises the entire functionality of the multi-agent element within the IDIMS system by coordinating the activities performed by the Application Interface System Agent, User Interface System Agent, Middle Agents and Information Agents. It consists of a set of mobile agents[4] that handle the request-response process between the different agent systems in IDIMS.

The *Middle Agents* form a collection of software agents capable of performing various knowledge management activities at the request of other agents within the IDIMS multi-agent system. For instance, these agents maintain the knowledge, data and case bases by storing new information or updating existing information/knowledge when a request was generated by an Application Agent through the Information Management Centre. Similarly, they also retrieve requested information as instructed by the User Interface System Agents. The Middle Agents do not necessitate a graphical user interface, as they are able to reach their objectives by performing the requested actions in the background.

[4] Mobile agents have the ability to move around a network in a self-directed way (Nwana 1996, Bradshaw 1997).

The *Information Agents* exploit the vast amount of information available in wide area networks such as the Internet and retrieve specific required information. These agents will reach their true potential when the web will be semantically enabled. The current trend of research suggests that the World Wide Web will be upgraded to a new environment i.e. Semantic Web where data will be defined and linked in such a way that it can be used by people and processed by machines (Berners-Lee 1998; Decker et al. 2000; Fensel 2000; Ramsdell 2001; Berners-Lee et al. 2001; Dumbill 2001; Hendler et al. 2002).

The *Application Interface System Agent* includes software agents integrated with the applications frequently used by designers in a distributed environment. The objective of these agents is to capture design data specific to design applications (e.g. 3D modelling tools, FEA tools) and forward this data for storage in the ontological knowledge base so that information and knowledge can be automatically made available to the distributed design teams. These agents can act without user interaction operating behind the applications used by the designer. They are able to cooperate with the Middle Agents through the Information Management Centre for knowledge management activities. Currently, the development process of the Application Interface System Agent focuses on the implementation of the Pro-Engineer Agent whose task is to manage the data generated by the CAD models manipulated in the Pro-Engineer CAD application.

The *User Interface System Agent* addresses the user needs in a DDE through user interface agents with objectives set by the designer and/or by other agents. Examples of such agents include the *Main User Agent, the Browser Agent* and the *Search Agent.*

*3.2.1. Main User Agent*

The Main User Agent represents the main interface between the IDIMS system and the designer. It has the ability of performing tasks received from its user or from other agents (through cooperation based on FIPA ACL). One of its main tasks is to retrieve the designer's profile. This profile is based on the agent learning from the designer's preferences and interaction with the IDIMS system, and then adapting its behaviour accordingly. Ongoing work focuses on capturing this profile of the designer and dynamically adapting the graphical user interfaces of all other agents within the User Interface System Agent (e.g. Browser Agent, Search Agent) so as to suit user preferences (learning and creativity are supported).

*3.2.2. Browser Agent*

Figure 6 shows the Browser Agent displaying general information on a selected assembly from the knowledge base. (Note: This screenshot is not indicative of the eventual interface to the IDIMS architecture). The Browser

Agent serves incoming messages that request, for example, functional and structural information on a specific product. Furthermore, this agent offers the functionality to browse all the products available in the knowledge base (ontology). The agent's graphical user interface displays general information on a product and also specific information regarding the product's components, joints and fasteners.

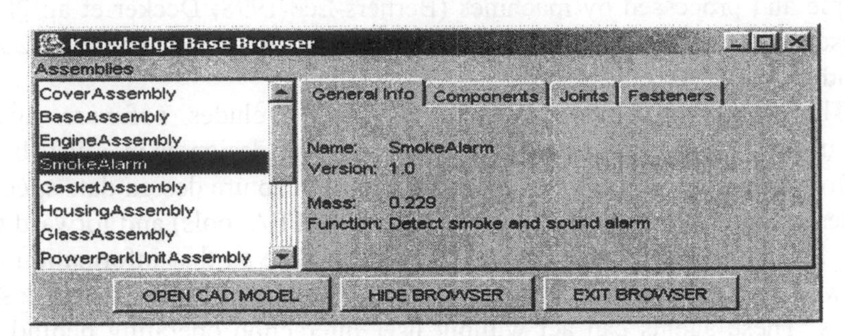


*Figure 6.* Browser Agent's graphical user interface: The General Info Tab

The components tab displays the subassemblies and parts that are components of the currently selected assembly. These components are displayed in a tree structure whose root is the parent assembly of the currently selected one and the first child node is represented by the selected assembly. The user can select a subassembly or part to access more design information. When a subassembly is selected from the tree structure, the Browser Agent will automatically select it in the list and display similar design information. Figure 7 shows the Components tab and how the agent reacts when the user requires more information on a part of the currently selected assembly.

Part information includes name, version, mass, function, parent assembly, material, process, label, features and parameters. The user can get more information on the features and parameters of a part (e.g. id, description, unit, value) by selecting it from the list. Additionally, more complete information on the material (e.g. density, colour, texture, fatigue, impact strength, tensile strength, Youngs Modulus, environmental indicator, hazard, recyclability, sustainability, biodegradability) is available at just one mouse click. The Joints and Fasteners tabs display a table containing all the joints respectively fasteners of the currently selected assembly. Figure 8 displays the Joints tab. Each joint is a triplet containing the two components that are joined together and the fastener used to join them. Selecting a joint from the table can access further information like the disassembly time for the joint and the disassembly tool.

The Main User Agent and the Search Agent can cooperate with the Browser Agent by sending request messages whenever the requirement of displaying product information arises. Table 1 presents all the 'REQUEST' type messages managed by the Browser Agent.

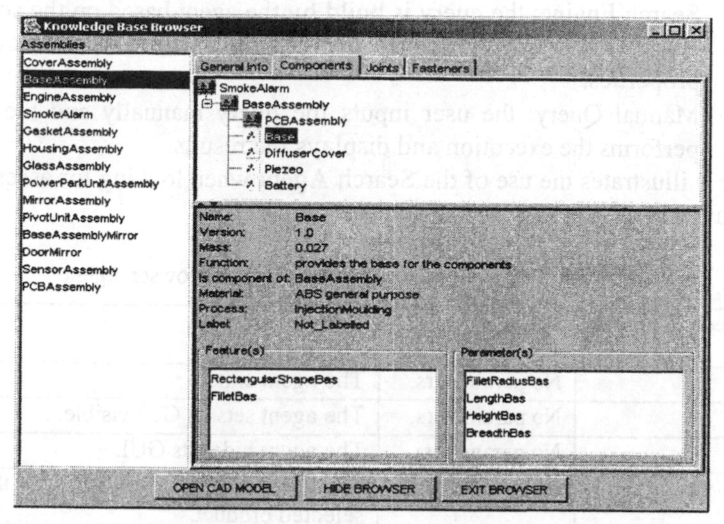


*Figure 7.* Browser Agent's graphical user interface: The components tab

Knowledge Base Browser

Assemblies: CoverAssembly, BaseAssembly, EngineAssembly, SmokeAlarm, GasketAssembly, HousingAssembly, GlassAssembly, PowerParkUnitAssembly, MirrorAssembly

General Info | Components | Joints | Fasteners

Joints table

| Description | First component | Second component | Fastener(s) |
|---|---|---|---|
| Base-DiffuserCover | Base | DiffuserCover | SnapFit2 |
| Battery-Base | Battery | Base | SnapFit5 |
| Base-PCBAssembly | Base | PCBAssembly | FlatHeadScrewWithInsert1 |
| Battery-PCBAssembly | Battery | PCBAssembly | SolderWeld0 |
| Piezeo-PCBAssembly | Piezeo | PCBAssembly | SolderWeld2 |

OPEN CAD MODEL | HIDE BROWSER | EXIT BROWSER

*Figure 8.* Browser Agent's graphical user interface: the Joints tab

The information displayed by the Browser Agent at the request of the user is the result of the cooperation process between the user's agent and the Middle Agents able to work directly with the ontology library in order to handle knowledge management activities.

### *3.2.3 Search Agent*

The Search Agent, Figure 9 provides the functionality to perform queries for products with specified constraints. Having the objective set by the user, the Search Agent cooperates with the Middle Agents who are able to retrieve the requested information from the ontology library. The RDF representation of the engineering design ontology is browsed using RDQL with the support of the Jena toolkit. The designer can use the Search Agent to query the knowledge base in two main ways as follows:

- Search Engine: the query is build by the agent based on the selected concepts and properties and the conditions specified for one or more properties.
- Manual Query: the user inputs the query manually and the agent performs the execution and displays the results.

Figure 9 illustrates the use of the Search Agent when looking for assemblies that have the mass between 0.02 and 0.1.

TABLE 1. Request messages served by the Browser Agent

| Message Content | Parameters | Action |
|---|---|---|
| exit | No parameters. | The agent exits. |
| show | No parameters. | The agent sets its GUI visible. |
| hide | No parameters. | The agent hides its GUI. |
| show-info | No parameters. | The agent displays information on the selected product. |
| show-components | No parameters. | The agent displays the components of the selected product (structure information). |
| show-joints | No parameters. | The agent displays the joints of the currently selected product. |
| show-fasteners | No parameters. | The agent displays the fasteners used in the selected product. |
| display | The name of the assembly to be displayed in browser. | The agent displays general information on the assembly with the given name (if exists). |

## 4. Conclusions and Future Work

This paper presented a preliminary work on an ontological oriented Multi-agent System based approach to knowledge management within a distributed design environment. Because the Distributed Design environment is extremely complex and knowledge oriented, there is a definite need for appropriate knowledge management strategies and support tools. The IDIMS system presented in this paper is intended to support and facilitate the management of information and knowledge for the purposes of achieving successful design solutions throughout the DDE. The ontology library coupled with multiple agents within the IDIMS system, enables the communication, cooperation and negotiation with each other in order to support distributed designers collocated within the distributed design environment. The Information Management Centre of the IDIMS system facilitates the seamless integration of multiple agents with the design knowledge base (ontology). This is achieved by defining design information and knowledge in ontological terms in order to semantically unify agent and designer communication, throughout the distributed design environment.

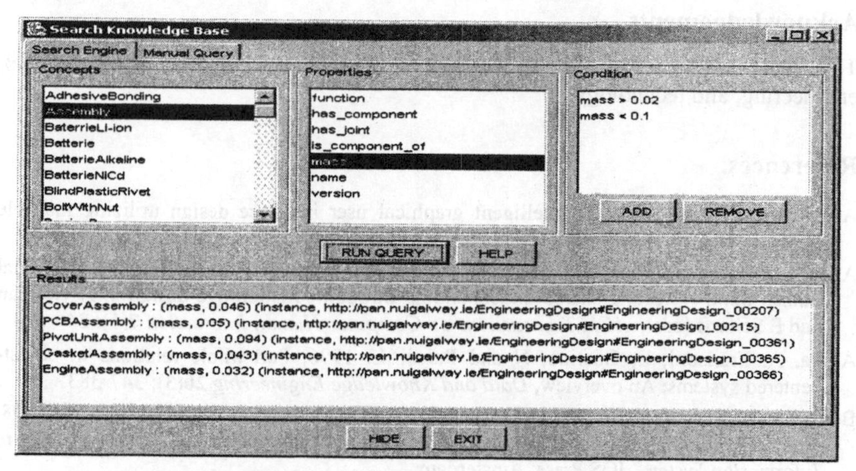


*Figure 9*. Search Agent's graphical user interface: Search Engine

Current and future work on the IDIMS system is focusing on developing a more robust ontology for representing the engineering design domain. This includes ontological alignment to current (and possible future) design standards. The multi-agent system development work is currently focusing on the Application Interface System Agents, which are being integrated into a Pro-E CAD environment through Jlink (J-Link is a java language toolkit for Pro/Engineer that makes it possible for the development of java programs that access the internal components of a Pro/Engineer session). The application agents will capture specific data and information from a Pro/E geometric model and translate it to an internal format as defined by the ontology's classes and slots. On going development work on the User Interface System Agents will ensure that user interfacing between participants of the DDE will be fully supported. Future work on the IDIMS system will also focus on the testing, validation and improvement of the proposed multi-agent system architecture and ontological knowledge base. IDIMS system work has also seen the commencement of research into human-computer interaction, collaborative virtual environments, information visualisation and particularly the user interface design.

While standard technologies usually offer pinpoint solutions (because they are dealing with data and rarely with information or knowledge), ontologies and MAS systems represent a novel approach to address the knowledge management problems that exist within highly knowledge-oriented environments such as distributed design. Capable of managing and processing not only data and information but also knowledge, the multi-agent solution takes a more holistic approach to the design environment by supporting the designer's decision making process towards achieving a 'global' optima.

## Acknowledgements

This work is supported by the IRC (Irish Research Council) set for science, engineering, and technology.

## References

Agah, A and Tanie K: 2000, Intelligent graphical user interface design utilizing multiple fuzzy agents, *Interacting with Computers* **12**(5): 529-542.

Arias, E, Eden, H, Fischer, G, Gorman, A and Scharff, E: 2000, Transcending the individual human mind - creating shared understanding through collaborative design *in* A Gorman and E Scharff (eds), *ACM Transactions on Computer-Human Interaction* **7**(1): 84 - 113.

Artala, A, Franconi, E, Guarino, N and Pazzi, L: 1996, Part-whole relations in object-centered systems: An overview, *Data and Knowledge Engineering* **20**(3): 347-383.

Bergamaschi, S, Castano, S, De Capitani di Vimercati, S, Montanari, S and Vincini, M: 1998, An intelligent approach to information integration *in* N Guarino (ed), *Formal Ontology in Information System,* IOS Press, Amsterdam.

Berners-Lee, T: 1998, Semantic web road map. http://www.w3.org/DesignIssues /Semantic.html, World Wide Web Consortium.

Berners-Lee, T, Hendler, J and Lassila, O: 2001, The semantic web, *Scientific American* **284**(5): 34-43.

Bigus, JP, Schlosnagle, DA, Pilgrim, JR, Mills, WN and Diao, Y: 2002, ABLE: A toolkit for building multiagent autonomic systems, *IBM Systems Journal* **41**(3): 350-371.

Bradshaw, JM: 1997, An introduction to software agents, *in* JM Bradshaw (ed), *Software Agents,* MIT Press, Cambridge.

Brennan, A: 1996, A graphical user interface design tool to facilitate managerial learning, *PhD Thesis*, CIMRU, Univeristy College Galway.

Chaib-draa, B and Dignum, F: 2002, Trends in agent communication language, *Computational Intelligence* **18**(2).

Cooper, RG and Kleinschmidt, E J: 1995, Benchmarking the firm's critical success factors in new product development, *Journal of Product Innovation Management* **12**(5): 374-391.

Coyne, RD, Rosenman, MA, Radford, AD, Balachandran, M and Gero, JS: 1989, *Knowledge Based Design Systems,* Addison Wesley, Boston, MA.

Cutkosky, MR, Engelmore, RS, Fikes, RE, Genesereth, MR, Gruber, TR, Mark, WS, Tenenbaum, JM and Weber, JC: 1993, PACT: An experiment in integrating concurrent engineering systems, *IEEE Computer Special Issue on Computer Supported Concurrent Engineering* **26**(1): 28-37.

Cross, N: 1994, *Engineering Design Methods: Strategies for Product Design*, John Wiley & Sons, Chichester, England.

Decker, S, Melnik, S, Van Harmelen, F, Fensel, D, Klein, M, Broekstra, J, Erdmann, M and Horrocks, I: 2000, The semantic web - on the respective roles of XML and RDF, *IEEE Internet Computing*, November, pp. 63-74.

Dong A and Agogoino AM: 1997, Text analysis for constructing design representations, *Artificial Intelligence in Engineering* **11**(2): 65-75.

Dumbill, E: 2001, Building the semantic web, *http://www.xml.com/pub/a/ 2001/03/07/buildingsw.html*, XML.com.

Edmonds, EA and Candy, L : 1994, Support for collaborative design: Agents and emergence, *Communications of the ACM* **37**(7).

Fensel, D: 2000, *Ontologies: A Silver Bullet for Knowledge Management and Electronic Commerce*, Springer, Berlin.

Finger, S and Dixon, JR: 1989, A review of research in mechanical engineering design - part 1- descriptive, prescriptive and computer based models of the design process, *Research in Engineering Design*, Springer **1** pp. 51-67.

Finkelstein, L and Finkelstein, A: 1983, Review of design methodology, *IEE Proceedings, Part A* **130**: 213--222.

Fischer, G and Ostwald, J: 2001, Knowledge management: Problems, promises, realities and challenges, *IEEE Intelligent Systems*, pp. 60-72.

Gaines, B: 1997, Editorial: Using explicit ontologies in knowledge-based system development, *International Journal of Human-Computer Systems* **46**: 181.

Griffin, A: 1997, PDMA research on new product development practices: Updating trends and benchmarking best practices, *The Journal of Product Innovation Management* **14**(6), 429-458

Gomez-Perez, A: 1999, Ontological engineering: A state of the art, *Expert Update, Ontono* **2**(3): 38-43.

Gruber, TR: 1993, A translation approach to portable ontology specification, *Knowledge Acquisition* **5**(2): 199-220.

Hales, C: 1987, Analysis of the engineering design process in an industrial context, *PhD Thesis*, Department of Engineering, Cambridge, University of Cambridge.

Hendler, J, Berners-Lee, T and Miller, E: 2002, Integrating applications on the semantic web, *Journal of the Institute of Electrical Engineers of Japan* **122**(10): 676-680.

http://www.fipa.org, Foundation for Intelligent Physical Agents.

Hubka, V and Eder, WE: 1996, *Design Science*, Springer-Verlag, London.

Huget, MP: 2002, Desiderata for agent oriented programming languages, University of Liverpool.

Jennings, NR: 2000, On agent-based software engineering, *Artificial Intelligence* **117**(2): 277-296

Jennings, NR, Sycara, KP and Wooldridge, M: 1998. A roadmap of agent research and development, *Journal of Autonomous Agents and Multi-Agent Systems* **1**(1): 7-38.

Labrou, Y, Finin, T and Peng, Y: 1999, Agent communication languages: The current landscape, *IEEE Intelligent Systems* **14**(2):45-52.

Larsson, A, Torlind, P, Mabogunje, A and Milne, A: 2002, Distributed design teams: Embedded one-on-one conversations in one-to-many, *Common Ground Conference*, London UK.

Lawson, B: 1990, *How Designers Think: The Design Process Demystified*, 2nd Ed, Architectural Press, London.

Lindemann, U, Anderl, R, Gierhardt, H and Fadel, G.M: 2000: 24 hr design and development - an engine design project, *Proceedings of CoDesigning*, New York, Springer.

Mena, E, Kashyap, V, Illarramendi, A and Sheth, A: 1998, Domain specific ontologies for semantic information brokering on the global information infrastructure, *in* N Guarino (ed), *Formal Ontology in Information Systems*, IOS Press, Amsterdam.

Nonaka, I and Konno, N: 1998, The concept of "Ba": Building a foundation for knowledge creation, *California Management Review* **40**(3): 40-54.

Nonaka, I and Takeuchi, H: 1995, *The Knowledge Creating Company: How Japanese Companies Create the Dynasties of Innovation*. New York, Oxford University Press.

Nvbuomwan, N, Sivaloganathan, S and Jebb, A: 1996, A survey of design philosophies, models, methods and systems, *Part B Journal of Engineering Manufacture* **210**: 301-19.

Nwana, HS: 1996, Software agents: An overview, *Knowledge Engineering Review* **11**(3): 1-40.

Odell, J (ed.): 2000, Agent technology - green paper, OMG - *Agent Platform Special Interest Group*.

Oliveira, E, Fischer, K and Stepankova, O: 1999, Multi-agent systems: which research for which applications, *Robotics and Autonomous Systems* **27**: 91-106.

Pahl, G and Beitz W: 1996, *Engineering a Systematic Approach*, London, Springer.

Patel, U, D'Cruz, MJ and Holtham, C: 1997, Collaborative design for virtual team collaboration: A case study of jostling on the web, *ACM Press*, New York, pp. 289-300.

Pena-Mora, F, Hussein, K, Vadhavkar S and Benjamin, K: 2000, CAIRO: A concurrent engineering meeting environment for virtual design teams, *Artificial Intelligence in Engineering* **14**: 202-219.

Polanyi, M: 1966, *The Tacit Dimension*, Garden City, New York: Doubleday & Co.

Poslad, S, Buckle, P and Hadingham, R: 2000, The FIPA-OS agent platform: Open source for open standards, *Proceedings of the 5th International Conference and Exhibition on the Practical Application of Intelligent Agents and Multi-Agent*, UK.

Protégé: 2000, http://protege.stanford.edu/.

Pugh, S: 1991, Total design, *Addison Wesley*, Wokingham, England.

Ramsdell, JD: 2001, A foundation for a semantic web, *MITRE Corporation Publication*.

Roche, T: 1999, Development of a design for the environment workbench, *Report*, CIMRU, Industrial Engineering Dept. Galway, UCG.

Toye, G, Cutkosky, MR, Leifer, LJ, Tenenbaum, JM and Glicksman, J: 1992, SHARE: A methodology and environment for collaborative product development, *Post-Proceedings of the Second Workshop on Enabling Technologies: Infrastructure for Collaborative Enterprises (IEEE)*.

Uschold, M and Gruninger, M: 1996, Ontologies: Principles, methods and applications, *The Knowledge Engineering Review* **11**(2): 93-136.

Vadhavkar, S: 2001, Team interaction space effectiveness for globally dispersed teams, theory and case studies, *ScD Thesis*, MIT Cambridge, MA.

Vianni, V, Parodi, A, Alty, JL, Khail, C, Angulo, I, Biglino, D, Crampes, M, Vaudry, C, Daurensan, V and Lachaud, P: 2000, Adaptive user interface for process control based on multi-agent approach, *AVI 2000 Conference Proceedings*, Palermo, Italy, pp. 201-204.

Wood, WH and Agogino, AM: 1996, A case-based conceptual design information server, *Computer Aided Design* **28**(5): 361-369.

Wooldridge, M and Ciancarini, P: 2001, Agent-oriented software engineering: The state of the art, *in* P Ciancarini and M Wooldridge (eds), *Agent-Oriented Software Engineering*, Springer-Verlag Lecture Notes, AI Volume 1957.

JS Gero (ed), *Design Computing and Cognition'04*, 479-497


# MULTIJADE

## *A Domain Independent Multiagent Active Design Documents Environment*

FLÁVIO M VAREJÃO, RODRIGO L GUIMARÃES, CLAUDIA S BRAUN
*Universidade Federal do Espírito Santo, Brazil*

and

FERNANDA A GAMA
*Companhia Siderúrgica de Tubarão, Brazil*

**Abstract.** In this paper, we present MultiJADE, a domain independent multiagent active design documents environment. MultiJADE uses multiagent technology to support activities in concurrent and distributed design systems. It is able to deal with various domains and allows multiple designers to work simultaneously. MultiJADE is based on the Active Design Documents (ADD) approach. It uses the ADD's potential for capturing and generating design rationale for supporting conflict mitigation whenever conflicts between designers arise during a concurrent design. This paper shows how MultiJADE enables the construction of concurrent and distributed design systems and how they may be used for mitigating conflicts during the design process. We also exemplify the use of MultiJADE with a prototype system that supports the decision about human resource allocation costs on a concurrent and distributed design.

## 1. Introduction

A concurrent and distributed design process involves people with different domain expertises working in diverse locations on different portions of the overall design solution. Since there are intersections among the design parts, conflicts often occur. The designers need to collaborate with each other to solve these conflicts. In order to optimize the conflict mitigation, designers need to understand the rationale underneath the decisions of the other designers involved in the conflict. They generally get this rationale during personal meetings. However, this type of procedure usually delays conflict mitigation once designers may not be always available for participating on the meetings.

Active Design Document (ADD) systems (Garcia and Howard 1993) use the metaphor of a computational design assistant to support designers on the development and documentation of a design. ADD systems use a domain specific design process model to capture information about the design decisions during the design process. The computational model and the information captured during the design enable ADD systems to generate the rationale about the design over demand. Even though ADD systems are a powerful tool for supporting designers on their work, they are not able to deal with distributed and concurrent design problems because they are developed to assist a sole designer.

Multiagent systems (MAS) are a collection of distributed autonomous systems capable of accomplishing complex tasks through interaction, coordination, collective intelligence and emergence of patterns of behavior. The integration of the multiagent approach with the ADD system features resulted on the MultiADD (Multiagent Active Design Document) approach. MultiADD systems support cooperative development on a concurrent and distributed design problem (Garcia and Vivacqua 1997).

MultiADD represents an evolution of the ADD approach. But, likewise ADD systems, MultiADD systems are domain dependent. Therefore, for each design application, it is necessary to construct a new MultiADD system from scratch. This increases significantly the cost of using a MultiADD system and limits the wide dissemination of the MultiADD approach.

This paper presents MultiJADE (Multiagent Java-based Active Design Documents Environment), a multiagent environment that makes the construction of MultiADD systems easier on different design problem domains.

MultiJADE uses the ADD features to support participants of a concurrent and distributed design on making their individual design decisions and on negotiating their conflicts. MultiJADE supports the creation of different design models that may be used by distinct designers on various designs. MultiJADE also improves the communication between designers since they may use it to send messages by e-mail or chat and supports quick conflict mitigation because the designers may better understand the rationale of other designers, without requiring personal meetings.

In the remainder of this paper, we provide an overview of MultiJADE. Section 2 briefly reviews the Active Design Documents and the MultiADD approaches. Section 3 presents MultiJADE describing its logical and physical architecture. Section 4 presents an example of the application of MultiJADE. Section 5 summarizes the MultiJADE benefits and outlines the future directions of research.

## 2. Active Design Documents

An ADD system uses a design process model of a particular domain to support the designer. It observes the designer work, makes suggestions and checks if the designer decisions are consistent with its model. During this process, ADD also documents the design rationale. ADD can capture and generates the design rationale because its computational model allows the interpretation of the designer decisions.

The ADD approach is based on the metaphor of having a computational apprentice supporting and following the designer decisions. The apprentice uses its design process model for making decisions (when asked by the designer) or for checking the designer decisions (when the designer is making decisions by him/herself). In the latter case, whenever the apprentice expectations for the designer decisions are correct, it means that the apprentice model is compatible with the design process. Otherwise, the system asks the designer to check its proposed decisions or adjust the apprentice model in order to reflect its reasoning. At the end of the design process, the apprentice will be able to generate rationale for explaining the design from its model.

### 2.1. REPRESENTATION LANGUAGE

An ADD computational model is represented by a parametric dependency network. Parameters are related to each other by dependency links. According to Garcia et al (1997), there are three types of parameters: primitive, derived and decided.

Primitive parameters are used to represent the input specifications of a design case. Derived parameters apply functions over parameter values. Decided parameters choose between their associated alternatives, by the analysis of criteria and restrictions. There is also a utility function to balance the relative importance of each criterion.

Figure 1 shows an illustrative example of this representation. In this example, the designer uses an ADD system to help him/her decide which material should be used to construct a bridge in the sea level.

“Design Load” (expected load), “Height” and “Width” are primitive parameters used for specifying the design case. The user inputs their values during the design process. “Pressure” is a derived parameter. It is computed by applying the formula:

$$(\text{Design Load} * 9.8) / (\text{Height} * \text{Width}) \quad (1)$$

over the “Design Load”, “Height” and “Width” values. The value 9.8 represents the gravity acceleration in $m/s^2$.

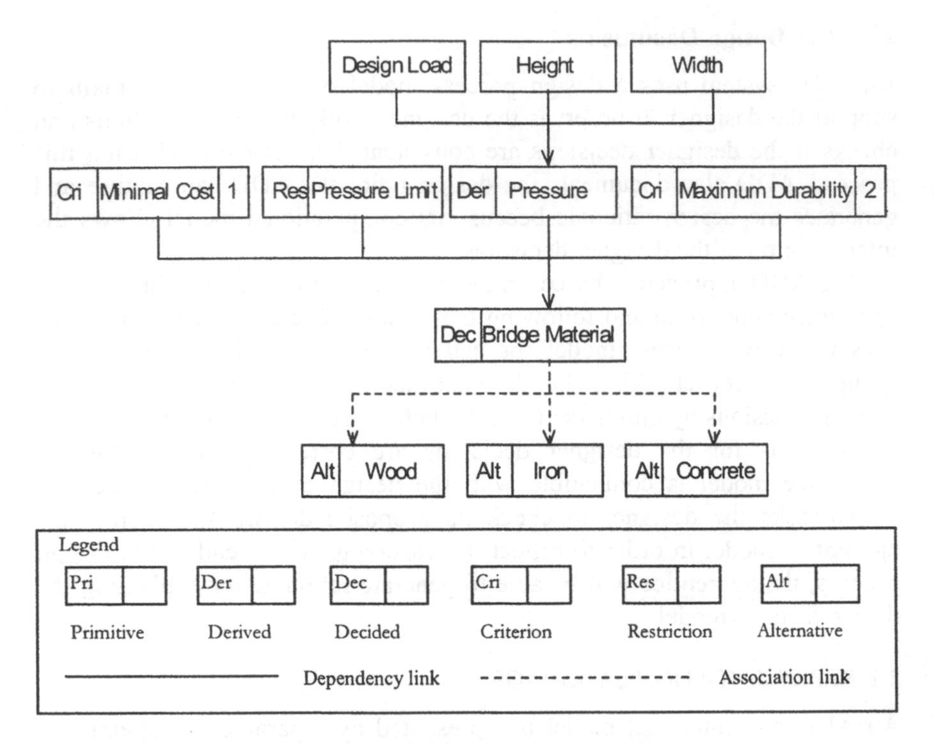


*Figure 1.* Bridge decision network

The last parameter of this model is "Bridge Material", a decided parameter. There are three material alternatives associated to this parameter: "Wood", "Iron" and "Concrete". Each alternative has associated properties, criteria and restrictions.

Table 1 summarizes the properties considered by this decision problem.

TABLE 1. Material properties

| Bridge material | Durability | Cost (per square meter) | Supported Pressure |
|---|---|---|---|
| Iron | 50 years | US$ 100,00 | 100 kN/m$^2$ |
| Concrete | 30 years | US$ 70,00 | 60 kN/m$^2$ |
| Wood | 10 years | US$ 25,00 | 20 kN/m$^2$ |

"Durability" represents the material durability in years, "Cost" represents the cost per square meter in dollars and "Supported Pressure" represents how much load the material may support in N/m$^2$ (Newton per square meter). Figure 1 also shows that the "Bridge Material" parameter has associated a "Pressure Limit" restriction and the "Minimal Cost" and "Maximum Durability" criteria. The "Pressure Limit" restriction indicates that the "Pressure" value must be smaller than the "Pressure" property value of the alternative. The "Minimal Cost" criterion indicates that the decided parameter should evaluate better the alternatives with smaller prices. The "Maximum Durability" criterion indicates that more durable alternatives also have better evaluations. The associated number put along the criteria (1 to "Minimal Cost" and 2 to "Maximum Durability") are weights. They indicate that the "Maximum Durability" criterion is twice as important as "Minimal Cost".

## 2.2. JADE

An active design document is a domain dependent system. Whenever an active design document is needed on a domain, a new system must be built from scratch. Varejão and Pessoa (2002) developed JADE (Java-based Active Documents Environment), a computational tool that may be easily used for creating ADD systems on different domains. JADE also increases the types of interaction between the human and the computational system and makes the computational environment extensible and customizable.

JADE enables the reuse of the ADD parametric representation language. It also provides domain independent design, knowledge acquisition and explanation modules and interfaces that may be used on different active design document systems. Therefore, it speeds up and makes easier the whole construction of an ADD system. JADE is an object-oriented system based on a class framework shown in Figure 2. It provides classes describing each type of parameter of the parametric representation language. It is only necessary to create instances of these classes for constructing a new parametric network.

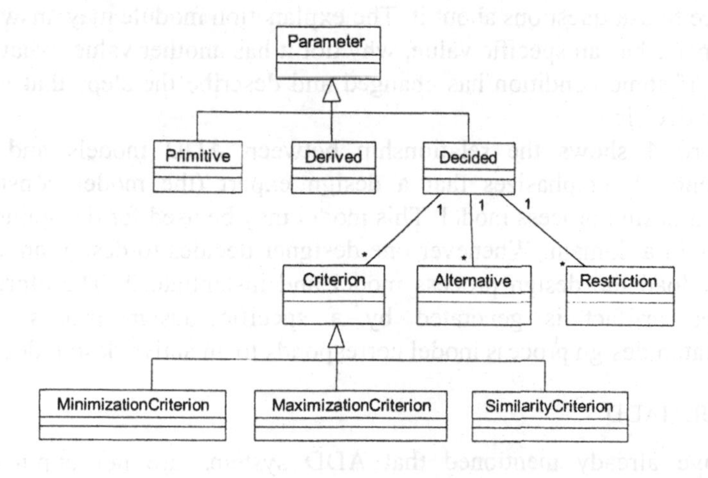


*Figure 2.* JADE object-oriented framework

Figure 3 shows the JADE architecture. The knowledge engineer (or a designer) uses the knowledge acquisition interface to create an ADD design process model. This knowledge base is constructed on the receptacle module. After constructing the model, the engineer saves it in the model library. Whenever a designer wants to design an artifact based on the saved model, he may ask JADE to load the model. In other words, the designer may create an ADD system by only loading a design process model. JADE automatically generates the user interfaces (for design, knowledge acquisition and explanation) of the new ADD system.

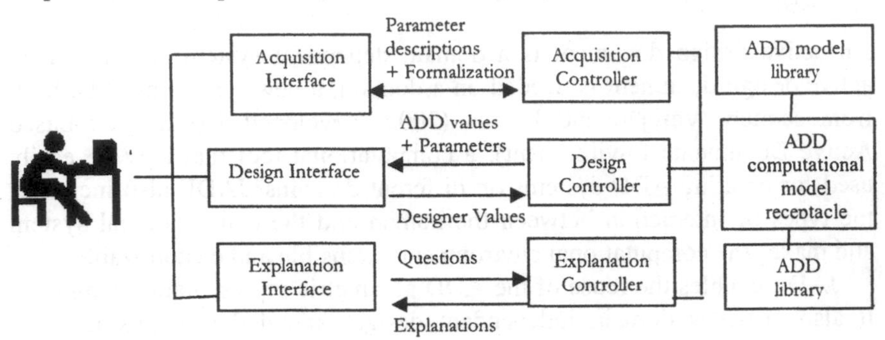


*Figure 3.* JADE architecture

The designer uses the design interface to work on a design case getting support from the ADD system inference capabilities. Once the designer starts the design process, an active design document is created and may be saved on the ADD library. Whenever the design case rationale is needed, the user may ask JADE to load the active document and utilize the explanation interface to ask questions about it. The explanation module may answer why a parameter has an specific value, why not it has another value, what would happen if some condition has changed and describe the steps that led to a specific decision.

Figure 4 shows the relationship between ADD models and active documents. It emphasizes that a design expert (the model constructor) creates a design process model. This model may be used for designing many artifacts in a domain. Whenever one designer decides to design an artifact, he may load the design process model and instantiate it. Therefore, each designed artifact is generated by a specific design process. Every instantiated design process model corresponds to an active design document.

## 2.3. MULTIADD

We have already mentioned that ADD systems are not applicable to concurrent and distributed design problems. They are single user systems. However, a large part of real design problems involve distributed and collaborative systems, which are beyond the capabilities of an individual agent. The capacity of an intelligent agent is limited by its knowledge, its resources, and its perspective. This bounded rationality is one of the underlying reasons for creating problem-solving organizations (Sycara 1998).

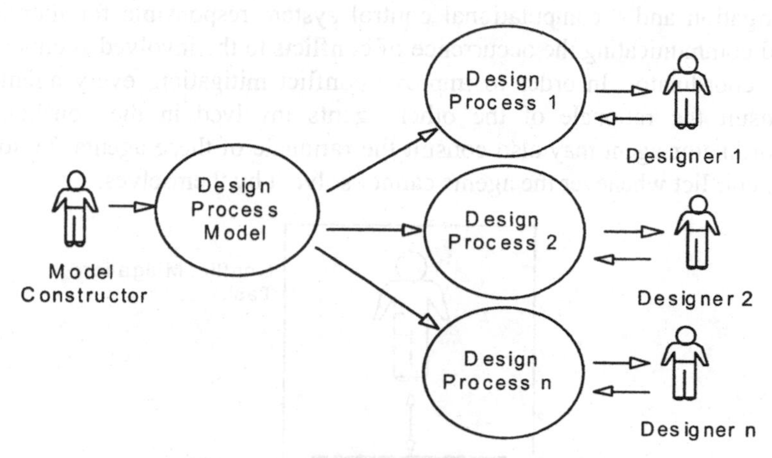


*Figure 4*. ADD model instantiation

Over the past few years, multiagent systems have become a crucial technology not only for effectively exploiting the increasing availability of diverse, heterogeneous and distributed on-line information sources, but also as a framework for building large, complex and robust distributed information processing systems which exploit the efficiencies of organized behavior (Lesser 1999).

Therefore, the expansion of the ADD approach with multiagent technology has a large potential of generating powerful tools for supporting concurrent and distributed design. In addition to the common advantages of multiagent technology, those tools have the capabilities of ADD systems for making decisions and generating the design rationale. Following this perspective, Garcia and Vivacqua (1997) proposed the MultiADD approach and used it for constructing a multiagent ADD system for supporting the concurrent and distributed design of process plants of offshore oil platforms.

Figure 5 shows the MultiADD architecture. MultiADD is composed by many design agents. Every design agent is a pair composed of an ADD system and a human designer. The ADD system have a specific parametric network model for representing part of the whole design process knowledge. The human designer use the respective ADD system to support his/her individual assignments. Even though most part of a design agent work is independent from the others, there are interactions between them. An agent decision may be used as input for other agents or there may be a global constraint to be jointly satisfied by some design agents. Therefore, a design agent must interact with the others to reach a feasible solution for the global problem.

MultiADD also contains a special coordinator agent (compound by an ADD system and a human design team manager) to oversee conflict mitigation and a computational control system responsible for identifying and communicating the occurrence of conflicts to the involved agents and to the coordinator. In order to improve conflict mitigation, every agent may consult the rationale of the other agents involved in the conflict. The coordinator agent may also consult the rationale of these agents for solving the conflict whenever the agents cannot solve it by themselves.

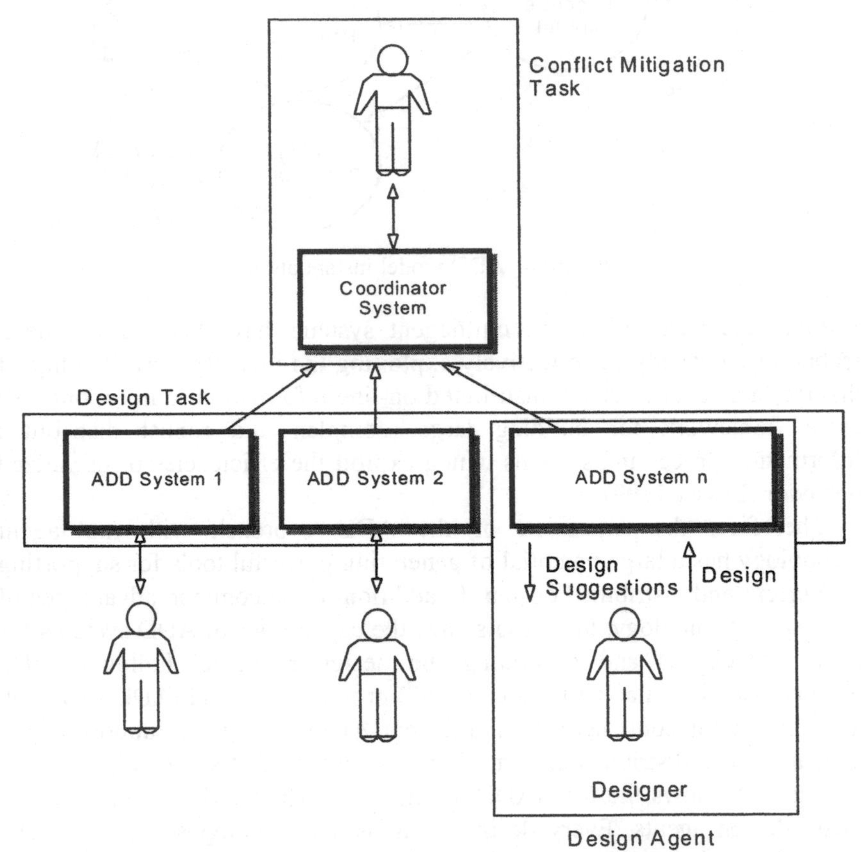


*Figure 5.* MultiADD architecture (adapted from Garcia and Vivacqua (1997))

Likewise ADD systems, MultiADD systems are domain dependent. This feature obstructs its wide dissemination and use. To solve this limitation we have developed MultiJADE, a computational environment for constructing and using multiagent active design documents.

## 3. MultiJADE

JADE (Varejão and Pessoa 2002) is a domain independent computational environment for constructing, using and reusing ADD systems. It doesn't deal with problems involving concurrent and distributed design. MultiJADE integrates the JADE tool with the MultiADD approach to support resolution of these kinds of problems.

MultiJADE use involves two main steps: MultiJADE model construction and MultiJADE model instantiation. Figure 6 illustrates the MultiJADE model construction. Initially, a knowledge engineer creates a shared design model. This model specifies the number of design agents and their possible interactions. It may also specify parts of parametric networks common to two or more design agents.

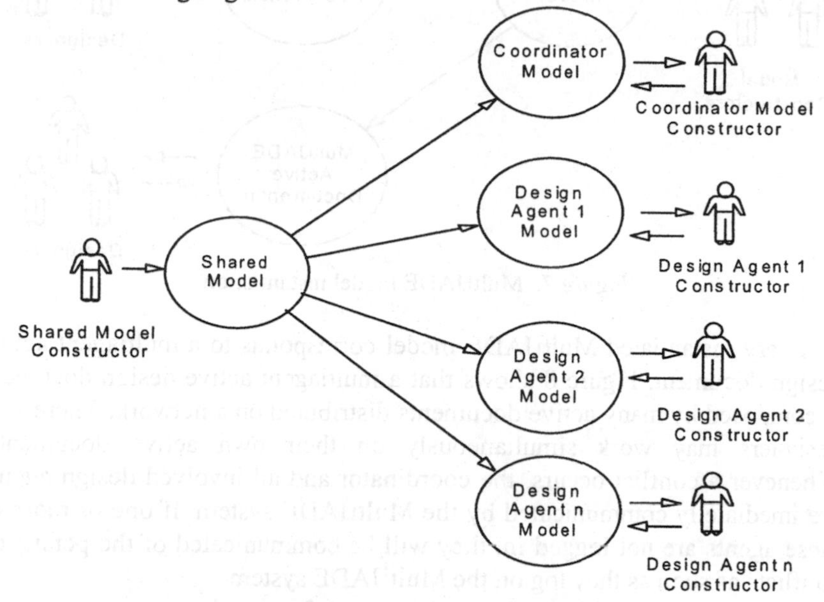


*Figure 6.* MultiJADE model construction

After the shared model is finished, the coordinator agent model and the design agent models may be created from the shared model. Usually, these models will be complemented or changed by design experts (the agent model constructors). Whenever all agent models are finished, the model is completed and may be used for creating multiagent active document systems.

Figure 7 illustrates that many multiagent active documents may be generated from the same MultiJADE model. For instantiating a MultiJADE model it is only necessary to choose a MultiJADE model from a model library, set some coordinator parameter values and assign the human designers for each agent of the concurrent and distributed design process.

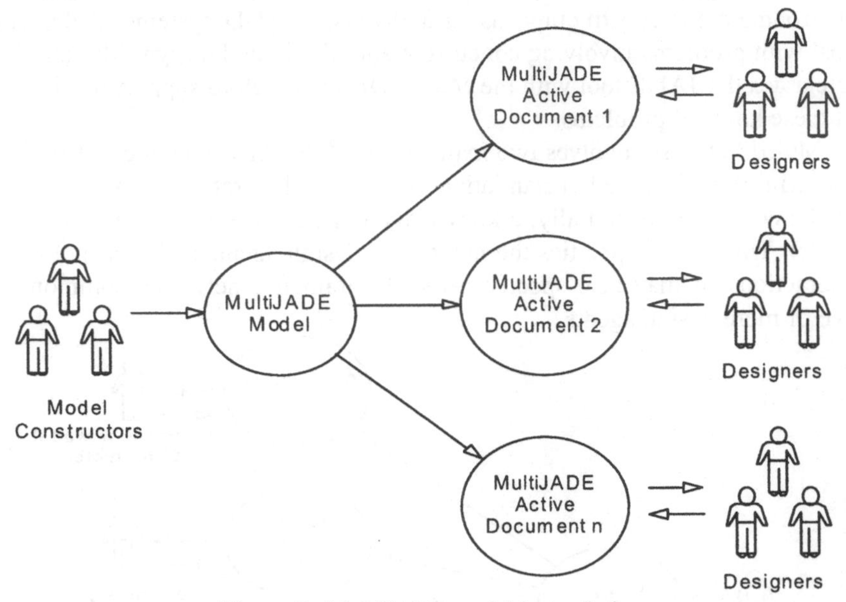


*Figure 7.* MultiJADE model instantiation

Every instantiated MultiJADE model corresponds to a multiagent active design document. Figure 8 shows that a multiagent active design document is composed by many active documents distributed on a network. Therefore, designers may work simultaneously on their own active documents. Whenever a conflict occurs, the coordinator and all involved design agents are imediatelly communicated by the MultiJADE system. If one or more of these agents are not logged in, they will be communicated of the persisting conflicts as soon as they log on the MultiJADE system.

MultiJADE also allows the assignment of more than one designer to work on the same active document, but they may not do it at the same time. Once an active document is opened by an assigned designer, its access is blocked to the other designers. Moreover, MultiJADE does not provide support for conflict mitigation of designers inside the same design agent.

### 3.1. LOGICAL ARCHITECTURE

Concurrent design implies that some decisions of one agent will impact some decisions of other agent. For modeling these situations, the parameters of a MultiJADE system may be classified as communicating parameters. These parameters are used by two or more agents and their values must obey some constraints. If one constraint is violated, there is a conflict. Figure 9 shows an example of communicating and non communicating parameters:

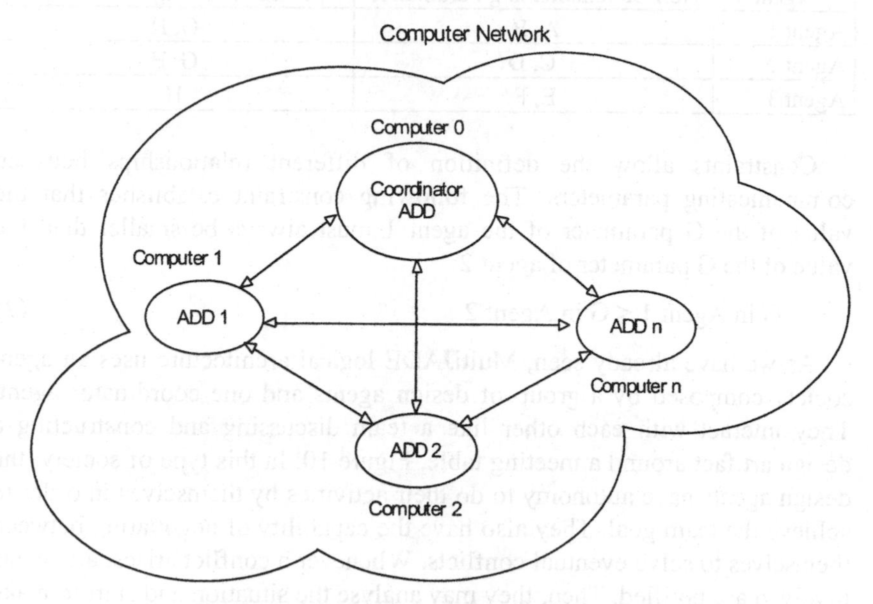


*Figure 8.* Multiagent active design document

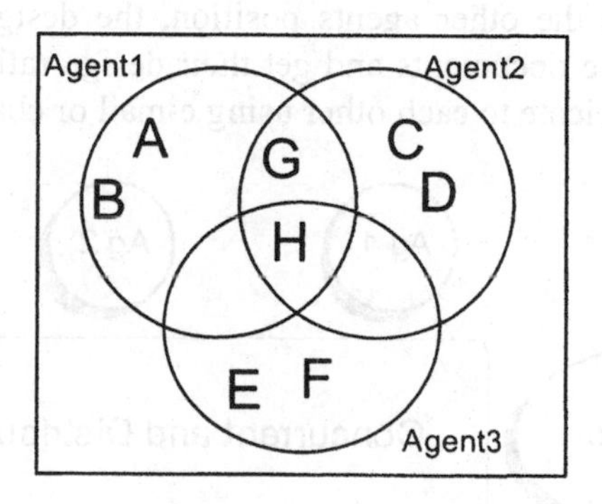


*Figure 9.* Agent parameters

The agent 1 design process model has the A, B, G and H parameters. Agent 2 model has the C, D, G and H parameters and agent 3 has the E, F and H parameters. The G and H parameters are communicating parameters. They are replicated only on the design models they belong. Table 2 identifies the communicating and non communicating parameters of each design agent.

TABLE 2. Agent parameters

| Agent | Non Communicating Parameters | Communicating Parameters |
|---|---|---|
| Agent 1 | A, B | G, H |
| Agent 2 | C, D | G, H |
| Agent 3 | E, F | H |

Constraints allow the definition of different relationships between communicating parameters. The following constraint establishes that the value of the G parameter of the agent 1 must always be smaller than the value of the G parameter of agent 2:

$$G \text{ in Agent } 1 < G \text{ in Agent } 2 \qquad (2)$$

As we have already seen, MultiJADE logical architecture uses an agent society composed by a group of design agents and one coordinator agent. They interact with each other like a team discussing and constructing a design artifact around a meeting table, Figure 10. In this type of society, the design agents have autonomy to do their activities by themselves in order to achieve the team goal. They also have the capability of negotiating between themselves to solve eventual conflicts. Whenever a conflict arises, all agents involved are notified. Then, they may analyse the situation and start to make changes on their active documents in order to solve the conflict. Whenever these changes do not solve the conflict, they need to negotiate. In order to better understand the other agents position, the designer may consult the other agents active documents and get their design rationale. The designers may also communicate to each other using e-mail or chat.

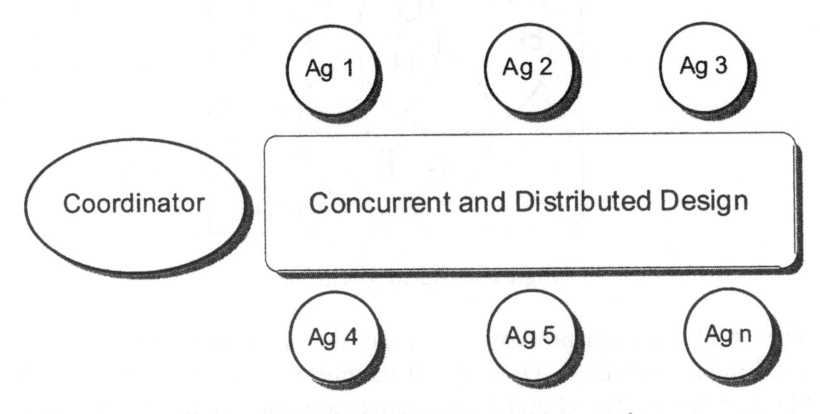


*Figure 10.* MultiJADE meeting table metaphor

A conflict solution is achieved when the values defined by all involved agents satisfy the design constraint. Every time an agent assigns a value to a parameter involved on a constraint, the MultiJADE control system checks if the constraint is not violated.

If a conflict persists for a certain time period, the control system requires the coordinator agent to choose a conflict solution. The coordinator solution must be accepted by all agents involved in the conflict since the agents cannot refuse it. MultiJADE automatically sets the solution values to all agents. The existence of a coordinator is important to guarantee that a conflict will not stop the project development.

The specific time limit to solve a conflict is defined by the coordinator agent based on the conflict complexity and on the design process schedule. While the computational part of the coordinator agent may support this type of decision, we expect that the human part will be in charge of setting the time limit on most cases due to the opportunistic and difficult nature of these decisions. Figure 11 shows the steps taken once a conflict has happened.

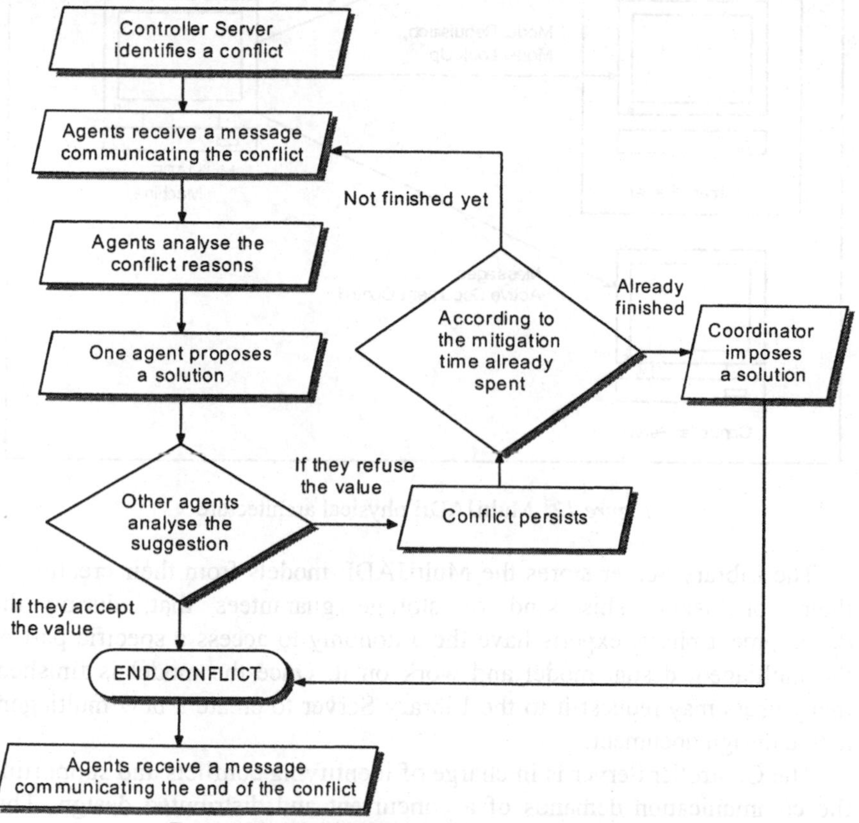


*Figure 11.* MultiJADE conflict resolution strategy

### 3.2. PHYSICAL ARCHITECTURE

The computational infrastructure needed to support MultiJADE is composed by three servers and many MultiJADE user clients. It is illustrated on Figure 12. The User Server is in charge of controlling the permissions to access the MultiJADE environment and its resources, such as the Library Server and the Controller Server.

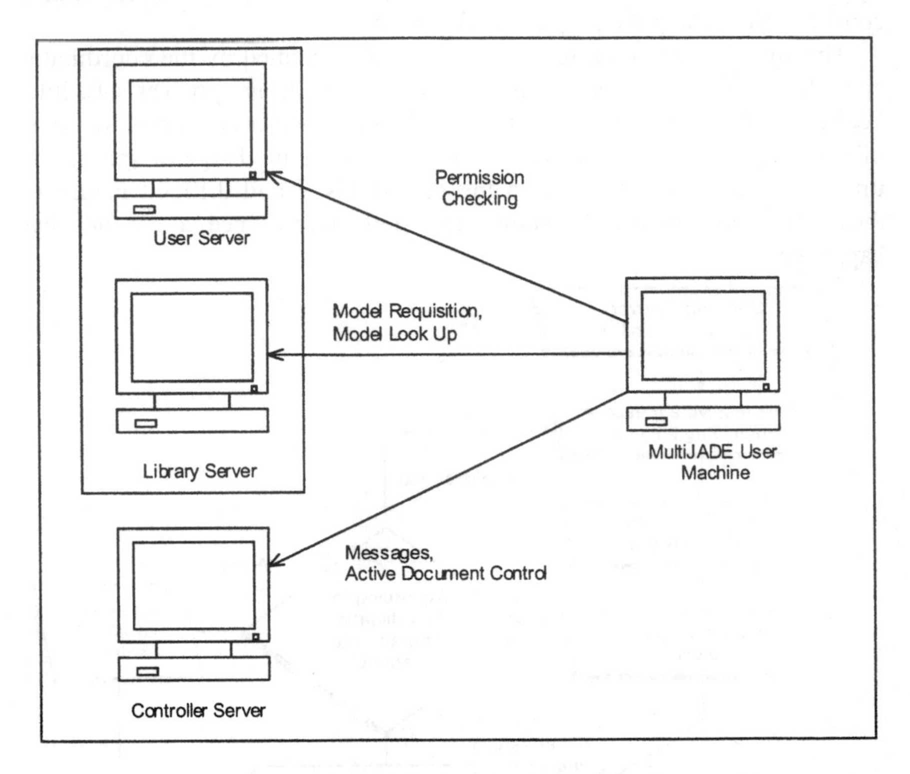


*Figure 12.* MultiJADE physical architecture

The Library Server stores the MultiJADE models from their creation to their conclusion. This kind of storage guarantees that, during the development phase, experts have the autonomy to access a specific part of the multiagent design model and work on it. Once the model is finished, many users may request it to the Library Server to create a new multiagent active design document.

The Controller Server is in charge of identifying conflicts and supporting the communication demands of a concurrent and distributed design. This support includes storing the current values of each communicating parameter and checking the constraints whenever one of these values are changed. The Controller Server also stores the design rationale of each design agent. This rationale is consulted by other agents for understanding the reasons of one decision during conflict mitigation. In addition, the Controller Server manages all communications among agents. MultiJADE communication processes include messages, chat and e-mail. These options give flexibility to the designers for using the most appropriate type of communication whenever necessary. Chat sessions and e-mail messages are stored by the Controller Server and may be consulted by other agents. The Controller Server also communicates the coordinator agent about an expired conflict time limit.

## 4. MultiJADE Application

We have chosen a simple application related to the task of controlling human resource allocation costs to illustrate the use of a MultiJADE system. Financial resources are limited and frequently required by distinct projects at the same period on the same organization. The distribution of resources is a problem that always involves conflicts.

This application intends to help project managers from a department to plan and control the financial resources in different projects. Figure 13 shows the agents involved in the problem.

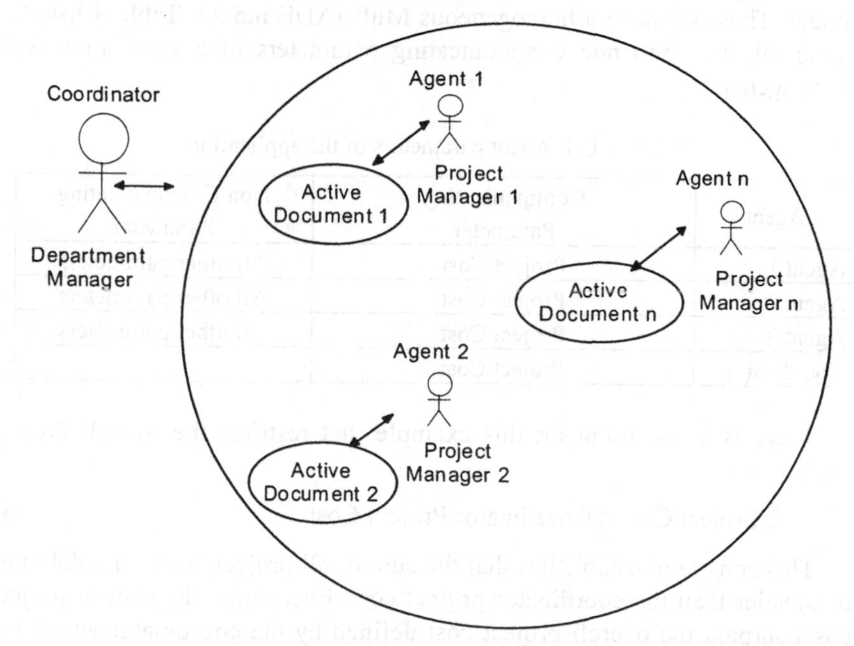


*Figure 13*. Agents in the human resource allocation cost application

In this example, the department is composed by a department manager and project managers. Each project manager controls one or more projects.

The MultiJADE model associates the department manager with the coordinator agent and the project managers with the design agents. Table 3 explains the parameters of a project manager model.

TABLE 3. Parametric network

| Parameters | Type | Definition |
|---|---|---|
| SystemAnalyst | Primitive | Number of System Analysts in a project |
| Developer | Primitive | Number of Developers in a project |
| SystemAnalystCost | Primitive | System Analyst cost per working day |
| DeveloperCost | Primitive | Developer cost per working day |
| SystemAnalystAllocation | Primitive | System Analyst allocation in days |
| DeveloperAllocation | Primitive | Developer allocation in days |
| SystemAnalystTotalCost | Derived | SystemAnalyst * SystemAnalystAllocation |
| DeveloperTotalCost | Derived | DeveloperCost * DeveloperAllocation |
| ProjectCost | Derived | SystemAnalystTotalCost + DeveloperTotalCost |

This type of problem have all design agents with the same design process model. Thus, we have a homogeneous MultiJADE model. Table 4 lists the communicating and non communicating parameters of a application with three agents.

TABLE 4. Agent parameters of the application

| Agent | Communicating Parameter | Non Communicating Parameter |
|---|---|---|
| Agent 1 | Project Cost | All other parameters |
| Agent 2 | Project Cost | All other parameters |
| Agent 3 | Project Cost | All other parameters |
| Coordinator | Project Cost | - |

There is a constraint on this example that restricts the overall Project Cost:

$$\Sigma \text{ Project Cost} < \text{Coordinator Project Cost} \quad (3)$$

This constraint establishes that the sum of all projects in the model must be smaller than the coordinator project cost. Every time the sum of project costs surpass the overall project cost defined by the coordinator agent, the MultiJADE control system identifies the conflict and communicate it to the design agents. The design agents may review their demands and change the number of professionals or the number of working days allocated to their project in order to reduce the cost.

The design agents have a time period to solve the conflict. Sometimes, it is possible to share a professional for two or more projects. MultiJADE supports the task of identifying this type of professional sharing by allowing the design agents to consult the other agents rationale. If the conflict persists after the time period has expired, MultiJADE asks the coordinator to interfere and solve the conflict. The coordinator may increase the overall project cost or arbitrarily reduce one or more project costs.

The complexity of this problem may be significantly increased if we consider the deadlines for closing the projects. Once you reduce the project number of allocated professionals, its duration grows. Therefore, the design agents and the coordinator agent also have to negotiate deadline dates.

## 5. Conclusion

MultiJADE highly reduces the required effort and cost for constructing MultiADD systems. The model constructors only need to specify the number of design agents, create the parametric networks and specify the communicating parameters to create a new model. The design and explanation interfaces and the multiagent architecture are automatically created by MultiJADE. Moreover, it is only necessary to select a MultiJADE model and start to use it for creating a new multiagent active design document.

One important feature of MultiJADE is the immediate communication to the designers of every decision that affects them. Thus, they may try to solve the conflict as fast as they happen. This is much better than the personal meeting approach that requires a lot of design backtracking because the conflict identification is delayed until the meeting day. Another distinguishing feature of MultiJADE is to allow the designers consult the rationale of the other designers, without requiring personal meetings, in order to make conflict mitigation easier. This feature only may be provided by systems capable of capturing, recording and generating design rationale such as the active design document systems.

Whenever the designers cannot solve the conflict by themselves, MultiJADE communicates this fact to the coordinator. Then, the coordinator analyses the conflict and defines a solution. MultiJADE communicates this solution to the involved designers who must accept it. This seems to be a very strict position. Future versions of MultiJADE should give a chance to the design agents for contesting the coordinator proposed solution.

Currently, MultiJADE only distributes scalar parameters. This may also be restrictive once there is concurrent problems that need to distribute more structured datatypes. In addition, MultiJADE requires that all communicating parameters should be settled a priori and remain fixed along the design process. This may not be appropriate on concurrent design problems where the conflict types may change dynamically.

It would be possible to use some multiagent tools already developed and available like JADE (Bellifemine, Poggi and Rimassa 1999), JATLite (Jeon, Petrie and Cutkosky 2000) or Zeus (Collis 1999) to create a multiagent active design document. However, these tools would require a lot of work for constructing individual active design document agents from scratch. We have chosen to reuse the JADE tool (Varejão and Pessoa 2002) and construct our own multiagent environment.

The actual MultiJADE version implemented a centralized architecture where one coordinator plays a central role. We are designing extensions to support other architectures like the hierarchical architecture and total interaction architecture. The hierarchical architecture replicates the centralized architecture in many levels. Each centralized architecture defines a group of agents that act as a small team to achieve the team goal. The communication between groups is made by coordinators. The total interaction architecture proposes MultiJADE systems without coordinators. In this case, the design agents have total autonomy and should solve the conflict by themselves. This architecture demands more negotiation among the agents and may produce design deadlocks whenever conflicts are not solved.

Another interesting future work is to turn MultiJADE web-based like the domain dependent collaborative system Wink (a web-based system for collaborative project management in virtual enterprises) developed by Bergamachi et al.(2003).

## References

Bellifemine, F, Poggi, A and Rimassa, G: 1999, JADE – A FIPA-compliant agent framework, *CSELT internal technical report*. Part of this report has been also published in *Proceedings of PAAM'99*, London, pp.97-108. Available at http://sharon.cselt.it/projects/jade.

Bergamachi, S, Gelati, G, Guerra, F and Vincini, M: 2003, Wink: A web-based system for collaborative project management in Virtual Enterprises, *Fourth International Conference on Web Information Systems Engineering*, pp.176-185.

Garcia, ACB, Andrade, JC, Rodrigues, RF and Moura, R: 1997, ADDVAC: Applying active design documents for capture, retrieval and use of rationale during offshore platform VAC design, *Proceedings of the 14th National Conference on Artificial Intelligence*, AAAI Press, Rhode Island,Providence, pp. 986-991.

Garcia, ACB. and Howard, HC: 1993, Active design documents: From information archives to design model construction, *Proceedings of the Artificial Intelligence in Engineering Conference*, Toulose, France, pp. 233-244,

Garcia, ACB. and Vivacqua, A: 1997, MultiADD: A multiagent active design document model to support group design, *Proceedings of the 14th National Conference on Artificial Intelligence*, AAAI Press, Rhode Island, Providence, pp. 1066-1071

Jeon, H, Petrie, C and Cutkosky, MR: 2000, JATLite: A Java agent infrastructure with message routing, *IEEE Internet Computing 1089-7801/00/ IEEE* **4**(2): 87-96. Available at http://www-cdr.stanford.edu/ProcessLink/papers/jat/jatlite.html

Lesser, VR: 1999, Cooperative multiagent systems: A personal view of the state of the art, *IEEE Transactions on Knowledge and Data Engineering* **11**(1): 133-142.

Nwana, HS, Ndumu, DT, Lee, LC and Collis, JC: 1999, ZEUS: A toolkit and approach for building distributed multi-agent systems, *Proceedings of the Third International Conference on Autonomous Agents*, Seattle, WA, USA, ACM Press, pp. 360--361

Sycara, KP: 1998, Multiagent systems, *AI magazine* **19**(2): 79-92.

Varejão, FM and Pessoa, RM: 2002, JADE: A computational environment for constructing, using and reusing active design documents, *Proceedings of the Fifth WG 5.2 Workshop on Knowledge Intensive CAD*, Borg, JC & Farrugia, PF (eds), pp. 99-114.

JS Gero (ed), *Design Computing and Cognition'04*, 499-517


# A COMPUTATIONAL FRAMEWORK FOR THE STUDY OF CREATIVITY AND INNOVATION IN DESIGN: EFFECTS OF SOCIAL TIES

RICARDO SOSA, JOHN S GERO
*University of Sydney, Australia*

**Abstract.** This paper describes a socio-cognitive framework to study the interaction between designers and social groups. Experimentation with situational factors of creativity is presented. In particular, social ties in a population of adopters are shown to shape the way in which designers are considered as change agents of their societies.

## 1. Introduction

Creative design is widely recognised as one of the foundations for social change (Gero 1996). This paper explores some fundamentals of the relationship between designers and social groups. Our motivation is to understand how individual actions can be determined by collective conditions and in turn trigger structural changes. Conventional research focuses on distinct units of analysis, i.e. personal or social processes separately (Conte et al 2001).

An increasingly accepted approach to the study of creativity is based on the relation between individual-generative and group-evaluative processes. Under this view, creativity is seen as a social construct (Saunders and Gero 2001) or communal judgment (Gardner 1993) where the creative individual is considered not in isolation but in interaction with an environment of physical and social dimensions. Being socially constructed, standards of what constitutes creative solutions evolve (Simonton 2000). This requires a broader inquiry of design that extends the unit of study outside the cognitive dimension to include the social aspect of creativity (Amabile 1983).

The term *creativity* is polysemous and ambiguous. In the literature it refers to aesthetic appeal, novelty, quality, unexpectedness, uncommonness, peer-recognition, influence, intelligence, learning, and popularity (Runco and Pritzker 1999). In this paper creativity is defined by a set of complementary processes including adoption of a solution by a population, nomination by specialists or gatekeepers and colleague recognition.

Innovation is defined by the diffusion of a solution across a population of adopters (Rogers 1995). This paper presents an artificial society of adopters and designers as an empirical test-bed where qualitative generalisations about the nature of creative behaviour can be grounded.

## 2. Socio-Cognitive Agent Architecture

The dominant architecture of rational agency divides the system into two explicit parts: agent and environment (Wooldridge 2000). The rational agent is the sole causal determinant of behaviour whilst changes in the environment may reflect the impact of actions by other agents or external effects. Under such view, social interaction is contained as part of the external state. In the prototypical ant colony, behaviour is determined by a fixed reaction to pheromone levels present in the environment. However, the behaviour of more complex and social individuals is not expected to be hardwired as a reaction to environmental stimuli.

For a social agent it is important to perceive more than environmental expressions. Interpretations of who laid the pheromone trail and with what intentions become key determinants of social behaviour. Individual learning helps but it is not feasible for every agent to learn at all times in a complex environment. Individual determinants are complemented by reliance on group support. Societies provide collective cues that are not necessarily expressed in the physical environment. The main difference between a physical and a social environment is that the former may exist independently from the agents that inhabit it. Agents and physical environments can be explicitly defined and treated as separate entities. However, a social environment is a function of the aggregate effect of agent interaction over time. Whilst no single agent has control over its society, the emergent structure of a social environment feeds back and co-determines individual states and behaviours.

Social norms and conventions can clearly co-determine individual behaviour in design. In our agent architecture there is a place for a range from individual to social mechanisms mediating the agent-environment divide. Figure 1 shows a schematic definition of our architecture where agent-environment interaction is mediated by layers that range from individual to collective characteristics.

Agent behaviour **M** in an environment **E** can thus be defined as:

$$\mathbf{M} = \sum \{ m_{i\text{-}n} [\mathbf{S}(m_{i\text{-}n} \wedge \mathbf{E'})] \} \tag{1}$$

where individual behaviour **M** is determined by the sum of internal state $m_{i\text{-}n}$ and construed situation **S**. Internal states consists of components $m_{i\text{-}n}$ such as perceptions, goals, preferences, skills, knowledge, and actions. Environment **E** is perceived by a bounded agent as interpreted external state **E'**. Situation

**S** is therefore a function of the combination of internal and interpreted external state. For instance, a perceived external state may be a measure of group pressure. However, perceived group pressure by itself has no meaning, i.e., it is not a situation but a passive contextual feature. Perceived group pressure becomes part of a situation when construed in combination with a relevant internal state such as a certainty or extroversion threshold. Equivalent group pressure perceived by agents with different internal states may indeed lead to the construction of entirely different situations, eg. compliance or assertiveness. Equivalent contexts may thus generate different behaviours within different situations. When situational factors are strong determinants, agent behaviour is normalised whereas if personal factors $m_i$ dominate, behaviour is more differentiated across a population. Under this view, individual differences in isolation are insufficient to explain behaviour.

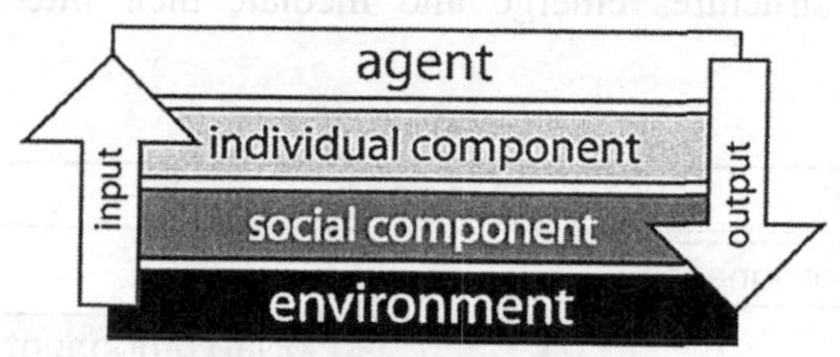


*Figure 1.* The social agent: behaviour is determined by a combination of individual and shared components

The Asch compliance paradigm (1951) illustrates the nature of our agent architecture. In this widely replicated experiment test subjects that respond correctly in isolation tend to be influenced by associates of the experimenter when placed within certain social settings. In some situations subjects tend to comply with group judgements even when clearly wrong. Within this paradigm, personal factors partially determine behaviour. If a test subject has certain individual characteristics such as high extroversion (Eysenck 1991) it tends to avoid compliance. Situational factors complement the determination of behaviour. If a test subject is asked to respond earlier in the task or if a previous participant differs and breaks unanimity the subject tends to differ and provide an independent response. Resulting behaviour can only be explained by a combination of individual and situational conditions.

Different insights were extracted from the verbal account of yielding subjects from the original experiment after conditions were revealed (Asch 1951). Influence effects fell into three categories: distorted perception, distorted judgement, and distorted action. These sources of behaviour can be mapped onto internal components $m_{i\text{-}n}$ of complier subjects as a function of their appraisal of the situation:

$\mathbf{M} = \sum \{ \qquad A : perception_i[\mathbf{S}(\mathbf{E'}_i)]$

$$B : judgement_i[\mathbf{S}(\mathbf{E}'_i)]$$
$$C : action_i[\mathbf{S}(\mathbf{E}'_i)] \quad \} \quad (2)$$

where $\mathbf{E}'_i$ represents a perceived collective state of unanimity and **S** the complier's appraisal of a situation within which sufficient conditions exist for distortion. Individuals from group A have their perceptions distorted and their behaviour **M** is subsumed by group influence. Judgements and actions are distorted in groups B and C respectively. The first two resemble informative influence whilst the third is a type of normative influence.

2.1. MULTI-AGENT SYSTEM

The defining characteristics of social agents would have no relevance separate from a social group. Figure 2 shows a diagram of our multi-agent system where components $m_{n,i}$ become part of the group structure. As agents interact group structures emerge and mediate their interaction with the environment.

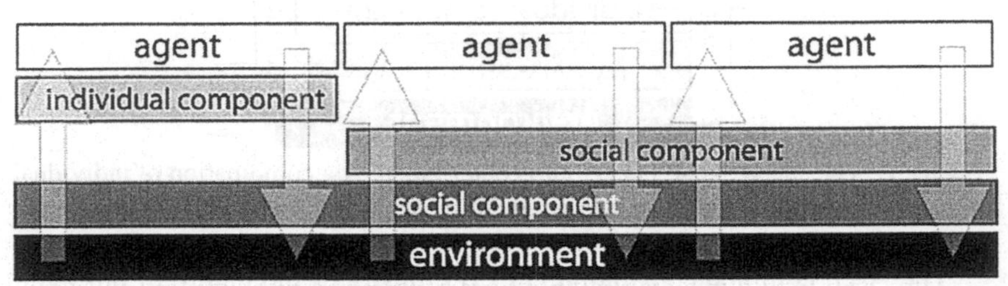


*Figure 2.* Socio-cognitive architecture where behaviour components become part of emergent group structures

These structures are shared by agents at different times causing them to exhibit different degrees of normalised behaviour. For instance, perceptions may become collectively biased, preferences may be emphasised by groups at different times, and socially permissible actions may be established. The collective state of a society $N$ can thus be defined as:

$$N = \{ \mathbf{M}_{i\text{-}n}[\mathbf{S}_{i\text{-}n}] \} \quad (3)$$

where state $N$ is a function of agents' behaviour $\mathbf{M}_{i\text{-}n}$ codetermined by internal states and their situation $\mathbf{S}_{i\text{-}n}$. A situation can be defined at the individual level as shown in eq. 1 and it can also be shared by a group as defined in eq. 3 above. A shared situation is perceived by a group of agents as a result of the combination of internal states and perceived external state. Extending the previous example, at the individual level a situation may be one of compliance whilst at the group level it may be a situation of unanimity. These are corresponding interpretations of one common collective structure, i.e., group pressure.

Individual behaviour under this view is defined as a function of the agent and the situation. This approach supports equivalent agents acting differently within different situations, and different agents acting similarly within similar situations.

Social agents inhabit more than one social space, i.e. they have adjacency relations to other agents in multiple social environments simultaneously. For instance, an individual may have different positions in kinship and work structures. Other approaches like cellular automata conflate physical and social location into a notion of neighbourhoods (Schelling 1971). Each social space can be modelled with different parameters: social tie strength and number of ties are the structural properties addressed in this paper.

This architecture is used to implement a model of design beyond internal thinking processes. Key actors including consumers or adopters of artefacts, opinion leaders and designers are modelled as social agents. Individual and situational factors are included in the design of experiments to inspect phenomena related to creativity and innovation.

## 3. Adoption Framework

A multi-agent system is implemented based on a population of adopter agents and their social interaction. Adopter behaviour consists of evaluating available solutions and deciding to adopt or abstain. Solutions are formulated in a simple linear representation as shown in Figure 3(a). This representation is chosen because it provides a number of intuitive geometric features as well as multiple interpretations and emergence. The objective is to support compromise in a multi-objective decision making and negotiation of requirements, which are typical of design problems (Goel 1994).

Adopters evaluate artefacts according to individual perception **V** and preference **F** values. Variation of perception enables different interpretations across a population as shown in Figure 3(b). Variation of preferences ($0.0 \leq \mathbf{F} \leq 1.0$) enables different decisions based on shared interpretations.

Perception is implemented by a shape-recognition algorithm executed independently by every individual adopter. Every adopter agent executes a search for closed shapes with a branch limit **V**. This search renders a set of closed shapes **G** that stands for the artefact's features perceived by each adopter. Perception traits **V** are assigned from a Gaussian distribution. This reasoning mechanism is computationally expensive and is scheduled at intervals of adoption. Adopters do not learn individually at every time-step but base their decisions on approximations that they update regularly. This representation refers to the idea that in human populations there may be a number of distinct but overlapping views of an artefact's features, i.e. adopter segmentation. Variation of percepts across a population is controlled by the standard deviation of **V**. Different studies may consider different percept variation assuming more subjective or more normalised interpretation across a target population.

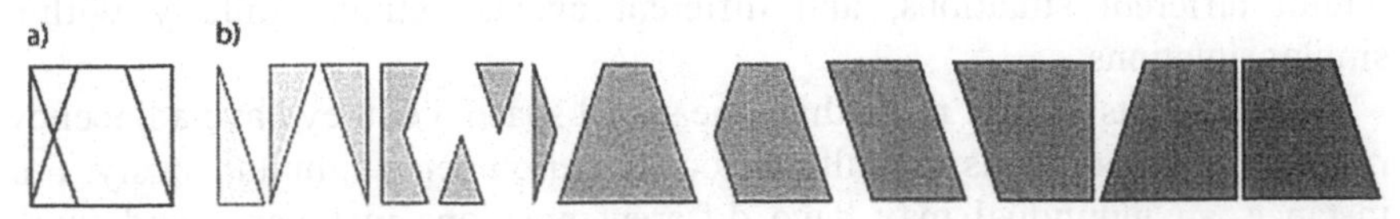


*Figure 3.* (a) Sample artefact and (b) a range of shapes that adopters may perceive

### 3.1. ADOPTION DECISION

The adoption decision process consists of a multivariate evaluation where adopters seek to maximise conflicting criteria. Criteria attached to this representation include factors such as number of shapes, ratio of shapes aligned along horizontal and vertical axes, preferred number of sides, intersection or overlapping of shapes and similar shape bounds. The performance of an artefact is estimated by each adopter following:

$$\mathbf{P} = \{ [ (\mathbf{G}/\mathbf{G}_{min}) + (\mathbf{V}_{(+-2)}/\mathbf{G}) + (alignX/\mathbf{G}) + (alignY/\mathbf{G}) + (ints/\mathbf{G}) + (bounds/\mathbf{G})] \;/\; (\mathbf{G}^2\text{-}\mathbf{G}) \} \tag{4}$$

where artefact performance $\mathbf{P}$ provides a measurement based on an individualised set $\mathbf{G}$ of perceived features. The sum of number of shapes (size of $\mathbf{G}$), ratio of shapes with $\mathbf{V}_{(+-2)}$ number of sides, ratio of shapes horizontally aligned (*alignX*), ratio of shapes vertically aligned (*alignY*), ratio of shape intersection (*ints*) and ratio of similar bounds (*bounds*) normalised over the combinatorial size of $\mathbf{G}$.

Ratings $\mathbf{P}$ of artefacts under evaluation are compared by each adopter reaching an adoption decision where $\mathbf{G}_{max}$ is the artefact with preferred features as follows:

$$\mathbf{G}_{max} = \{ \mathbf{C}_{i\text{-}n} [(\mathbf{P}_{max} - \mathbf{P}_{mean}) (\mathbf{B}_{i\text{-}n})] \} \tag{5}$$

where adoption choice $\mathbf{G}_{max}$ refers to the artefact's perceived features that lead in the criterion $\mathbf{C}_{i\text{-}n}$, i.e. with the largest performance differentiation ($\mathbf{P}_{max}$ - $\mathbf{P}_{mean}$). Biases $\mathbf{B}_{i\text{-}n}$ are weights between 0.0 and 1.0 that adopters incorporate in the form of individual preferences or can be externally manipulated by the experimenter. The shape of $\mathbf{G}_{max}$ shows an implicit 'novelty' criterion by which adopters tend to choose artefacts that they perceive to have the highest differentiation from others. Adoption is therefore a function of how competing artefacts compare at that time. To be adopted an artefact needs to perform well in a criterion that other artefacts do not meet and it helps if such criterion is positively biased by adopters' preferences. Preferences evolve over time following a mechanism of habituation $\mathbf{U}$. As adopters choose the best available artefact, their preference for the criterion best satisfied by that artefact is gradually increased.

$$\mathbf{U} = \{ \mathbf{F}_i \mathrel{+}= \mathbf{C}_i \} \tag{6}$$

3.2. VERIFICATION

This adoption framework provides a method to manipulate perceptions **V** and preferences **F** for verification purposes. Figure 4 shows a verification run where a set of artefacts is randomly generated and made available to a population of 100 adopters.

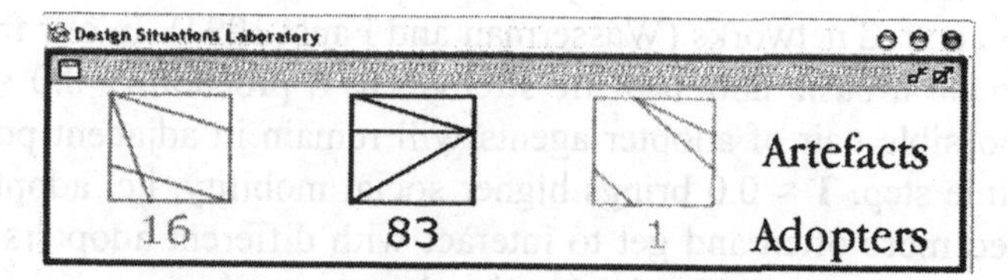


*Figure 4*. A verification run where the weight of criterion $\mathbf{F}_{\text{alignX}}$ is externally increased. As a result, a majority (83 of a 100 population) chooses the artefact that performs best in the horizontal alignment criterion.

The group's preference for shapes aligned in the horizontal axis is externally increased by assigning extra weight to the criterion $\mathbf{F}_{\text{alignX}}$. As a result of this bias, adopters tend to choose the artefact with the highest performance in this criterion, i.e. where all shapes are horizontally aligned. However, not all adopters' decisions converge since perceptions **V** are not homogeneous.

3.3. ADOPTION SATISFACTION

Adopter satisfaction is computed as a measure of quality. In the adoption decision, if the choice criterion $\mathbf{C}_{i\text{-}n}$ equals the leading preference of an adopter, its satisfaction level is set to a maximum discrete value. Else, if the choice criterion is one standard deviation above the mean of the adopter's preferences, then the satisfaction level is set to a medium level. Otherwise, the agent has adopted an artefact that performs best in a criterion that is of little relevance to the adopter and its satisfaction level is set to a minimum. An adopter may abstain from adoption if no difference is perceived between artefacts, i.e. iff $\mathbf{P}_{i\text{-}n}$ is equal for all artefacts and $\mathbf{G}_{max} = \varnothing$.

3.4. SOCIAL INTERACTION

At every iteration step, adopters rely on social interaction to validate their perceptions, spread preferences and in general to conduct their adoption decisions. To this end different social spaces **L** are defined where adopters interact. At initial time adopter agents are randomly assigned a location on each space. These social spaces have different rules of interaction and development. Two aspects addressed in this paper are social tie strength (**T**) and neighbourhood size (**H**).

Ties are interaction links between nodes in a social network and represent the relationship between adopter agents (nodes) in a social space (Wasserman and Faust 1994). **T** is determined by the probability that associated nodes may interact over a period of time (Granovetter 1973). Strong social ties exist between nodes in a kinship network, whilst weak ties exist in networks where casual encounters occur between strangers or acquaintances. **H** is determined by the number of links from a node -also called ego-centred networks (Wasserman and Faust 1994). In our framework we implement a basic notion of tie strength as a probability $0.0 \leq \mathbf{T} \leq 1.0$ that any possible pair of adopter agents will remain in adjacent positions at the next time step. $\mathbf{T} \approx 0.0$ brings higher social mobility, i.e. adopter agents are shuffled more often and get to interact with different adopters over any given period. In contrast, $\mathbf{T} \approx 1.0$ bonds adopter together causing a decrease in social mobility, i.e. adopter agents remain in their same neighbourhoods interacting with the same agents for longer periods of time.

### 3.5. INFLUENCE DOMINANCE

A social space $\mathbf{L}_1$ in this framework is set where adopters exchange preferences **F**. Within a second social space $\mathbf{L}_2$ percepts **G** are traded. A third space is set where agents exchange adoption decisions $\mathbf{G}_{max}$. In all spaces **H** has an initial value of 2 that varies during a system run according to the influence that an adopter exerts on others. More influential adopters have larger neighbourhoods. Adopters can be located in social spaces with particular assumptions according to the hypothesis under inspection. For instance, the purchase of cars may be shaped by influence interaction in kinship networks whilst that of mobile phones may be strongly influenced by peer networks. Figure 5 shows a sample influence structure where an adopter with high dominance **D** has a large neighbourhood $\mathbf{H} = 6$.

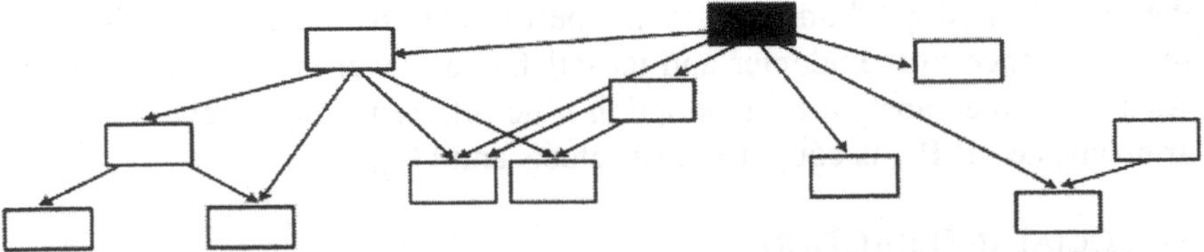

*Figure 5.* Influence structure in a sample space. Adopters are represented by rectangles, influence dominance by arrows. Vertical axis plots influence dominance **D**. Neighbourhood size increases with dominance.

The distribution of influence dominance **D** in an adopter population is measured by the Gini coefficient, a summary statistic of inequality. The Gini coefficient $\gamma$ is used in studies of wealth distribution where group resources are limited and exchanged among members of a population. Influence can be seen as analogous to wealth in that it is generated by the interaction between two agents where one may increase its share at the expense of another. When $\gamma \approx 1.0$ influence is concentrated by a few adopters and more stable dominance hierarchies exist. In contrast, when $\gamma \approx 0.0$, influence is more distributed among adopters. More formally,

$$\gamma = \{ \sum [(|d_i - d_{i+1}|) / \mathbf{D}_{mean}] / (2\, \mathbf{D}^2) \} \tag{7}$$

where the difference of every possible pair of dominance values ($d_i - d_{i+1}$) is divided by the mean of the entire dominance set of the population ($\mathbf{D}_{mean}$). The relative mean difference ($\gamma$) is obtained by dividing the sum of pair differences by the square of the size of the dominance set ($\mathbf{D}^2$) (Dorfman 1979).

At initial time agents are randomly assigned extroversion thresholds **X** in every social space (Eysenck 1991). An adopter agent is assigned different **X** in different social spaces. Extroversion values are not fixed during a system run but change as a result of exerting influence over other agents.

Exchange between any pair of adopters starts by a comparison of their extroversion thresholds **X**. In the social space where preferences **F** are exchanged, the adopter agent with the higher extroversion of the pair influences the less extrovert adopter on the criterion with the highest preference. A negotiation process occurs by which the influenced adopter increases its preference by half the difference between their preferences. However, if the chosen artefact of both adopters is the same and their preferences too similar, the more extrovert changes its focus of attention by shifting its preference to another criterion. This is a way to implement uniformity-avoidance and novelty-seeking behaviour, i.e. "$p_i$ is an adopter's top preference until it perceives that $p_i$ is commonplace". Within other social spaces different content is exchanged following a similar approach. More formally influence **I** between adopters *i* and *j* is of the form:

$$\mathbf{I}_{i\text{-}j} = |\mathbf{X}_i - \mathbf{X}_j| \; \{ \mathbf{F}_j \mathrel{+}= [(\mathbf{F}_i - \mathbf{F}_j)\,(0.5)] \} \tag{8}$$

where the more extrovert adopter *i* influences less extrovert *j*. Negotiation occurs as the target preference **F** of agent *j* approaches agent *i* by a ratio of their difference. The exchange of percepts **V** and adoption choices $\mathbf{G}_{max}$ in their corresponding social spaces takes place in the same form.

Whilst the details of this interaction can be fine-tuned to match a theory or observations, the key idea is that adopter agents exchange building blocks of their adoption process. This way even if an influential adopter is successful in spreading its preferences and percepts, the adoption decisions of influenced adopters need not converge. Namely, adopters with equal top preferences may still perceive artefacts differently and therefore reach different adoption decisions as shown in Figure 4.

In ergodic systems such as 2-dimensional cellular automata, a population converges from any initial random configuration. In contrast, when exchange occurs in more than one social space, the population may not converge as time $\rightarrow \infty$ due to random walks being transient (Sosa and Gero, 2003).

### 3.6. OPINION LEADERSHIP AND GATEKEEPING

Promoters are opinion leaders whose dominance **D** increases at least one standard deviation above the mean as a result of social interaction. At initial time the set of promoters **R** is empty. As a result of social interaction over time, adopter populations form social structures of hierarchy. These structures can be determined by various exchange processes. An adopter population may have characteristics that enable many agents to gain opinion leadership temporarily or may generate only a limited number of stable opinion leaders. Whilst in the former $\gamma \approx 0.0$ supports social mobility, the latter exhibits social stability and $\gamma \approx 1.0$. Namely,

$$\mathbf{R} = \{ d_i > (\mathbf{D}_{mean} + \mathbf{D}_{stdev}) \} \quad (9)$$

where a promoter **R** consists of every adopter whose dominance is greater than one standard deviation above the mean of group dominance **D**. The role of promoters in this framework is to form a two-way bridge between adopters and designers. Firstly, they serve as adoption models providing designers with positive feedback for reinforcement learning. Secondly, promoters become gatekeepers of the field given their ability to nominate artefacts for entry into the artefact repository **Y**, i.e. a collection of artefacts that defines the material culture of a population (Feldman et al 1994).

Since the number of promoters is by definition a small ratio of the adopter population, they are more likely to spend more real and computational resources in analysing available artefacts. With an adopter background, promoters follow the standard adoption decision process described above but also gain access to more detailed evaluation criteria.

The artefact repository **Y** is initialised with an entry threshold $\varepsilon = 0$. During a system run $\varepsilon$ is increased supporting a notion of group progress by which the entry bar is raised with every entry. Two possible entry modes are addressed in this paper. Promoters can nominate artefacts that either increase the population's threshold of entry $\varepsilon$ or perform well in different criteria than existing entries. Promoters evaluate artefacts using geometric descriptions like orthogonal rotation, uniform scale, and reflective symmetry. The nomination of artefacts by promoters occurs at a control rate specified by the experimenter. Figure 6 shows sample repository entries as nominated by promoters based on their geometric relationships.

Entry threshold ε to repositories has a decay mechanism **A** of the form:

$$\mathbf{A} = \{ \varepsilon -= (0.05\varepsilon) \} \tag{10}$$

where ε decays marginally over time. **A** is executed when promoters fail to nominate qualified entries above ε.

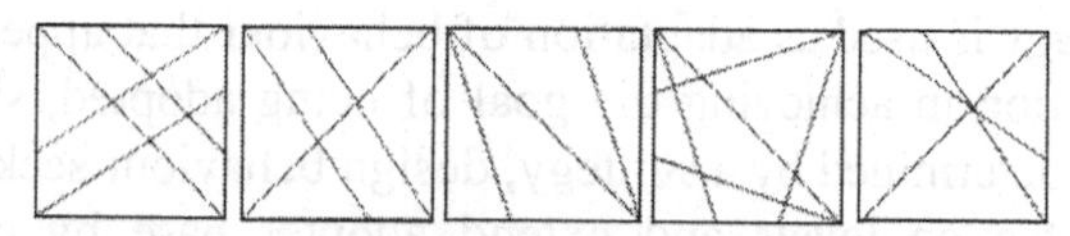

*Figure 6*. Sample entries to the repository. Geometric relationships can be recognised within artefact shapes including scale, rotation, and symmetry.

Adopters and promoters provide the first elements of our definition of creativity. A creative design must be recognised and adopted by a population. Cumulative adoption of artefacts addresses a notion of popularity (Simonton 2000). It must also be selected by gatekeepers, i.e. experts representative of their social group. This selection is based on rules of entry that evolve as artefacts and societies change. Critics' choice addresses the idea that creativity is judged by relevant arbiters (Gardner 1993; Feldman et al 1994). Lastly, adopter categories enable classification on the basis of when they choose an artefact (Rogers 1995).

## 4. Design Behaviour

At initial step, the size of a group of designer agents is determined as a ratio of the adopter population. Initial artefacts are configured and assigned to each designer. Designer agents are given a set of standard constraints to which their artefacts must comply. Designers' knowledge and adopter bases, recognition levels, and repository entries are all set to zero at the beginning of a system run. Knowledge base refers to domain rules that designer agents generate and apply during a simulation. Adopter base is defined by cumulative adoption. Recognition is given by peer designers that imitate features of an existing solution. At initial time the role of designer agents is to present their artefacts for adopters' assessment. The details of the design task are determined by the adopter group decisions and the ability of competing designers to generate solutions. The goal of designers in this system is to consistently generate artefacts that are chosen by adopters, are selected by critics, and are imitated by peers.

The execution of design behaviour can be parameterised according to the hypotheses under consideration. Namely, design update and adoption rates can be assumed to be periodical or synchronous. In this paper we assume that design occurs at adoption intervals during which adopters execute their decisions and interact socially. Variations of these assumptions are required to model different product markets and industries, requiring particular experimentation scenarios.

Designers may engage in different types of behaviour at any given time depending on a number of internal and external factors. Contingent design strategies can be seen as the product of the confluence of these conditions. The term *strategy* is used as adaptation of behaviour that appears to serve an important function in achieving the goal of being adopted, short-listed and influential. As determined by a strategy, design behaviour seeks to increment adopters' satisfaction levels and extend adopter base by capitalising on relative superiority (competition), by exploiting weaknesses of competitors, or by seeking artefact differentiation or diversification.

Designer agents seek a type of contingent strategy where they learn a *design rule*, i.e. an instance of domain knowledge tied to the artefact representation. In this case a design rule is made of artefact feature, target criterion and target perception. Rules are generated based on the designer's model of the population's adoption process. In order to do this, designers establish contracts with promoters that adopt their artefacts. This is a way to implement positive feedback since otherwise a designer would not have access to target criteria and target perception, i.e. a promoter may be an adopter of a competing artefact or may be abstaining from adopting. Having access to a framework of adoption, a designer can emulate the collective decision process by generating hypotheses of possible alternative artefacts.

Designers evaluate and change the configuration of their artefacts in order to improve performance along the modelled adoption criteria provided by promoters. Namely, designers sort the lines of their artefacts according to their contribution to the formation of perceived shapes. Designers are able to delete or generate new lines as a function of adopter perception **V**. Hypotheses consist of informed changes to current artefacts. In particular features that do not contribute to good performance are randomly replaced. Hypotheses are then evaluated following the multi-criteria adoption function of equations 4 and 5. A design rule $\lambda$ consists of artefact changes that increase its performance along a target criterion.

$$\lambda = \{ \mathbf{G}_h = \Delta\mathbf{P} : \mathbf{C}_i \} \tag{11}$$

where a hypothesised feature $\mathbf{G}_h$ results in an increment of artefact performance **P** in a presumed criterion $\mathbf{C}_i$. The positive value of $\Delta$ stands for the improvement ratio of $\lambda$.

Processing abilities and synthetic abilities are assigned at initial time to each designer. Processing refers to the capacity of designers to generate and retrieve domain rules, whilst synthesis controls the number of hypotheses that designers can generate before having to transform their artefacts. In this paper designers are assigned constant abilities at initial time. However, abilities gradually increase as a function of design behaviour. These two parameters determine individual differences across designers. This enables experimentation with the impact of individual factors on creativity, but is beyond the scope of this paper.

If a designer is not able to generate new domain knowledge, it seeks a strategy to apply existing design rules $\lambda$. Here two assumptions can be implemented: domain knowledge may have private or public access. If private, every designer agent only has access to their own rules, whilst in public mode all designers have access to all existing rules.

$$\text{Apply} = \{ \lambda = \Delta\mathbf{P} : \mathbf{C}_i \} \tag{12}$$

where an existing rule $\lambda$ that improves performance **P** in a target criterion **C** is applied to an artefact. If a designer is not able to generate or apply relevant knowledge, the last option is to imitate other designers. Imitation is the simplest form of collective learning, i.e. blind learning since information about features, criteria, and perception is missing. Imitation is defined as the transfer of a random feature **G.** Designers whose artefacts have low adoption rates imitate artefacts with higher rates. This is acknowledged as peer recognition given to the designer of the source artefact. Recognition from colleagues indicates the influence of a designer and further extends its processing and synthetic abilities.

Designers may address the perceived group's choice criterion $\mathbf{C}_i$ or they may determine an alternative target criterion. This choice is a function of perceived adopter preferences **F** and estimated artefact performance **P**. If a designer considers that its artefact's performance is competitive (i.e. equal or above mean adopter preference) capitalisation is chosen and alternatives are sought to improve performance on the choice criterion (exploit relative superiority). If estimated performance is instead low on adopter preferences then designers seek to differentiate their artefacts from the highly competitive industry by selecting their best performing criterion.

$$\text{Competition} = \mathbf{P} \geq \mathbf{F}_{mean} \tag{13}$$

$$\text{Differentiation} = \mathbf{P} < \mathbf{F}_{mean} \tag{14}$$

Designer agents in this system are not equipped with creative abilities per se. The aim is not to introduce special traits to assess the effects of agents' creativeness as assessed by the experimenter. Instead, all designers are assigned equivalent sets of mechanisms. No extraordinary process within the individual is hardwired but in time agent interaction renders a social self-organised construct of how a designer may exhibit behaviour considered creative within its society.

Strategic Differentiation Index (SDI) is an index estimated collectively by adopters that reflects the perceived differentiation across the available artefacts (Nattermann 2000). With a design system initialised in a converged state, SDI = 0.0. As designers seek to generate artefacts that differ from other available artefacts SDI > 0.

$$SDI = \{ \sum (\mathbf{P}_{var} / \mathbf{C}_{i\text{-}n}) \} \quad (15)$$

where SDI is the sum of mean performance variance for all criteria as estimated by the adopter population.

This social framework encapsulates some of the characteristics of design problems including ill-structuredness and interpretation; decomposability; incremental solutions; hypothesis generation; negotiable and nomological constraints; no right or wrong answers; and delayed feedback (Goel 1994).

Design behaviour complements our definition of creativity. Adoption rate is a trend measure used to determine what designer is imitated at a particular time step. Peer-recognition is considered a necessary element in the creativity literature (Runco and Pritzker 1999). The contribution of each designer to domain knowledge is interpreted as transformation of the design space (Gero 1996), learning, experience, and the ten-year rule (Runco and Pritzker 1999). Extension and exhaustion of the design space (i.e., rule generation and application) refer to exploratory-transformational creativity (Boden 1999). The number of hypotheses generated resembles idea productivity. The number of entries selected by gatekeepers gives a measure of a designer's contribution to the repository or domain (Feldman et al 1994). The limit on hypothesis generation and constraints on representation addresses the relationship between constraints and creativity (Amabile 1983).

A designer is considered creative by its social group when it reaches large adopter groups, its artefacts are entered into the repository, other designers imitate its artefacts, it transforms the design space by formulating knowledge, and its adopters have high satisfaction levels. In general experimentation with this framework consists of exploring the effects that both individual and situational factors have on determining the creativity of designers.

## 5. Experiment: Social Ties

This experiment addresses the role of social ties in the formation of influence structures in a population and the associated effects on creativity and innovation. Tie strength **T** is implemented as the frequency of contact between adopters (Marsden and Campbell, 1984). A series of simulations are run where the initial configuration of adopters and designers is kept constant and the strength of social ties **T** is the independent variable. Monte Carlo runs are conducted to explore the range $0.0 \leq \mathbf{T} \leq 1.0$ over 7500 iterations in populations of 10 agents. This explores the range where agents remain in their social location at all times to where agents change social locations at all times, respectively. Preliminary runs showed that dependent variables stabilise between 2500 and 5000 iterations. The resulting dataset is then filtered in order to exclude outliers, i.e. values 1.5 standard deviations from the mean. All the following results represent means of 10 simulation runs. Each simulation run is initialised in a converged state to avoid biases in the form of random initial artefact configurations. Therefore at iteration step 0, adopters perceive no differentiation between artefacts and all abstain from adopting. It is only after a designer first modifies an artefact that adoption commences.

## 5.1. DOMINANCE HIERARCHIES

As **T** increases, social mobility is seen to decrease causing agents to interact more often with a stable group of neighbours. As a result, influence is more concentrated ($\gamma \approx 1.0$), i.e. a few adopters exert dominance over others. In contrast, as **T** decreases, social mobility increases and agents have contact within a varying neighbourhood. In such conditions, influence structures of dominance are more distributed ($\gamma \approx 0.0$), i.e. hierarchies become more flat. Figure 7 shows a scatter plot and power-law relation of tie strength **T** and Gini coefficient $\gamma$.

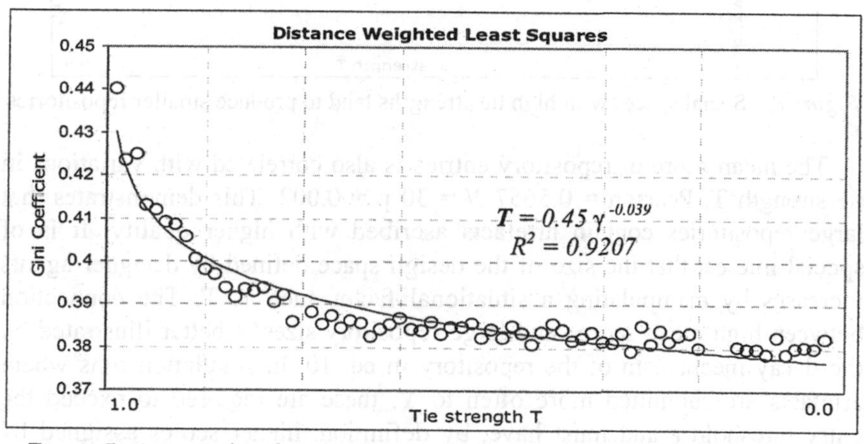


*Figure 7.* A power law function is empirically demonstrated for tie strength and Gini coefficient ($\mathbf{T} = 0.45\ \gamma^{-0.039}$)

## 5.2. GATEKEEPING EFFECTS

The formation of dominance structures shows unexpected effects in adoption and design behaviour. On the one hand an inverse correlation is shown between tie strength **T** and number of entries to the repository **Y**. Lower **T** is correlated with larger repositories as shown in Figure 8, Pearson = 0.6706 $N$ = 30 p = 0.001.

This phenomenon may be due to the nature of the promoter role and is particularly insightful in relation to gatekeeping (Feldman et al 1994). In societies with rigid influence hierarchies (**T** ≈ 1.0) there is less variation in adopters that play the promoter role. Therefore interpretations that serve to evaluate artefacts for entry remain constant over time. In contrast, in societies with lower **T** and therefore where influence is distributed rather than concentrated there is a higher change rate of gatekeepers. Consequently, more diverse evaluations of artefacts mean more artefacts are submitted to the repository. As an effect, designers in general tend to receive more recognition for their work.

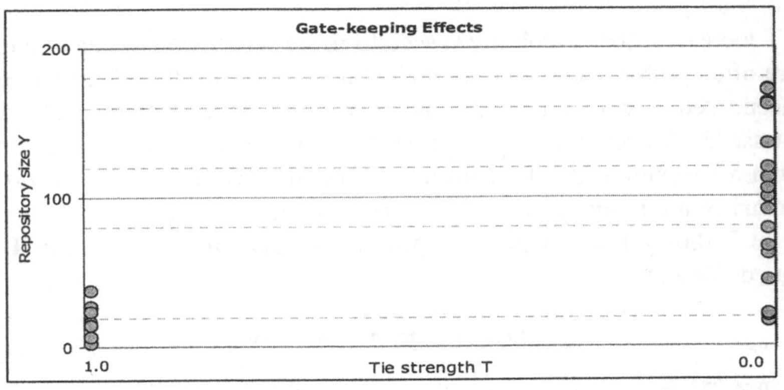


*Figure 8.* Social spaces with high tie strengths tend to produce smaller repositories

The mean score of repository entries is also correlated with variations in tie strength **T**, Pearson = 0.5657 $N$ = 30 p = 0.002. This demonstrates that large repositories contain artefacts ascribed with higher quality. It is of special interest that the size of the design space defined by designer agents increases by manipulating a situational factor such as **T**. The connection between high mean scores and large repository sizes is better illustrated by the decay mechanism of the repository in eq. 10. In simulation runs where artefacts are submitted more often to **Y**, these are required to exceed the entry threshold $\varepsilon$ and must have, by definition, higher scores assigned by promoters.

### 5.3. DIFFERENTIATION EFFECTS

The differentiation of design artefacts is measured through the Strategic Differentiation Index (SDI) (Nattermann 2000). These experiments show that SDI decreases with higher **T** and has the opposite effect as **T** decreases.

In other words, designers operating on tight social spaces where influence structures are rigid tend to generate more similar artefacts. The same designers operating on better distributed influence social spaces have a tendency towards higher differentiation, Pearson = 0.5755 $N$ = 30 p = 0.004.

## 5.4. PROMINENCE EFFECTS

Lastly, effects on the size and nature of adopter groups are addressed. Monte Carlo runs where tie strength **T** is the independent variable consistently show that **T** is positively correlated with adopter group size, Pearson = 0.608 $N$ = 26 p = 0.001. This illustrates that low tie strengths increase abstention. On the other hand, **T** is also correlated with distribution of adoption defined as the ratio of smallest and largest adopter groups as shown in Figure 9, Pearson = 0.6796 $N$ = 26 p = 0.001. Namely, in social spaces where **T** ≈ 0.0 and influence is more distributed, adopters tend to abstain more and their choice tends to be more closely distributed across designers. In contrast, **T** ≈ 1.0 increases total adoption and concentration of choices. In other words, the competitiveness between designers and their prominence can be determined by the way in which their evaluating groups organise.

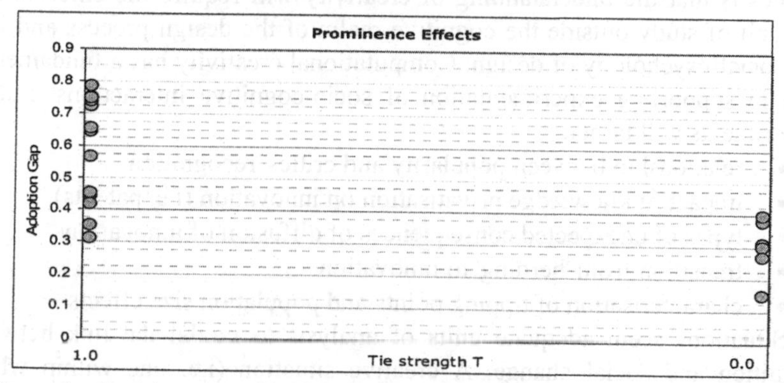


*Figure 9.* Social spaces with high tie strengths (**T** ≈ 1.0) tend to produce higher variation between adopter groups' sizes

## 5.5. SUMMARY

The experiments presented in this section suggest that creativity transcends the individual domain. Patterns of creative figures show that characteristics external to the individual may indeed determine who and how is considered creative in a society. Graham, Einstein, Picasso and Freud have been characterised as extraordinary creators. Whilst their personality traits and abilities have little in common (Gardner 1993), similarities exist between the structures of the fields within which they operated. Namely, a few powerful critics rendered influential judgements about the quality of their work (Feldman et al 1994).

Our experiments illustrate a fundamental idea about the nature of creativity and innovation, i.e. that a situational factor that regulates the way in which adopters interact may have a significant effect on how both designers and adopter groups operate. The implications are that by observing the performance of designers alone it is not possible to put forward conclusions about their individual characteristics. Instead, the cause of behaviour could be a situational factor that defines not the designers but their evaluators.

## 6. Discussion

In this paper a social framework for the study of creativity and innovation in design has been introduced and used to experiment with a situational factor of creativity in design. Factors that regulate aggregate behaviour of a population of adopters are shown to affect the way designers operate and their impact as change agents of their societies. A corollary of these types of studies is that the understanding of creativity will require the extension of the unit of study outside the cognitive realm of the design process and into the social-psychology of design. Computational creativity has a fundamental role in supporting experimentation of socio-cognitive interactions. Future experiments will target:

- relationship between popularity and critics' recognition
- impact of knowledge privatisation on innovation (i.e. patents)
- types of unexpected consequences of diffusion of innovations
- transition from 'herding' to innovation
- characterisation of tipping points and population size effects

Situations seem adequate units of analysis to model the link between cognition and social change. A creative situation (i.e. one within which designers with different characteristics are likely to trigger a social change) could be typified to complement the dominance of studies that focus on the creative personality.

## Acknowledgements

This research is supported by an International Postgraduate Research Scholarship and a University of Sydney International Postgraduate Award. Computing resources are provided by the Key Centre of Design Computing and Cognition.

## References

Amabile, TM: 1983, *The Social Psychology of Creativity*, Springer-Verlag, New York.

Argyle, M, Furnham, A and Graham JA: 1981, *Social Situations,* Cambridge University Press, Cambridge.

Asch, SE: 1951, Effects of group pressures upon the modification and distortion of judgment, *in* H Guetzkow (ed), *Groups, Leadership, and Men*, Carnegie Press, Pittsburgh, pp. 177-190.

Boden, M: 1999, Computer models of creativity, *in* R Sternberg (ed), *Handbook of Creativity*, Cambridge University Press, Cambridge, pp.351-372.

Conte R, Edmonds B, Moss S and Sawyer RK: 2001, Sociology and social theory in agent based social simulation, *Computational and Mathematical Organization Theory* **7**(4): 183-205.

Dorfman, R: 1979, Formula for Gini coefficient, *Review of Economics Statistics* **61**: 146-9.

Eysenck, HJ: 1991, Dimensions of personality, *Personality and Individual Differences* **12**: 773-90.

Feldman, DH, Csikszentmihalyi, M, and Gardner, H: 1994, *Changing the World, A Framework for the Study of Creativity*, Praeger, Westport.

Gardner, H: 1993, *Creating Minds, An Anatomy of Creativity Seen Through the Lives of Freud, Einstein, Picasso, Stravinsky, Eliot, Graham and Gandhi*, Basic Books, New York

Gero, JS: 1996, Creativity, emergence and evolution in design: concepts and framework, *Knowledge-Based Systems* **9**(7): 435-448.

Goel, V: 1994, A comparison of design and nondesign problem spaces, *Artificial Intelligence in Engineering* **9**(1): 53-72.

Granovetter, MS: 1973, The strength of weak ties, *American Journal of Sociology* **78**(6): 1360-80.

Howe, MJA, Davidson, JW, and Sloboda, JA: 1999, Innate traits: reality or myth?, *in* SJ Ceci and WM Williams (eds), *The Nature-Nurture Debate, The Essential Readings*, Blackwell, Malden, pp. 258-289.

Marsden, PV and Campbell, K: 1984, Measuring tie strength, *Social Forces* **63**(2): 482-501.

Nattermann, PM: 2000, Best practice does not equal best strategy, *The McKinsey Quarterly* **2000**(2): 38-45.

Rogers, EM: 1995, *Diffusion of Innovations*, The Free Press, New York.

Ross, L and Nisbett, R: 1991, *The Person and the Situation*, McGraw-Hill, New York.

Runco, M and Pritzker, S (eds): 1999, *Encyclopedia of Creativity*, AcademicPress, San Diego.

Saunders, R and Gero, JS: 2001, The digital clockwork muse, *in* G Wiggins (ed), *AISB'01 Symposium on AI and Creativity in Arts and Science*, University of York, York, pp.12-21.

Schelling, K: 1971, Dynamic models of segregation, *Mathematical Sociology* **1**: 143-186.

Simon, HA: 2001, Creativity in the arts and the sciences, *The Kenyon Review* **23**(2): 203-221.

Simonton, DK: 2000, Creative development as acquired expertise: theoretical issues and an empirical test, *Developmental Review* **20**: 283-318.

Sosa, R and Gero, JS: 2003, Social change: exploring design influence, *in* D Hales, B Edmonds, E Norling and J Rouchier (eds), *Multi-Agent Based Simulation III*, Lecture Notes in Artificial Intelligence, 2927, Springer, Berlin, pp. 106-119.

Wasserman, S and Faust, K: 1994, *Social Network Analysis: Methods and Applications,* Cambridge University Press, Cambridge.

Wooldridge, M: 2000, *Reasoning About Rational Agents*, MIT Press, Cambridge.

# WORDS AND RATIONALE IN DESIGN

JS Gero (ed), *Design Computing and Cognition'04*, 521-540


# QUANTIFYING COHERENT THINKING IN DESIGN: A COMPUTATIONAL LINGUISTICS APPROACH

ANDY DONG
*University of Sydney, Australia*

**Abstract.** Design team conversations reveal their thinking patterns and behaviour because participants must communicate their thoughts to others through verbal communication. This article describes a method based on latent semantic analysis for measuring the coherence of their communication in a conversation mode and how this measurement also reveals patterns of interrelations between an individual's ideas and the group's ideas. While similar studies have been done on design documentation, it was unclear whether computational techniques that have been applied to communication in text could be successfully applied to communication in a conversational mode. Transcripts of four engineering/product design teams communicating in a synchronous, conversational mode during a design session were studied. Based on the empirical results and the proposition that a team's verbal communication offers a fairly direct path to their thinking processes, the article proposes the link between coherent conversations and coherent thinking.

## 1. Introduction

### 1.1. VERBAL COMMUNICATION IN DESIGN

Social activities such as information exchange, compromise and negotiation figure prominently into phenomenological descriptions of design. The field of social cognition in design studies these social activities and their reciprocal influence on a design team's ability to create collective sense-making of the function, behaviour, structure and meaning of a product. Knowing how teams develop a shared, organised understanding and mental representation of knowledge about the designed artefact is critical for educating designers and for building design support environments.

One of the accepted methods for studying and understanding social cognition in design is the observation of designers (Lawson 1997). In design teams, because designers must communicate their thoughts to others, verbal communication offers a fairly direct path to the thinking processes of designers. "Think aloud" transcripts present a vehicle for peering into the thinking processes occurring during designing. The widely adopted method of verbal protocol analysis converts the "think aloud" transcripts into data (Ericsson and Simon 1993). The methodological roots for a rich body of literature on the cognitive processes of designers lay in verbal protocol analysis (Cross et al. 1996).

Recently, there has been the growing recognition to understand design in group situations. This appreciation has led to more in-depth analyses of design groups, that is, designers working together rather than individual designers working alone. Analyses of design team discussions using protocol analysis have led to interesting insights into the thinking behaviour of design groups such as strategies for reaching shared understanding (Valkenburg and Dorst 1998) and the patterns of basic cognitive processes in design groups (Stempfle and Badke-Schaub 2002). Computational studies of design team documentation (communication in text) illustrated the link between textual coherence and shared understanding (Hill et al. 2001), team performance (Dong et al. 2004), and variations in the design direction that a design group may pursue over a long period of time (Song et al. 2003). Being able to deduce aspects of design team behaviour automatically from their ongoing conversations would have a beneficial impact on studies in design workspaces which have recognised that verbal dialog constitutes a major part of design session activity (Tang and Leifer 1998) and advanced design research laboratories (Milne and Winograd 2003).

Increasingly, design teams rely on computer-mediated communication and collaboration tools such as email, chat, bulletin boards, "blogs" and virtual environments. Human based protocol studies would be unable to scale to the volume of information and communication generated by designers in real-world settings. Could computational methods assist in seeing the "haystacks" of behaviour before humans go about finding the "needles?" There is also an increasing call to encourage an ethic of self-reflection (Schön 1987). Placing teams under the watchful eye and lens of social science researchers, while of theoretical analytical value, may offer little for constructive in-process improvement. Could autonomous agents with the capability to assess the team communication supplement or confirm the teams' own awareness of breakdowns in communication?

This article concerns with the broader question of how cognitive processes during design could be assessed computationally, that is without reliance on hand coded protocol analysis. The approach taken to address one aspect of this question, coherent thinking in design teams, is based on computational linguistics. The premise of the article is that the psychological similarity between thoughts is reflected in the semantic coherence between words in the way they co-occur in dialog and other language-based communication. Latent semantic analysis (LSA) is a computational linguistics tool which provides one way to model this "psychological similarity between thoughts" based on language (Landauer 1999). The article demonstrates how to extend latent semantic analysis to analyse design communication in a conversation mode and develops metrics to measure the duration of coherent discussions in design teams. Based on the theory of LSA as a representation of knowledge (Landauer and Dumais 1997), the article proposes the alignment between coherent conversations and coherent thinking in teams.

For the purposes of this research, coherent thinking is not equated with thinking alike. That would leave the notion of coherent thinking at the level of "groupthink." Instead, coherent thinking is characterised by the alignment of thoughts among the design team participants that leads to the joint construction of knowledge about the designed artefact. Second, the article assumes that the designers' conversation is goal-oriented. They are not just "chatting," but are, for example, resolving a problem or cooperatively designing an artefact. The remainder of the article describes the computational linguistics method and what the empirical results imply about the coherent thinking of the design teams.

## 2. Methodology

### 2.1. LATENT SEMANTIC ANALYSIS IN DESIGN RESEARCH

Latent semantic analysis (LSA) (Landauer et al. 1998) is a text analysis method that characterizes the semantic similarity between words in texts using a high-dimensional semantic space. The mathematical foundation for LSA lies in singular value decomposition (SVD), a matrix approximation method for reducing the dimensions of a matrix to the most significant vectors. The principal advantage of LSA over other computational linguistics techniques such as lexical chain analysis and anaphora resolution is LSA's examination of context instead of individual word meaning. All other computational linguistics tools degrade in performance when the data set is noisy or semantic understanding is insufficient. Examining context instead of semantics removes obfuscation created by "noise" in the oral and written communication, and scales up to deal with very large corpora.

LSA has been applied to a wide range of problems: full-text information retrieval (Foltz et al. 1998); assessing learning (Landauer et al. 2000; Shapiro and McNamara 2000); analysing the cognitive processes underlying communication (Landauer 1999); and assessing the outcome of design teams (Dong 2004). The problem, though, is the direct application of the LSA method to analysing text from a conversation. If one were to attempt to examine the coherence of a discussion over time based on the methods described by Song (2003) or Hill (2002), one would not obtain much useful information because of the high variation in context between one utterance and the next. If we were to assume, instead, that the topical focus of a design conversation should be coherent, then there should exist a mostly orderly relationship between semantic similarity and distance between utterance boundaries. An utterance boundary is defined by the turn-taking between speakers, that is, when one speaker stops talking and the next speaker begins. This orderly mapping can be revealed by looking at the coherence between any two utterances (communicative acts) as a function of the "distance" between the utterance boundaries. That is, instead of computing and plotting the coherence between adjacent utterances as suggested by Song (2003), compute the average coherence between utterances which are "one" utterance away, the average coherence between utterances which are "two" utterance boundaries away, and so forth, to expose the structuring of language over the entire conversation.

## 2.2. LSA FOR ANALYSING DESIGN TEAM DISCUSSIONS

The basic method for analysing text with LSA follows established procedures. We refer the reader to the cited sources above for detailed discussions on applying LSA. First, the utterances from the design team's conversation are entered into a database, recording an utterance identifier, the name of the speaker, and the utterance itself. The utterances may consist of multiple complete "sentences" or incomplete thoughts such as single word statements within an utterance boundary. Then, the utterances are parsed into principal phrases. Next, a term frequency matrix is constructed. The term frequency matrix counts the occurrences of the principal phrases in each utterance boundary. This term frequency matrix **X** is of dimension n (rows) words $w_1, w_2, \ldots, w_n$ and m (columns) utterance boundaries $d_1, d_2, \ldots d_m$ where each matrix entry indicates the total frequency of occurrence of term $w_n$ in utterance $d_m$ and $m < n$.

Then, the singular value decomposition (SVD) of the matrix **X** is computed. The SVD of **X** is defined as $\mathbf{X} = \mathbf{USV}^T$ where **U** (n x m) and **V** (m x m) are the left and right singular matrices (orthonormal), respectively, and **S** (m x m) is the diagonal matrix of singular values. SVD yields a simple strategy to obtain an optimal approximation for **X** using smaller matrices. If the singular values in **S** are ordered descending by size, the first k largest may be kept and the remaining smaller ones set to zero. The product of the resulting k-reduced matrices is a matrix $\widetilde{X}$, which is approximately equal to **X** in the least squares sense, and is of the same rank. That is, $\widetilde{X} \approx \mathbf{X} = \mathbf{USV}^T$. The number of singular dimensions to retain is an open issue in the latent semantic analysis literature. Based on prior research (Hill et al. 2002), retaining the first 100 dimensions or dimensions 2 to 101 resulted in satisfactory performance. In this study, we retained the first 100 dimensions. We refer the reader to the literature (Deerwester et al. 1990) for the proof that the rows of the matrix VS are coordinates of the utterance in this k-reduced space. These utterance coordinate row-vectors form the basis for the coherence analyses. The final step is to compute the coherence of the utterances.

### 2.3. MEASURING THE COHERENCE OF UTTERANCES

The standard definition of the (cosine) coherence between any two adjacent utterances represented by the LSA vectors $\mathbf{d}_m$ and $\mathbf{d}_{m+1}$ is the dot product of the utterance vectors normalized by the product of their norm.

$$\chi_m^{m+1} = \frac{d_m \cdot d_{m+1}}{\|d_m\| \|d_{m+1}\|} \tag{1}$$

The ultimate objective is to consider how the topical focus (coherence) of a conversation changes during a design session. In the design sessions studied, each team's conversation was goal-directed; the teams had to complete the design of an artefact that satisfied a prescribed set of goals and constraints. To examine the change in focus, the average coherence between utterances which are "one" utterance away, the average coherence between utterances which are "two" utterance boundaries away, and so forth, were computed and plotted.

## 3. Experimentation

### 3.1. DATA SETS

Two data sources were examined in this study. The first was a transcript from the mountain bike backpack design problem at the 1994 Delft Protocols Workshop with human analysis using the reflective practice as an observation method (Valkenburg and Dorst 1998). The transcript contains 2190 "raw" utterances among three design students over a 118 minute period. The second set comes from the Bamberg Study (Stempfle and Badke-Schaub 2002) of design thinking in teams based on their communication. There are three teams in the Bamberg Study, denoted by 1102, 2202 and 2302. Because the Java natural language parser (Stanford 2004) used for this research only parses English, the Bamberg transcripts were translated from German to English by native-language speakers with technical language proficiency and edited by the author for consistency in terminology by the two translators.

### 3.2. RESULTS

Figure 1 to Figure 4 display the average coherence between utterances as a function of distance between utterance boundaries for all transcripts. In the plots, the "distance" is the number of utterance boundaries between two utterances. The solid dots are the raw data; the open circles are a curve fit of the data (discussed below). The plots are intriguing in that they appear fairly regular with a non-zero slope initially, but approach an asymptotic limit. Outside a certain distance between utterances, the coherence 'drops off' and is highly scattered.

Dissecting the conversations by speaker, we can then look at the coherence of each speaker's utterances, as in Figure 5 and Figure 6. The solid dots are the group's utterance coherence (as before) and the other dots represent each speaker. For the Delft Protocol Study Team, (Figure 5), even though each speaker's coherence is roughly constant, in additive, they increase the coherence of the conversation. It is as if the contribution of each person to the group's conversation led to a more coherent conversation by the group; this appears to confirm the prior assessment of this team by the protocol analysts of the establishment of a shared understanding. Bamberg Team 2302 exhibited a similar conversation pattern as shown in Figure 7.

The Bamberg 1102 and 2202 Teams differed. Illustrating the per-speaker coherence for the Bamberg 1102 Team in Figure 6, one notes that each individual's coherence is "scattered" within a band which also contains the group's overall coherence, as shown by the solid red dots. That is, the contribution of each group member to the group's discussion does not increase the coherence of the overall conversation.

The smoothness of the curves of Figure 1 to Figure 4 motivated the interest to investigate the regularity in the data set. Polynomial curve fits on the data were conducted to obtain the best polynomial curve fit in the least sum square of errors sense. The curve fit would offer a functional relationship between number of utterances and the coherence of those utterances from which metrics of coherent discussions could be derived. It was found that a 3$^{rd}$ degree polynomial curve (Eq. 2) best fit the data.

$$y = m_1x^3 + m_2x^2 + m_3x + m_4 \tag{2}$$

Using these curve fits, it then becomes possible to quantify coherent discussions via analytic metrics. Two methods are proposed: the loss of information and the rate of decay of coherence.

The first metric is based on the notion of measuring the loss of coherence using a decibel scale. The decibel scale is a convenient way to compare amplitudes. In this case, the amplitudes are the level of coherence of the discussions. The equation for the decibel scale is shown in Eq. 3 where $c_0$ is the average coherence for adjacent utterances and $c$ is the coherence of an

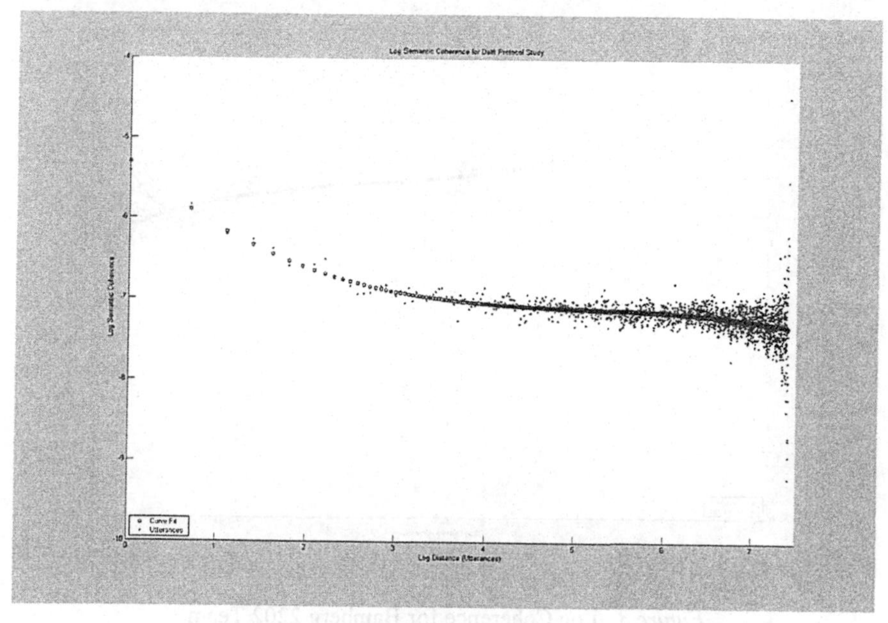


*Figure 1.* Log Coherence for Delft Protocol Team

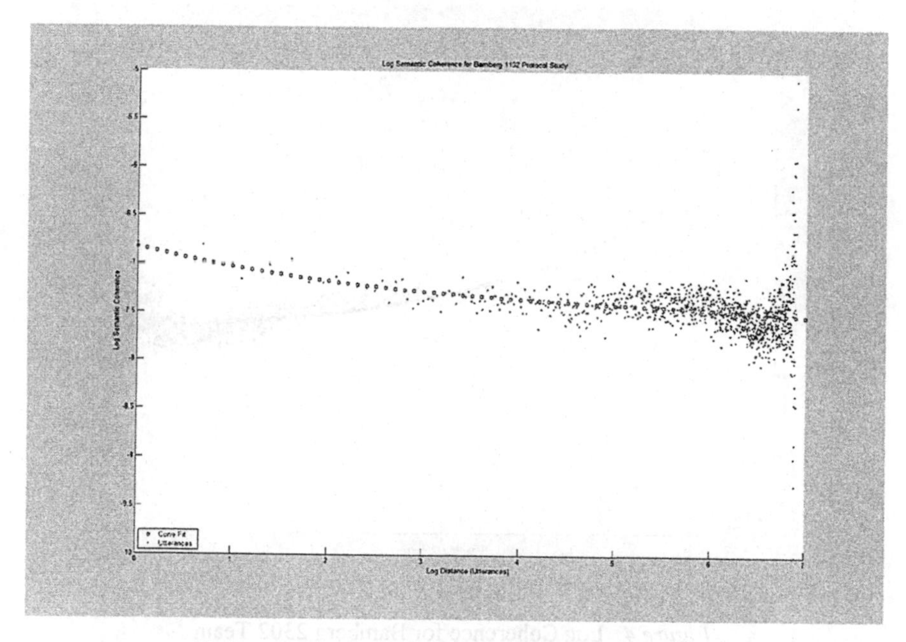


*Figure 2.* Log Coherence for Bamberg 1102 Team

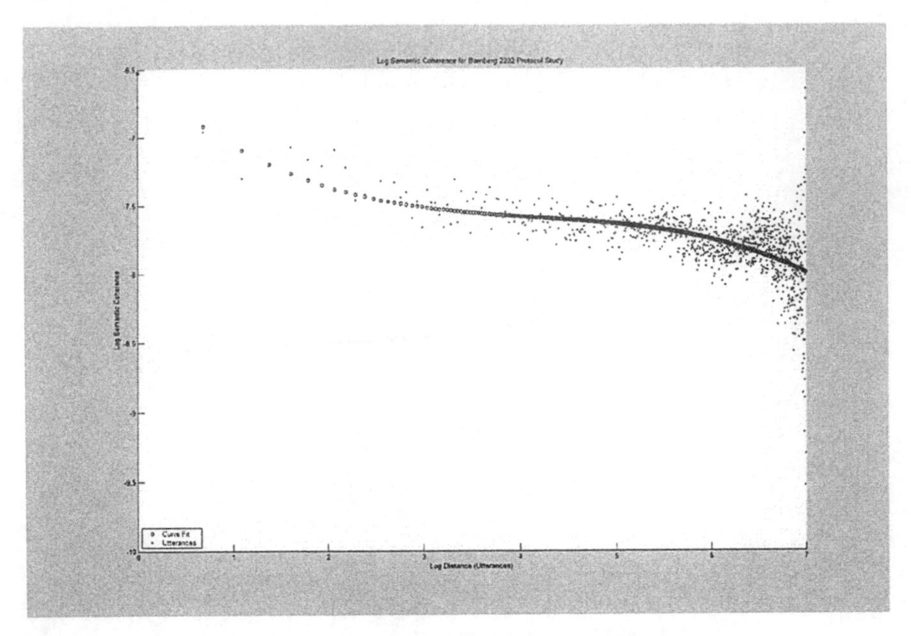


*Figure 3.* Log Coherence for Bamberg 2202 Team

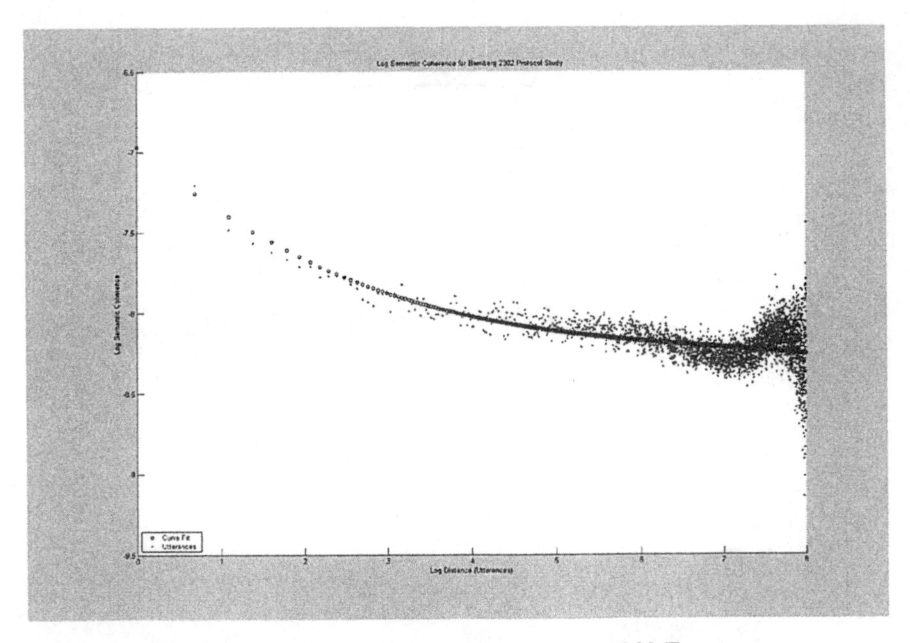


*Figure 4.* Log Coherence for Bamberg 2302 Team

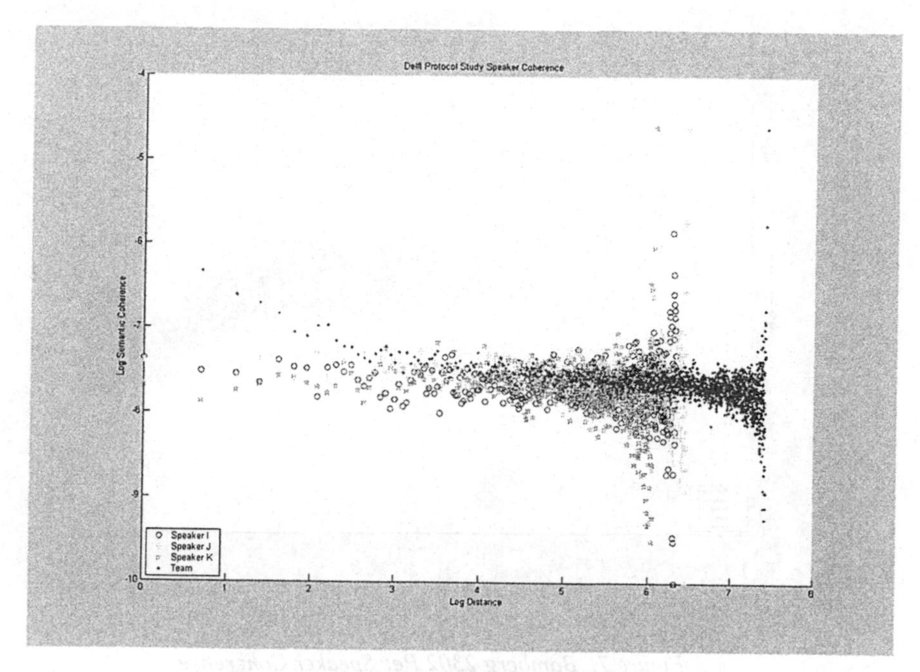


*Figure 5.* Delft Per Speaker Coherence

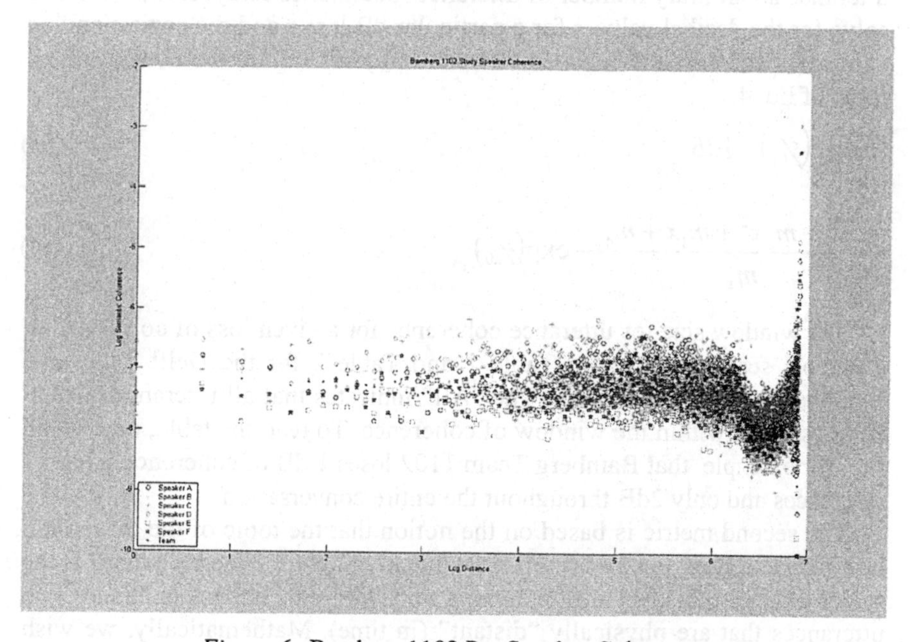


*Figure 6.* Bamberg 1102 Per Speaker Coherence

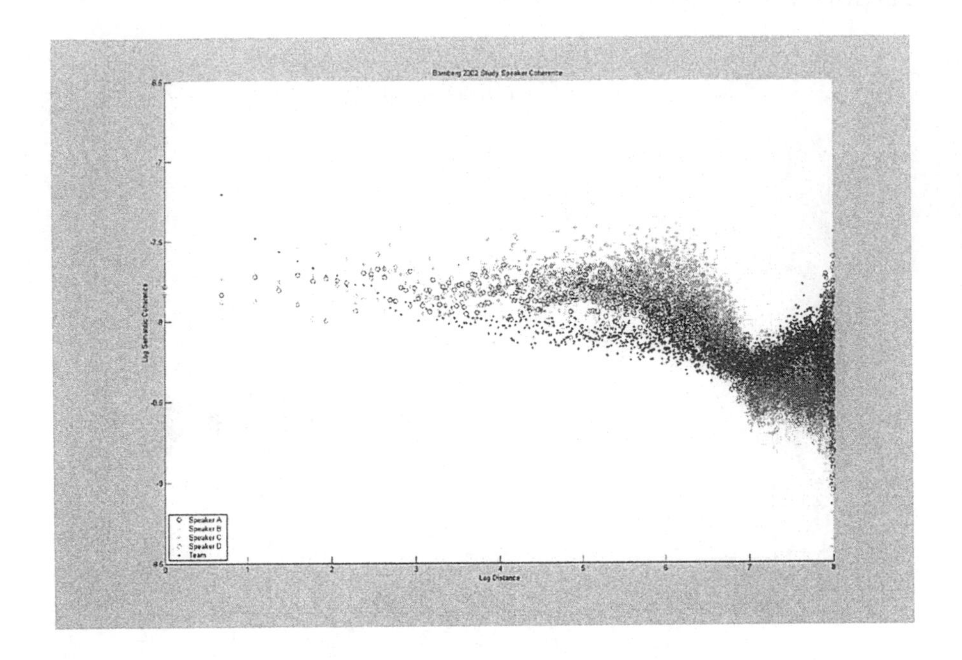

*Figure 7. Bamberg 2302 Per Speaker Coherence*

utterance an arbitrary number of utterance boundaries away. It is possible to solve for the decibel value $c$ for a particular dB loss $z$ and the corresponding value of $x$, the distance between utterances, by solving for the non-zero, real roots of Eq. 4.

$$20\log\left(c/c_0\right)=1dB \qquad (3)$$

$$\frac{m_1x^3+m_2x^2+m_3x+m_4}{m_4}-\exp\left(z/20\right) \qquad (4)$$

The window size for utterance coherence for a given loss of coherence in a decibel scale is shown in Table 1 and Table 2 for the Delft Team and Bamberg Teams, respectively. The "All" indicates that all utterances would be considered within the window of coherence. To read the table, one would say, for example, that Bamberg Team 1102 loses 1 dB of coherence within 7 utterances and only 2dB throughout the entire conversation.

The second metric is based on the notion that the topic of the discussion will likely shift during a session. How quickly might the topic change? If the topics change quickly, an utterance is very likely to be less coherent with utterances that are physically "distant" (in time). Mathematically, we wish to find the slope of the best linear fit of the data within the incline region of the "raw" coherence data as in Figure 1. To do this, a linear curve fit was conducted from the first data point of the coherence data plotted as in Figure 1, which corresponds to the average coherence of adjacent utterances, to the point on the 3rd order curve fit where the first derivative is closest to 0. This point is defined as the asymptotic limit. The rate of decay of coherence of utterances is shown in Table 3 for the Delft and Bamberg Teams. To read these tables, one would say the Delft Team loses topical similarity at a rate of 19%.

TABLE 1. Delft dB Coherence

| dB | Utterances |
|---|---|
| 1 | 1.3 |
| 2 | 1.9 |
| 3 | 3 |
| 5 | 13 |

TABLE 2. Bamberg dB Coherence

| dB | Utterances (Team 1102) | Utterances (Team 2202) | Utterances (Team 2302) |
|---|---|---|---|
| 1 | 7 | 2 | 2 |
| 2 | All | 4 | 8 |
| 3 | All | 57 | 111 |
| 5 | All | All | All |

TABLE 3. Decay of Coherence

| Team | Utterances |
|---|---|
| Delft | -19% |
| Bamberg 1102 | -6% |
| Bamberg 2202 | -17% |
| Bamberg 2302 | -10% |

The rate of decay of the Bamberg 1102 Team is 66% less than the Bamberg 2302 and almost three times lower than the Bamberg 2202 Team.

## 4. Discussion

### 4.1. DISCUSSION OF EXPERIMENTAL RESULTS

The computational analyses revealed two patterns of conversations in the data sets: one in which the speakers built upon each other's utterances resulting in an increased level of coherence as exhibited by the Delft Team and Bamberg Team 2302, which we'll term *constructive dialog*, and one in which there was little in the way of building upon each other's expressions of ideas as demonstrated by the Bamberg 1102 and 2202 teams, which we'll term *neutral dialog*. The data sets examined do not reveal a third possibility, *destructive dialog*, in which each other's expressions of ideas negate one another.

Theories of socio-linguistics offer many explanations, as evidenced by the wide body of empirical literature that focuses on when and whether people adapt their discourse to each other, why these design teams might exhibit one of these three types of dialog patterns. Early on, Bahktin (Wertsch 1991) theorised that effective group communication occurs when the group shares a voice. The collective voice is dynamic; nevertheless, effective group communication occurs when all members are able to borrow from and relate to this combined group voice. The constructive dialog pattern supports this assertion. Conversely, when speakers lack "grounding," they may not initiate the next relevant contribution to the conversation (Clark and Schaefer 1987). The neutral and destructive dialog patterns offer empirical evidence of this situation.

From the design research standpoint, the challenge is not necessarily to understand the socio-linguistic factors that affect discourse, though those are important. What is of interest here is what the patterns of conversations imply about coherent thinking in the design team. If the objective of designers' conversations is to pool their resources to negotiate different design perspectives and specialties to solve a design problem, then the greater challenge is to infer what the empirical data depict about designers' individual and distributed mental processes while they are simultaneously involved in the collective activity of conversation and designing (which would involve drawing on paper or making simple models from available workspace materials.)

These three types of conversational coherence patterns revealed by the computational analysis could be related to behaviour associated with coherent thinking in design teams:

Constructive Dialog: By direct application of experience and each designer's knowledge to solve the design problem, indirect relations among components of knowledge stored in each designer's mind augment and expressed through conversation amplify coherent thinking by the group.

Neutral Dialog: While each individual designer may have a coherent plan "in mind" for solving the design problem, there is little attempt to reconcile each designer's "object world." (Bucciarelli 1994) The designers do not contribute to the conversation based on what has taken place before (i.e., what a previous designer has said/expressed).

Destructive Dialog: The design team is unable to reconcile viewpoints; the conversation breaks down due to disagreements. Collaboration and design suffer; thinking is incoherent.

For the DelftTeam, a move occurred on average every 28 utterances which, according to the computational data, is about 5½dB of loss of coherence. According to the Bamberg study, the teams "spend 8.3 communicative acts on content-related communication before switching to process-related communication" (Stempfle and Badke-Schaub 2002, p. 486). The switching could be thought of as changing the content of communication, or thinking coherently about the "content" and then about the "process." The value of 8.3 is a bit over 1dB loss of coherence for Team 1102, almost 2.5dB loss of coherence for Team 2202, and about 2dB loss of coherence for Team 2302. At this time, there is insufficient data or theory to make statements about the correlation between the computational metrics of coherent discourse and the moves and communicative acts that were ascertained by the protocol analysts. However other insights are available. Looking across the rows of Table 2, there is only a loss of 2dB for Team 1102 throughout their entire conversation whereas both Team 2202 and 2302 lose 5dB. The researchers reported that "proceeding in the design task can be labelled from "chaotic" (group 3) to "planned" (group 1)" (Stempfle and Badke-Schaub 2002, p.484). Yet, the above graphs showed that "group 3" (2302) exhibited a constructive dialog pattern. This apparent contradiction might be explained by the researchers' observation (Stempfle 2004; Stempfle and Badke-Schaub 2002) that this team experienced disagreements and challenges of ideas which nonetheless lead to careful analyses and selection of a design idea. Even though there is some of loss of coherence throughout their (2302) conversation, on the whole, these designers built upon each other's ideas. Thus, the computational analyses suggest that it would be faulty to automatically dismiss a team's conversation as being incoherent (and their thinking incoherent) when topics change quickly or where there appears to be high levels of disagreement. Rather, if the argumentation style builds upon prior evidence, then the resulting complete dialog may be constructive.

Also, the rate of decay is not apparently directly linked with the type of design team dialog. Even though Bamberg Team 2302 had a much higher rate of decay than Bamberg Team 1102, Figure 7 shows the dialog of Bamberg Team to be constructive whereas Bamberg Team 1102 is neutral.

Likewise, although the Delft Team had the highest rate of decay, this team has been shown to exhibit communication and design strategies to increase the level of shared understanding of the team. Again, rapid changes on topic of dialog during a design discussion may not necessarily indicate an incoherent conversation. In summary, the empirical results demonstrate the need to consider both the analytic metrics and the patterns.

The analysis of the conversation of the coherence of design team conversations was more complex than originally expected. Unfortunately, it was not possible to "assign a number" to the level of coherence of the conversation and correlate that to successful outcomes as had been done for communication in text. What are revealed by this computational analysis are aggregate patterns of how individuals contributed to the group conversation. That is, the analysis transforms the communicative residue of teams into devices for quantifying and visualising the level and quality of communication interactions. The analysis also provided accounting metrics of the nature of group conversations. Given that ethno-methodological methods and discourse analysis are beyond the capacity of design teams, automatically deducing characteristics of their communication may lead to feedback mechanisms which support them in assessing the state of their design sessions and adjusting their behaviour as required.

### 4.2. LANGUAGE, COMMUNICATION AND HUMAN COGNITION

What interpretations about coherent cognition do the results on coherent discussions provide? To answer this question, we must consider the relation between language, communication and cognition.

The intersection of language and human cognition is regarded as a fundamental insight to human cognition. The field of sociolinguistics examines how language works to convey information and to connect individuals to cognitive systems. Researches in human developmental psychology and language development, such as Vygotsky's (1978) body of work on the social origins of language and thinking, Bahtkin's (1981) thesis on the development of shared discourse through the assimilation of others' voices, and Wertsch's (1991) socio-cultural framework for mediated action, situate language expression and human cognition within social contexts and connect theories of human cognition to the physical and social world. In all, language acquisition, whether individual or social, is regarded as a key factor for understanding mental functioning.

The study of distributed cognition in design concerns with the generation, transmission and evaluation of information and knowledge to create collective sense-making of the function, behavior, structure and meaning of a product. This is based on Hutchins's (1995) *distributed cognition approach* which promotes the need to look at individuals and the artefacts and tools they use and at the social organisation factors that influence cognition.

Finally, a predominant model of how knowledge is stored in memory asserts that certain types of knowledge are stored as semantic memory (Quillian 1968).

The premise of the article was that the psychological similarity between stored knowledge as semantic memory in each designer's mind is reflected in the semantic coherence between words in the way they co-occur in dialog and other language-based communication.

Therefore, whereas verbal communication portrays the cognitive processes of the designers and whereas certain types of design knowledge are stored as semantic memory and distributed throughout each designer, we can hypothesise that the contribution of individual knowledge to the formation of the group's knowledge manifests in group verbal communications.

Given this hypothesis, one conservative interpretation of the empirical data is that some groups are more able to engage in a constructive dialog. But these teams aren't just talking for the sake of talking. They're engaging in a conversation to design an artifact. They have a goal-oriented conversation. Thus, a more radical interpretation of the data is that the patterns of utterance coherence portray the overlap of their knowledge and how the knowledge of each individual designer contributes to the group knowledge. The metrics of loss and decay of coherence appear to indicate how quickly the teams change directions in thinking rather than not thinking coherently.

Cognitive coherence is an important issue in design cognition research. The coherence of the collective mental functioning facilitates the coordinated action that is required for successful team-based design. Of the basic cognitive processes in design (Stempfle and Badke-Schaub 2002), exploration, generation, selection and comparison, exploration and generation serve to widen a problem space whereas comparison and selection narrow a problem space. Changes in coherent thinking should mirror the changing episodes of design cognition. That is, exploration and generation occur during periods of decreasing coherence (divergent thinking) whereas selection and comparison occur during periods of static or generally increasing coherence (focus). If these episodes could be identified automatically, then it would provide an effective means by which to analyse the frequency, duration and activities during these cognitive episodes. Other research (Valkenburg 2000) has shown that during these episodes design teams create shared understanding and that the nature of the design activities in these episodes can assist (or detract from) the establishment of shared understanding in design. A high rate of decay of coherence or loss of coherence might be an issue if the design team needs to stay on topic; however, the high rate might be desirable if they need to explore a wide variety of design solutions. Thus, the significance of the metrics and of the patterns is based on the design situation.

## 5. Design Cognition Viewer

While it may be possible to scale-up tools and methodologies from linguistics, cognitive psychology and anthropology to study the inter-relationship between communication and cognition, it is not *a priori* obvious how this could be done for any authentic scale design problems. Computational systems for characterising and critically reflecting upon cognitive performance based on communication could resolve the scale-up issue. For this reason, a prototype design cognition viewer, as illustrated in Figure 8, has been developed in conjunction with this research. The design cognition viewer is part of the "Team Thinking" module of Team Agora, a computational system being developed by the author for supporting reflective and productive collaboration in teams by rendering awareness of the social mechanisms that come into play in team collaboration. The team thinking module supports two main functions: analysis of periods of coherent cognition and a cognitive actions analysis (to be reported in a future paper.)

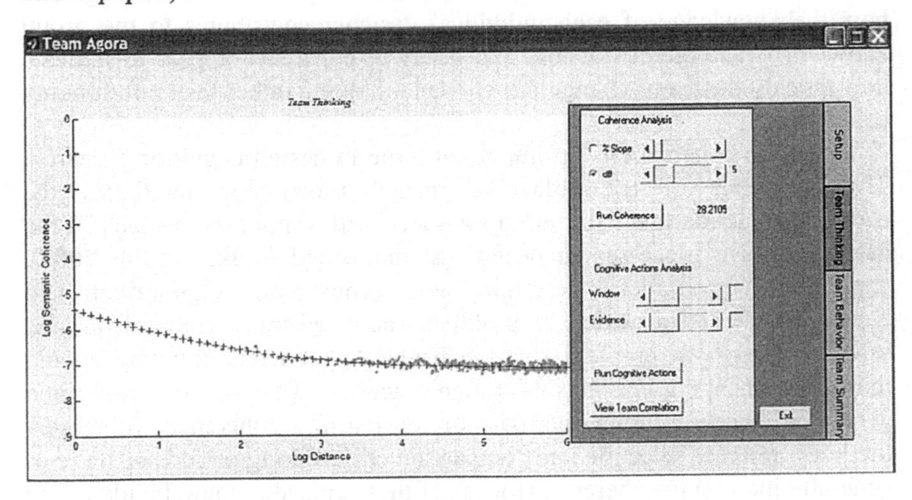


*Figure 8.* Design Cognition Viewer

As shown in the figure, the user can select the method (by dB or slope) by which to compute the duration of coherent thinking, displayed to the right of the "Run Coherence" button, by the team for a given data set. More importantly, though, the graph shows the rate of change and the variation in coherence. Recalling Figure 1 and Figure 2, there is a visible difference in the rate of decay of coherent conversation (thinking) between the Delft Team and the Bamberg Team 1102. While studies have not yet been conducted to ascertain whether this system actually encourages more reflective analysis of their thinking processes, it demonstrates an application of computational linguistic analysis for analysing design team conversations to make visually transparent team collaboration through communication in an accessible way to academics, researchers and people interested in reflection-upon-action of team processes.

Ideally, the viewer would be as rigorous and carefully constructed as "hand done" verbal protocol analysis. However, because the auto-generated visualisations of design cognition are faster to generate, one can utilise the results as a rough indicator or guide by which to segment verbal protocols to search for interesting areas. The design researcher can then conduct a more in-depth analysis of the dialog segments to ascertain boundaries between different cognitive episodes in design.

For the designer or the design teams, the design cognition viewer is intended to a means for reflection upon action by rendering transparent the history of their cognitive performance as demonstrated by their dialogs. Given the ethical issues associated with monitoring dialogs and the impression that such a tool subjects humans to microscopic surveillance, the intent is not to suggest realism in modeling cognitive performance but rather to encourage an atmosphere in which design teams continually self-monitor their performance to reflect upon action. Such a design tool is not a mechanical aid for improving the efficiency of specific design processes but rather intended to improve the effectiveness of the cognitive activities which take place during designing.

## 6. Concluding Remarks

The purpose of this paper was to address the methodological shortcoming of applying latent semantic analysis to the study of design communication in a conversation mode rather than in a text mode. As a practical method, LSA produces useful visual representations of the coherence of the discourse of design teams that enable design researchers to understand the nature of their conversation. Because LSA is nearly automatic, it permits the rapid analyses of a large number of design conversations in near real-time. This capability creates the potential to embed design assessment techniques into design environments and information and communication technology tools purported to *measurably* enhance the communication of design teams. There is much reported research that certain collaborative design environments support and improve communication in design teams and that "more" communication necessarily leads to improved team performance. A more objective assessment might be to measure whether the CSCW tools increase, decrease, or do not affect the ability of the design team to think coherently. Likewise, it would be interesting to study how computer-mediated interaction (such as in design collaboration environments) alters their conversations.

Computational methods such as the one described in this article make steps towards achieving the goal of including design cognition assessment techniques into computer-supported collaborative design environments. This type of analysis may lead to larger studies, in terms of the number of teams and the duration of the analysis of the teams' design period, of designers than is practical with protocol analysis. Clearly, not being able to quantify the influence of facial cues, gestures, and diagrams in the communication is a shortcoming. There exists nascent research on facial gesture recognition and computational shape emergence analyses which may lead to additional insights about the mental functioning of design teams.

In summary, a method for objectively measuring the cognitive coherence of design teams based on synchronous communication in a conversation mode has been presented. The impact of the research lays in the inclusion of these intelligent design cognition assessment techniques into design environments and information and communication technology tools intended to *measurably* increase the cognitive capacity of design teams. An interesting area of investigation would be to measure the effect of various media and computer-supported design and collaboration tools on the cognitive performance of design teams and the influence of such systems. One such question that could be addressed is *how much* digital communication environments alter the cognitive behaviour of design teams. Computational systems for characterizing and critically reflecting upon cognitive performance based on communication is part of continuing research in design cognition performance and management systems. The work contributes to an emerging research area on computational methods for studying the behaviour of designers.

## Acknowledgements

The author would like to thank Rianne Valkenburg for providing the Delft protocol study transcript, Joachim Stempfle for the Bamberg transcripts, and Udo Kannengiesser and Dirk Schwede for translating the Bamberg transcripts.

## References

Bakhtin, MM: 1981, *The Dialogic Imagination: Four Essays by M.M. Bakhtin*, Holquist, M (ed), Holquist, M and Emerson, C (trans.), University of Texas Press, Austin, TX.

Bucciarelli, L: 1994, *Designing Engineers*, Cambridge, MA, MIT Press.

Clark, HH and Schaefer, EF: 1987, Collaborating on contributions to conversation, *Language and Cognitive Processes* **2**: 19-41

Deerwester, S, Dumais ST, Furnas, GW and Landauer, TK: 1990, Indexing by latent semantic analysis, *Journal of the American Society For Information Science* **41**(6): 391-407.

Ericsson, KA and Simon, HA: 1993, *Protocol Analysis: Verbal Reports as Data*, MIT Press, Cambridge, MA.

Dong, A, Hill, A and Agogino, A: 2004, A document analysis technique for characterizing design team performance, *Journal of Mechanical Design (to appear)*.

Foltz, PW, Kintsch, W and Landauer, TK: 1998, The measurement of textual coherence with latent semantic analysis, *Discourse Processes* **25**(2&3): 285-307.

Hill, A, Song, S, Dong, A and Agogino, AM: 2001, Identifying shared understanding in design using document analysis, *Proceedings of the 2001 ASME Design Engineering Technical Conferences*, Pittsburgh, PA, DETC2001/DTM-21713.

Hill, A, Dong, A and Agogino, AM: 2002, Towards computational tools for supporting the reflective team, *in* Gero, JS (ed), *Artificial Intelligence in Design '02*, Kluwer, Dordrecht, Netherlands, pp. 305-325.

Cross, N, Christiaans, H and Dorst, K: 1996, *Analysing Design Activity*, New York, Chichester.

Hutchins, E: 1995, *Cognition in the Wild*, MIT Press, Cambridge, MA.

Landauer, TK, Foltz, PW and Laham, D: 1998, Introduction to latent semantic analysis, *Discourse Processes* **25**: 259-284.

Landauer, TK, Laham, D and Foltz, PW: (2000, The intelligent essay assessor, *IEEE Intelligent Systems and Their Applications* **15**(5): 27-31.

Landuaer, TK: 1999, Latent semantic analysis: A theory of the psychology of language and mind, *Discourse Processes* **27**(3): 303-310.

Lawson, B: 1997, *How Designers Think*, Architectural Press, Oxford.

Mabogunje, A and Leifer, LJ: 1997, Noun phrases as surrogates for measuring early phases of the mechanical design process, *Proceedings of the 9th International Conference on Design Theory and Methodology*, Sacramento, CA.

Milne, A and Winograd, T: 2003, The iLoft project: A technologically advanced collaborative design workspace as research instrument, *Proceedings of the 13th Annual Intenational Conference on Engineering Design (ICED03)*, Stockholm, Sweden.

Quillian MR. 1968, Semantic Memory, *in* J Minsky (ed), *Semantic Information Processing*, Cambridge, MA, MIT Press.

Schön, DA: 1987, *Educating the Reflective Practitioner*, Jossey-Bass, San Francisco.

Shapiro, AM and McNamara, DS: 2000, The use of latent semantic analysis as a tool for the quantitative assessment of understanding and knowledge, *Journal of Educational Computing Research* **22**(1): 1-36.

Song, S, Dong, A and Agogino, AM: 2003, Time variation of "story telling" in engineering design teams, *in* A Folkeson, K Gralen, M Norell & U Sellgren (eds), *Research for Practice: Innovation in Products, Processes and Organisations*, *Proceedings of the 14th International Conference on Engineering Design*, Stockholm, Sweden, The Design Society.

Stanford JavaNLP Project. Retrieved from http://www-nlp.stanford.edu/javanlp/ on February 2, 2004.

Stempfle, J (js.beratung.training@gmx.de) (2004, 22 January), Group 3, E-mail to Andy Dong (adong@arch.usyd.edu.au).

Stempfle, J and Badke-Schaub, P: 2002, Thinking in design teams – an analysis of team communication. *Design Studies* **23**: 473-496.

Tang, JC and Leifer, LJ: 1998, A framework for understanding the workspace activities of designers, *Proceedings of the 1988 ACM Conference on Computer-Supported Cooperative Work*, ACM Press, New York pp. 244-249.

Valkenburg, R and Dorst, K: 1998, The reflective practice of design teams. *Design Studies* **19**: 249-271.

Vygotsky, LS: 1978, *Mind in Society,* Harvard University Press, Cambridge, MA.

Wertsch, J: 1991, *Voices of the Mind, A Sociocultural Approach to Mediated Action*, Harvard University Press, Cambridge, MA.

JS Gero (ed), *Design Computing and Cognition'04*, 541-556


# WORD GRAPHS IN ARCHITECTURAL DESIGN

BAUKE DE VRIES, JORAN JESSURUN, NICOLE SEGERS, HENRI ACHTEN
*Eindhoven University of Technology, The Netherlands*

**Abstract.** In architectural design words are an unemployed source of information. Through a series of case studies we deduced a design annotation data model. All entities in this model can be captured from the design draft, except one: the word-relation. Therefore a system was developed that can generate word graphs using single words from the draft as input. The system searches for semantic relations between words and for new in-between words that can connect two existing words. Filters were developed to filter out only those graphs that were considered interesting by the designers. The envisioned applications of word graphs in the context of Computer Architectural Aided Design are the reduction of fixation during the design process and enhancing creativity by inferring novelty.

## 1. Introduction

Words have not enjoyed the same amount of attention as the support of graphic representations in research on Computer Aided Architectural Design (CAAD). Research on CAAD has been focused since the introduction of the first drawing systems mainly on the development of intelligent drawing objects. CAAD is firmly established for the production of the final design, and is now moving towards support of the early design phase, for example in the area of sketch recognition (Leclercq 2001; Do et al. 2000). The input of words in CAAD is treated as graphical entity or at most reinterpreted into the character format.

In architectural design sketches, annotations are used for clarification of, or commenting on, the design at hand. In our view, if architects write in the act of designing, then words should be considered as part of the design process just like sketches. Words are complementary to the sketch and provide information about the design (Lawson and Loke 1997). Other research even demonstrated that there is a relationship between the design quality and the use of words (Wong and Kvan 1999). Words are a valuable, yet unemployed source of information that can help us better understand the design and thereby provide better support for the designer.

Based on this observation, we have set out to develop a method that can interpret words while designing and process them into a semantic representation. The form in which this has been implemented, is by presenting captured words in graphs to the designer. Relationships between annotated words are inferred and new words are generated that are associated to these words. We have investigated which filtering and presentation of generated graphs is necessary to produce output in a format that can be interpreted by the designer. Using a test implementation we explored the possibilities of word graphs in the context of architectural design. The findings show that it is possible to interpret annotated words and to positively stimulate the architect by structuring these words and by offering new related words.

The outline of the article is as follows: First we briefly discuss the outcome of design case studies. The outcome is organized in a design annotation data model that includes the design entities that make up a design. In the next section the word graph system and the technologies that were use in the implementation are described in detail. The feedback filtering section discusses which output of the words graph system is considered interesting and how the filters operate. A list of typical examples demonstrates how word graphs can shift the interpretation of words. Finally we discuss the potentials of word graphs in the context of CAAD.

## 2. Design Annotations

### 2.1. CASE STUDIES

Although we had indications from our own experience and from other research (Wong and Kvan 1999; Lawson and Loke 1997) that words play an important role during the design process, we did not know what this role is, why it is important, and how it fits within the whole design process. Therefore we decided to organize a short series of case studies, namely observed design sessions, introspective design sessions, and student assignments, to analyze the use of words in the early phase of architectural design. The cases studies are extensively reported in (Segers 2004). Here we will only describe the results that contribute to the understanding of annotated words. From the case studies we have deduced the entities that are used in designs: Words, Sketches, Images, and Marks.

#### *2.1.1. Words*

Words are used for different purposes and in different combinations. Types of writing that were found were annotations, a list of items (with numbers or bullets), a diagram-like placement of keywords, or words that comprise complete sentences. Annotations often have as function to clarify or comment on a sketch, image, or mark. A list of items is used to make a list of attributes to an idea, while a diagram-like placement of keywords is often an abstraction. A statement consisting of complete sentences explains ideas more thoroughly.

*2.1.2. Sketch*

In the studies we found that there are three different types of sketches: a small icon or diagram representation, an isometric or perspective representation, and a projection (façade), section (vertical), or plan (horizontal section) representation. Sketches never seem to be finished. Architects edit them in a later phase, or use them as underlying sketch to trace some of the old lines and add new ones through transparent sheets.

*2.1.3. Images*

Images are pictorial examples that are retrieved from an outside source. Some images are included in the assignment, such as photos from the site or maps, and other images are taken from books, magazines, or the Internet. Images are used to clarify or illustrate an idea, to give inspiration to the architect, or to trace over.

*2.1.4. Marks*

Marks are any graphic element that are not meant as word, sketch or image, but that indicate a relationship between two entities or that single out a particular entity. The type of the marks varies from arrows, lines, and encircling to framings. Marks appear to have a different function, depending on what they connect, and how and where they are placed. The arrow with a single arrowhead for instance mostly points at conclusions, solutions, questions or other important issues. In or near a sketch, this arrow indicates an entrance, a line of sight, or a movement.

Now that we have identified the design entities that are used in the design draft, we can construct a data model in the next section to specify how the design entities – including words – are related to each other.

## 2.2. WORD ASSOCIATIONS IN DESIGN

In Figure 1, a UML class schema (Fowler and Scott 2000) is given how design-data is structured as classes. For each design entity holds that a unique time- and place-stamp attribute is assigned to the entity with each design-action, i.e. when an entity was created, moved, related, or re-used. Extra classes have been added to construct the complete design annotation model, namely paper, page, content, container, mark-relation, and word-relation. When an architect uses pen and paper to work, the paper can be a roll of transparent paper, or a workbook with white empty pages. Turning to a new page has several functions. A new empty sheet prevents the designer to be distracted from earlier design draft, and to have more space for drafting, when continuing the design draft with the same or a new idea. On each page are multiple objects: containers such as papers and images, and contents such as words, marks, and sketches. A container is an abstract class for images and paper, which is a placeholder for (new) content. Content is an abstract class for words and sketches. Content is generated during a design session, whereas a container is imported into the design. The mark-relation and the word-relation classes represent relationship types. In case of the mark-relation the relationship type can sometimes be deduced from the mark, such as an arrow with the meaning 'leads to', but architects do not have a univocal symbol language. Word-relations are often not explicitly defined. Sometimes marks serve this purpose. However, implicit relationships will be inferred from the design by the designer during the design process, but they don't become part of the design data since they are not explicitly stated.

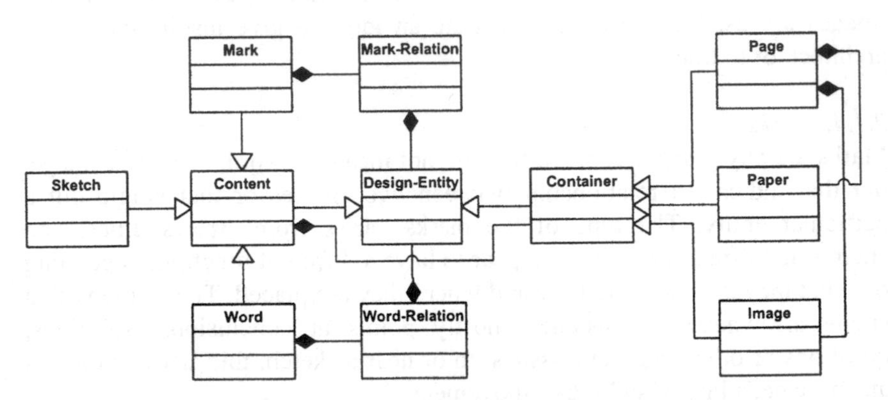


*Figure 1.* Design annotation data model

In this research project we have focused on the role of words and we have investigated the possibility to infer relationships between words. More specifically we have researched lexical associations as a method to stimulate architectural design by associative reasoning. Silberman describes two types of associations: semantic and episodic (Silberman 1999). Semantic associations are theoretical: there is a common understanding about the relationship. An example of a semantic association is for instance 'tree' with 'branch'. A tree and a branch are related to one another in theory; it is a relation not depending on an individual's perception. Episodic associations are associations that are not theoretical, but exist because something happened in time or space: an episode. An example of an episodic association is 'yellow' with 'submarine'. The relation exists because the Beatles made a song called 'yellow submarine'. Silberman suggests that showing semantic associations causes people to come up with more semantic and episodic associations.

## 3. Word Graphs

Episodic associations occur through personal experience and are therefore difficult to capture, and nearly impossible to infer "after the fact." For this reason, we have focused on the generation of semantic associations to describe the words-relation explicitly. We decided to develop a system to test whether the generation of semantic associations can fill in the word-relation class of Figure 2 and thereby contribute to the understanding of the design process.

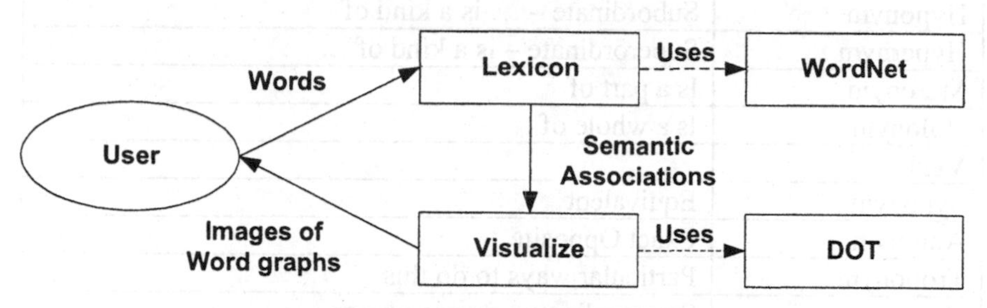


*Figure 2.* Wordgraph test system

The test system has been implemented as a stand-alone system that can be used by the designer on demand. The system reads words that are typed by the user and outputs word graphs. In this procedure, a Lexicon component processes the words and searches for semantic associations. The words and their semantic associations are transferred into a graph structure by the Visualize component. For the implementation, two existing software libraries were used, namely WordNet and DOT, as explained in the following sections.

### 3.1. LEXICON

In the case studies we have tested existing lexical systems (e.g. WordNet browser: www.cogsci.princeton.edu/cgi-bin/webwn) to investigate the potential added value in a design context. The words that the architect writes down during the early phase of the design process do not include much jargon. Therefore, an existing lexical ontology was investigated instead of developing a design ontology specifically. We confronted the designers with the result that is generated when entering the words that were written down during a design session. During this process of re-interpretation the designers indicated that they were interested in the ability of these systems to bring forward *new related words*. What was missing however is the possibility to *structure the words* that were entered, in a meaningful way. Therefore we set out to develop a Lexicon component that can generate such structures that are intrinsically semantic, namely by the exploiting semantic relationships between words that are specified in lexical libraries. The relationships that are subject to our area of research are listed in Table 1.

TABLE 1. Relationships

| **Noun** | |
|---|---|
| Synonym | Equivalent |
| Antonym | Exact Opposite |
| Hyponym | Subordinate – … is a kind of |
| Hypernym | Superordinate – is a kind of … |
| Meronym | Is a part of |
| Holonym | Is a whole of |
| **Verb** | |
| Synonym | Equivalent |
| Antonym | Exact Opposite |
| Troponym | Particular ways to do this |
| Hypernym | Superordinate – is one way to… |
| Entailment | Cannot without, … entails doing |
| **Adjective** | |
| Synonym | Equivalent |
| Antonym | Exact Opposite |
| Similar | Not exactly the same |
| Pertainym | Relational Adjective |
| **Adverb** | |
| Synonym | Equivalent |
| Antonym | Exact Opposite |
| | |

The structure that emerges from the words entered by the designer can be conceptualized as a graph structure with the words as the nodes and the relations between words that were found by the system as the edges of the graph. The system will take all words that are written down on one page (i.e. a sheet of paper) into consideration. Consequently, every time when a new word is entered the system will search for new graph structures. These graph structures do not necessarily have to form one graph, but can also consist of subgraphs.

The Lexicon component makes use of the WordNet library (Miller et al. 1990). The basic principle in WordNet is the so-called synset. A synset is a synonym set, a set of words that are interchangeable while maintaining the same meaning. Each sense of a word is in a different synset, and each synset has it's own set of semantic pointers, indicating a relation between synsets (word meanings). Because of the fact that one word mostly has several synsets with each their own set of semantic pointers, there are usually many words related to one word.

Next to direct word relations, the Lexicon component also searches for words that connect two other words that are entered into the system. These intermediate words are especially meaningful in the case of hypernym and holonym relationships. To discriminate these relationships from the original ones we will add the index 2. To give an example of what is exactly a hypernym$^2$ relation, suppose that the architect types in 'brother' and 'sister'. The system does not only find them related as antonyms, but also with one intermediary word ('relative') as a hypernym$^2$ relation. The system will show the graph with all three words included, generating the word 'relative', since the architect did not enter that word in himself.

### *3.1.1. Word Graph Structure and Processing*

Figure 3 shows a UML diagram of the generated graph structure. A WordGraph contains WordNodes and RelationEdges. A WordNode is a reference to one word in a WordNet synset. A RelationEdge makes a connection between two WordNodes. The relation type can be any type in WordNet, but also synonym, which is implicitly defined in the synset. Consequently two word nodes can point to the same synset. WordNodes and RelationEdges have the properties ignored and fresh. The ignored property indicates that a node or edge should be ignored in the graph. This property is used by the filtering system explained below. The fresh property indicates that a node or edge has just been added. This is used to indicate which new nodes and edges where generated since the last input. The generated property of the WordNode indicates that it is a word generated by the system and not entered by the user. The WordGraph structure also keeps track of all SubGraphs. In the SubGraph class, nodes and edges are grouped that are connected together by traversing linked edges and other nodes. If there is no connection between two nodes, then the nodes are in different SubGraphs. A SubGraph is fresh when at least one node or edge is fresh.

When a new word is entered into the system, the following happens; the Lexicon component looks up all synsets for the word in WordNet. Every synset is added as a fresh word node into the graph. If a new word is entered that is already in the SubGraph as a generated word, then the generated word is replaced by a fresh non-generated version. Next the Lexicon component searches for relations between each fresh word and all other words. It does this by taking the two words and then checking if there is a relation between those words. The system will check for synonym, antonym, hypernym, holonym and entailment relations. If a relation is found a fresh edge with that relation type is created between those two words. For the hypernym$^2$ and holonym$^2$ relation types the system searches for an intermediate word that has a hypernym or holonym relation with both original words. This intermediate word is added as a fresh generated word and the system searches again for relations between the added word and all other words in the graph as described above, but this time it does not search for hypernym$^2$ and holonym$^2$ relations again. When this process ends, the system locates all fresh sub graphs with more than one word. These graphs will be presented to the user.

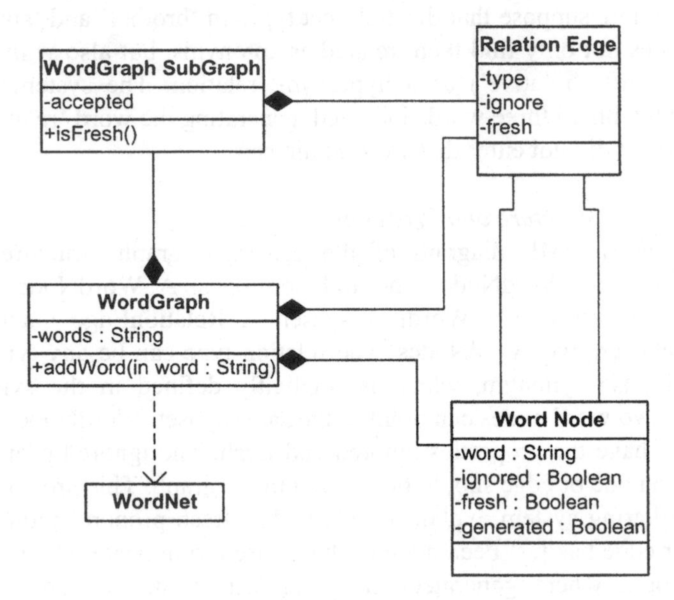


*Figure 3.* WordGraph data model

Hypernyms are the reverse relations of hyponyms. When searching for a hypernym relation between two words, both relation types can be used. In WordNet however, a word has much more hyponyms than hypernyms. It takes less time to find a relation between words when looking at the hypernyms relations in WordNet. The same is true for meronyms and holonyms. In this case it is quicker to search for relations when following the holonyms relations, Figure 4.

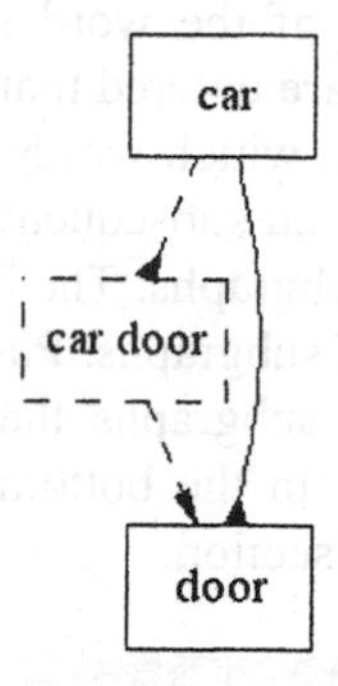


*Figure 4.* Finding a holonym relation via a hypernym relation

When checking for a holonym relation between two words, the system needs also to traverse hypernyms. The reason for that is because only direct holonym relations are contained in WordNet, but a holonym relation can be connected to another hypernym that is thereby also a holonym of the original word. For example a 'door' is part of a 'car'. To find this relation the system first traverses the hypernym relation between 'door' and 'car door', followed by the holonym relation between 'car' and 'door'.

The Lexicon component generates graph structures in a textual format. The following section describes the processing of the textual information into a graphical representation.

3.2. VISUALISATION

For a fast and easy understandable representation we developed a set of symbols that visualizes the (sub)graphs. Words written by the architect are displayed in a white textbox and words generated by the system are displayed in a yellow text box. Additional information is obtained from the type of word at hand, i.e. a noun, verb, adjective, or adverb. For this purpose the shape of the word-objects is used: a rectangle indicates a noun, an ellipse a verb, and a parallelogram an adverb or adjective. To indicate the type of relation, the links between the nodes are used. These are displayed as arrows: a black arrow indicates a hypernym or hypernym$^2$, or a hyponym or hyponym$^2$. The position of the arrow-head specifies whether is it a hypernym$^{(2)}$ or a hyponym$^{(2)}$. Color is used for less frequently occurring relations: Blue stands for entailment, red means antonym, and green means synonym. For the implementation of the Visualization component we used the DOT library [Koutsofios and North 1993]. DOT creates hierarchical layouts of directed graphs. A DOT language file specifies the objects in a directed graph and makes an optimal output layout of the objects.

Figure 5 displays the interface of the word graph test system. At the bottom part is a field where words are entered that should be included in the word graph. The left column shows which words have been entered. In the main window the subgraphs are added subsequently every time when a new word generates one or more new subgraphs. The horizontal scrollbar allows for browsing through the history of subgraphs. Pushing the Complete Graph button results in a sheet with all subgraphs that are generated from the complete list of words. The filters in the bottom left that can by checked on/off will be discussed in the next section.

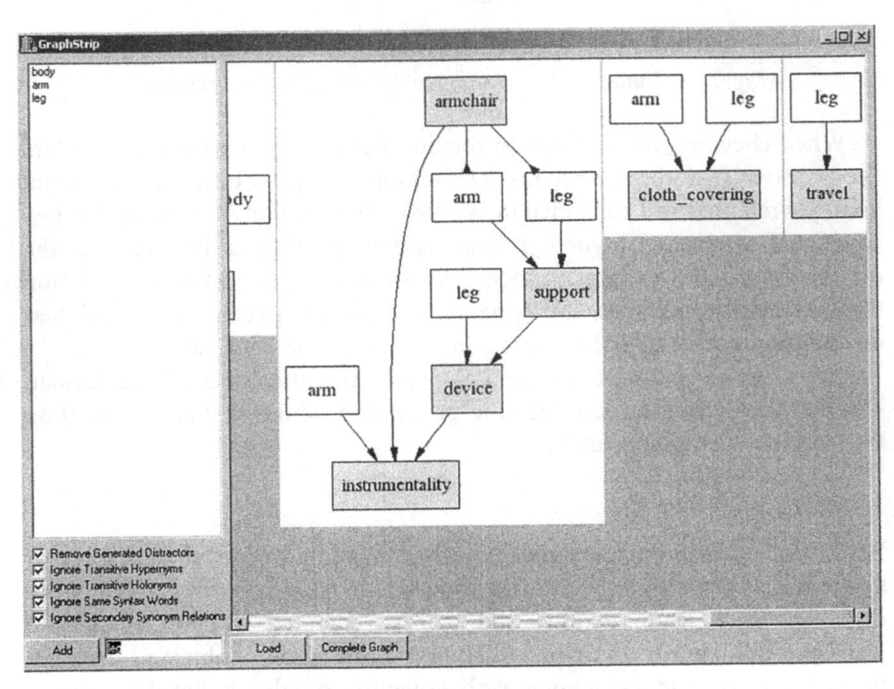


*Figure 5.* Word graph test system

## 4. Feedback Filtering

When words are added, the process described in the previous section generates all possible semantic relations. Some of these relations are not interesting or can easily be deduced from other relations. The displayed graphs become too large which makes them too complicated to quickly understand. For this reason it is necessary to implement filters that reduce the size of the graph and which present only interesting relations and new words. We can distinguish between two classes of filters: redundancy filters and relevance filters, Table 2. The redundancy filters called "ignore transitive hypernyms," "ignore transitive holonyms," and "ignore secondary synonym" remove relations that can be deduced from other relations. The relevance filters "remove generated distractors" and "ignore same syntax words" remove relations and words that do not seem to be interesting for the user. The relevance filters have been defined in post-design sessions with participants who were presented generated feedback and had to indicate which relations were (not) interesting.

TABLE 2. Feedback examples

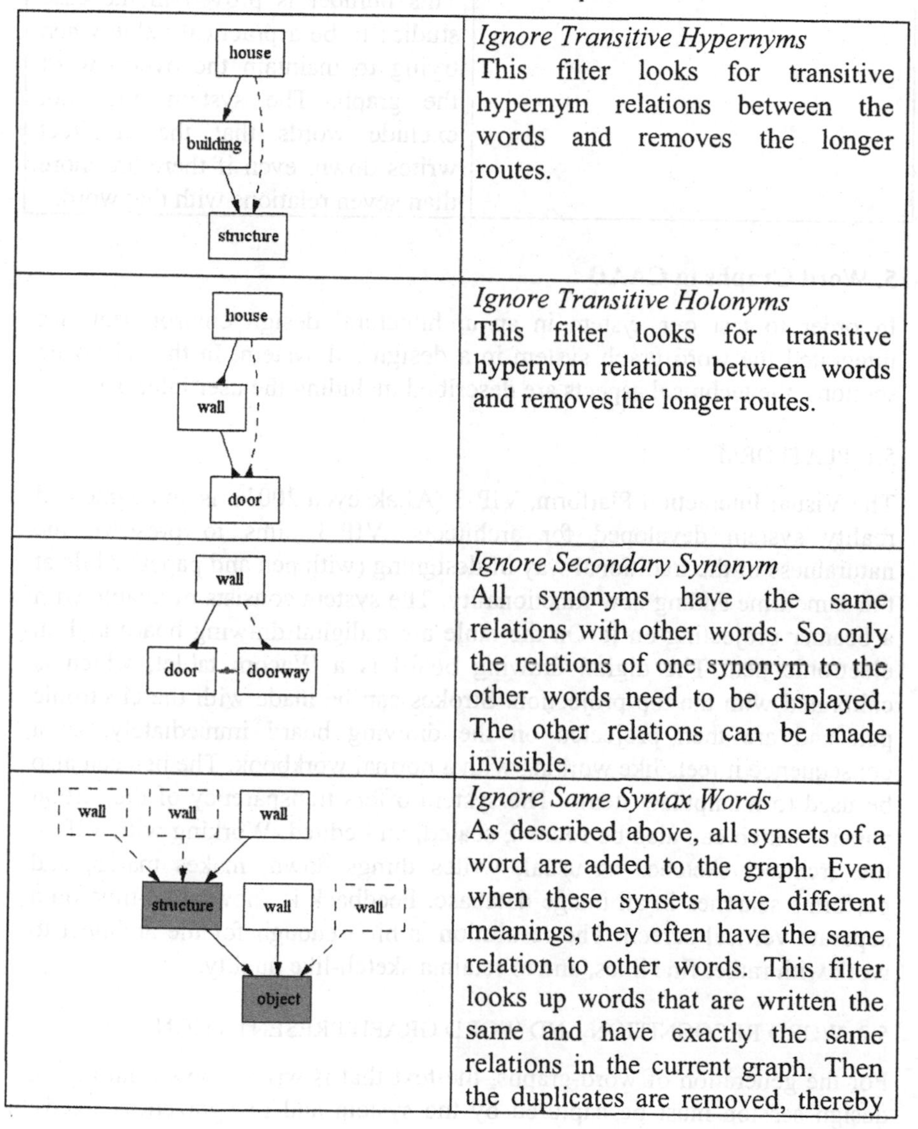


| | |
|---|---|
| | *Ignore Transitive Hypernyms*<br>This filter looks for transitive hypernym relations between the words and removes the longer routes. |
| | *Ignore Transitive Holonyms*<br>This filter looks for transitive hypernym relations between words and removes the longer routes. |
| | *Ignore Secondary Synonym*<br>All synonyms have the same relations with other words. So only the relations of onc synonym to the other words need to be displayed. The other relations can be made invisible. |
| | *Ignore Same Syntax Words*<br>As described above, all synsets of a word are added to the graph. Even when these synsets have different meanings, they often have the same relation to other words. This filter looks up words that are written the same and have exactly the same relations in the current graph. Then the duplicates are removed, thereby |

| | |
|---|---|
| | creating a much more readable graph. The information that there where different meanings of that word is now lost. |
| object | *Remove Generated Distractors*<br>This filter looks for generated words with more then 7 relations and removes them from the diagram.<br>This number is proved in the case studies to be a practical value when trying to maintain the overview of the graph. The system does not exclude words that the architect writes down, even if there are more than seven relations with that word. |

## 5. Word Graphs in CAAD

In order to test our system in an architectural design environment, we integrated the word graph system in a design aid system. In the following sections, the technical aspects are described including the user interface.

### 5.1. PLATFORM

The Visual Interaction Platform, VIP-3 (Aliakseyeu 2001), is an augmented reality system developed for architects. VIP-3 aims to preserve the naturalness of the traditional way of designing (with pen and paper) while at the same time adding new functionality. The system consists of a table with a beamer projecting on it. On the table are a digital drawing board and an electronic pen. The digital drawing board is a Wacom tablet, which is calibrated with the top projection. Strokes can be made with the electronic pen and are then projected on the drawing board immediately; as a consequence it feels like working with a normal workbook. The pen can also be used to manipulate items. The system offers transparency of the virtual papers, which can also be rotated, scaled, and edited. Working with VIP-3, the architect sketches as usual, writes things down, makes marks, and performs searches in an image database. Feedback is shown real-time on a separate vertical screen. The resolution is high enough for the architect to work with rather fine lines, and to retain a sketch-like quality.

### 5.2. WORD RECOGNITION AND WORD GRAPH PRESENTATION

For the generation of word-graphs, the text that is written down during the design session must be captured by the system and recognized as words.

Although handwriting recognition software has improved rapidly over the last years, it failed for our purpose for the following reasons:

- The quality of text in design drafts is often very poor.
- Not just a specific set, but all words need to be recognized.
- Text in design drafts is written under various angles.
- Text must be isolated from other strokes such as marks and sketches.

It is found that no such software is available. A pragmatic solution to this problem of word recognition is the Human Handwriting Recognition Server module, in short: the HHRS module. The HHRS module basically functions by a human observer who recognises the words that are made in the design draft. The observer is located in a different area from the design system. All strokes appear on the observer's monitor. The observer selects the strokes that are part of a word, as displayed in Figure 6, and types in the word for the CAAD system.

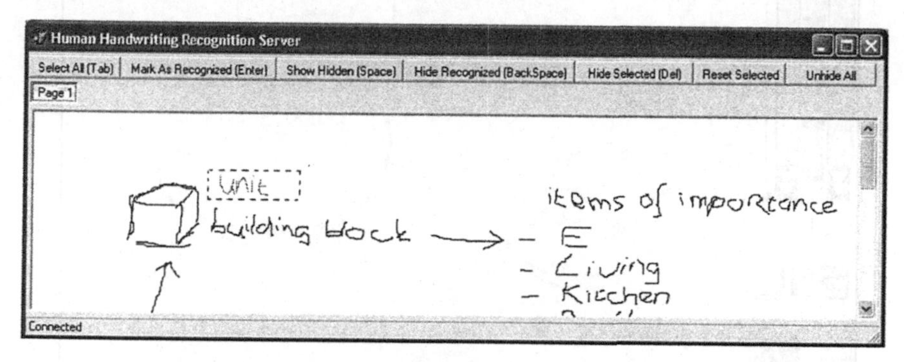


*Figure 6.* Human handwriting recognition server module

As soon as a new word is entered, possibly new word graphs are generated and displayed to the system user on the vertical screen. Word graphs are displayed on the vertical field in a strip-like manner. An average of five word graphs is visible on the vertical screen, to prevent displaying of too much information. New word graphs appear on the right and the old ones 'slide' to the left. With the 'word graph menu', the architect can see all word graphs by scrolling back and forth through the strip of word graphs. One word graph from the strip is highlighted, indicating the selected word graph. With the button 'Get Graph' the selected word graph is copied to the horizontal work field as an image to become part of the design draft.

In the snapshot from the VIP desktop of Figure 7 we see in the top-left corner a vertical slider bar with images, in the lower-left corner a drawing tool menu, some sketches and writings in the working area and an included word graph in the top-right corner.

The word graph in the design draft was selected from all generated word-graphs that have been presented to the user. The words that had been written down are: eye, hair and light. Using these words the system generates (amongst others) the word graph as shown in Figure 8. Next to the various relationships presented by arrows, the following intermediate-words have been generated: mammal, body-part and process. As can be seen in Figure 7, the architect encircled the word 'mammal' and it inspired her to make some sketches that look like animals.

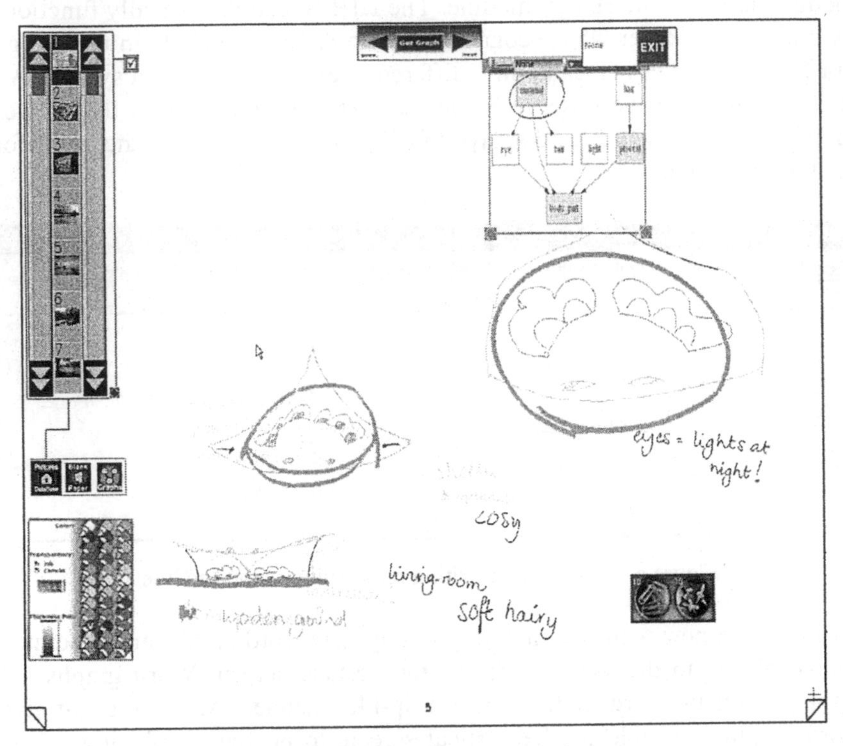


*Figure 7.* Snapshot from design desktop

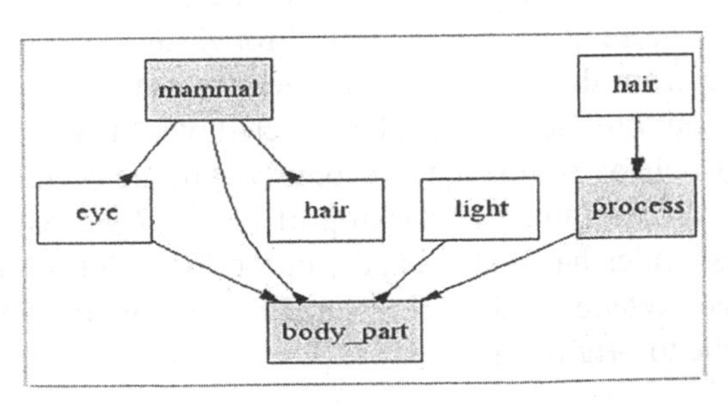


*Figure 8.* Included word graph in design draft

## 6. Discussion

In the previous sections we have explained how word graphs can be generated and presented. The aim of the research is to use the unemployed potentials of words in designs in the context of Computer Aided Architectural Design systems. Now that we have a system that can generate semantic graphs from words, they can be exploited in many ways. In this section we propose two application areas we consider promising, namely: fixation reduction and creativity enhancement.

### 6.1. FIXATION REDUCTION

A commonly known problem that architects face from time to time is fixation. Fixation occurs when a designer sticks too long to the same idea. Purcell and Gero indicate that fixation appears to be associated with the absence of domain specific knowledge and a reliance on everyday knowledge activated by exposure to a picture of a familiar example (Purcell and Gero 1996). They state that when the familiarity effect is removed and the example is innovative, designers do not suffer from fixation any longer, except maybe from the fixation on 'being different.' Insights about other possible uses or meanings of ideas can support overcoming fixation. Finke et al. (Finke et al. 1992) state that 'contextual shifting' or considering an idea in a new or different context is a way to achieve this.

Through the generation of new semantic associations from the words that are used in a design, word graphs can provide means to enable contextual shifting. Restructuring of the words in design and reflection on the new structure provide a different view on the design using the common knowledge that is stored in lexical relationships.

### 6.2. ENHANCING CREATIVITY

Creativity is a human characteristic, which is notoriously difficult to define. Consequently it is even more difficult to prove creativity enhancement. Creativity is related to characteristics that we can understand better, namely novelty. According to Saunders and Gero a novel situation is one that is similar enough to previous experiences to be recognized as a member of a class but different enough from the other member of that class to require significant learning (Saunders and Gero, 2001). Words that are written by the designer can be used to infer new output in the form of semantic relationships and new words. Some relationships will be coherent with the users expectations. Other relationships will come unexpected since they were not intended while writing. Nevertheless, these unexpected relations are meaningful because of the lexical basis. The word graph system generates new in-between words that are at a distance from what the designer has written. In the word graphs this means that there is one relation between the original words and the new in-between words. A word distance of one (new) word that connects two words written by the designer is not too far to understand the lexical relationship. This word distance seems right for being recognized as a novelty.

6.3. CONCLUSION

CAAD systems aiding in the creative design process should not only provide modeling features but also structure design information and trigger new ideas by shifting the users attention. The notion of design annotations provides a framework for describing the role of words in design by means of relations to sketches, marks and images. Meaningful relations between words in a graph as demonstrated in this paper through the word graph test system and design aid system, can play an important role in triggering new ideas during the design process.

## References

Aliakseyeu, A: 2003, *A Computer Support Tool for the Early Stages of Architectural Design*, PhD-thesis, Eindhoven University of Technology, Eindhoven, The Netherlands.

Do, EY, Gross, M, Neiman, B and Zimring, G: 2000, Intentions in and relations among design drawings, *Design Studies* **21**(5): 483-503.

Finke, RA, Ward, TB and Smith, SM: 1992, *Creative Cognition – Theory, Research, and Applications*, MIT Press, Cambridge, Massachusetts.

Fowler, M and Scott, K: 2000, *UML Distilled, A Brief Guide to the Standard Object Modelling Language*, Addison-Wesley, Amsterdam.

Koutsofios, E and North, SC: 1993, *Drawing Graphs with DOT*, Technical Report, AT&T Bell Laboratories, Murray Hill, NJ.

Lawson, B and Loke, SM: 1997, Computers, words and pictures, *Design Studies* **18**(2): 171-183.

Leclercq, P: 2001, Programming and assisted sketching, *in* B de Vries, J van Leeuwen and H Achten (eds), *CAADFutures 2001*, Kluwer, Dordrecht, pp. 15-31.

Miller, GA, Beckwith, R and Fellbaum, C: 1990, Introduction to WordNet: An on-line lexical database, *International Journal of Lexicography* 3(4): 235-244.

Purcell, TA and Gero, JS: 1996, Design and other types of fixation, *Design Studies* **17**(4): 363-383.

Saunders, R and Gero, JS: 2001, Designing for interest and novelty: motivating design agents, *in* B de Vries, J van Leeuwen and H Achten (eds), *CAADFutures 2001*, Kluwer, Dordrecht, pp. 725-738.

Segers, NM: 2004, *Computational Representations of Words in Architectural Design: Development of a System Supporting Creative Design*, PhD-thesis, Eindhoven University of Technology, Eindhoven, The Netherlands (forthcoming).

Silberman, Y, Miikkulainen, R and Bentin, S: 2001, Semantic effect on episodic associations, *in* JD Moore and K Stenning (eds) *Proceedings 23rd Annual Conference of the Cognitive Science Society*, Cognitive Science Society, Edinburgh.

Wong, W and Kvan, T: 1999, Textual support of collaborative cesign, *in* O Ataman and J Burmudez (eds), *ACADIA '99*, Salt Lake City, pp. 168-176.

JS Gero (ed), *Design Computing and Cognition'04*, 557-575


# AN INTEGRATED APPROACH FOR SOFTWARE DESIGN CHECKING USING DESIGN RATIONALE

JANET E BURGE, DAVID C BROWN
*Worcester Polytechnic Institute, USA*

**Abstract.** *Design Rationale* (DR), the reasons behind decisions made while designing, offers a richer view of both the product and the decision-making process by providing the designer's intent behind the decisions. DR is also valuable for checking to ensure that the intent was adhered to throughout the design, as well as pointing out any unresolved (or undocumented) issues that remain open. While there is little doubt of the value of DR, it is typically not captured during design. SEURAT (Software Engineering Using RATionale) is a system we have developed to explore uses of design rationale. It supports both the display of and inferencing over the rationale to point out any unresolved issues or inconsistencies. SEURAT is tightly integrated with a software development environment so that rationale capture and use can become integrated into the software development process.

## 1. Introduction

For a number of years, members of the Artificial Intelligence (AI) in Design community have studied *Design Rationale* (DR), the reasons behind decisions made while designing. DR offers a rich view of both the product and the decision-making process by providing the designer's intent behind the decisions. DR is also valuable for checking to ensure that the intent was adhered to throughout the design as well as pointing out any unresolved (or undocumented) issues that remain open.

An area where rationale for past decisions is especially useful is during software maintenance. One reason for this is that the software lifecycle is a long one. Large projects may take years to complete and spend even more time out in the field being used (and maintained). Maintenance costs can be more than 40 percent of the cost of developing the software in the first place (Brooks 1995). The panic over the "Y2K bug" highlighted the fact that software systems often live on much longer than the original developers intended. Also, the combination of a long life cycle and the typically high personnel turnover in the software industry increases the probability that the original designer is unlikely to be available for consultation when problems arise.

All these reasons argue for as much support as can be provided during maintenance. Semi-automatic maintenance support systems, such as Reiss's constraint-based system (Reiss 2002), that work on the code, abstracted code, design artifacts, or meta-data, assist with maintaining consistency between artifacts. Design Rationale, however, assists with maintaining consistency in designer reasoning and intent.

### 1.1. DIFFICULTIES WITH RATIONALE

While rationale has great potential value, rationale is not in widespread use. One difficulty, despite much research, is the capture of design rationale. Recording all decisions made, as well as those rejected, can be time consuming and expensive.

Documenting the decisions can impede the design process if decision recording is viewed as a separate process from constructing the artifact (Fischer et al. 1995). Designers are reluctant to take the time to document the decisions they did not take, or took and then rejected (Conklin and Burgess-Yakemovic 1995). A real danger is the risk that the overhead of capturing the rationale may impact the project schedule enough to make the difference between a project that meets its deadlines and is completed, versus one where the failure to meet deadlines results in cancellation (Grudin 1995). One way to mitigate these risks is to provide tools for rationale capture and use that are tightly integrated with those used during the designing process so that capturing and using the rationale becomes part of the standard process, not an extra task that needs to be performed with its own set of tools and standards.

### 1.2. USES OF RATIONALE

The key to making the capture worthwhile, as well as providing requirements for DR representation, is the use for, *and usefulness of*, the rationale. In this paper, we describe the SEURAT (Software Engineering Using RATionale) system, which integrates tools for rationale capture, visualization, and use into a standard software engineering environment. SEURAT addresses a number of uses for rationale:

- *Design verification* – using rationale to verify that the design meets the requirements and the designer's intent.
- *Design evaluation* – using rationale to evaluate (partial) designs and design choices relative to one another to detect inconsistencies.
- *Design maintenance* – using rationale to locate sources of design problems, to indicate where changes need to be made in order to

modify the design, and to ensure that rejected options are not inadvertently re-implemented.
- *Design assistance* – using rationale to clarify discussion, check impact of design modifications, and perform consistency checking.

This paper is structured as follows: in section 2, we describe related work. In section 3, we describe the overall approach. Section 4 describes the rationale representation developed for SEURAT and section 5 presents the Argument Ontology, a key component of the rationale representation. Section 6 describes inferences to be performed over the rationale and section 7 gives the conclusions and outlines future work.

## 2. Related Work

Work on design rationale has focused on three main issues: capture, representation, and use. While SEURAT supports capture by providing the capability to enter the rationale, capture is not a main focus of the work. The related work on representation is presented as part of the representation discussion in section 4. In this section, we describe related work on rationale use.

### 2.1. RATIONALE USE

There are a number of systems that focused on uses for rationale for both engineering and software design. JANUS (Fischer et al. 1995) critiques the design and provides the designers with rationale to support the criticism SYBIL (Lee 1990) and InfoRat (Burge and Brown 2000) both check that the rationale behind each decision is complete. C-Re-CS (Klein 1997) performs consistency checking on requirements and recommends a resolution strategy for detected exceptions.

Co-MoKit (Dellen, et al. 1996) uses a software process model to obtain design decisions and causal dependencies between them. WinWin (Boehm and Bose 1994) aims at coordinating decision-making activities made by various "stakeholders" in the software development process. Bose (Bose 1995) defined an ontology for the decision rationale needed to maintain the decision structure. The goal was to model the decision rationale in order to support decision maintenance by allowing the system to determine the impact of a change and propagate modification effects. Chung et al. (2000) developed an NFR Framework that uses non-functional requirements to drive the software design process, producing the design and its rationale.

### 2.2. EVALUATING USEFULNESS

While the usefulness of rationale has not been studied in as much detail as the capture and representation, there have been some experiments performed. Field trials performed using itIBIS and gIBIS (Conklin and Burgess-Yakemovic 1995) indicated that capturing rationale was found to be useful during both requirements analysis and design, and that the process also helped with team communication by making meetings more productive. Karsenty (1996) studied how DR could be used to evaluate a design. In this study, 50% of the designers' questions were about the rationale behind the design and 41% of those questions were answered using the recorded rationale.

## 3. Approach

For the SEURAT system we have chosen to focus our efforts on software engineering and concentrate on how rationale could be used during software maintenance, one of the most difficult and expensive phases of the software life cycle. Our goal is to create a system that can be tightly integrated with existing development tools so that rationale capture and use can become a part of the development process, not something that is done after the fact.

We are currently building the SEURAT system as a plug-in to the Eclipse Tool Platform (www.eclipse.org) so that it can be tightly integrated with a Java IDE (Interactive Development Environment) and other design tools that plug into Eclipse. This allows us to connect the rationale with the code and design artifacts that it explains. It ensures that the software maintainers are aware of and use the rationale.

SEURAT will present the relevant DR when required and allow entry of new rationale for the modifications. The new DR will then be verified against the existing DR to check for inconsistencies. There are several types of checks that should be made: structural inferences to ensure that the rationale is complete, evaluation, to ensure that it is based on well-founded arguments, and comparison to rationale collected previously for similar modifications to see if the same reasoning was used. In the latter, the previous rationale could be used as a guide in determining the rationale for the new modification.

Figure 1 shows SEURAT as part of the Eclipse Java IDE. SEURAT participates in the development environment in three ways: a Rationale Explorer (lower left pane) that shows a hierarchical view of the rationale and allows display and editing of the rationale; a Rationale Task List (lower right pane), that shows a list of errors and warnings about the rationale; and Rationale Indicators that appear on the Java Package Explorer (upper left pane) and in the Java Editor (upper right pane) to show where rationale is available for a specific Java element. The examples in this paper come from a meeting scheduling system. Note that the screenshots are in color, making the icons much easier to distinguish than when reproduced in black and white.

The software developer enters the rationale to be stored in SEURAT while the system the rationale describes is being developed. SEURAT supports the entry by providing rationale entry screens for each type of rationale element.

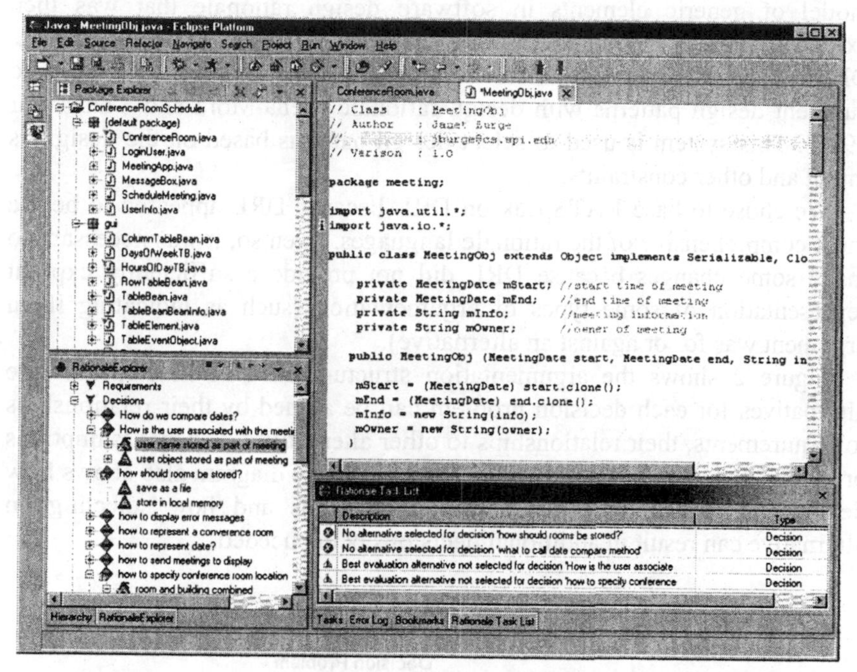


*Figure 1.* SEURAT and eclipse

## 4. Representation

A DR representation needs to be formalized and well structured, as opposed to just free text, so that computer-based checking and inferences are possible. We have generated a rationale representation, called RATSpeak, and have chosen to use a semi-structured argumentation format because we feel that argumentation is the best means for expressing the advantages and disadvantages of the different design options considered. Argumentation formats date back to Toulmin's representation (Toulmin 1958) of datums, claims, warrants, backings and rebuttals. This is the origin of most argumentation representations. More recent argumentation formats include Questions, Options, and Criteria (QOC) (MacLean et al. 1995), Issue Based Information System (IBIS) (Conklin and Burgess-Yakemovic 1995), and Decision Representation Language (DRL) (Lee 1991). Each argumentation format has its own set of terms but the basic goal is to represent the decisions made, the possible alternatives for each decision, and the arguments for and against each alternative.

Argumentation has been used in rationale representations that were created specifically for software design. Potts and Bruns (1988) created a model of generic elements in software design rationale that was then extended by Lee (1991) in creating DRL, the language used in SIBYL. DRIM (Design Recommendation and Intent Model) was used in a system to augment design patterns with design rationale (Peña-Mora and Vadhavkar 1996). This system is used to select design patterns based on the designers intent and other constraints.

We chose to base RATSpeak on DRL because DRL appeared to be the most comprehensive of the rationale languages. Even so, it was necessary to make some changes because DRL did not provide a sufficiently explicit representation of some types of argumentation (such as indicating if an argument was for or against an alternative).

Figure 2 shows the argumentation structure used in RATSpeak. The alternatives for each decision problem can be argued by their relationships to requirements, their relationships to other alternatives, and by assumptions or claims that support or deny the alternatives. The diagram also shows how decisions can be subdivided into sub-decisions and how selecting an alternative can result in additional decisions being needed.

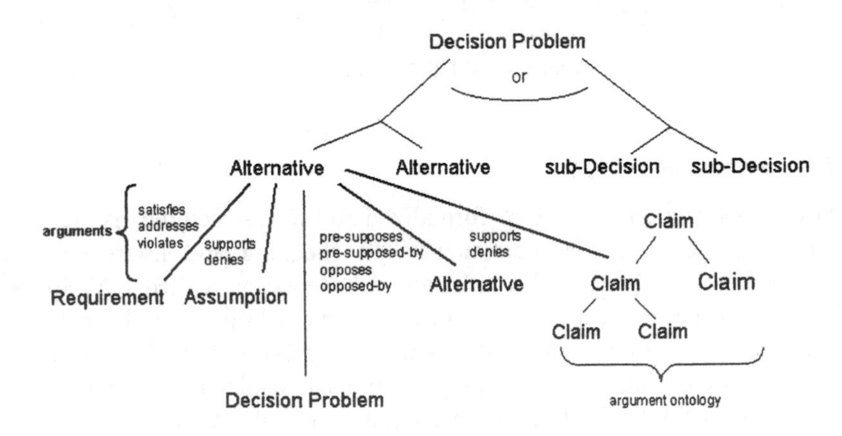


*Figure 2.* RATSpeak argumentation structure

RATSpeak uses the following elements as part of the rationale:

- *Requirements* – these are the requirements, both functional and non-functional. These can either be represented explicitly in the rationale or be pointers to requirements stored in a requirements document or database. For the purposes of our examples, we will show them as part of the rationale. Requirements serve two purposes in RATSpeak, one is as the basis of arguments for and against alternatives. This allows RATSpeak to capture cases where an alternative supports or violates a requirement. The other purpose is so that the rationale for the requirements themselves can be captured.

- *Decision Problems* – these are the decisions that must be made as part of the development process. They are expressed in the form of questions.
- *Questions* – these are questions that need to be answered before the answer to the decision problem can be defined. The question can include the procedures or programs that need to be run or simple requests for information. While questions are not a standard argumentation concept, they can augment the argumentation by specifying the source of the information used to make the decisions, which is useful during software maintenance.
- *Alternatives* – these are alternative solutions to the decision problems. Each alternative will have a status that indicates if it is accepted, rejected, or pending.
- *Arguments* – these are the arguments for and against the proposed alternatives. They can either contain requirements (i.e., an alternative is good or bad because of its relationship to a requirement), claims about the alternative, assumptions that are reasons for or against choosing an alternative, or relationships between alternatives (indicating dependencies or conflicts). Each argument is given an *amount* (how much the argument applies to the alternative, i.e., how flexible, how expensive) and an *importance* (how important the argument is to the overall system or to the specific decision).
- *Claims* – these are reasons why an alternative is good or bad. Each claim maps to an entry in an *Argument Ontology* of common arguments for and against software design decisions. Each claim also indicates what *direction* it is in for that argument. For example, a claim may state that a choice is *NOT* safe or that an alternative *IS* flexible. This allows claims to be stated as either positive or negative assertions. Claims also contain an *importance*, which can be inherited or overridden by the arguments referencing the claim.

- *Assumptions* – these are similar to claims except that it is not known if they are always true. Assumptions do not map to items in the Argument Ontology.
- *Argument Ontology* – this is a hierarchy of common argument types that serve as types of claims that can be used in the system. These are used to provide the common vocabulary required for inferencing. Each ontology entry contains an *importance* that can be overridden by claims that reference it.
- *Background Knowledge* – this contains *Tradeoffs* and *Co-Occurrence Relationships* that give relationships between different arguments in the Argument Ontology. This is not considered part of the argumentation but is used to check the rationale for any violations of these relationships.

Figure 3 shows the relationships between the different rationale entities.

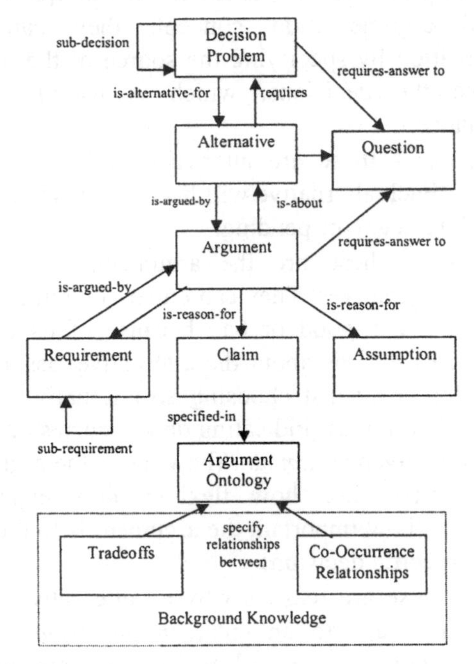


*Figure 3.* Relationships between rationale entities

## 5. Argument Ontology

One key element in the RATSpeak representation is the Argument Ontology. Our work on InfoRat showed the importance of providing a common vocabulary to support inferencing over the content of the rationale as well as over its structure. To support this, we have developed an ontology of reasons for choosing one design alternative over another. This ontology forms a hierarchy of terms with abstract reasons at the root and increasingly detailed reasons towards the leaves.

RATSpeak provides the ability to express several different types of arguments for and against alternatives. One type of argument is if an alternative satisfies or violates a requirement. Other arguments refer to assumptions made or dependencies between alternatives. Another type of argument involves claims that an alternative supports or denies a Non-Functional Requirement (NFR). These NFRs, also known as "ilities" (Fillman 1998) or quality requirements, refer to overall qualities of the resulting system, as opposed to functional requirements, which refer to specific functionality. As we describe in (Burge and Brown 2002), the distinction between functional and non-functional is often a matter of context. RATSpeak also allows NFRs to be represented as explicit requirements.

There have been many ways that NFRs have been organized. CMU's Quality Measures Taxonomy (SEI 2000) organizes quality measures into Needs Satisfaction Measures, Performance Measures, Maintenance Measures, Adaptive Measures, and Organizational Measures. Bruegge and Dutoit (2000) break a set of design goals into five groups: performance, dependability, cost, maintenance, and end user criteria. Chung et al. (2000) provides an unordered list of NFRs as well as specific criteria for performance and auditing NFRs.

For the RATSpeak argument ontology, we took a bottom-up approach by looking at what characteristics a system could have that would support the different types of software qualities. This involved reviewing literature on the various quality categories to look for how a software system might choose to address these qualities. For example, one quality attribute that is a factor in design decisions is scalability. We looked to see what might contribute toward scalability in a software design and added these attributes to the ontology. For example, one way to increase scalability is to minimize the number of connections a system must set up, another is to avoid using fixed data sizes that may limit the capacity of the system. Our aim was to go beyond the idea of design goals or quality measures to look at *how* these qualities might be achieved by a software system.

In maintenance, the maintainers are more likely to be looking at the lower-level decisions and will need specific reasons why these decisions contribute to a desired quality of the overall system. It is probable that decisions made at the implementation level are likely to correspond to detailed reasons in the ontology, while higher level decisions are more likely to use reasons at the more abstract levels.

After determining a list of detailed reasons for choosing one alternative over another, an Affinity Diagram (Jiro 2000) was used to cluster similar reasons into categories. These categories were then combined again. The more abstract levels of the hierarchy were based on a combination of the NFR organization schemes listed earlier (the CMU taxonomy as well as Bruegge and Dutoit's design goals). Also, NFRs from the Chung list were used to fill in gaps in the ontology.

Figure 4 shows the first two levels of the Argument Ontology displayed in SEURAT.

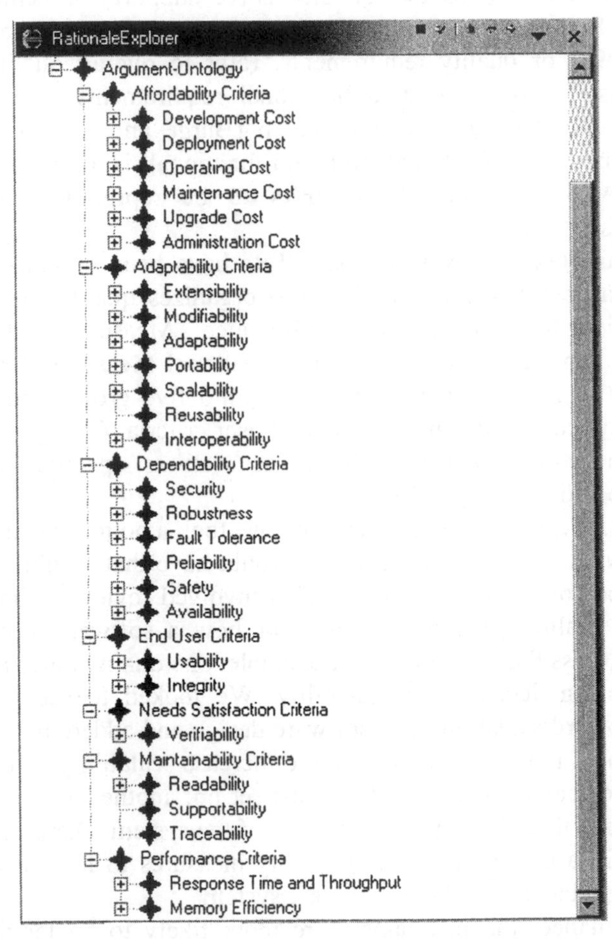


*Figure 4.* Top levels of argument ontology

Each of these criteria then have sub-criteria at increasingly more detailed levels. As an example, Figure 5 shows the sub-criteria for Usability as displayed in SEURAT. The ontology terms are worded in terms of arguments: i.e., <*alternative*> is a good choice because it <*ontology entry* >, where *ontology entry* starts with a verb. The SEURAT system has been designed so that the user can easily extend this ontology to incorporate additional arguments that may be missing. With use, the ontology will continue to be augmented and will become more complete over time. It is possible to add deeper levels to the hierarchy but that will make it more time consuming for the developer to find the appropriate item when adding rationale.

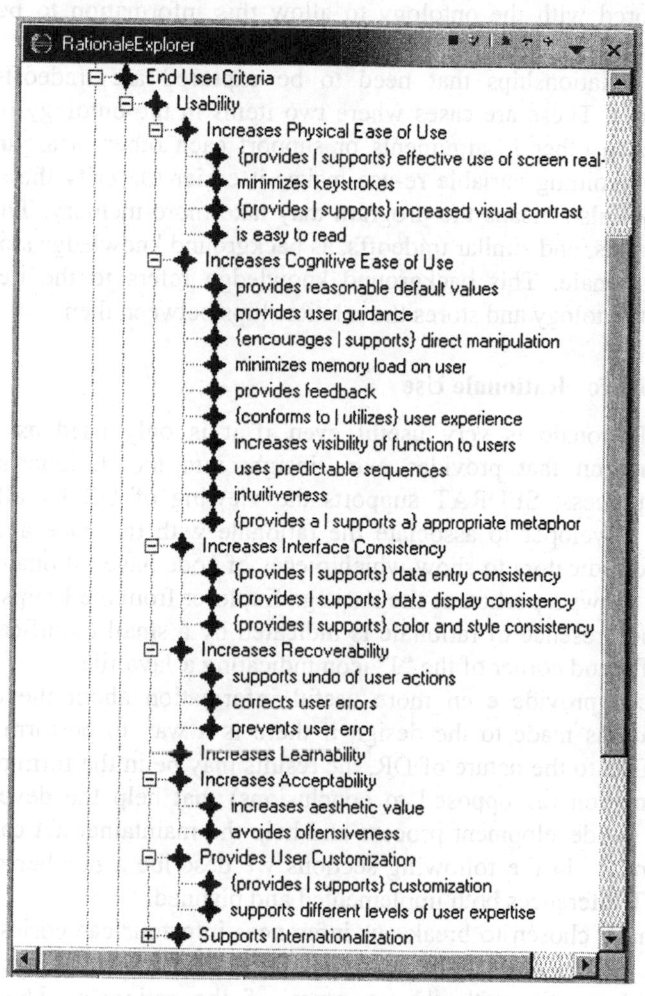


*Figure 5.* Sub-criteria for usability

Similar hierarchies have been developed for the other categories in Figure 4. One thing to note is that it is not a strict hierarchy—there are many cases where items contributing toward one quality also apply to another. One example of this is the strong relationship between scalability and performance. Throughput and memory use, while primarily thought of as performance aspects, also impact the scalability of the system. In this case, and others that are similar, items will belong to more than one category.

The argument ontology also includes a default importance for each item. These are present so that SEURAT users can specify this information for a particular system. This is used in weighing the different arguments during inferencing. The importance can be overridden for each claim or argument but is stored with the ontology to allow this information to be global if desired.

Other relationships that need to be captured are tradeoffs and co-occurrences. These are cases where two items in the ontology often either oppose each other in arguments or support each other in arguments. For example, avoiding variable re-use makes it easier to verify the software is correct but also means the program may take more memory. The user can represent this, and similar tradeoffs, as background knowledge stored as part of the rationale. This background knowledge refers to the items in the argument ontology and stores the relationships between them.

## 6. Support for Rationale Use

Design Rationale is very useful even if it is only used as a form of documentation that provides extra insight into the designer's decision-making process. SEURAT supports the viewing of DR by allowing the software developer to associate the rationale with the code and by using Rationale Indicators to show which pieces of code have rationale available. Figure 6 shows a portion of the Package Explorer from the Eclipse Java IDE where the presence of rationale is indicated by a small modification to the upper left hand corner of the "J" icon indicating a Java file.

DR can provide even more useful information about the design and modifications made to the design if there is a way to perform inferences over it. Due to the nature of DR, the results may be in the form of warnings or information (as opposed to conclusions) that help the developer keep track of the development process and help the maintainer act carefully and consistently. In the following sections we describe a number of different SEURAT inferences both implemented and planned.

We have chosen to break our inferences into four categories: syntactic, semantic, queries, and historical. Syntactic inferences are those that are concerned mostly with the structure of the rationale. They look for information that is missing. Semantic inferences require looking into the content of the rationale to evaluate the consistency of the design reasoning and point out cases made where less-supported decisions were made. Rationale queries give the user the ability to ask questions about the rationale, and historical inferences use a history of rationale changes to help the user learn from past mistakes, rather than repeating them.

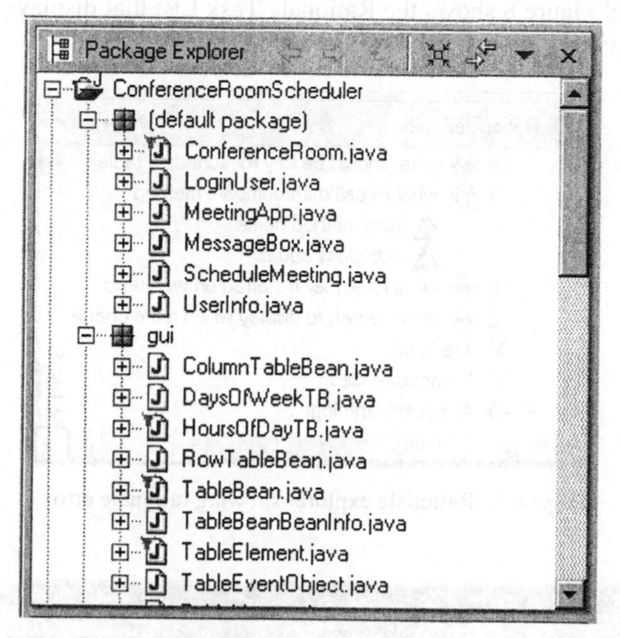


*Figure 6.* Package explorer showing rationale associations

## 6.1. SYNTACTIC INFERENCING

Syntactic inferencing is primarily concerned with the structure of the rationale – ensuring that the rationale is complete. This is a significant aid to the developer to make sure they do not leave any unresolved issues behind when building the system. These inferences include the following:

- Checking for decisions with no selected alternatives;
- Checking for decisions with more than one selected alternative when there should be only one;
- Checking for selected alternatives with no arguments in their favor;
- Checking for selected alternatives with only arguments in opposition;
- Checking for biased arguments where some alternatives have many arguments (for and/or against) while others have few or none.

SEURAT currently displays which alternative has been selected for each decision and indicates that there is an error if no alternative is selected. Errors are shown in two places: an error indicator on the rationale item in the Rationale Explorer and an error description on the Rationale Task List. Figure 7 shows the Rationale Explorer with an error indicator showing that no alternative was selected for the decision "what to call date compare method" and Figure 8 shows the Rationale Task List that displays that error, and others. Errors are indicated by a red icon containing an "X".

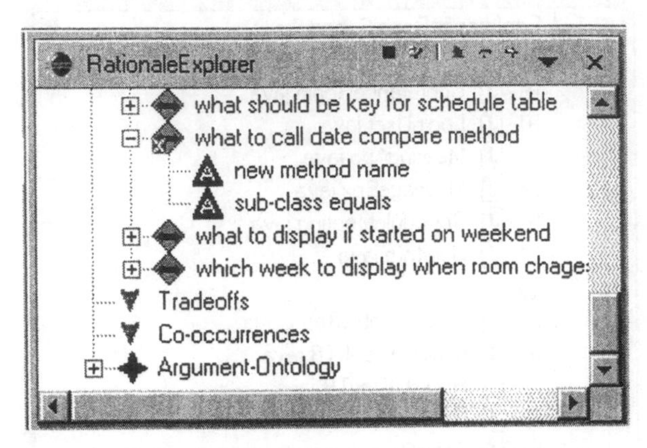


*Figure 7.* Rationale explorer showing rationale error

Rationale Task List

| Description | Rationale | Type |
|---|---|---|
| No alternative selected | how should rooms be stored? | Decision |
| No alternative selected | what to call date compare method | Decision |
| Best evaluated alternative not selected for decision | How is the user associated with the meeting | Decision |

*Figure 8.* Rationale task list

6.2. SEMANTIC INFERENCES

While syntactic inferences look at the structure of the rationale, the semantic inferences look at the content. This allows a more in-depth look for any inconsistencies in reasoning that are captured in the rationale. The syntactic inferences implemented in SEURAT include the following:

- Determining if the best supported alternatives were selected;
- Checking for contradictory arguments by using the argument ontology to compare claims;
- Checking for violated requirements;
- Checking for violations of the tradeoff and co-occurrence relationships captured in the rationale.

Some of these results are shown as errors, such as when a requirement is violated, while others are warnings. Figure 9 shows how a warning is indicated when the inferencing shows that the best alternative was not selected for the decision "How is the user associated with the meeting." This warning also shows up on the Rationale Task List shown in Figure 8. The warning is displayed as a yellow triangle icon with an exclamation point inside it.

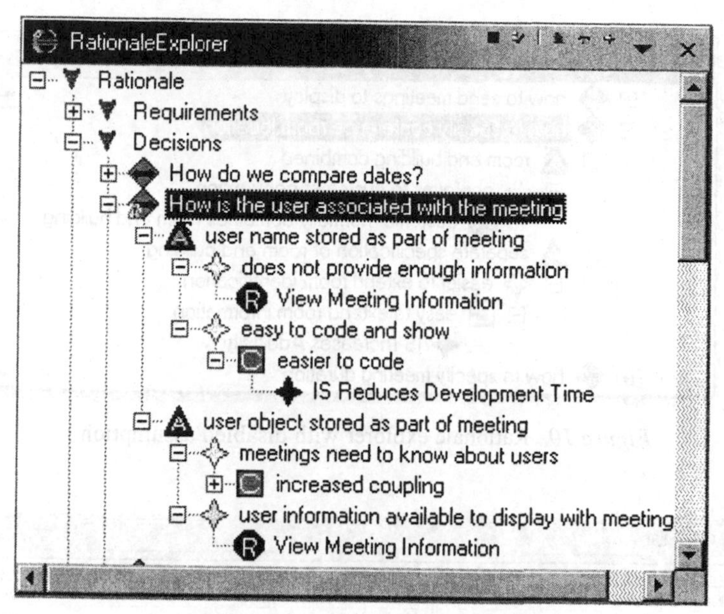


*Figure 9.* Rationale explorer with warning indicator

The semantic inferencing also allows the user to do some "what-if" reasoning by making changes to the rationale and seeing what effect that has on the decisions that have been made. For example, SEURAT provides the ability to disable requirements or assumptions and re-compute the evaluation of each decision. Figure 10 shows the Rationale Explorer after an assumption, "customer normally combines room and building" has been disabled. The assumption, denoted by an icon containing an "A", is changed to have a "D" in the upper right hand corner showing it is disabled. When the decision is re-evaluated, a warning icon is shown because the selected alternative (denoted by an "S" in the upper right hand corner) is no longer the best supported. The new warning is added to the Rationale Task List shown in Figure 11.

Another way that semantic inferencing is useful is in evaluating the effect of changing priorities on the design. Arguments for and against alternatives can consist of requirements, other alternatives (in case of dependencies), arguments, and claims. Each argument has an importance associated with it that can either be set at the argument level of, in the case of assumptions and claims, inherited. Each claim is associated with an entry in the argument ontology, which also has an importance assigned. The user can change the importance at any of the three levels (ontology, claim, or argument) and will be able to examine how that change affects the evaluation of the rationale.

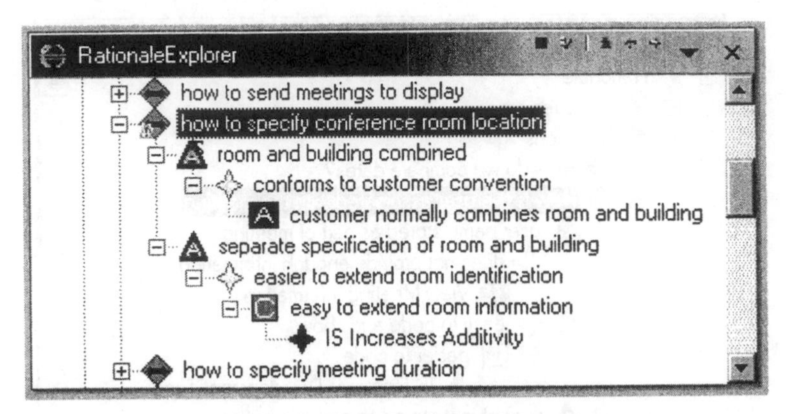


*Figure 10.* Rationale explorer with disabled assumption

Rationale Task List

| Description | Rationale | Type |
|---|---|---|
| No alternative selected | how should rooms be stored? | Decision |
| No alternative selected | what to call date compare method | Decision |
| Best evaluated alternative not selected for decision | How is the user associated with the meeting | Decision |
| Best evaluated alternative not selected for decision | how to specify conference room location | Decision |

*Figure 11.* Rationale task list with new warning

## 6.3. RATIONALE QUERIES

Rationale queries are inferences that are performed only upon request, not automatically when the rationale changes. These queries are used to obtain additional information about the rationale or the design. Rationale queries supported by SEURAT include the following:

- Listing all selected alternatives that address or satisfy a specific requirement;
- Listing all non-selected alternatives that address or satisfy a specific requirement;
- Listing which alternatives are argued (for or against) by a specific claim or specific ontology entry;
- Listing where there were importance overrides (from the default specified in the argument ontology) in the rationale;
- Listing all disabled entities (assumptions, claims, or requirements) in the rationale;

- Listing the most frequently referenced ontology entries (i.e., common arguments for and against alternatives).

The results of these queries will not result in errors or warnings about the rationale but will provide useful information to assist the developer in understanding the rationale and the system it describes.

### 6.4. HISTORICAL INFERENCING

The final component of SEURAT inferencing is inferences that take advantage of stored history about changes to the rationale. One of the motivating reasons for keeping track of rationale is to avoid repeating past mistakes by documenting alternatives that were attempted and then rejected. History-based inferences would be used to ensure that the developer does not select an alternative that was rejected before without being aware of the reasons for why it failed the first time. The rationale history is also used to keep track of which areas of the design have been the most volatile.

### 6.5. SEURAT IMPLEMENTATION

SEURAT is implemented as a Java Plugin for the Eclipse framework. This provides tight integration with the Eclipse Java IDE where the rationale associations are shown as part of the Java editor used to develop the code that the rationale describes and where the rationale and rationale status displays are all shown as windows within the IDE. The rationale is stored in a MySQL database. This provides scalability to large amounts of rationale and allows the use of SQL queries to assist in the inferencing. The database relationships can be used to propagate the results of rationale changes to other affected portions of the rationale. These links between the rationale act much like dependencies described in a truth maintenance system except that we are not asserting the truth of the statements.

## 7. Conclusions and Future Work

The SEURAT system provides a number of important innovations contributing towards effective use of rationale for software maintenance. The first is the argument ontology. This contributes in several ways: first, by creating an extensive list of reasons for making software design choices and secondly by using these reasons to support semantic inferencing to determine the impact of these choices on the software system and to promote consistency in the rationale. Another key contribution is the integration of SEURAT into a software development environment used by the developers and maintainers. This allows both the developers and maintainers to use the rationale without having to remember to invoke a separate utility or environment and lessens the disruption that can occur when switching from development to documentation.

Future work on SEURAT will involve expansion of the inference set and enhancements to the integration with the Eclipse Java IDE. These changes will increase both the functionality and usability of the SEURAT system. The system will be evaluated in a series of experiments with software developers of varying levels of expertise performing a series of maintenance tasks to determine the effectiveness of the rationale support.

We feel that the SEURAT system will be invaluable during development and maintenance of software systems. During development, SEURAT will help the developers ensure that the systems they build are complete and consistent. During maintenance, SEURAT will provide insight into the reasons behind the choices made by the developers during design and implementation. The benefits of DR are clear but only with appropriate tool support, such as that provided by SEURAT, can DR live up to its full potential as an aid for revising, maintaining, and documenting the software design and implementation.

## References

Boehm, B and Bose, P: 1994, A collaborative spiral software process model based on theory W, *Proc. 3rd International Conf. on the Software Process,* IEEE Computer Society Press, CA, pp. 59-68.

Bose, P: 1995, A model for decision maintenance in the win win collaboration framework, *Proc. of the Conf. on Knowledge-based Software Engineering*, IEEE Computer Society Press, CA, pp. 105-113.

Brooks, FP Jr.: 1995, *The Mythical Man-Month*, Addison Wesley, MA.

Burge, JE and Brown, DC: 2000, Inferencing over design rationale, *in* J Gero (ed), *Artificial Intelligence in Design '00*, Kluwer Academic Publishers, Netherlands, pp. 611-629.

Burge, JE and Brown, DC: 2002, *NFRs: Fact or Fiction?*, Technical Report WPI-CS-TR-02-01, Computer Science Department, WPI.

Bruegge D and Dutoit A: 2000, *Object-Oriented Software Engineering: Conquering Complex and Changing Systems,* Prentice Hall.

Chung, L, Nixon, BA, Yu, E and Mylopoulos, J: 2000, *Non-Functional Requirements in Software Engineering*, Kluwer Academic Publishers.

CMU: 2002, Quality measures taxonomy,
http://www.sei.cmu.edu/str/taxonomies/view_qm.html

Conklin, J and Burgess-Yakemovic, K: 1995, A process-oriented approach to design rationale, *in* T Moran and J Carroll (eds), *Design Rationale Concepts, Techniques, and Use*, Lawrence Erlbaum Associates, Mahwah, NJ, pp. 293-428.

Dellen, B, Kohler, K and Maurer, F: 1996, Integrating software process models and design rationales, *Proc. of the Conf. on Knowledge-based Software Engineering*, IEEE Computer Society Press, pp. 84-93.

Filman, RE: 1998, Achieving ilities, *Proc. of the Workshop on Compositional Software Architectures*, Monterey, CA, USA.

Fischer, G, Lemke, A, McCall, R and Morch, A: 1995, Making argumentation serve design, *in* T Moran and J Carroll (eds), *Design Rationale Concepts, Techniques, and Use*, Lawrence Erlbaum Associates, pp. 267-294.

Grudin, J: 1995, Evaluating opportunities for design capture, *in* T Moran and J Carroll (eds), *Design Rationale Concepts, Techniques, and Use*, Lawrence Erlbaum Associates, NJ, pp. 453-470.

Jiro, K: 2000, *KJ Method: A Scientific Approach to Problem Solving*, Tokyo: Kawakita Research Institute.

Karsenty, L: 1996, An empirical evaluation of design rationale documents, *Proceedings of the Conference on Human Factors in Computing Systems*, Vancouver, BC, April 13-18.

Klein, M: 1997, An exception handling approach to enhancing consistency, completeness and correctness in collaborativ requirements capture, *Concurrent Engineering Research and Applications*, **March:** 37-46.

Lee, J: 1990, SIBYL: A qualitative design management system, *in* PH Winston and S Shellard (eds), *Artificial Intelligence at MIT: Expanding Frontiers*, MIT Press, Cambridge, MA, pp. 104-133.

Lee, J: 1991, Extending the Potts and Bruns model for recording design rationale, *Proc. of the 13th International Conf. On Software Engineering,* Austin, TX, pp. 114-125.

MacLean, A, Young, RM, Bellotti, V, and Moran, TP: 1995, "Questions, options and criteria: elements of design space analysis", *in* T Moran and J Carroll (eds), *Design Rationale Concepts, Techniques, and Use,* Lawrence Erlbaum Associates, NJ, pp. 201-251.

Peña-Mora, F and Vadhavkar, S: 1996, Augmenting design patterns with design rationale, *Artificial Intelligence for Engineering Design, Analysis and Manufacturing* **11**: 93-108.

Potts, C and Bruns, G: 1988, Recording the reasons for design decisions, *Proc. of the International Conf. On Software Engineering*, Singapore, pp. 418-427.

Reiss, SP: 2002, Constraining software evolution, *Proc. of the International Conference on Software Maintenance*, Montreal, Quebec, Canada, pp. 162-171.

Toumlin, S: 1958, *The Uses of Argument*, Cambridge University Press

# INTERACTIONS: SENTIENCE AND SKETCHING

JS Gero (ed), *Design Computing and Cognition'04*, 579-592


# DESIGNING A UBIQUITOUS SMART SPACE OF THE FUTURE: THE PRINCIPLE OF MAPPING

TAYSHENG JENG
*National Cheng Kung University, Taiwan*

**Abstract.** In this paper, we present a conceptual framework for the development of a spatial system prototype that is oriented toward ubiquitous smart spaces. A key problem in developing ubiquitous smart spaces is how to map a physical space to the underlying computing infrastructure and the corresponding patterns of situated interaction in every life. This paper explores the design issues encountered in the development of ubiquitous smart spaces. Methods for mapping different aspects of ubiquitous smart spaces are described and illustrated by a number of brief examples from the implementation of the spatial system prototype.

## 1. Introduction

There is growing interest in developing ubiquitous computing technologies to build a smart space of the future with natural interaction interfaces that can be more deeply integrated with our daily life. Examples of ubiquitous smart spaces are Intelligent Environments (Coen et al. 1999), iRoom (Johanson et al. 2002), Classroom2000 (Abowd 2002), AwareHome (Kidd et.al. 1999), EasyLiving (Brumitt 2000), House_n (Intille 2002). The common goal is to augment physical spaces with ubiquitous computing technologies for supporting our daily lives more effectively. The goal is difficult to achieve because, in addition to the design problem, it must cross the boundaries of disciplines to define the future of computing.

Architecture is creating new challenges as computing moves into everyday life. As hardware components become smaller, faster, and cheaper, computational devices are becoming embedded in building elements such as doors, walls, floors, and furniture. These devices are largely invisible and linked together through wireless network. Buildings would largely contain smart spaces and appliances that provide access to computational resources at any place and time. Architecture becomes an interaction interfaces between humans and computers. A key problem in developing for such ubiquitous smart spaces is to map physical spaces to the underlying computing infrastructure and the corresponding patterns of situated interaction in everyday life. Many ubiquitous smart spaces projects have emphasized the computational capabilities, with less concern for the *logical mappings* between user experiences, digital infrastructures, and physical interfaces in a broader context of design.

Mapping is an essential part of design process. In this paper, we do not attempt to identify the technological issues or detail the functionality of ubiquitous smart spaces, but instead focus on the principle of mapping in designing a successful ubiquitous smart space system. Our interest is in the inter-disciplinary approach to the method of mapping, in support of the design of ubiquitous smart space of the future. The development of ubiquitous smart spaces, involving possibly unique sets of criteria that are also ill-structured. Many types of design are of this sort; our particular interest is the inter-relationships between human experience, physical infrastructure, and interaction design.

This paper is organized as follows. First, we provide a conceptual framework offering an integrative view of the key issues encountered in the design of ubiquitous smart spaces. We use a practical example of our laboratory throughout the paper to illustrate the design process. In section 3 and 4, we describe an approach to the method of mapping according to the conceptual framework. In section 5, we set forth a set of criteria for architectural computing at the boundaries between humans and spaces. Some of its benefits, current status and some open questions about the development of ubiquitous smart spaces are discussed.

## 2. A Conceptual Framework

In order to determine if the existing methods can support the design of ubiquitous smart spaces, it is first necessary to understand the patterns of everyday life, what system infrastructures underpin our everyday lives, where and what physical interfaces afford it. In this section, we present a conceptual framework for structuring the problem domain. The design issues encountered in the development of ubiquitous smart spaces are identified.

### 2.1. THE FRAMEWORK

As a ubiquitous smart space is designed, the issues concerning it fall within three general areas:

- **Patterns of everyday life:** A variety of patterns are implicated in our daily lives. A generic pattern is that *people* interact with *things* in *places*. In order to design a place for living, we must understand how people interact with things when they are at work, at home, at school, on the town, and on the move.

- **Technological Infrastructure:** People are no longer using a singular desktop computer to get their work done. They interact with heterogeneous computational devices such as tablet PCs, PDAs, and wearable and embedded computers. The computational devices can be either wearable computers carried by people moving from one place to another, or interactive devices embedded in physical spaces such as digital desks, wall-sized interactive whiteboards, etc. This require a technological infrastructure to accommodate a new way of interacting with computation. The technological infrastructure is composed of multiple layers, including interaction multimodalities, computational devices, electronic services, and wireless networks.
- **Physical Spaces:** Using multiple computational devices correspondingly creates different forms of situated interactive experiences in the physical world. The physical world can be conceptualized in terms of cities, sites, and spaces. In addition to the underlying technological infrastructure support, we are concerned with the design of building sites, physical structures, skins, services, furniture, helping improving the quality of life for everyone in the physical world.

We are concerned with the design of ubiquitous smart spaces with respect to three levels of design: patterns of everyday life, technological infrastructure, and physical spaces. The idea of ubiquitous smart spaces encompasses many different kinds of user experience, computational devices, and spatial settings. An important aspect is to relate the patterns of everyday life to the technological infrastructure and the physical environments necessary to support our everyday lives, as shown in Figure 1.

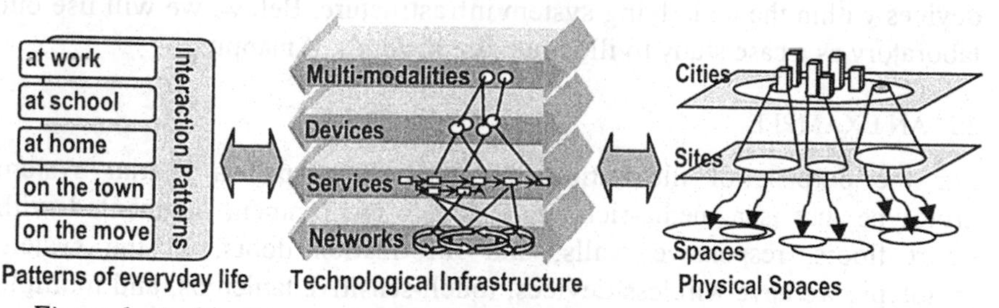


*Figure 1.* A ubiquitous smart space involves three levels of design: patterns of everyday life, technological infrastructure, and physical environments

In accordance with the conceptual framework, several functional requirements quickly arise for the development of ubiquitous smart spaces. The challenge before us is:

*to map the patterns of everyday life to the underlying technological infrastructure, and the corresponding physical spaces.*

This challenge is the focus of ubiquitous smart spaces. Even now, the challenge is making the incremental transition to study architectural computing cross the boundaries of disciplines. We will discuss the future of architectural computing in section 5.

Mapping is an essential part of design process. We have proposed two methods of mapping. The first method is to map the patterns of everyday life to the underlying technological infrastructure. We must understand human experience and how they interact with surroundings so as to invent the technological infrastructure to support it. An important aspect is that the technological infrastructure needs to be maintained to reflect the patterns of everyday life. The technological infrastructure must support dynamic configuration to fit the everyday practices, in a manner allowing adaptation of a spatial configuration in a particular set of cognitive and social context.

The second method is to map the technological infrastructure to the physical spaces. This method is accomplished by embedding a variety of computational devices (e.g. displays, sensors, e-tags) in the architectural elements such as walls, floors, doors, and furniture. The sensors make the computer aware of human gesture, voices, and motions. It is quite possible the conceptualization of how buildings and spaces are used and perceived relies on how interaction multimodalities, services, and network facilities are provided by the technological infrastructure.

The usability of a ubiquitous smart space will only be realized when a specific user community experiment it. In order for us to fully reveal how a ubiquitous smart space will be used in the future, we need a full-scale spatial system prototype that can instrument a variety of computational devices within the underlying system infrastructure. Below, we will use our laboratory as a case study to illustrate the methods of mapping.

### 2.2. AN EXAMPLE

For the purpose of illustration, we have developed a spatial system prototype that is a media-rich multi-device environment augmented with smart floors, responsive walls, and information doors. In the system prototype, we have wireless devices, touch-sensitive tabletops, and multiple large wall-size displays augmented with interactive whiteboards (e.g. Mimio). A set of web cameras are used for image recognition and motion detection within the space. A view of the spatial system prototype is shown in Figure 2.

The spatial system prototype was initially designed for design presentation. We designed an interactive CAD platform that is spatially-aware in a sense that the display reacts immediately in response to changes in user's position (Jeng and Lee 2003). We embedded a transparent screen in the wall-sized glass door, allowing students interacting between inside and outside. The spatial system prototype has been installed in our Information Architecture Laboratory, which is more thoroughly presented in CAADFutures conference in 2003. We will describe the details of the prototype in Section 4.

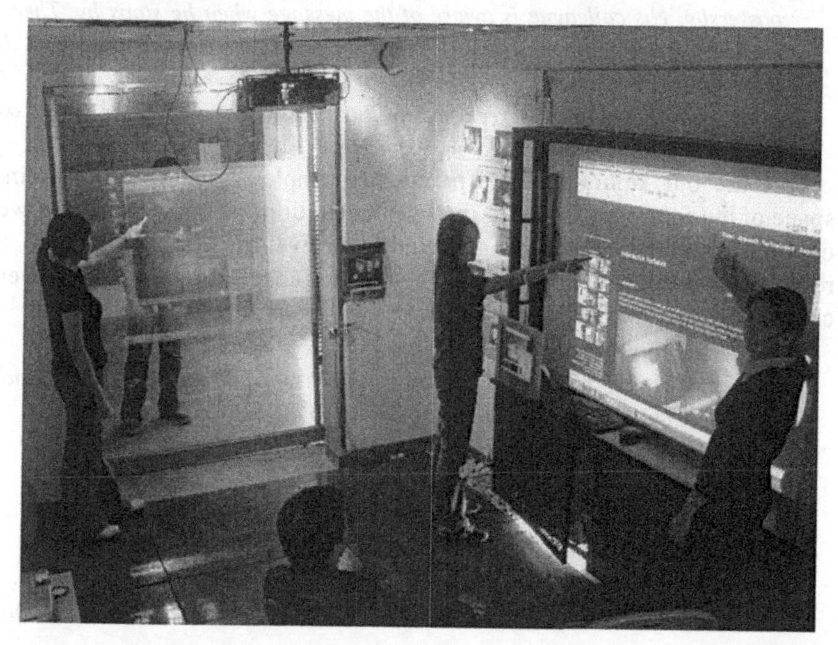

*Figure 2.* A view of the spatial system prototype in the information architecture lab

## 3. Mapping Patterns of Everyday Life to System Infrastructure

In order to reveal the underlying system infrastructure support for new kinds of interaction, we begin with a scenario that suggests a ubiquitous smart space to support the existing pattern of design work. The example is drawn from the practical actions of individuals in our laboratory that is commonly used for group meeting and studio discussion. There may be a variety of actions and activities in the laboratory but we will focus on a small part of human-space interaction interfaces for purposes of discussion. Let us consider the following scenario:

> *Jeng steps on the smart floor at the entrance every morning. His status is identified by the electronic tags mounted on his shoes. The e-tag reader triggers the display of the personal schedule on the side wall. Jeng checks and reviews design work in progress. He marks up his calendar on the touch screen whiteboard for the next meeting. In the afternoon, Jeng moves on to a weekly*

*design meeting. In the design presentation, he picks up a project display tablet and plugs it into the smart wall. This automatically brings up a project web page with related links on the wall screen. Jeng holds a laser pointer to remotely move the digital links over the screen. Later, in the coffee break, Jeng presents his ideas on an information door that is connected to the web server in the building. The information door broadcasts design information to passersby. His colleague is aware of the message when he stops by. Then he leaves messages from the outside of the information door for commentary. The messages are captured and added to the project home page. With the transparency and touch-screen capabilities, the information door serves as a medium for situated and social interactions.*

We use our laboratory as a practical example of the ubiquitous smart space of the future and observe the practical actions of individuals that were organized through reoccurring patterns of events. In the scenario, the reoccurring events are underlined. The typical reoccurring patterns of events can be classified as follows:

1. *personal identification – step at the entrance, login to the computer,*
2. *check work in progress – display personal schedule, mark on the calendar, any other forms of communication prior to the meeting.*
3. *information processing – processing information individually or collaboratively.*
4. *presentation – multimodal presentations, laser pointing, whiteboard sessions.*
5. *meeting – brainstorming, discussion, understanding, data organization.*
6. *situated and social interaction – coffee break, commentary, broadcasting.*

We attempt to identify the patterns (e.g. identification, check work in progress, information processing, presentation, meeting, situated and social interaction) that are independent from the types of everyday life (e.g. at work, at school, at home, on the town, on the move). Each pattern is mapped on a particular set of sensors and actuators, which reveals technological usage and infrastructural support, Figure 3.

**Reoccurring Events:** The pattern of everyday life can be defined through reoccurring events. A composite event can be defined in a higher level of abstraction when we use a new form of interaction interfaces anchored to a particular place. For example, in the course of "identification", the user may be identified at the entrance and then get access to information on the door when he steps in the workspace. This implies that the patterns of events such as "identification" and "check work in progress" will merge into a composite event when the patterns are tied to a particular place. Of course, it requires continuous interaction support from the underlying computing infrastructure.

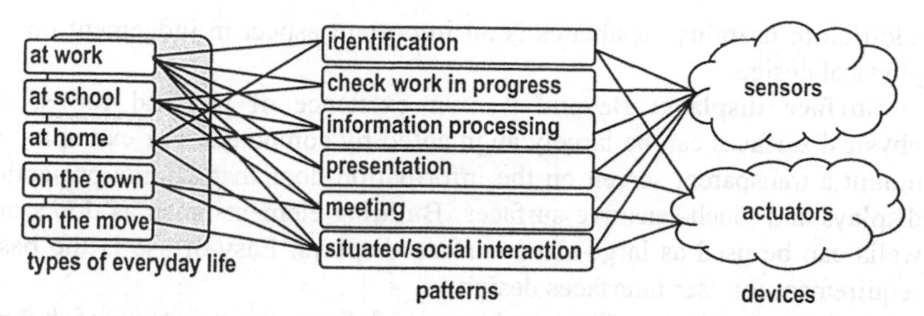


*Figure 3.* Mapping the patterns of everyday life to the technological infrastructure

**Continuous Interaction:** Identifying the patterns of events is not our primary focus. Rather, we are interested in the relationships between the patterns of events that reveal spatial usage and technological infrastructure support. For example, the user may leave in the middle of the meeting, then continue the meeting in the home office immediately after he is back home. Clearly, the technological infrastructure should support the continuous interaction when one moves from one place to another. A building can provide a variety of natural interfaces that may be situated to cope with unstructured activities as like our everyday lives. This implies the need for building technological infrastructure in such a way that computation should move from a localized facility to a constant presence.

**Pattern Language:** Alexander proposed patterns as a method for capturing common knowledge and shared solutions to recurring problems (Alexander et al. 1977). We adopt the idea of patterns as a means of capturing user experiences through reoccurring patterns of events in everyday practices. Such patterns of events are implicated in the daily lives, which provide a means of understanding the context of interactive technology usage. The main purpose of identifying patterns is to allow them to be used as part of design process.

## 4. Mapping System Infrastructure to Physical Spaces

If we think of a space as an information appliance, we will design a building for interaction, Figure 4. The challenge is not simply to design a building to be a container of computational devices, but rather create the user experience through embodied interaction within the physical space. Given the range of technological infrastructure, there are rules defining how the physical interfaces must be composed to make a meaningful ontology in the physical environment. We highlight three interrelated component concepts for physical spaces. The concepts are:

**Physical forms:** The physical components are used to construct and enclose a void space include floors, walls, ceilings, doors, and furniture. In addition to usability, aesthetics is an important aspect in judgement on the physical design.

**Surface displays:** Beyond it own existence of physical forms, the physical surfaces can be largely augmented by computers. For example, we mount a transparent screen on the information door that affords projection displays and touch-sensitive surfaces. Building elements such as doors and walls can be used as large-sized surface displays. Easy-of-use is the basic requirement for user interfaces design.

**Activity territory:** The void space defines the territory of human activities and the invisible sensing area for interaction with computation. For example, the information door affords social interaction on inside and outside i.e. between the laboratory and the hallway. The sensors are hidden behind the door and the cameras are mounted on the ceiling. Of course, any computational devices impose a set of constraints on its own setting such as the display clearance imposed by video projectors, the range for sensing, the limited distance for laser pointer interaction.

Let us consider the settings in our laboratory as a practical example. We implement some computer-augmented architectural components generally applied to the design of ubiquitous smart spaces. These architectural components provide three specific types of physical computing for interaction:

- **Smart floors:** We mount a RFID reader under the floor in the laboratory, which defines a spatial territory for personal identification. The embedded sensors signify the presence of human gesture and motions using computer vision technology, allowing software agents to interpret the information and send signals to actuators so as to produce real world actions.
- **Responsive walls:** We design a variety of wall-sized displays in the laboratory. Next to the wall-sized display is a RFID reader mounted in the exhibit side wall of the laboratory. When the user places a RFID-mounted acrylic table on the exhibit side wall, the wall-size display will automatically show the project-specific information corresponding to the particular tablet.
- **Information doors:** The information door serves as a medium between laboratory and the hallway. We mount a transparent screen on the laboratory door. A video projector is mounted over the ceiling and a web camera for image recognition. We adopt a mouse as an input device attached on the touch-screen door surface at eye level. Passerby will use the mouse-like device to interact with information on the door. The information door provides a projection surface for practical services such as posting design work, personal scheduling, and organizational announcements. The user looks in the transparent door and places a finger on the glass, believing that she is touching

the digital information and interacting the person directly in the other side of the door.

***Smart floors:*** *We mount a RFID reader under the floor in the laboratory, which defines a spatial territory for personal identification.*

***Responsive walls:*** *When the user places a RFID-mounted acrylic table on the embedded side wall, the wall-size display will automatically show the project-specific information corresponding to the particular tablet*

***Information doors:*** *The user looks in the transparent door and places a finger on the glass, believing that she is touching the digital information and interacting the person directly in the other side of the door.*

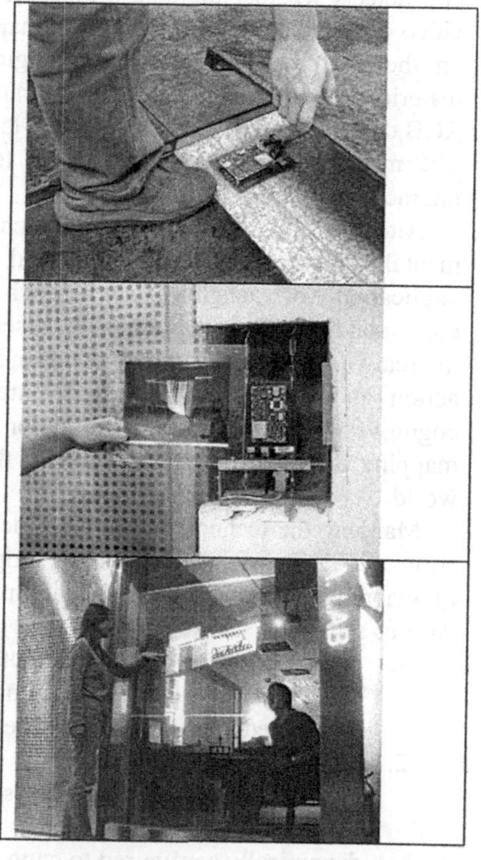

*Figure 4.* Mapping the technological infrastructure to the physical space such as smart floors, responsive walls, and information doors

Once we have built a physical space with the system infrastructure, the next level of principles is to map physical representations to digital representations. The principle of mapping in the technological level can be characterized in two parts. The mapping may be *external*, imposed by technical mechanisms or the vision and perception technologies, as inherent features in the use of computational devices. In the ubiquitous smart space, for example, the vision recognition system signifies the presence of human bodies, gesture, motions, or movements in a camera view. Once a visualization pattern is matched within the camera view, it would respond a reaction to the physical world. In short, the digital representation is mapped onto the physical interaction, and vise versa. Consider, for example, the application of laser pointer interaction. When one holds the laser pointer and points at an object, the vision recognition system captures a live video and detects whether the spot's location is within a certain area. The live video captured by the camera can be mapped onto the real-world projection on the wall display. In terms of digital representations, the real-world material properties (e.g. laser spot) is transformed into digital attributes (e.g. RGB color). It triggers an action if the laser spot is within the specified area. The mapping mechanism contributes to the application of laser pointer interaction.

Alternatively, the mapping may be characterized *internally*, as the user's mental representation of the physical and social context. Consider an application of tangible media where moving a physical token correspondingly alters the digital representation. The human senses perceive the real-world change and map the user's action to perception. Coupling the action and perception creates an internal representation in the user's cognitive world. In the course of action and perception, it reveals multiple mapping between the physical world, the digital world, and the cognitive world.

Mapping the technological infrastructure to the physical space is an ill-structured design problem. We have sought out to exploit the basic principle of mapping to convey an understanding of the dynamic configuration of various settings.

1. different subset of the computational devices can be isolated or simultaneously used. This implies the physical space must be partitioned for different purposes of usage.
2. computational devices may be dynamically added to the infrastructure when people step in the space with wearable computers. This implies that the physical environment must be dynamically configured to cope with different kinds of composition.

Our experiments show that an attempt to define a single physical space that can encompass the whole universe of computational devices is doomed to failure. However, the relationship between the technological infrastructure and the physical space is not one-to-one mapping. In our experimental study, we found that multiple devices in physical spaces can be more than the sum of the functions they perform. The interaction of devices creates new form of spaces. It demands a shift in the way we think about the manifestation of computation in the work and design environments from graphical user interfaces to a new physical computing infrastructure support.

Our response to the mapping problem is to define modular domains of augmented reality for physical interaction, and to provide the means to integrate combinations of these modules into custom ubiquitous smart spaces. Of course, new devices can added at any time and it should be possible to reconfigure the setting in the space through an alternative physical interface. In order to develop such a modular space for adaption, an often-used strategy is to design different layers of architectural components out of a building constructed form. We refer to it as *an information skin* of the space. The concept of information skins facilitates the design of ubiquitous smart spaces.

## 5. Architectural Computing at the Boundaries between Humans and Spaces

Having outlined the methods of mapping, the following questions are: Where is the sensor data go? How is it interpreted? Where is the interpreted data reasoned about? What is the intelligence in the spatial system? To answer these questions, we propose a human-space interaction model as shown in Figure 5.

Human-space interaction is dramatically different from traditional models of human-computer interaction. It is different in that multiple devices are aggregated and contained within the space. Each person no longer interacts with a single computer. One may use multiple computational devices at a time. On the other hand, a single device may be shared by a group of people. Interaction with computation soon becomes an environmental experience, rather than a desktop one.

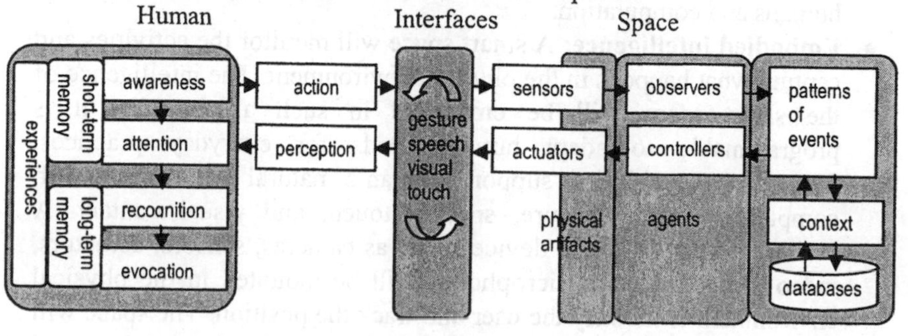


*Figure 5.* A human-space interaction model

The human-space interaction model being proposed here is based on three key elements: *human, interfaces,* and *spaces*. In the "space" part of the model, physical elements are embedded with sensors and actuators. When an event is detected by the sensor, the sensors convey message to the observer agents and produce a pattern of events in the space memory. The agents contain observers and controllers to receive and dispatch events correspondingly. The observer agent transforms sensor data to a meaningful event, which is compared to the pre-defined pattern of events in the context databases. An action is triggered by the controller agent when the event matches one of the patterns of the events.

To make physical space intelligent, we need to map context databases to human memories and experience. In the "human" part of the model, humans can be considered as a cyborg when they are augmented with wearable computers. They use gesture, speech, touch, or visual contact to interact with computation without independent sets of input and output channels. Human has action and perception. The human senses perceive the real-world change and map the user's action to perception. Coupling the action and perception creates an internal representation in the user's cognitive world. Underlying the action and perception is the short-term and long-term memories that map onto the user experiences in varied physical and social context.

In accordance with this perspective, we set forth a set of criteria for adaptive design of ubiquitous smart spaces. Ubiquitous smart spaces should provide three specific types of assistance for users:

- **Physical interfaces:** Physical spaces offer natural interfaces to provide services to human needs. Increasingly, architectural elements such as floors, walls, doors, and furniture are augmented to support intuitive effective non-intrusive interaction with computation. When broadly considered, architecture becomes an interface between humans and computation.
- **Embodied intelligence:** A smart space will monitor the activities and capture what happens in the physical environment. The intelligence of the smart space will be embodied in such a way that it is programmable to adapt human needs for everyday practices. Perceptual capabilities support human's natural interaction with computation using gesture, speech, touch, and visual contact. A variety of computational devices such as cameras, sensors, actuators, video projectors, and microphones will be mounted in the physical environment to identify the user and track the position. The space will provide services to human needs in accordance with the cognitive and social context.
- **Situated interaction:** A situatedness-based support for social construction of user experiences is required in the smart space. Increasingly people work on the move. Mobile devices will be context-aware in such a way that the embedded program is affected by where, how, and what they interact. A smart space will dynamically link the mobile devices and construct its memory to store user experiences with the computational devices. The context memory eventually serves as a knowledge base for a smart space to change its function and behavior in context.

## 6. Discussion and Conclusion

In this paper, we present a conceptual framework for the development of a spatial system prototype that is oriented toward ubiquitous smart spaces. In terms of conceptual framework, we can see an emerging vision of the future of architectural computing where people would interact continually with computation embedded in physical spaces. There is likely a profound change to our future work, design, and living environments. Our goal is to support the features of ubiquitous smart spaces in studios, classrooms, laboratories, hallways, and all spaces of our physical environments.

In response to such a shifting paradigm, we addresses the principle of mapping that is essential to realize the design and research issues as well as the underlying structure of ubiquitous smart spaces. Now, we are in the transitional state where efforts are being made to develop the cutting-edge media technology inherent in ubiquitous computing for future practices. One approach has addressed the integrated technologies that address human needs, either perceptual or vision recognition, allowing automatic detection of human intent and triggering physical operations. Another line of our work has addressed the development of different forms of physical computing, either handheld or embedded in the space, more deeply integrated into our physical environment.

This paper sets forth a set of methods and basic principles that are essential to realize the mapping process of ubiquitous smart spaces. The methods of mapping support the formulation of design solutions for integrating new technologies in ubiquitous smart spaces. The conceptual framework and the implementation of the prototype have served as a logical basis to elaborate broad design concepts and technical mechanisms that may be carried out toward the ubiquitous smart space of the future.

## Acknowledgements

This work was supported by the Taiwan National Science Council, grant No. NSC92-2211-E-006-087.

## References

Abowd, GD, Mynatt, ED and Rodden, T: 2002, The human experience, *IEEE Pervasive Computing* **2**: 48-57.

Alexander, C, Ishikawa, S, Silverstein, M, Jacobson, M, Fiksdahl-King, I and Angel, S: 1977, *A Pattern Language,* New York: Oxford University Press.

Brumitt, B, Meyers, B, Krumm, J, Kern, A and Shafer, S: 2000, EasyLiving: Technologies for intelligent environments, *Proceedings of Handheld and Ubiquitous Computing*.

Coen, M, Phillips, B, Warshawsky, N, Weisman, L, Peters, S, and Finin, P: 1999, Meeting the computational needs of intelligent environments: The metaglue system, *Proceedgins of 1st International Workshop on Managing Interactions in Smart Environments (MANSE'99)*, pp.201-212.

Dourish, P: 2001, *Where the Action is: The Foundations of Embodied Interaction*, MIT Press, Cambridge, MA.

Fitzmaurice, GW: 1993, Situated information spaces and spatially aware palmtop computers, *Communications of the ACM* **36**(7): 39-49.

Intille, SS: 2002, Designing a home of the future, *Pervasive Computing* **1**(2):76-82

Kidd, C et al: 1999, The AwareHome: A living laboratory for ubiquitous computing research, *Proceedings of 2nd International Workshop Cooperative Buildings*, Springer-Verlag.

Jeng T and Lee, C: 2003, *iCube*: Ubiquitous media spaces for embodied interaction, *Proceedings of CAADFutures*, Taiwan, Kluwer Academic Publishers, pp. 225-234.

Jeng T and Lee, C: 2003, Tangible design media: Toward an interactive CAD platform, *International Journal of Architectural Computing* (**1**)2: 153-168

Johanson, B, Fox, A, and Winograd, T: 2002, The interactive workspaces project: Experiences with ubiquitous computing rooms, *IEEE Pervasive Computing* **1**(2): 67-74.

McCullough, M: 2001, On typologies of situated interaction, *Human-Computer Interaction* **16**: 337-349.

Winograd, T: 2002, Interaction spaces for twenty-first-century computing, *in* JM Carroll (ed), *Human-Computer Interaction in the New Mullennium*, pp. 259-274.

Weiser, M:1991, The computer for the 21st Century, *Scientific America* **265**(3): 66-75.

JS Gero (ed), *Design Computing and Cognition'04*, 593-612


# ELEMENTS OF SENTIENT BUILDINGS

ARDESHIR MAHDAVI
*Vienna University of Technology, Austria*

**Abstract.** This paper describes the concept and elements of sentient buildings. A sentient building is one that possesses a sensor-supported, dynamic, and self-updating internal representation of its own components, systems, and processes. It can use this representation, amongst other things, toward the full or partial self-regulatory determination of its indoor-environmental status.

## 1. Introduction

### 1.1. THE PROBLEM

Buildings face a growing number of requirements. Specifically, multiple environmental control systems must operate in a manner that is energy-effective, environmentally sustainable, economically feasible, and occupationally desirable. However, a considerable number of buildings fail to meet these requirements. This in part due to design flaws in buildings and their systems, and in part due to the way systems are operated.

Requirements associated with the day-to-day operation of buildings are not high on the priority list of the primary designers of buildings. Building service systems are often perceived as technical add-ons, to be taken care of by specialists in the later phases of design. This attitude is not conducive to conceiving well-engineered and integrated building control systems.

Currently, controllers (for heating, cooling, ventilation, lighting, acoustics, etc.) operate within a distributed hierarchy of individual system components at several levels. While this allows, in principle, for a distributed implementation of control logic, it also leads to the development of environmental systems as isolated sub-systems. Most commercially available environmental control systems for buildings do not offer control logic and software that consider the impact of the interactions and interferences of multiple devices. Thus, devices affecting the same control zone are seldom integrated. Likewise, in most buildings, the level of vertical integration of local and central systems is insufficient, Figure 1. These problems are aggravated due to the contingencies associated with the boundary conditions of system operation, e.g. dynamic changes in the outdoor conditions as well as people's activities and control actions.

These issues imply the need for better building control systems and better methods to design them. Toward this end, the present paper introduces the concept and features of "sentient" buildings (Mahdavi 2001a).

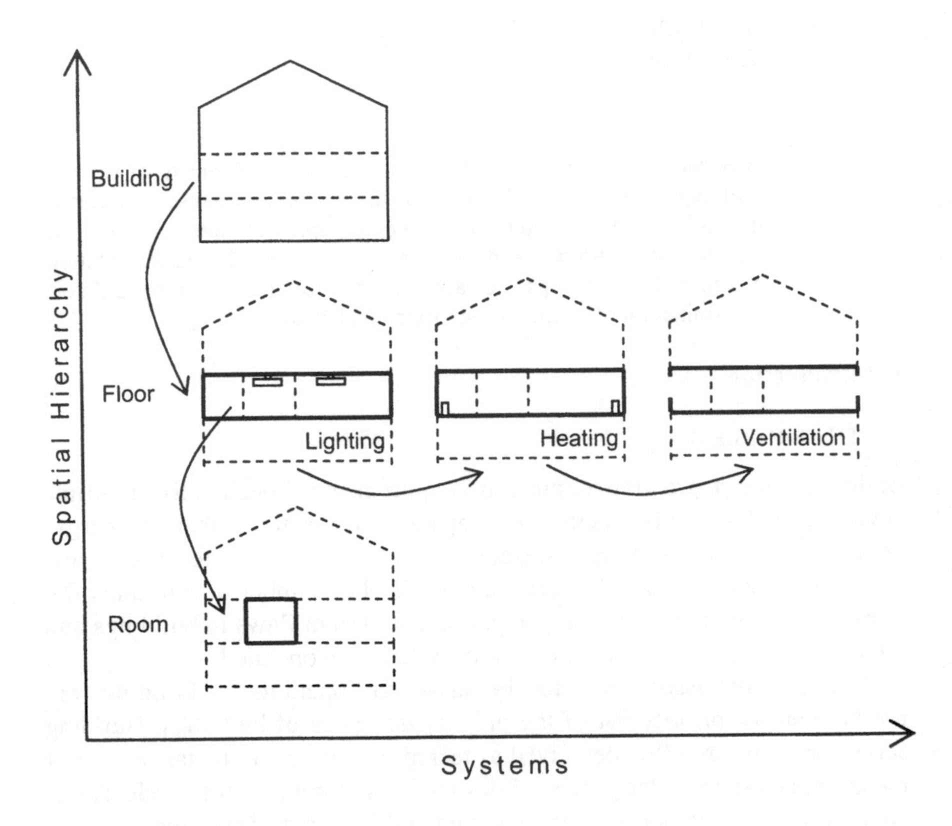


*Figure 1.* Illustrative representation of the requirement for the integrated operation of multiple building control systems across multiple levels of spatial hierarchy

## 1.2. APPROACH TO A POSSIBLE SOLUTION

The approach elucidated in this paper is based on the assumption that building operation functionality can benefit from the presence of an explicit and dynamic model of building constitution, status, and context. Such a representational system can be conceived in the building design phase and subsequently carry over into the operation phase. Thereupon, the model must continuously update itself to reflect changes and, thus, provide a reliable informational basis for various management and control applications. An example of such an application concerns the decision-making functionality of the building's indoor environmental systems. This functionality can use the model to perform exploratory behavioral simulations of alternative control scenarios and evaluate the results toward preferable course of action. The main concepts involved in this approach are as follows:

**Sentient buildings** – A sentient building is one that possesses an internal self-organizing representation of its own components, systems, and processes. It can use this representation, amongst other things, toward simulation-based (full or partial) self-regulatory determination of its own status.

**Self-organizing building models** – A self-organizing building model is a complex, dynamic, sensor-supported, self-updating, and self-maintaining building representation with instances for building context, structure, components, systems, processes, and occupancy. As such, it can serve as the internal representation of a sentient building toward real-time building operation support (building systems control, facility management, etc.).

**Simulation-based building control** – Within the framework of a simulation-based control strategy, control decisions are made based on the comparative evaluation of the simulated implications (predicted future results) of multiple candidate control options (Mahdavi 2001b).

Note that in this contribution the terms "sentient" and "self-organizing" are used in a "weak" sense and are not meant to imply ontological identity with biological systems in general and human cognition in particular.

## 2. Sentient Buildings

A sentient building, as understood in this paper, involves the following constituents, Figure 2:

i) Occupancy – Inhabitants, users, and the visitors of the building;
ii) Components, systems – Physical constituents of building as a technical artifact (product);
iii) Self-organizing building model – The core representation of the building's context, components, systems, and processes. To be operationally effective, it is updated fairly autonomously based on pervasive sensor-supported data collection and algorithms for the interpretation of such data.
iv) Model-based executive unit – The decision-making agency of the sentient building. Simulation-based control strategies are part of this unit's repertoire of tools and methods for decision making support.

Depending on the specific configuration and the level of sophistication of a sentient building, occupants may directly manipulate the control devices or they may request from the executive unit desired changes in the state of the controlled entity. Likewise, the executive unit may directly manipulate control devices or suggest control device manipulations to the users.

## 3. Self-Organizing Models

### 3.1. REQUIREMENTS

To serve effectively as the representational core of a sentient building, a self-organizing model must fulfill at least two requirements. First, such a model must incorporate and integrate both a rather static building product view and a rather dynamic behavioral view of the building and its environmental systems. Second, to provide real-time building operation support, the model must be easily adaptable, i.e. it must respond to changes in occupancy, systems, and context of the building. Ideally, the model should detect and reflect such changes automatically, i.e. it must update itself autonomously.

### 3.2. BUILDING AS PRODUCT

Numerous representational schemes (product models) have been proposed to describe building elements, components, systems, and structures in a general and standardized manner. Thereby, one of the main motivations has been to facilitate hi-fidelity information exchange between agents involved in the building delivery process (architects, engineers, construction people, manufacturers, facility managers, users). An instance of such a representation or a shared building model, Figure 3, was developed in the course of the SEMPER project, a research effort toward the development of an integrated building performance simulation environment (Mahdavi 1999; Mahdavi et al. 1999a). Such a shared building model can be adapted as part of a self-organizing building model and provide, thus, a sentient building with the requisite descriptions of building elements, components, and systems.

### 3.3. PERFORMANCE AS BEHAVIOR

The representational stance of most building product models is decompositional and static. In contrast, simulation allows for the prediction of buildings' behavior over time and can, thus, provide a kind of dynamic representation. In order to suffice the requirements of a sentient building, the underlying representation must combine detailed building product information with building controls process modeling.

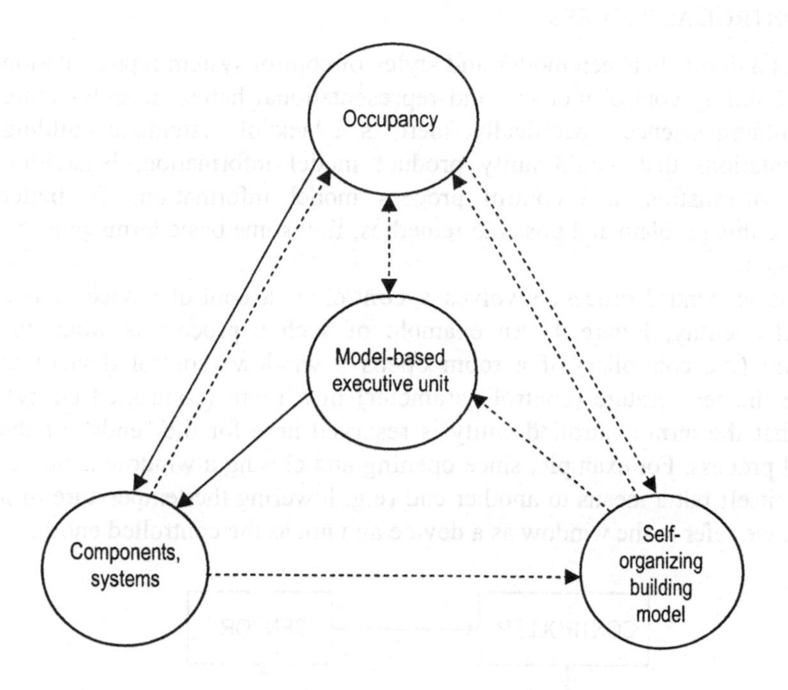


*Figure 2.* Scheme of the constitutive ingredients of a sentient building (continuous lines: control actions; dashed lines: information flows)

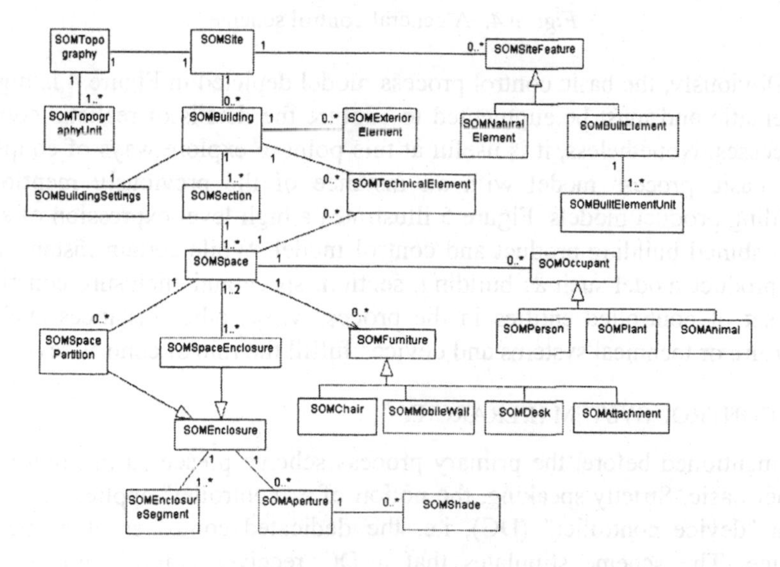


*Figure 3.* SEMPER's shared object model (SOM)

3.4. CONTROL AS PROCESS

There is a divide between modes and styles of control system representation in the building control industry and representational habits in architecture and building science. Specifically, there is a lack of systematic building representations that would unify product model information, behavioral model information, and control process model information. To better illustrate this problem and possible remedies, first some basic terms must be mentioned.

A basic control process involves a controller, a control device, and a controlled entity, Figure 4. An example of such a process is when the occupant (the controller) of a room opens a window (control device) to change the temperature (control parameter) in a room (controlled entity). Note that the term controlled entity is reserved here for the "ends" of the control process. For example, since opening and closing a window is not an end on itself but a means to another end (e.g. lowering the temperature of a space), we refer to the window as a device and not as the controlled entity.

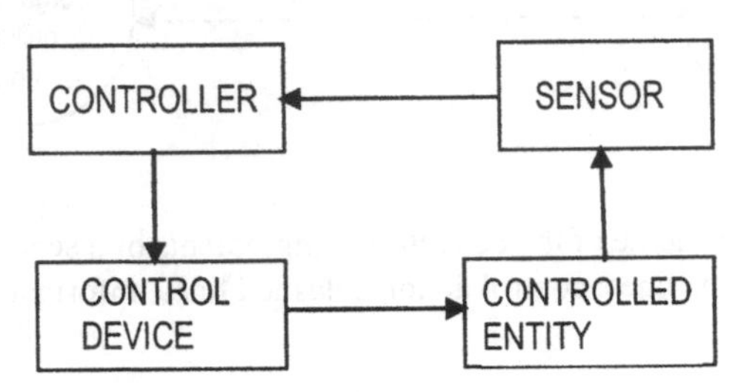


*Figure 4.* A general control scheme

Obviously, the basic control process model depicted in Figure 4 is highly schematic and must be augmented to capture the details of realistic control processes. Nonetheless, it is useful at this point to explore ways of coupling this basic process model with an instance of the previously mentioned building product models. Figure 5 illustrates a high-level expression of such a combined building product and control model. While certain instances of the product model such as building, section, space and enclosure constitute the set of controlled entities in the process view, other instances such as aperture or technical systems and devices fulfill the role of control devices.

3.5. CONTROL SYSTEM HIERARCHY

As mentioned before, the primary process scheme presented in Figure 5 is rather basic. Strictly speaking, the notion of a "controller" applies here only to a "device controller" (DC), i.e. the dedicated controller of a specific device. The scheme stipulates that a DC receives control entity's state information directly from a sensor, and, utilizing a decision-making functionality (e.g. a rule or an algorithm that encapsulates the relationship between the device state and its sensory implication), sets the state of the device. Real world building control problems are, however, much more complex, as they involve the operation of multiple devices for each environmental system domain and multiple environmental system domains (e.g., lighting, heating, cooling, ventilation).

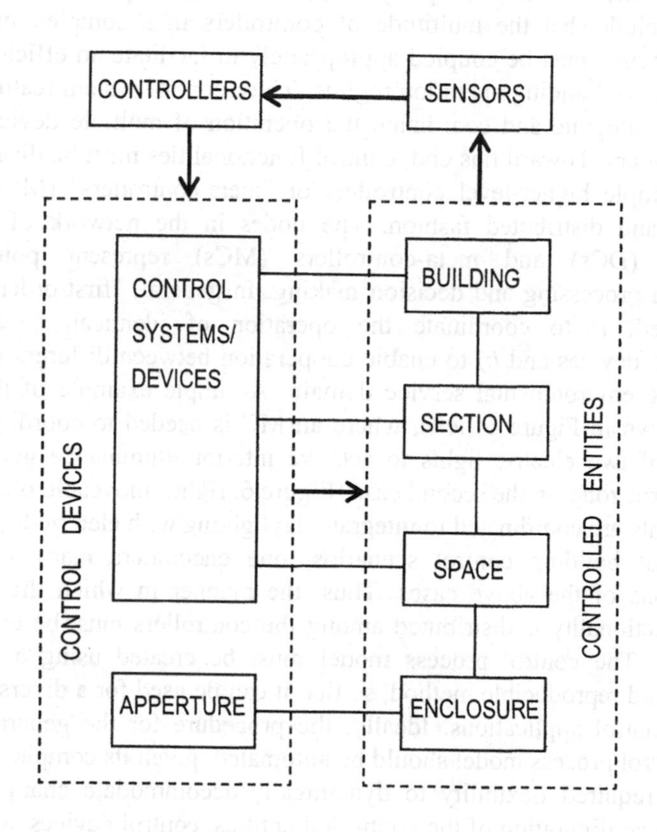


*Figure 5.* A high-level building product and control process scheme

As such, the complexity of building systems control could be substantially reduced, if distinct processes could be assigned to distinct (and autonomous) control loops. However, controllers for various systems and components are often interdependent. A controller may need the information from another controller in order to devise and execute control decisions. For example, the building lighting system may need information on the buildings thermal status (e.g. heating versus cooling mode) in order to identify the most desirable combination of natural and electrical lighting options.

Moreover, two different controllers may affect the same control parameter of the same impact zone. For example, the operation of the window and the operation of the heating system can both affect the temperature in a room. In such cases, controllers of individual systems cannot identify the preferable course of action independently. Instead, they must rely on a higher-level controller instance (i.e., a "meta-controller"), which can process information from both systems toward a properly integrated control response.

We conclude that the multitude of controllers in a complex building controls scheme must be coupled appropriately to facilitate an efficient and user-responsive building operation regime. Thus, control system features are required to integrate and coordinate the operation of multiple devices and their controllers. Toward this end, control functionalities must be distributed among multiple higher-level controllers or "meta-controllers" (MCs) in a structured and distributed fashion. The nodes in the network of device controllers (DCs) and meta-controllers (MCs) represent points of information processing and decision making. In general, "first-order" MCs are required: *i)* to coordinate the operation of identical, separately-controllable devices and *ii)* to enable cooperation between different devices in the same environmental service domain. A simple example of the first case is shown in Figure 6 (left), where an MC is needed to coordinate the operation of two electric lights to achieve interior illuminance goals in a single control zone. In the second case (Figure 6, right), moveable blinds and electric lights are coordinated to integrate daylighting with electric lighting.

In actual building control scenarios, one encounters many different combinations of the above cases. Thus, the manner in which the control system functionality is distributed among the controllers must be explicitly configured. The control process model must be created using a logical, coherent, and reproducible method, so that it can be used for a diverse set of building control applications. Ideally, the procedure for the generation of such a control process model should be automated, given its complexity, and given the required flexibility to dynamically accommodate changes over time in the configuration of the controlled entities, control devices, and their respective controllers.

### 3.6. AUTOMATED GENERATION OF CONTROL REPRESENTATION

We have developed and tested a set of constitutive rules that allow for the automated generation of the control system model (Mahdavi 2001a; 2001b). Such a model can provide a template (or framework) of distributed nodes which can contain various methods and algorithms for control decision making. Specifically, four model generation rules are applied successively to the control problem, resulting in a unique configuration of nodes that constitute the representational framework for a given control context. The first three rules are generative in nature, whereas rule 4 is meant to ensure the integrity of the generated model. The rules can be stated as follows:

i) Multiple devices of the same type that are differentially controllable and that affect the same sensor necessitate a meta-controller.
ii) More than one device of different types that affect the same sensor necessitates a meta-controller.
iii) More than one first-order meta-controller affecting the same device controller necessitates a second-order (higher-level) meta-controller.
iv) If in the process a new node has been generated whose functionality duplicates that of an existing node, then it must be removed. Any resulting isolated nodes must be re-connected.

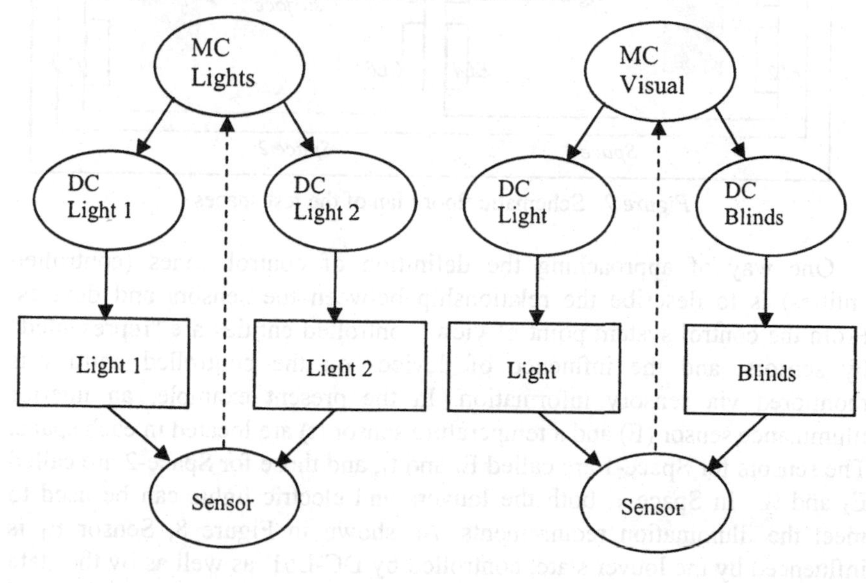


*Figure 6.* Left: Meta-controller for individually controllable identical devices; Right: Meta-controller for different devices addressing the same control parameter

The following example illustrates the application of these rules (Mertz and Mahdavi 2003). The scenario includes two adjacent rooms, each with four luminaires and one local heating valve, which share a set of exterior moveable louvers, Figure 7. Hot water is provided by the central system, which modulates the pump and valve state to achieve the desired water supply temperature. In each space, illuminance and temperature is to be maintained within the set-point range. This configuration of spaces and devices stems from an actual building, namely the Intelligent Workplace

(IW) at Carnegie Mellon University, Pittsburgh, USA (Mahdavi et al. 1999b).

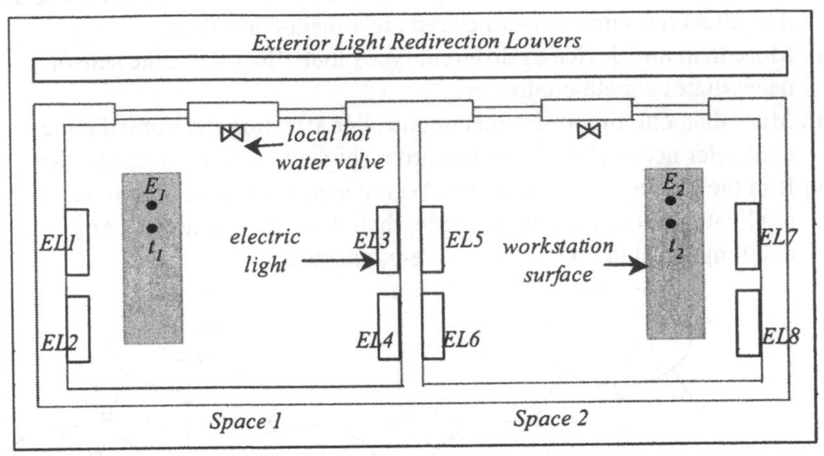


*Figure 7.* Schematic floor plan of the test spaces

One way of approaching the definition of control zones (controlled entities) is to describe the relationship between the sensors and devices. From the control system point of view, controlled entities are "represented" by sensors, and the influence of devices on the controlled entities is monitored via sensory information. In the present example, an interior illuminance sensor (E) and a temperature sensor (t) are located in each space. The sensors for Space-1 are called $E_1$ and $t_1$, and those for Space-2 are called $E_2$ and $t_2$. In Space-1, both the louvers and electric lights can be used to meet the illumination requirements. As shown in Figure 8, Sensor $E_1$ is influenced by the louver state, controlled by DC-Lo1, as well as by the state of four electric lights, each controlled by a DC-EL. Similarly, both the local valve state and the louver state influence the temperature in Space-1 ($t_1$). Analogous assumptions apply to Space-2.

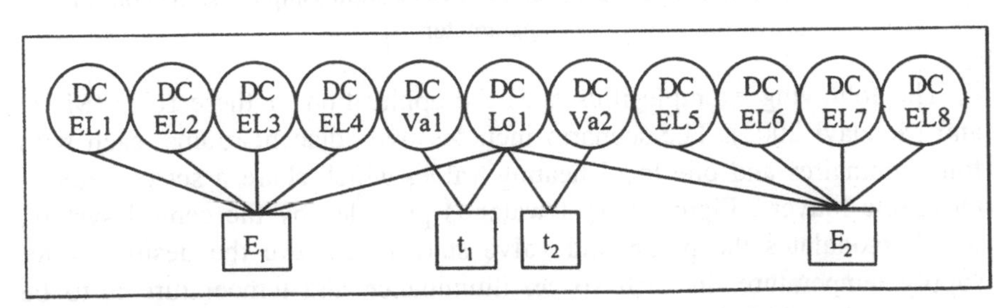


*Figure 8.* Association between sensors and devices in Figure 7

Once the control zones (controlled entities) have been defined, the generation rules can be applied to the control problem, resulting in the representation of Figure 9. As to the application of rule 1, four nodes, namely DC-EL1, EL2, EL3, and EL4 are of the same device type and all impact sensor $E_1$. Thus, an MC is needed to coordinate their action: MC-EL_1. Similarly, regarding the application of rule 2, both DC-Lo1 and DC-Va1 impact the temperature of Space-1. Thus, MC-Lo_Va_1 is needed to coordinate their action. As to rule 3, four MC nodes control the DC-Lo1 node. Thus, their actions must be coordinated by an MC of second order, namely MC-II EL_Lo_Va_1. In this example, Rules 1, 2, and 3 were applied to the control problem to construct the representation. Using this methodology, a scheme of distributed, hierarchical control nodes can be constructed. In certain cases, however, the control problem contains characteristics that cause the model not to converge toward a single top-level controller. In these cases, Rule 4 can be applied to ensure convergence.

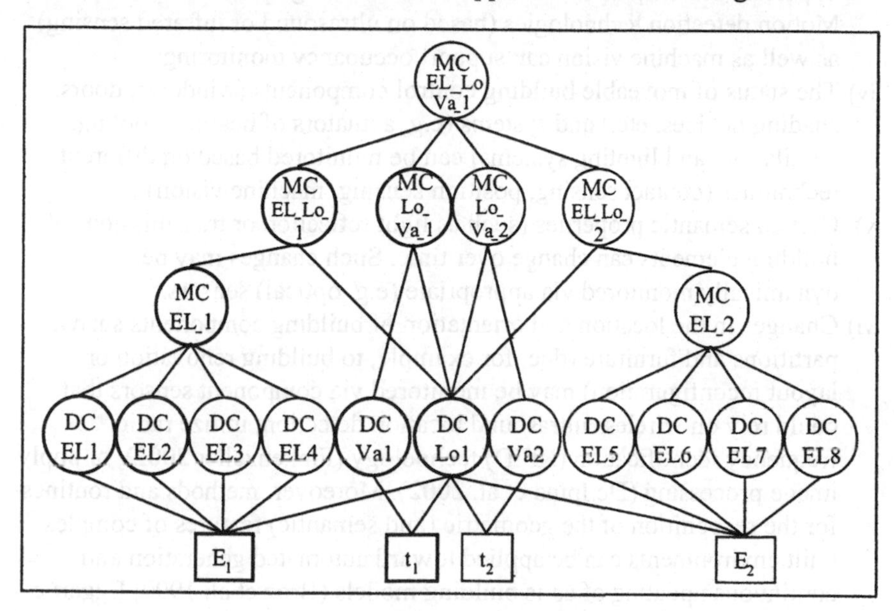


*Figure 9.* An automatically generated control model from Figures 7 and 8

## 3.7. REAL-TIME MODEL UPDATING

Once a building model is available with instances for building context, structure, systems, status, processes, and occupancy, it can be used to support the real-time building operation (building systems control, facility management, etc.). However, given the complexity of such a model, it needs to be self-organizing, i.e. it must maintain and update itself fairly autonomously. Depending on the type and the nature of the entity, system, or process to be monitored, various sensing technologies can be applied to continuously update the status of a building model:

i) Information about critical attributes of external micro-climate can be gained via a number of already existing sensor technologies (Mahdavi et al. 1999b; Wouters 1998). A compact and well-equipped weather station is to be regarded as a requisite for every sentient building.
ii) The success of indoor environmental control strategies can be measured only when actual values of target thermal and visual performance variables are monitored and evaluated. Further advances in this area are desirable, particularly in view of more cost-effective solutions for embodied high-resolution data monitoring infrastructures.
iii) Knowledge of the presence and activities of building occupants is important for the proper functionality of building operation systems. Motion detection technologies (based on ultrasound or infrared sensing) as well as machine vision can support occupancy monitoring.
iv) The status of moveable building control components (windows, doors, shading devices, etc.) and systems (e.g. actuators of heating, cooling, ventilation, and lighting systems) can be monitored based on different techniques (contact sensing, position sensing, machine vision).
v) Certain semantic properties (such as light reflection or transmission) of building elements can change over time. Such changes may be dynamically monitored via appropriate (e.g. optical) sensors.
vi) Changes in the location and orientation of building components such as partitions and furniture (due, for example, to building renovation or layout reconfiguration) may be monitored via component sensors that could rely on wireless ultrasound location detection, utilize radio frequency identification (RFID) technology (Finkenzeller 2002), or apply image processing (De Ipina et al. 2002). Moreover, methods and routines for the recognition of the geometric (and semantic) features of complex built environments can be applied toward automated generation and continuous updating of as-is building models (Broz et al. 1999; Eggert et al. 1998; Faugeras et al. 1998).

## 4. A Simulation-Based Control Strategy

### 4.1. INTRODUCTION

We argued that the nodes in the network of DCs and MCs in a building's control scheme represent points of information processing and decision making. An important challenge for any building control methodology is to find effective methods of knowledge encapsulation and decision making in such nodes. There are various ways of doing this. The simulation-based control method is discussed below (Mahdavi 2001b; 1997). This method can be effectively applied, once the main requirement for the realization of a sentient building is met, namely the presence of a unified building product and process model that can update itself dynamically and autonomously.

The use of performance simulation tools for building design support has a long tradition. The potential of performance simulation for building control support is, however, less explored. Conventional control strategies may be broadly said to be "reactive". A thermostat is a classical example: The state of a control device (e.g. a heating system) is changed incrementally in reaction to the measured value of a control parameter (e.g. the room temperature). Simulation-based strategies may be broadly characterized as "proactive". In this case, a change in the state of a control device is decided upon based on the consideration of a number of candidate control options and the comparative evaluation of the simulated outcomes of those options.

### 4.2. APPROACH

Modern buildings allow, in principle, for multiple ways to achieve desired environmental conditions. For example, to provide a certain illuminance level in an office, daylight, electrical light, or a combination thereof can be used. The choice of the system(s) and the associated control strategies represent a non-trivial problem since there is no deterministic procedure for deriving a necessary (unique) state of the building's control systems from a given set of objective functions (e.g., desirable conditions for the inhabitants, cost-effectiveness of the operation, minimization of environmental impact).

Simulation-based control has the potential to provide a remedy for this problem. Instead of a direct mapping attempt from the desirable value of an objective function to a control systems state, the simulation-based control adopts an "if-then" query approach. In order to realize a simulation-based building systems control strategy, the building must be supplemented with a multi-aspect virtual model that runs parallel to the building's actual operation. While the real building can only react to the actual contextual conditions (e.g., local weather conditions, sky luminance distribution patterns), occupancy interventions, and building control operations, the simulation-based virtual model allows for additional operations: *a)* the virtual model can move backward in time so as to analyze the building's past behavior and/or to calibrate the program toward improved predictive potency; *b)* the virtual model can move forward in time so as to predict the building's response to alternative control scenarios. Thus, alternative control schemes may be evaluated and ranked according to appropriate objective functions pertaining to indoor climate and occupancy comfort, as well as environmental and economic considerations.

### 4.3. PROCESS

To describe the simulation-based control process in simple terms, we shall consider four process steps, Table 1:

i) The first step identifies the building's control state at time $t_i$ within the applicable control state space (i.e. the space of all theoretically possible control states). For clarity of illustration, Table 1 shows the control state space as a 3-dimensional space. However, the control state space has as many dimensions as there are distinct controllable devices in a building.

ii) The second step identifies the region of the control state space to be explored in terms of possible alternative control states at time $t_{i+1}$.

iii) The third step involves the simulation-based prediction and comparative ranking of the values of pertinent performance indicators for the corpus of alternative identified in the second step.

iv) The fourth step involves the execution of the control action, resulting in the transition of control state to a new position at time $t_{i+1}$.

### 4.4. AN ILLUSTRATIVE EXAMPLE

To further illustrate the steps introduced in section 4.3, we consider a demonstrative experiment regarding daylight control in an office space in the previously mentioned Intelligent Workplace (see Figure 7). About 60% of the external wall of this space consists of glazing. The facade system includes a set of three parallel external moveable louvers, which can be used for shading and – to a certain degree – for light redirection. These motorized louvers can be rotated anti-clockwise from a vertical position up to an angle of 105°. As an initial feasibility test of the proposed simulation-based control approach, we considered the problem of determining the "optimal" louver position (Mahdavi 2001b).

*Step 1* – In this simple case, the control state space has just one dimension, i.e. the position of the louver. We further reduced the size of this space, by allowing only four discrete louver positions, namely 0° (vertical), 30°, 60°, and 90° (horizontal).

*Step 2* – Given the small size of the control state space in this case, we considered all four possible louver positions as candidate options.

*Step 3* – The lighting simulation application LUMINA (Pal and Mahdavi 1999) was used for the prediction of light levels in the test space. In the present case, measured outdoor irradiance values were used at every time step to generate the sky model in LUMINA for the subsequent time step. However, trend forecasting algorithms could be used to predict outdoor conditions for future time steps. For each time step the simulation results (mean illuminance and uniformity levels on a horizontal plane approximately 1 m above the floor) were used to rank and select the most desirable control scenario based on two illustrative objective functions:

TABLE 1. Schematic illustration of the simulation-based control process

| | | |
|---|---|---|
| Step 1 | | Control state at time $t_i$ |
| Step 2 | | Identification of candidate control states for time $t_{i+1}$ |
| Step 3 | | Simulation-based determination and evaluation of the performance implications of the control options identified in step 2 |
| Step 4 | | Transition to the most desirable control state at time $t_{i+1}$ |

The first function aims at minimizing the deviation of the average (daylight-based) illuminance level $E_m$ from a user-defined target level $E_t$.

The second objective function aims at maximizing the uniformity (Mahdavi and Pal 1999) of the illuminance distribution. At time interval $t_i$, the simulation tool predicted for four candidate louver positions the expected interior illuminance levels for the time interval $t_{i+1}$ (test space geometry and photometric properties, as well as the outdoor measurements at time interval $t_i$ were used as model input). Based on the simulation results and objective functions, it was possible to determine for each time step the louver position which was considered most likely to maximize the light distribution uniformity or to minimize the deviation of average illuminance from the target value.

*Step 4* – Device controller instructed the control device (louver) to assume the position identified in step 3 as most desirable.

## 4.5. CHALLENGES

### *4.5.1. Introduction*

In previous sections we described the simulation-based strategy toward building systems control and how this approach, supported by a self-organizing building model, could facilitate the operation of a sentient building. The practical realization of these methods and concepts, however, requires efficient solutions for various critical implementation issues.

There are two basic problems of the proposed approach which we briefly mention but will not pursue in detail, as they are not specific to simulation-based control methodology but represent basic problems related to simulation methods and technologies in general:

First, the reliability of simulation algorithms and tools is always subject to validation, and this has been shown to be a difficult problem in the building performance simulation domain. In the context of sentient building implementations, there is an interesting possibility to improve on the predictive capability of the simulation applications by "on-line" calibration of simulation results. This can be done by continuous real-time monitoring of the performance indicator values (using a limited number of strategically located sensors) and comparing those with corresponding simulation results. Using the results of this comparison, appropriate correction factors may be derived based on statistical methods and neural network applications.

Second, preparation of complete and valid input data (geometry, materials, system specifications) for simulation is often a time-consuming and error-prone task. In the context of self-organizing models, however, such data would be prepared mostly in an automated (sensor-based) fashion, thus reducing the need for human intervention toward periodic updating of simulation models.

In the following discussion, we focus on arguably the most daunting problem of the simulation-based control strategy, namely the rapid growth of the size of the control state space in all those cases where a realistic number of control devices with multiple possible positions are to be dealt with. Consider a space with *n* devices that can assume states from $s_1$ to $s_n$. The total number, z, of combinations of these states (i.e. the number of necessary simulation runs at each time step for an exhaustive modeling of the entire control state space) is given by:

$$z = s_1 \cdot s_2 \cdot \ldots \cdot s_n \tag{1}$$

This number represents a computationally insurmountable problem, even for a modest control scenario involving a few spaces and devices. To address this problem, multiple possibilities must be explored, whereby two general approaches offer themselves, namely, *i)* the reduction of the size of the control state space region to be explored; *ii)* the acceleration of the computational assessment of alternative control options.

### *4.5.2. The Control State Space*

At a fundamental level, a building's control state space has as many dimensions as there are controllable devices. On every dimension, there are as many values as there are possible states of the respective device. This does not imply, however, that at every time step the entire control state space must be subjected to predictive simulations.

The null control state space – Theoretically, at certain time steps, the size of the applicable control state space could be reduced to zero. Continuous time-step performance modeling is not always necessary. As long as the relevant boundary conditions of systems' operation have remained either unchanged or have changed only insignificantly, the building may remain in its previous state. Boundary conditions denote in this case factors such as outdoor air temperature, outdoor global horizontal irradiance, user request for change in an environmental condition, dynamic change in the utility charge price for electricity, etc.

Limiting the control state space – Prior to exhaustive simulation of the theoretically possible control options, rules may be applied to reduce the size of the control state space to one of practical relevance. Such rules may be based on heuristic and logical reasoning.

Compartmentalization – The control state space may be structured hierarchically, as we saw in section 3. This implies a distribution of control decision making across a large number of decision making nodes. At every level, a control node accesses the control alternatives beneath and submits a ranked set of recommendations above. For this purpose, different methods may be implemented in each node, involving rules, tables, simulations, etc.

Simulation routines thus implemented, need not cover the whole building and all the systems. Rather, they need to reflect behavioral implications of only those decisions that can be made at the level of the respective node.

"Greedy" navigation and random jumps – Efficient navigation strategies can help reduce the number of necessary parametric simulations at each time step. In order to illustrate this point, consider the following simple example. Let D be the number of devices in a building and P the number of states each device can assume. The total number z of resulting possible combinations (control states) is then given by equation 2.

$$z = P^D \tag{2}$$

To reduce the size of the segment of the control state space to be explored, one could consider, at each time step, only three control states for each device, namely the status quo, the immediate "higher" state, and the immediate "lower" state. While this would mean a sizable reduction of the number of simulation, it is still too high to be of any practical relevance. Thus, to further reduce the number of simulations, we assume the building to be at control state A at time $t_1$. To identify the control state B at time $t_2$, we scan the immediate region of the control state space around control state A. This we do by moving incrementally "up" and "down" along each dimension, while keeping the other coordinates constant. Obviously, the resulting number of simulations in this case is given by:

$$z = 2D+1 \tag{3}$$

This would result obviously in a much more manageable computational load. However, this "greedy" approach bears the risk that the system could be caught in a performance state corresponding to a local minima (or maxima). To reduce this risk, stochastically-based excursions to the more remote regions of the control state space can be undertaken (Mahdavi and Mahattanatawe 2003).

*4.5.3. Efficient Assessment of Alternative Control Options*

Our discussions have so far centered on performance simulation as the main instrument to predict the behavior of a building in response to alternative control actions. The advantage of simulation is that it offers the possibility of an explicit analysis of various forces that determine the behavior of the building. The downside is that detailed simulation is computationally expensive. We now briefly discuss some possible remedies.

Customized local simulation – As mentioned earlier, simulation functionality may be distributed across multiple control nodes. These distributed simulation applications can be smaller and run faster, thus reducing the overall computational load of the control system.

Simplified simulation – The speed of simulation applications depends mainly on their algorithmic complexity and modeling resolution. Simpler models and simplified algorithms could reduce the computational load. Overt simplification could of course reduce the reliability of predictions.

Simulation substitutes – In certain cases detailed simulation applications may be replaced by computationally more efficient regression models or neural network copies of simulation applications. Regression models are derived based on multiple runs of detailed simulation programs and the statistical processing of the results. Likewise, neural networks may be trained by data generated by simulation programs. Such modeling techniques lack the flexibility of explicit simulation, but, if properly engineered, can match the predictive power of detailed simulation algorithms. Multiple prototypical designs of hybrid control systems that utilize both simulation and machine learning have been designed and successfully tested (Chang and Mahdavi 2002). Rules represent a further class of – rather gross – substitutes for simulation-based behavioral modeling. In certain situations, such rules could define the relationship between the state of a device and its corresponding impact on the state of the sensor. Rules can be derived based on experience, measured data, or logical reasoning (Mahdavi 2001b).

## 5. Conclusion

This paper described the concepts of sentient buildings and self-organizing building models. It explored the possibility of integrating primary models of buildings' composition and behavior in higher-level building control systems. It demonstrated how computational performance simulation codes and applications could become an integral part of the methodological repertoire of building operation and controls systems, such that a larger set of indoor environmental performance indicators and a richer set of indoor-environmental control options could be considered. An important feature of the proposed model-based strategy lies in its potential to support concurrent consideration of multiple control agenda. Thus, complex control strategies could be formulated to simultaneously address habitability, sustainability, and feasibility requirements in building operation.

## Acknowledgment

The research presented in this paper is supported in part by FWF (Austrian Science Foundation), Grant number: P15998-N07.

## References


Broz, V, Carmichael, O, Thayer, S, Osborn, J and Hebert, M: 1999, ARTISAN: An integrated scene mapping and object recognition system, *American Nuclear Society 8th Intl. Topical Meeting on Robotics and Remote Systems*, American Nuclear Society.

Chang, S and Mahdavi, A: 2002, A hybrid system for daylight responsive lighting control, *Journal of the Illuminating Engineering Society* **31**(1): 147 - 157.

De Ipina, DL, Mendonca, P and Hopper, A: 2002, TRIP: A low-cost vision-based location system for ubiquitous computing, *Personal and Ubiquitous Computing Journal* **6**(3): 206-219.

Eggert D, Fitzgibbon A and Fisher, R: 1998, Simultaneous registration of multiple range views for use in reverse engineering of CAD models, *Computer Vision and Image Understanding* **69**: 253-272.

Faugeras O, Robert L, Laveau S, Csurka G, Zeller C, Gauclin C and Zoghlami I: 1998, 3-D reconstruction of urban scenes from image sequences, *Computer Vision and Image Understanding* **69**: 292-309.

Finkenzeller, K: 2002, RFID-Handbuch, Hanser, ISBN 3-446-22071-2.

Mahdavi, A: 2001a, Aspects of self-aware buildings, *International Journal of Design Sciences and Technology* **9**(1): 35-52.

Mahdavi, A: 2001b, Simulation-based control of building systems operation, *Building and Environment* **36**(6): 789-796.

Mahdavi, A: 1999, A comprehensive computational environment for performance based reasoning in building design and evaluation, *Automation in Construction* **8**: 427 – 435.

Mahdavi, A: 1997, Toward a simulation-assisted dynamic building control strategy, *Proceedings of the Fifth International IBPSA (International Building Performance Simulation Association) Conference* **1**: 291 - 294.

Mahdavi, A and Mahattanatawe, P: 2003, A computational environment for performance-based building enclosure design and operation *in* Carmeliet, Hens, and Vermeir (eds), *Research in Building Physics,* Swets & Zeitlinger, Lisse, The Netherlands, pp. 815 – 824.

Mahdavi, A and Pal, V: 1999, Toward an entropy-based light distribution uniformity indicator, *Journal of the Illuminating Engineering Society* **28**(1): 24 - 29.

Mahdavi, A, Ilal, ME, Mathew, P, Ries, R, Suter, G and Brahme, R: 1999a, The architecture of S2, *Proceedings of Building Simulation '99, Sixth International IBPSA Conference,* Kyoto, Japan, Vol. III, pp. 1219 - 1226.

Mahdavi, A, Cho, D, Ries, R, Chang, S, Pal, V, Ilal, E, Lee, S and Boonyakiat, J: 1999b, A building performance signature for the "intelligent workplace": Some preliminary results, *Proceedings of the CIB Working Commission W098 International Conference: Intelligent and Responsive Buildings*, Brugge, pp. 233 - 240.

Mertz, K and Mahdavi, A: 2003, A representational framework for building systems control, *Proceedings of the Eight International IBPSA Conference*, Eindhoven, Netherlands, Vol. 2, pp. 871 – 878.

Pal, V and Mahdavi, A: 1999, Integrated lighting simulation within a multi-domain computational design support environment, *Proceedings of the 1999 IESNA Annual Conference*, New Orleans, pp. 377 - 386.

Wouters, P: 1998, Diagnostic techniques, *in* Allard (ed), *Natural Ventilation in Buildings; A Design Handbook*, James & James.

JS Gero (ed), *Design Computing and Cognition'04*, 613-632


# THREE R'S OF DRAWING AND DESIGN COMPUTATION

*A Drawing Centered View of Design Process*

MARK D GROSS, ELLEN YI-LUEN DO
*University of Washington, USA*

**Abstract.** A drawing centered view of design process focuses on the interplay between designer expertise, domain knowledge and media manipulation. We report on our computational sketching software systems that support design recording, reasoning, and resolving.

## 1. Introduction

### 1.1. MOTIVATION AND RELATED WORK

Ivan Sutherland's Sketchpad system first demonstrated the power and promise of interactive, intelligent, pen based graphics to support engineering design (Sutherland 1963). Among other innovative ideas, Sketchpad employed what is now known as object-oriented programming, graphics scene-graph hierarchy, interaction with a stylus, on-screen menus, two-handed interaction, and a relaxation-based constraint solver to maintain user-specified design constraints on a drawing. In addition to these technological innovations, Sketchpad demonstrated how computers might support semantically attached interactive design drawing, yet for many years little further work was done.

Several factors fueled a resurgence of interest in pen-based computing, beginning in the 1990s. Low cost LCD screens and tablet-stylus technology led to pen-based PDAs such as the Apple Newton, Palm Pilot, and others, as well as early pen computers and recently the Tablet PC. This new hardware combined with advances in pattern recognition rekindled interest in drawing-based systems for design. Over the past few years a great deal of new work has appeared on sketch design systems.

Many have noted the affordances of sketching in design. Sketching, for example, allows quick exploration of ideas at a high level of abstraction, avoids early commitment to a particular solution, allowing many alternatives to be explored (Fish and Scrivener 1990; Ullman, Wood et al. 1990). Some experienced designers postpone using computer-aided design tools because they feel that these tools require a degree of precision and commitment that is inappropriate for the early phases of design. They prefer to develop conceptual design sketches with pencil and paper, moving to computer-aided tools at a later stage of designing (Cross et al. 1996; Suwa and Tversky 1997).

Many systems aim to understand drawing to provide design feedback. For example, SILK enables a designer to sketch and test a user interface design prototype (Landay and Myers 2001). EsQUIsE interprets architectural sketches for design performance evaluation (Leclercq and Juchmes 2002). An increasing number of researchers consider sketch understanding a knowledge-based task. They argue that computers can provide domain knowledge for the task through sketching. For example, the recognition of device symbols and connections among them in a motor driven conveyor system diagram could infer motion by reasoning from domain knowledge (Kurtoglu and Stahovich 2002). The ASSIST mechanical engineering sketch design system can simulate and create movies of mechanical movements such as a car rolling down a hill (Davis 2002). Forbus et al, working on military "course of action" diagrams minimize sketch recognition's role. They argue that it is difficult for computers and easy for people, and the same information can be obtained from other sources such as speech and deictic references (Forbus, et al. 2001). Their system therefore attempts only basic recognition of spatial pointers (arrows) combined with speech recognition.

There is also a growing recognition of the value of ambiguity and tolerance for multiple interpretations (Mankoff et al. 2000; Gaver et al. 2003). Saund's "perceptual sketching" systems (Saund and Moran 1994; Saund et al. 2003) identify patterns in the lines of a sketch that enable users to select and work with the (emergent) objects that they see—regardless of how the sketch was initially constructed. Oviatt & Cohen's QuickSet system (Oviatt and Cohen 2000) combined sketch and speech recognition, showing that cues from sketch recognition can improve speech recognition performance, and vice-versa. Other work on computational support for drawing focuses on using sketches to produce more refined three-dimensional models (Igarashi and Hughes 2001; Contero González et al. 2003).

### 1.2. DESIGN: EXPERTISE, KNOWLEDGE, AND MEDIA

Observing designers at work, we find them simultaneously engaged in three rather different kinds of tasks. At a practical level they work with various media, for example making drawings and models and using these media in communicating with others about the design. They engage domain knowledge, for example making predictions and exercising judgments about how a design will perform. And designers allocate their time to different tasks in the design process, at some points searching for information, at other points generating new ideas, testing and evaluating alternatives, communicating with colleagues and clients, or developing basic concepts into specific detailed proposals. Hence we propose to look at design in terms of three distinct capacities: as the manipulation of different *media* with *domain knowledge*, governed by *design expertise*.

This understanding implies three fundamental components of computational support for design. First, computational tools must support the media that designers use to create, edit, view and review, and exchange designs and design ideas. These media include drawings and text as well as speech, audio, video, gesture, and collage. Designers are trained in the use of media to create, capture, consider, and convey ideas. Although conventional computer media for design, i.e., CAD drafting and modeling, has primarily focused on hard-line drawings and three-dimensional geometry, a great deal of creative work in design is actually carried out in the medium of freehand sketches and diagrams.

Second, computational tools must integrate domain knowledge into the design decision-making process. Designers know a lot about their domains, and they exercise a great deal of intelligence in making design decisions. On occasion, they refer to handbooks, case studies or well-known examples, previous similar projects; they call for technical analyses of projected design performance; they seek expert advice. Computational support for domain knowledge in design has been weighted heavily toward end-stage analysis and evaluation based on detailed design representations constructed with structured editors, rather than sketches and freehand drawings.

Third, computational tools must embody expertise in managing the design process. That is, independent of any specific design domain, the tools must help a designer manage the design process — ideation and retrieval of relevant precedents, generation and evaluation of alternatives, consulting experts, balancing stakeholder values, etc. Our work so far has concentrated on the relationship of design media and domain knowledge, and most of the work outlined here is in this territory. We return to consider the question of design expertise in the final section of this paper.

### 1.3. DRAWING—THE CENTRAL REPRESENTATION IN DESIGN

We take a drawing-centered view of designing. In domains such as mechanical engineering and architectural design where the product is a physical object, the drawing is typically the single representation that the designer uses throughout the designing process, from initial rough sketch to final fabrication drawing. Although other representations (such as specifications, component lists, and schedules) also play roles, the drawing remains the focus of design activity.

Our view of design is colored in the first place and most strongly by our own training and experience as practicing designers and by first-hand observation of colleagues and students in architecture schools where we have worked and in other disciplines such as mechanical engineering. Although we have done a few studies of design activity (Do et al. 2001) we largely rely on empirical work by Goldschmidt, Tversky, and others for insights about the cognitive roles and functions of drawing in design (Goldschmidt 1994; Schön and Wiggins 1992; Tversky 2002).

### 1.4. THREE R'S OF DRAWING AND DESIGN

Figure 1 outlines various activities that designers carry out using drawing. We group these activities into three categories (the three R's of design): recording, reasoning, and resolving. *Recording* activities concern design and knowledge capture—from the user's pen input at the stroke level to capturing speech and text during designing, including design rationale and other annotations that play a role in collaboration and negotiation. *Reasoning* activities engage domain knowledge in various ways—deducing the performance behavior of a proposed design, retrieving relevant design precedents or cases from a library or database, stimulating creative thought through reasoning-by-analogy, and so on. The third R, *Resolving*, concerns the development of drawings over the design process from rough, sketchy, and abstract representations to specific, definite, and well-defined design proposals.

### 1.5. OUR MULTIFARIOUS EFFORTS TO SUPPORT DESIGN DRAWING

Over the past decade we have engaged in a research program and developed computational support to investigate various aspects of designing. We began with a recognition-based sketching program written in Lisp called the Electronic Cocktail Napkin, which formed the platform for several initial forays into the territory. These efforts focused principally on how to recognize and interpret freehand drawings and extract and embed semantic content. We later extended this work with several components focused on other aspects of designing. For the most part, we developed each module separately, following the exigencies of student interest and funding. The result is a collection of apparently diverse proof-of-concept software projects published in disparate venues. Taken together they outline the principal components of a drawing-centered design system. The current paper assembles this body of work in relation to our view of design and decision support. In the interest of providing a larger picture, we omit technical details, which can be found in the reports we cite on the individual projects. We provide a commentary and framework to relate all these software systems in the context of drawing and design computation.

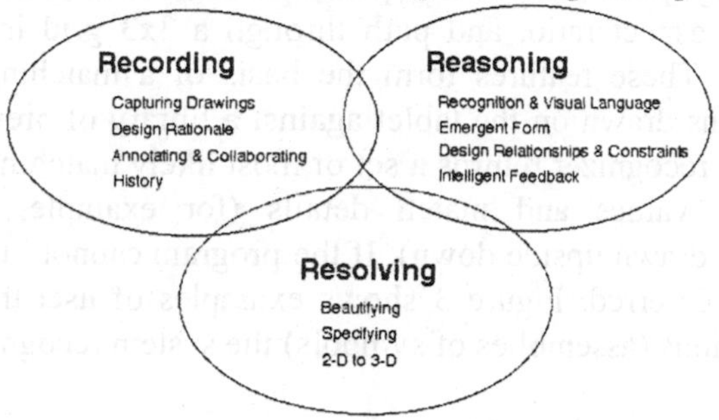


*Figure 1.* Computational sketching systems must support diverse design activities

We feel compelled to point out that although most of the examples in the work described below are from architecture, the model of designing and the systems we have built are quite domain independent. We would be pleased to see computer support for architectural design built along the lines we outline, but the strategies presented here could be incorporated into a wide variety of design support systems.

## 2. Drawing Tools to Support the Three R's of Design Activities

### 2.1. RECORDING – KNOWLEDGE CAPTURE

The capture and management of drawings is central, but important additional information about a design often accompanies drawings in various media—speech, text, photographs, even physical samples of materials—and at various levels of formal representation.

Recording activities, Figure 2, concern design and knowledge capture. At the most basic level we record the user's pen input strokes, pressure and speed using the Electronic Cocktail Napkin as a platform for diagram parsing and understanding. The Design Amanuensis and Design Recorder projects extended the capture to include speech and text and provide synchronized history for reply and search. To support shared drawing, annotation of design rationale, collaboration and negotiation, we developed the Immersive Redliner and Space Pen projects for design drawing in 3-D and also synchronous shared whiteboard drawing with design history in NetDraw.

#### *2.1.1. Capturing Strokes and Recognizing Drawing Elements*

The Electronic Cocktail Napkin (Gross 1996) captures a designer's drawing strokes from a tablet. The program stores raw data as a time-stamped sequence of tuples (x, y, and pressure) which it analyzes to extract features such as drawing speed, corners and points of inflection, direction, bounding box, size, and aspect ratio, and path through a 3x3 grid inscribed in the bounding box. These features form the basis of a matching scheme that compares glyphs drawn on the tablet against a library of previously trained templates. The recognizer returns a set of most likely matching glyphs along with certainty values and match details (for example, a figure was recognized but drawn upside down). If the program cannot identify a figure, recognition is deferred. Figure 3 shows examples of user-trained symbols and configurations (assemblies of symbols) the system recognizes.

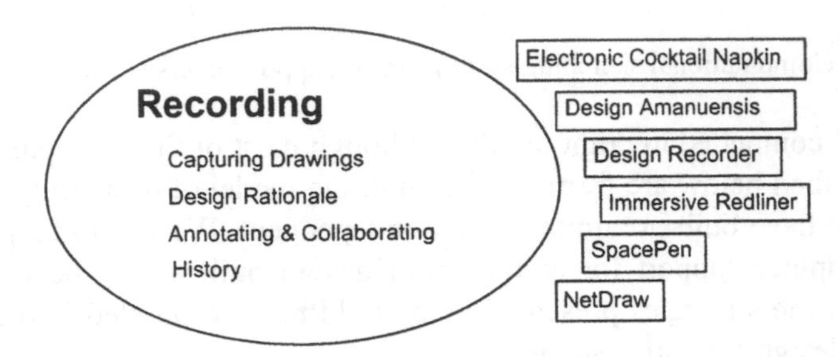


*Figure 2.* Recording activities concern design and knowledge capture

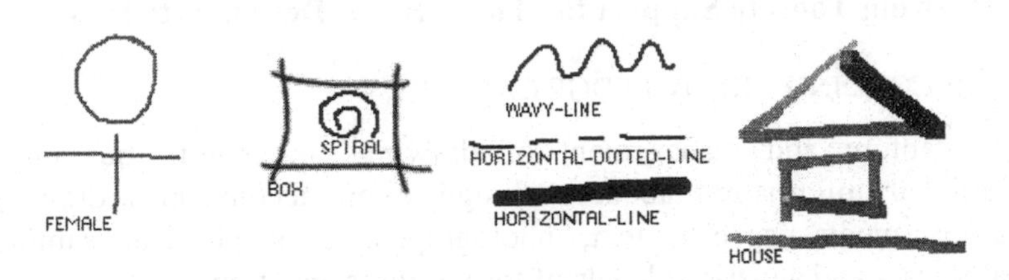


*Figure 3.* Electronic Cocktail Napkin recognizes basic symbols and configurations

*2.1.2. Recording Multi-Modal Design Conversations*

The Design Amanuensis combines the Napkin's stroke-capture with a continuous speech-to-text recognizer, storing the multimodal design conversation for later random-access playback and analysis (Gross et al. 2001). The text transcript enables search for key phrases ("where in the designing did wc talk about 'column size?' "). The search finds items in the spoken and text record, and highlights the drawing elements that were made at the time. The Napkin also enables search for particular figures—which similarly are linked through their timestamps to the associated text and speech in the design record. Figure 4 shows the drawing, control panel, and timeline views of the design record.

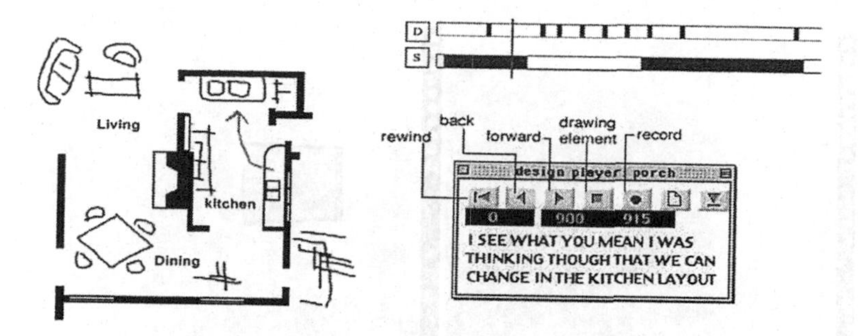


*Figure 4.* Design Amanuensis offers recording, playback, and search of the graphical and spoken design conversation

The Design Recorder is a later implementation that uses the Tablet PC to capture speech and drawing input from distributed collaborating designers. Figure 5 shows the chunking of speech and text displayed as selectable icons along a timeline.

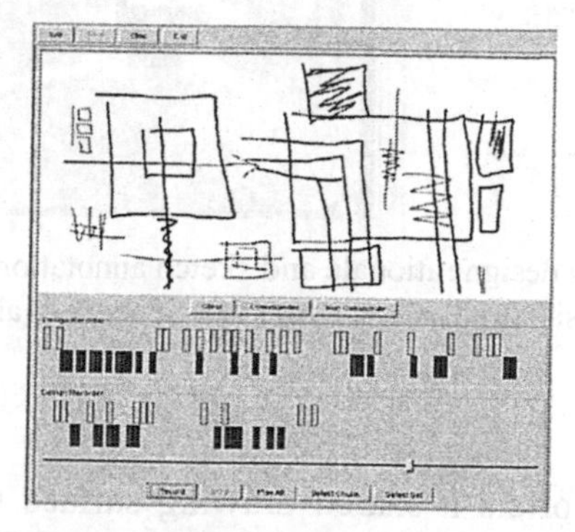

*Figure 5.* Design recorder: above: design drawing of two collaborating designers; below: 'chunks' of speech and drawing

### *2.1.3. Recording and Embedding Rationale in Designs*

The SpacePen (and a predecessor project, the Immersive Redliner) explored design annotation in three-dimensional models, Figure 6. Collaborating designers (or other stakeholders) sketch and attach text annotations to mark up a shared three dimensional model (Jung et al. 2002). A designer can propose adding a window by drawing directly on an existing wall in the 3-D model, Figure 6 top. The system recognizes symbols such as arrows and rectangles. Sketched lines and text annotations appear to subsequent viewers of the model. The system stores design critiques and rationale expressed as drawings or text which are located in the spatial context where they apply.

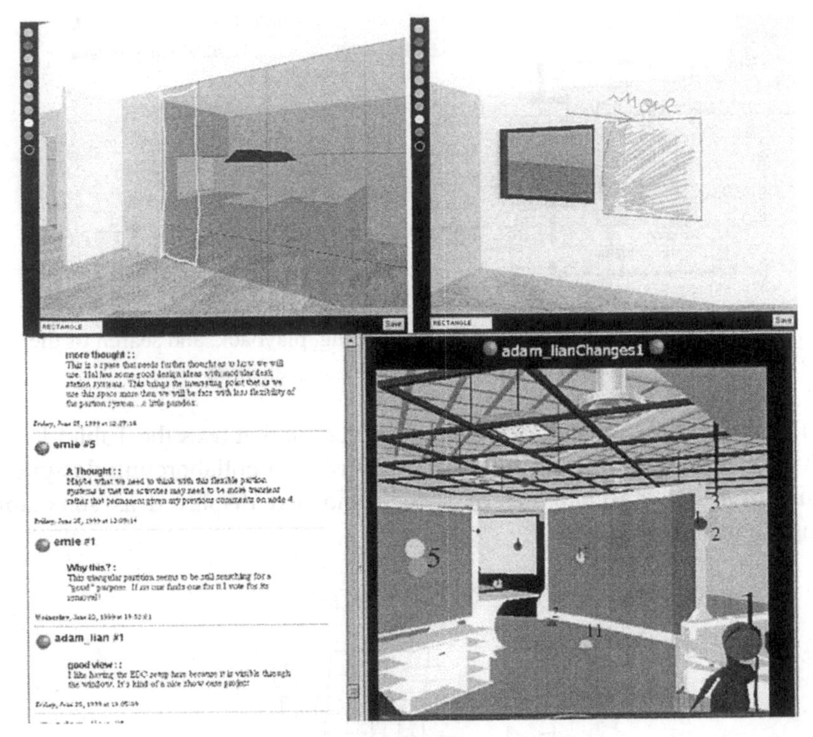


*Figure 6.* Recording design rationale and sketch annotations in a 3-D model – Immersive Redliner (below) and SpacePen (above)

### *2.1.4. Design History*

NetDraw, Figure 7, offers a shared drawing surface that enables several designers to work together simultaneously (Qian and Gross 1999). We also used NetDraw to explore several ideas related to synchronous collaboration. NetDraw stores and displays a history of the designing, so that a designer can return to a previous state and proceed to design from that point. NetDraw also offers an *ephemeral gesture* feature whereby a designer can sketch temporary marks over the design. These marks, useful for deictic references ("here is the main circulation path through the building") appear momentarily on other designers' drawings, then gradually fade away.

## 2.2. REASONING – DOMAIN SEMANTICS WITH INTELLIGENT SYSTEMS

The work described in the previous section treats drawings and attendant information simply as graphical data—one might say, as informal (human-readable) design representations. But much of our interest in drawing as a design representation hinges on knowledge that designers embed in drawings and the reasoning they apply to work on and with drawings. That is, we are interested in the intelligent processes of which design drawing is a part, and how we can use knowledge based computational systems to enhance design drawing.

Design drawings embody a designer's understanding of a domain, assumptions, and decisions about the design; a computer program can use recognition to extract this information and then reason about it.

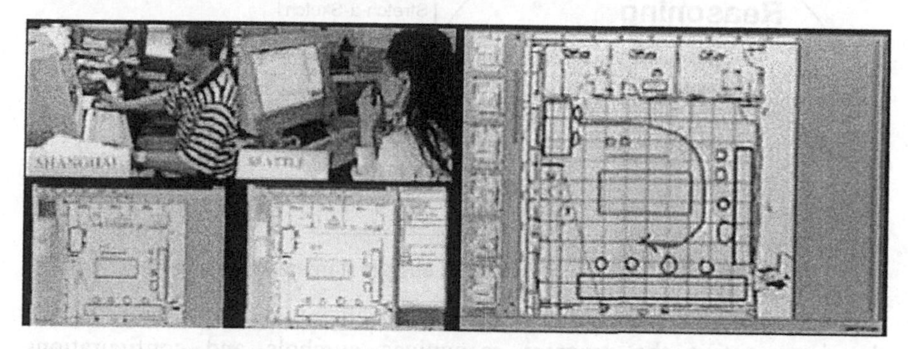

*Figure 7.* NetDraw: design history, ephemeral arrow gesture for deictic reference

Reasoning activities in design engage domain knowledge in various ways–including the performance behavior of a proposed design, retrieving relevant design precedents or cases from a library or database, stimulating creative thought through reasoning-by-analogy, and so on. We have implemented several software systems to support design reasoning in the form of connecting intelligent systems of domain knowledge to provide design feedback, Figure 8. The Napkin's visual language parsers identify the context of drawing elements and configuration. It also supports recognition of perceptual figures such as emergent shapes from overlapping figures. The Stretch-a-Sketch program maintains interactive behavior among drawing elements using constraint propagation. IsoVist, Design Evaluator and Napkin-Archie programs support intelligent feedback such as performance analysis, critiquing and retrieval of relevant cases from a database. Light Pen offers design decision support guided by lighting design practice rules. LAN-Tools activates a simulation of network communication and proposes modifications based on designer's hand drawn diagrams.

*2.2.1. Recognition and Visual Language, Emergence*

For computers to support design drawing as a knowledge-based activity, they must recognize and interpret drawing semantics. The Electronic Cocktail Napkin employs the symbol recognizer described earlier combined with a visual language parser to identify complex configurations built up hierarchically from simpler components. For example, it recognizes a building façade of windows, doors, and a roof composed in certain spatial relationships. The visual language allows for multiple alternative parses, to read a drawing in several different ways, grouping components into alternative assembly graphs.

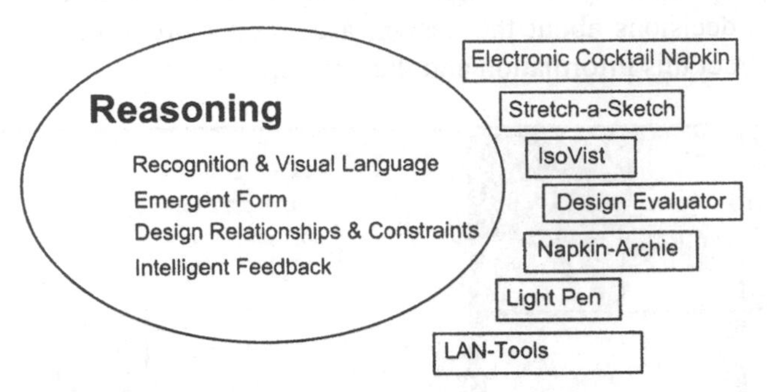


*Figure 8.* Reasoning activities engage domain knowledge

The Napkin's recognition scheme is contextual: depending on the drawing context the program recognizes symbols and configurations differently (Gross and Do 1996). For example, the same symbol may be recognized in an analog circuit diagram as an inductance and in a mechanical drawing as a spring. Conversely, when the Napkin identifies a symbol or configuration that is unique to a context (for example, a transistor symbol in an analog circuit diagram) then it uses this to determine context and thereby resolve pending recognition ambiguities.

A first task is recognizing drawing elements that the designer intended to represent—walls, columns, and windows in a floor plan, for instance, or levers, pulleys, and gears in a mechanical drawing. However, drawings often contain unintended figures formed by spatial relations among intentionally drawn components, termed emergent shapes or forms. These shapes stand out to the perceptive human designer and an intelligent drawing system must identify them also. We therefore developed an Electronic Cocktail Napkin component that searches the designer's drawing for symbols that emerge by combining strokes from two or more spatially proximate symbols and sub-strokes formed by intersecting and connecting strokes (Gross 2001). The system first generates a set of candidates; the symbol recognizer then selects those that match previously stored templates. For example, if the designer draws a diamond inscribed in a square, the program, previously trained to recognize triangles, identifies the four corner triangles, Figure 9.

### 2.2.2. *Constraints Bring Drawings to Life*

A designer sees more in a drawing than a static arrangement of arbitrary symbols; s/he sees its potential transformations. Domain semantics circumscribe the syntactic transformations of the diagram that are considered legal. For example, in transforming or editing a diagram of a mechanism—or a molecule—the designer maintains its essential spatial relationships. Which relationships in the diagram are essential and which are arbitrary depends on the domain. In an architectural plan geometry is essential; in an analog circuit diagram, it is not.

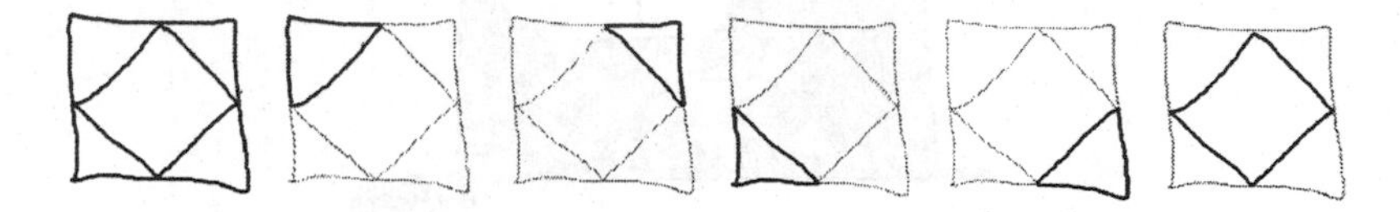

*Figure 9.* Sub-shapes identified in the first (leftmost) diagram

The Stretch-A-Sketch system, Figure 10, uses a constraint propagation engine to augment a drawing with interactive edit behavior (Gross 1994). First the system extracts the spatial relationships from the diagram—identifying, for example, elements that are connected, contained, adjacent, and so on. Then it instantiates these found relationships as constraints on the diagram elements. Subsequently as the designer edits the drawing, the system then maintains these constraints, so that elements remain connected, contained, adjacent etc.

Stretch-A-Sketch extracts constraint relationships from the drawing. A more sophisticated version might use domain knowledge—supplied by a module external to the sketching system—to select appropriate relationships to apply to the drawing.

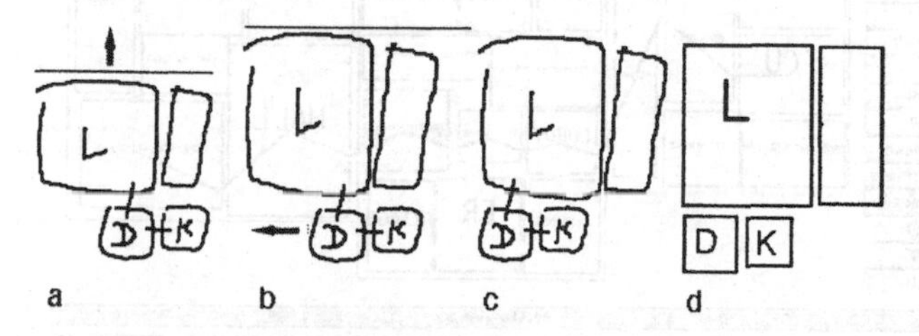


*Figure 10.* Stretch-a-Sketch interprets a floor plan (a) It infers an align-top constraint so that (b) stretching L also stretches the adjacent room, (c) rooms D and K retain their adjacency as D moves left, (d) the diagram rectified

### 2.2.3. *The Sketch as Locus for Intelligent Feedback*

In Stretch-A-Sketch, domain knowledge adds behavior to the designer's drawing through constraints. In this spirit, two other projects explore the idea of using the sketch itself as the locus of domain feedback. The IsoVist program, Figure 11, overlays design performance analysis on the designer's sketch (Do and Gross 1997). It calculates the area of a floor plan visible from a given vantage point (an "isovist"), and displays this polygon on the floor plan sketch. The designer moves the vantage point to display the isovist at various locations.

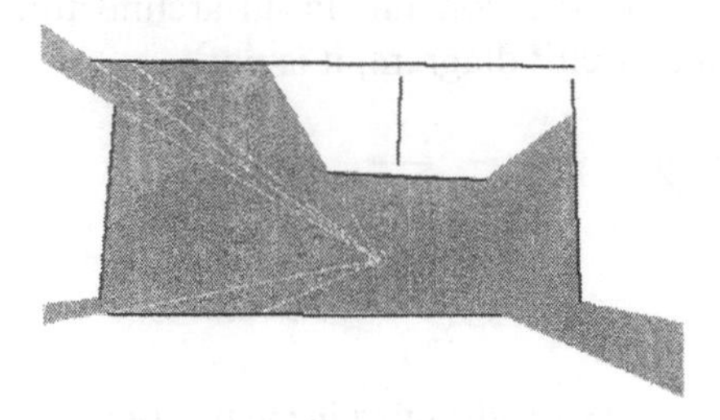

*Figure 11.* IsoVist overlays design performance evaluation on the sketch

Another example of domain based design performance evaluation is the Design Evaluator, Figure 12, a system for critiquing sketched designs, which analyzes a floor plan layout for spatial arrangement and circulation problems (Oh et al. 2004). When the system's rule checkers ('critics') detect problematic conditions, the Design Evaluator notifies the designer. Text messages link to a graphic overlay that marks the problem on the plan. For example, if the system finds a circulation problem, it will highlight the path on the floor plan drawing.

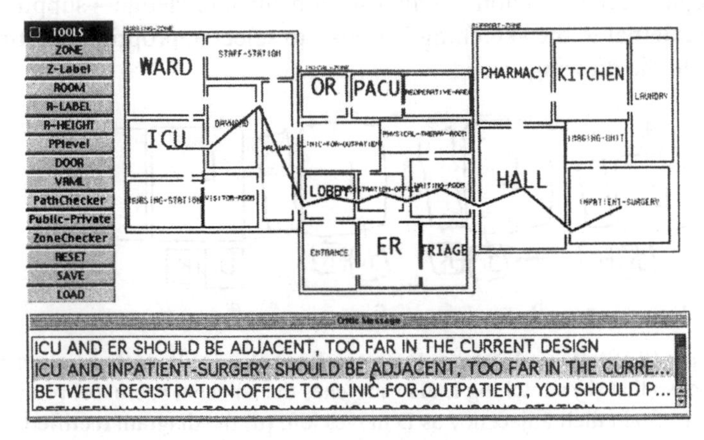


*Figure 12.* Design evaluator highlights designer's drawing (originally sketched, now rectified) with graphical critiquing. A critic highlights problematic path.

### *2.2.4. External Intelligent Systems*

A design drawing can serve as input to various external design aids or tools, which diagnose design problems and opportunities, suggest relevant or interesting references, offer advice, or evaluate design performance.

The Napkin-Archie system, Figure 13, links Cocktail Napkin with Archie, a case base of building designs with associated stories, problems, and solutions (Gross et al. 1994). A diagram index to the Archie case base enables a designer to retrieve design cases that match a given drawing. For example, Napkin-Archie might recognize a problematic entrance condition in the drawing and retrieve a relevant design case from the Archie library, illustrating the problem and potential solutions.

A similar scheme drives our "shape based reminding" system, in which drawing is used to retrieve visual images from a database (Gross and Do 1995). A diagram indexes each stored image. The system uses a multi-dimensional matching technique that compares the numbers, shapes, sizes, and spatial relationships of elements with a source drawing, to retrieve visually similar images. Depending on the image collection, the system can be used within a domain or across domains, as a creativity-stimulating scheme.

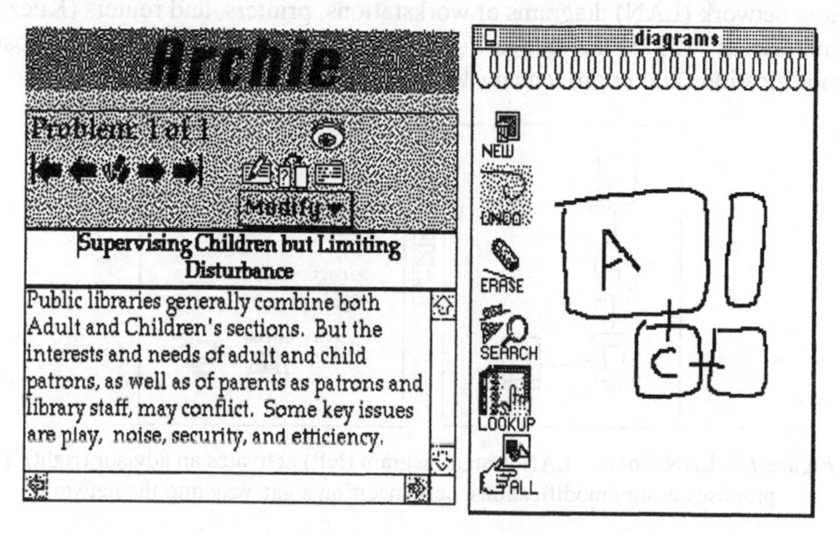


*Figure 13.* Napkin-Archie indexes a case based design aid with diagrams

The Light Pen system (Jung 2003), enhances the Space Pen program (Section 2.1.3) with an expert advisor for architectural lighting (Figure 14). The designer uses a "light pen" tool to sketch desired lighting patterns on the interior surfaces of a 3-D design model. The program analyzes the position, size, and intensity of these drawing marks and delivers this input to a rule based system for selecting lighting fixtures. The system proposes lighting fixtures and locations to produce the desired illumination. Light Pen shows how one might integrate knowledge-based advisors in a 3-D sketching system.

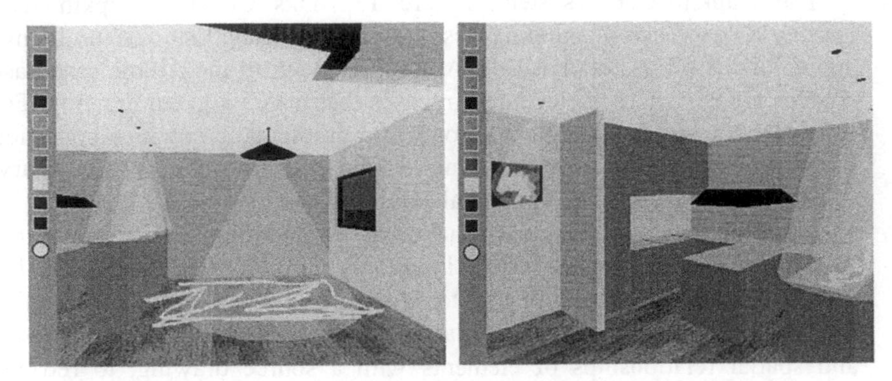

*Figure 14.* Light Pen offers lighting fixture suggestions based on sketches

The final example of linking with an external advisor is the LAN-Tools system, Figure 15. LAN-Tools employs the Cocktail Napkin to parse local area network (LAN) diagrams of workstations, printers, and routers (Kuczun and Gross 1997). A network design advisor module then proposes modifications and improvements to the designer's diagram.

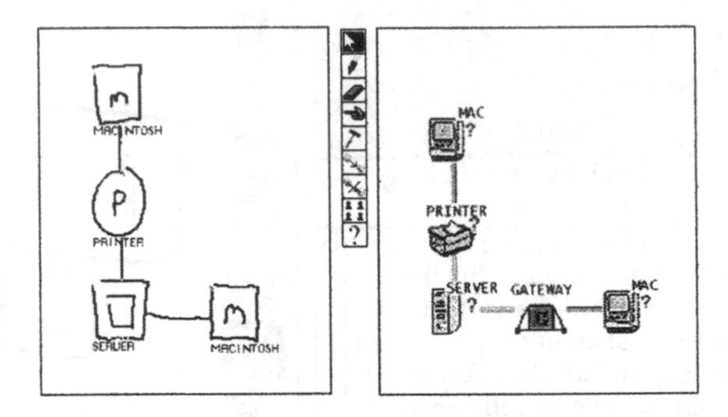


*Figure 15.* LAN-Tools – LAN design diagram (left) activates an advisor (right) that proposes design modifications, here inserting a gateway into the network

## 2.3. RESOLVING DESIGNS – FROM ABSTRACT TO SPECIFIC

The previous two sections have examined the first two R's of drawing and design — Recording and Reasoning — activities of recording design information and applying domain knowledge and expertise to the design drawing. We turn now to the third R, Resolving — the process of specifying a design, from initial concept to detailed specification. Although we have stressed the importance of retaining a sketchy appearance to accurately reflect the designer's level of commitment and decision-making in the early stages, sooner or later the designer must commit to decisions and proceed to specify the design. Computational drawing systems must recognize and support this trajectory—allowing vague, ambiguous, and abstract representations at the outset, supporting the move toward more detailed and definite ones, while allowing for some interplay and movement in both directions. Resolving concerns the development of drawing over the design process from rough, sketchy, and abstract representations to specific, definite, and well-defined design proposals. Drawing can be used to beautify and specify design (as in Electronic Cocktail Napkin and WebStyler) or translated into 3-D visualization with VR-Sketchpad, Figure 16. The incremental formalization includes recognition of drawing symbols and configurations, to beautification of rectified figures, including the substitution of the diagram elements to specification drawing illustrated in the Electronic Cocktail Napkin.

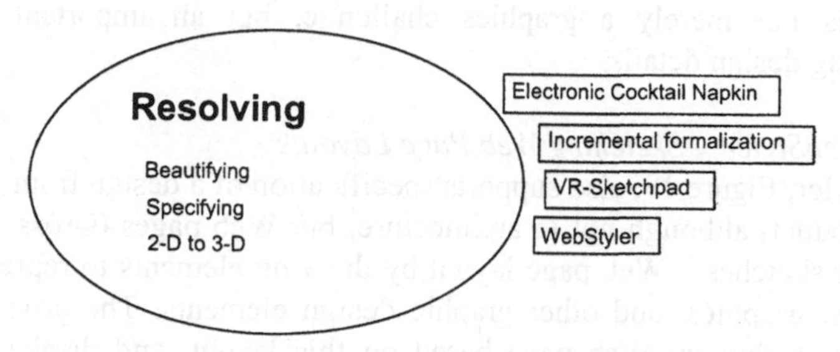


*Figure 16.* Resolving concerns drawing development & refinement during design

### *2.3.1. Beautification*

At the simplest level, the Cocktail Napkin's beautification scheme, in which each element type may have its own display method, moves the design along the path from crude sketch to precise drawing. When the program recognizes that the designer has drawn a stove, the beautified stove display method may replace the designer's sketchy stove with a dimensioned line drawing from a manufacturer's catalog, Figure 17. However, this step can involve additional selection criteria.

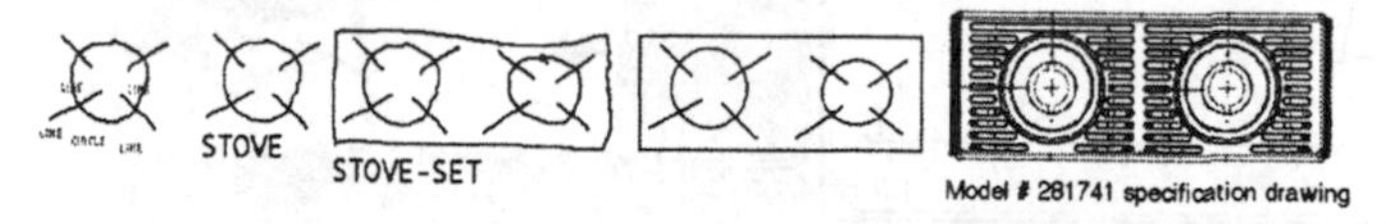


*Figure 17.* Incremental formalization, from diagram to CAD drawing replacement

### *2.3.2. Toward Specific Commitments, 2-D to 3-D*

The challenge of producing 3-D models from 2-D sketches has attracted much attention but it is often seen as a purely a problem in computer graphics. Rather, we see moving from a 2-D sketch to a 3-D model as a case of specifying a design. A 2-D design sketch is an abstraction of a 3-D model. Typically a 2-D sketch does not provide sufficient detail to produce a 3-D model, so information must be added in transforming the sketch to the model. For example our VR-Sketchpad program (Do 2002), Figure 18, transforms a floor plan sketch into a 3-D model. The program makes assumptions about the transformation: it extrudes walls and columns vertically, and it replaces floor plan symbols of furniture with 3-D models of furniture selected from the library.

Although the current VR-Sketchpad system just maps floor plan symbols one-to-one with library elements, an intelligent advisor could mediate this selection process, depending on characteristics of the floor plan, previously selected furniture, and so on. The transformation from 2-D sketch to 3-D model is not merely a graphics challenge, but an important case of specifying design details.

### *2.3.3. WebStyler —Sketching Web Page Layouts*

Web Styler, Figure 19, also supports specification of a design from sketch to final product, although not in architecture, but Web pages (Gross 1996). A designer sketches a Web page layout by drawing elements to represent text headings, graphics, and other graphic design elements. The program then generates a dummy Web page based on this layout, and displays it in a browser. The designer can attach actual content to the layout by selecting text files and graphics and associating them with elements of the Web page sketch. WebStyler then produces an HTML page layout with actual content.

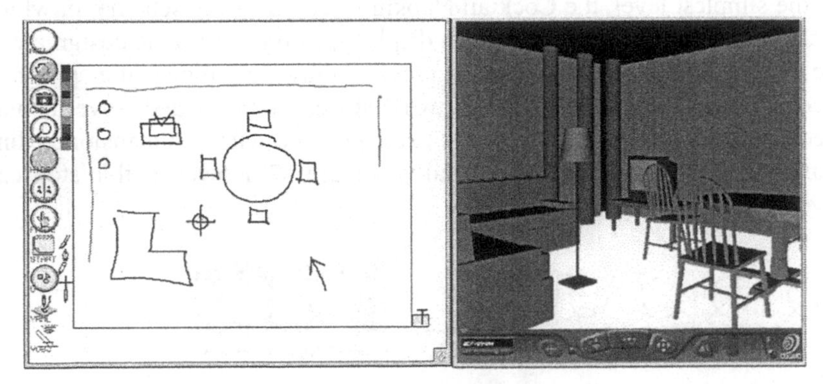

*Figure 18.* VR-Sketchpad generates specific 3-D models from rough 2-D sketches, extruding architectural elements and selecting furniture from a library

## 3. Discussion and Future Work

We have presented a raft of projects all centered on drawing in design. Individually, they explore topics from knowledge capture to visual analogy to simulating and evaluating performance. Taken together, the projects suggest various roles and uses for drawing, and ways of supporting these roles computationally, Figure 20. The specific computational support ranges from intelligent systems to management of informally expressed design rationale and annotations.

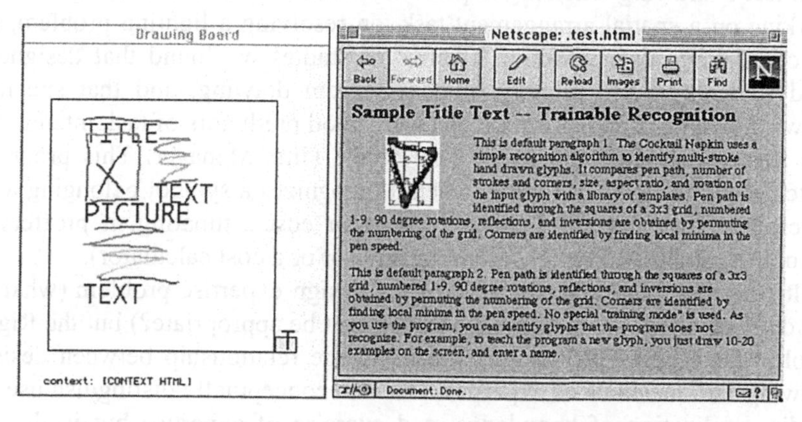


*Figure 19.* WebStyler generates web pages from sketched layouts

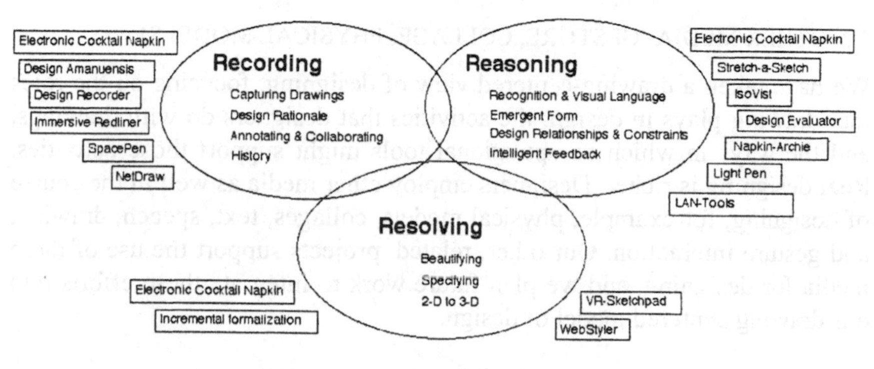


*Figure 20.* Software to support the three R's of design drawing.

## 3.1. EXPERTISE IN MANAGING DESIGN PROCESS

The bulk of our efforts so far have dealt with the integration of design knowledge and the medium of drawing. Expertise in managing design process, arguably represents the most fundamental question in the theory and methods of designing: how do designers decide what to work on, when, in the course of doing a design project? Apart from what designers know about their particular domain, what 'control' expertise do they exercise in deciding how to devote their efforts? And what kinds of computational support might aid in applying this control expertise?

The "Right Tool at the Right Time" project looked at whether monitoring designers' drawing acts could reveal useful information about their current purposes (Do 1998). The project posited that by monitoring an architect's drawing one might plausibly infer whether the designer was working on a spatial arrangement task, on resolving a lighting problem, or on calculating costs. Through a series of studies we found that designers (and students) could reliably infer task from drawing, and that specific drawing symbols and configurations were good predictors of task intentions. We then constructed the Right Tool Right Time Manager. This program watches the designer sketching: when it recognizes a symbol belonging to a specific subtask domain such as lighting or cost estimating, it proffers a supporting application (e.g., a lighting advisor or a cost calculator).

It considers only a small piece of the design expertise problem (what is the designer working on and what tools might be appropriate?) but the Right Tool at the Right Time project illustrates the relationship between design drawing and questions of expertise. One can conceptually distinguish use of media, application of knowledge, and exercise of expertise but in design practice the three are often intertwined.

### 3.2. OTHER MEDIA (GESTURE, COLLAGE, PHYSICAL MODELS)

We have taken a drawing-centered view of designing, focusing on the roles that drawing plays in design, the activities that designers do with drawings, and the ways in which computational tools might support those activities. Real designing is richer. Designers employ other media as well in the course of designing, for example, physical models, collages, text, speech, drawing, and gesture interaction. Our other, related, projects support the use of these media for designing, and we plan future work to integrate these efforts into our drawing centered model of design.

## Acknowledgements

This research was supported in part by the National Science Foundation under Grants IIS-96-19856, IIS-00-96138 and CCLI-0127579. The views and findings contained in this material are those of the authors and do not necessarily reflect the views of the National Science Foundation.

## References

Contero González, M, Naya Sanchis, F, Jorge, J and Conesa Pastor, J: 2003, CIGRO: A minimal instruction set calligraphic interface, *ICCSA 2003 Lecture Notes in Computer Science*, Springer, pp. 549-558

Cross, N, Christiaans H and Dorst K: 1996, *Analyzing Design Activity*, Wiley.

Davis, R: 2002, Sketch understanding in design: Overview of work at the MIT AI lab, *in* R Davis, J Landay and TF Stahovich (eds), *Sketch Understanding, 2002 AAAI Symposium*, Menlo Park, CA, AAAI: 24-31.

Do, EYL: 1998, The right tool at the right time: Investigation of freehand drawing as an interface to knowledge based design Tools, *PhD Thesis*, Georgia Tech.

Do, EYL: 2002, Drawing marks, acts, and reacts, toward a computational sketching interface for architectural design, *Artificial Intelligence for Engineering Design, Analysis and Manufacturing* **16**(3): 149-171.

Do, EYL and Gross, MD: 1997, Tools for visual and spatial analysis of CAD models, *in* R Junge, *Comp. Assisted Architectural Design Futures '97*, pp. 189-202.

Do, EYL, Gross, MD, Neiman, B and Zimring, C: 2001, Intentions in and relations among design drawings, *Design Studies* **21**(5): 483-503.

Fish, J, and Scrivener, S: 1990, Amplifying the mind's eye: Sketching and visual cognition, *Leonardo* **23**(1): 117-126.

Forbus, KD, Ferguson, RW and Usher, JM: 2001, Towards a computational model of sketching, *ACM Intelligent User Interfaces*, pp. 77-83.

Gaver, WW, Beaver J and Benford S: 2003, Ambiguity as a resource for design, *ACM Conference on Human Factors (CHI 2003)*, Fort Lauderdale, FL, ACM, pp. 233-240.

Goldschmidt, G: 1994, Visual analogy in design, *in* R Trappl (ed), *Cybernetics and Systems '94*, Singapore, World Scientific, pp. 507-514.

Gross, MD: 1994, Stretch-a-sketch, a dynamic diagrammer, *in* A Ambler (ed), *Proceedings of the IEEE Symposium on Visual Languages '94*, IEEE Press, pp. 232-238.

Gross, MD: 1996, The electronic cocktail napkin - working with diagrams, *Design Studies* **17**(1): 53-70.

Gross, MD: 2001, Emergence in a recognition based drawing interface, *in* JS Gero, B Tversky and T Purcell (eds), *Visual and Spatial Reasoning II*, Sydney Australia, Key Centre for Design Cognition and Computing, pp. 51-65.

Gross, MD and Do, EYL: 1995, Shape based reminding in creative design, *in* M Tan and R Teh (eds), *Global Design Studio: Computer Aided Architectural Design Futures '95*, Singapore, pp. 79-89.

Gross, MD and Do EYL: 1996, Ambiguous intentions, *ACM Symposium on User Interface Software and Technology*, Seattle, WA, ACM, pp. 183-192.

Gross, MD, Do EYL and Johnson BR: 2001, The design amanuensis: An instrument for multimodal design capture, *in* B d Vries, JP v Leeuwen and HH Achten (eds), *Computer Aided Architectural Design Futures 2001*, Kluwer, pp. 1-13.

Gross, MD, Lewin, J, Do, E, Kuczun, K, and Warmack, A: 1996, Drawing as an interface to knowledge based design, *Report*, Colorado Advanced Software Institute.

Gross, MD, Zimring, C and Do, E: 1994, Using diagrams to access a case base of architectural designs, *in* J Gero (ed), *Artificial Intelligence in Design '94*, Lausanne, Kluwer, pp. 129-144.

Igarashi, T and Hughes JF: 2001, A suggestive interface for 3D drawing, *ACM Symposium on User Interface Software and Technology (UIST), pp.* 173-181.

Jung, T, Do EYL and Gross MD: 2002, Sketching annotations in 3D on the Web, *Conference on Human Factors (SIGCHI)*, Minneapolis, ACM Press, pp. 618-619.

Kuczun, K and Gross, MD: 1997, Local area network tools and tasks, *ACM Conference on Designing Interactive Systems 1997*, Amsterdam, pp. 215-221.

Kurtoglu, T and Stahovich, TF: 2002, Interpreting schematic sketches using physical reasoning *in* R Davis, J Landay and T Stahovich (eds), *AAAI Spring Symposium on Sketch Understanding*, Menlo Park, CA, AAAI Press, pp. 78-85.

Landay, JA and Myers B: 2001, Sketching interfaces: Toward more human interface design, *IEEE Computer* **34**(3): 56-64.

Leclercq, P and Juchmes R: 2002, The absent interface in design engineering, *Artificial Intelligence for Engineering Design, Analysis and Manufacturing* **16**(3): 219 - 227.

Mankoff, J, Hudson, SE and Abowd, GD: 2000, Providing integrated toolkit-level support for ambiguity in recognition-based interfaces, *Human Factors in Computing (SIGCHI)*, ACM Press, pp. 368-375.

Oh, Y, Gross, MD and Do, EYL: 2004, Design evaluator: Critiquing freehand sketches, *Proceedings of Generative Computer Aided Design Systems*, Pittsburgh, Carnegie Mellon University, in press.

Oviatt, S and Cohen, P: 2000: Multimodal interfaces that process what comes naturally, *Communications of the ACM* **43**(3): 45-53.

Qian, D and Gross, MD: 1999, Collaborative design with netdraw, *in* G Augenbroe and C Eastman, *Computer Aided Architectural Design Futures '99,* Eastman, Atlanta, Kluwer, pp. 213-226.

Saund, E, Fleet, D, Larner, D and Mahoney, J: 2003, Perceptually-supported image editing of text and graphics, *User Interface Software Technology*, Vancouver, ACM, pp. 183-192.

Saund, E and Moran, TP: 1994, A perceptually-supported sketch editor, *User Interface Software and Technology*, Marina del Rey, CA, ACM Press, pp. 175-184.

Schon, DA, and Wiggins, G: 1992, Kinds of seeing and their functions in designing., *Design Studies* **13**(2): 135-156.

Sutherland, I: 1963, Sketchpad - a graphical man-machine interface, *PhD Thesis,* MIT.

Suwa, M and Tversky, B: 1997, What architects and students perceive in their sketches: A protocol analysis, *Design Studies* **18**: 385-403.

Tversky, B: 2002, What do sketches say about thinking? *in* T Stahovich, J Landay and R Davis (eds), *AAAI Spring Symposium Series -- Sketch Understanding,* Menlo Park, CA, AAAI, pp. 148-156

Ullman, D, Wood, S and Craig, D: 1990: The importance of drawing in the mechanical design process, *Computers and Graphics* **14**(2): 263-274.

## AUTHOR INDEX

## CONTACT AUTHORS' EMAIL ADDRESSES


| | |
|---|---|
| Akin, O | oa04@andrew.cmu.edu |
| Burge, JE | jburge@cs.wpi.edu |
| Cagan, J | cagan@cmu.edu |
| Chau, HH | hhchau@leva.leeds.ac.uk |
| Chira, C | camelia.chira@nuigalway.ie |
| Cruise, R | rbcruise@hotmail.com |
| Datta, S | sdatta@deakin.edu.au |
| de Assis Gama, F | fegama@tubarao.com.br |
| de Vries, B | B.d.Vries@tue.nl |
| Dong, A | adong@arch.usyd.edu.au |
| Gao, S | sgao@arch.hku.hk |
| Gómez de Silva Garza, A | agomez@itam.mx |
| Gross, MD | mdgross@u.washington.edu |
| Grzesiak- Kopeć, K | grzesiak@elf.ii.uj.edu.pl |
| Gu, ZY | zhenyu.gu@polyu.edu.hk |
| Heylighen, A | ann.heylighen@asro.kuleuven.ac.be |
| Jarratt, TAW | tawj2@eng.cam.ac.uk |
| Jeng, T | tsjeng@mail.ncku.edu |
| Koile, K | kkoile@csail.mit.edu |
| Krstic, D | george.krstic@ind.alcatel.com |
| Li, A | andrewili@cuhk.edu.hk |
| Liew, H | haldane@mit.edu |
| Mahdavi, A | amahdavi@tuwien.ac.at |
| McCormack, J | jonmc@csse.monash.edu.au |
| Mora, R | rodrigo_mora@sympatico.ca |
| Ostrosi, E | egon.ostrosi@utbm.fr |
| Shea, K | ks273@eng.cam.ac.uk |
| Ślusarczyk, G | Hliniak@softlab.ii.uj.edu.pl |
| Smith, GJ | g_smith@arch.usyd.edu.au |
| Sosa, R | rsos7705@arch.usyd.edu.au |
| Stouffs, R | r.stouffs@bk.tudelft.nl |
| Turner, A | a.turner@ucl.ac.uk |